VOLUME II   SPECIALIZED ASPECTS

MOLECULAR BIOLOGY OF THE GENE

FOURTH EDITION

James D. Watson   COLD SPRING HARBOR LABORATORY

Nancy H. Hopkins   MASSACHUSETTS INSTITUTE OF TECHNOLOGY

Jeffrey W. Roberts   CORNELL UNIVERSITY

Joan Argetsinger Steitz   YALE UNIVERSITY

Alan M. Weiner   YALE UNIVERSITY

The Benjamin/Cummings Publishing Company, Inc.

Menlo Park, California • Reading, Massachusetts • Don Mills, Ontario
Wokingham, U.K. • Amsterdam • Sydney • Singapore
Tokyo • Madrid • Bogota • Santiago • San Juan

Cover art is a computer-generated image of DNA interacting with the Cro repressor protein of bacteriophage λ. The image was prepared by the Graphic Systems Research Group at the IBM U.K. Scientific Centre.

Editor: Jane Reece Gillen
Production Supervisor: Karen K. Gulliver
Editorial Production Supervisor: Betsy Dilernia
Cover and Interior Designer: Gary A. Head
Contributing Designers: Detta Penna, Michael Rogondino
Copy Editor: Janet Greenblatt
Art Coordinator: Pat Waldo
Art Director and Principal Artist: Georg Klatt
Contributing Artists: Joan Carol, Cyndie Clark-Huegel, Barbara Cousins, Cecile Duray-Bito, Jack Tandy, Carol Verbeek, John and Judy Waller

Library of Congress Cataloging-in-Publication Data
Molecular biology of the gene.

  Rev. ed. of: Molecular biology of the gene / James D. Watson. 3rd ed. c1976.
  Bibliography
  Includes index.
  Contents: v. 2. Specialized Aspects.
  1. Molecular biology. 2. Molecular genetics.
I. Watson, James D., 1928–   . [DNLM: 1. Cytogenetics.
2. Molecular Biology. QH 506 M7191]
QH506.M6627 1987   574.87'328   86-24500
ISBN 0-8053-9613-6

DEFGHIJ-MU-898

The Benjamin/Cummings Publishing Company, Inc.
2727 Sand Hill Road
Menlo Park, California 94025

About the Authors

James D. Watson is the Director of the Cold Spring Harbor Laboratory. He spent his undergraduate years at the University of Chicago and received his Ph.D. in 1950 from Indiana University. Between 1950 and 1953 he did postdoctoral research in Copenhagen and Cambridge, England. While at Cambridge, he began the collaboration that resulted in 1953 in the elucidation of the double-helical structure of DNA. (For this discovery, Watson, Francis Crick, and Maurice Wilkins were awarded the Nobel Prize in 1962.) Later in 1953 he went to the California Institute of Technology. He moved to Harvard in 1955, where he taught and did research on RNA synthesis and protein synthesis until 1976. While at Harvard he also wrote the first, second, and third editions of *Molecular Biology of the Gene,* which were published in 1965, 1970, and 1976, respectively. He has been at Cold Spring Harbor since 1968, where his major interest has been the induction of cancer by viruses.

Nancy H. Hopkins is a Professor of Biology at the Massachusetts Institute of Technology. She graduated from Radcliffe College in 1964 and did graduate work at Yale and Harvard, receiving her Ph.D. in Molecular Biology and Biochemistry from Harvard in 1971. After postdoctoral work at the Cold Spring Harbor Laboratory, she joined the faculty at M.I.T., where she teaches and does research on the molecular biology of retroviruses. She is the primary author of Chapters 23 through 27 in Volume II of this edition of *Molecular Biology of the Gene.*

Jeffrey W. Roberts is a Professor of Biochemistry at Cornell University. He received a B.A. in Physics and Liberal Arts from the University of Texas in 1964 and a Ph.D. in Biophysics from Harvard in 1970. He was a postdoctoral fellow at Harvard and also did research at the MRC Laboratory of Molecular Biology in Cambridge, England, before going to Cornell in 1974. His current research interests are genetic regulation in bacteria and phages, in particular the regulation of transcription and the control of DNA repair functions. He is the primary author of Chapters 11, 12, 13, 16, and 17 in Volume I of this text.

Joan Argetsinger Steitz is a Professor of Molecular Biophysics and Biochemistry at Yale University. She graduated from Antioch College in 1963 and received a Ph.D. from Harvard in 1967. She did postdoctoral work at the MRC Laboratory of Molecular Biology before joining the Yale faculty in 1970. Her research interests have always focused on the structures and functions of RNA molecules; her current research is on gene expression in mammalian cells, with an emphasis on the roles of small RNA-protein complexes. A member of the National Academy of Sciences, she is a recipient of the National Medal of Science, among other awards. She is the primary author of Chapters 14, 15, 20, and 21 in Volume I of this text.

Alan M. Weiner is a Professor of Molecular Biophysics and Biochemistry at Yale University. He graduated from Yale College in 1968 and received his Ph.D. from Harvard in 1973. After postdoctoral work at Stanford University and M.I.T., he returned to Yale as a faculty member in 1976. His current research concentrates on the structure, function, and evolution of mammalian genes for small nuclear RNA species. He is the primary author of Chapters 22 and 28 in Volume II of this text.

Preface

Today no molecular biologist knows all the important facts about the gene. This was not the case in 1965 when the first edition of *Molecular Biology of the Gene* appeared. Then there were few practicing molecular biologists and not too many facts to learn. So what we knew about DNA and RNA could easily be explained to beginning college students. That year the final codons of the genetic code were being assigned, and everyone at the forefront of research could regularly assemble in the modest lecture hall at Cold Spring Harbor. Five years later, when the second edition appeared, our numbers were rising rapidly. Yet, despite the emerging popularity of molecular biology, it was still quite uncertain if the future would be as intellectually meaningful as the years just after the discovery of the double helix. The isolation of the first repressors and the demonstration that they bind specifically to control sequences in DNA seemed to some pioneers in DNA research to mark the end of the years of germinal discovery. With no means to isolate the genes of any higher organism, much less any way to know their nucleotide sequences, any pathway to understanding how genes guide the differentiation events that give rise to multicellular organisms seemed impossibly remote.

Happily, these worries did not last long. By the time the third edition of *Molecular Biology of the Gene* was published (1976), recombinant DNA procedures had given us the power to clone genes. Moreover, there was reason to believe that highly reliable methods to rapidly sequence long stretches of DNA would soon be available. As this new era of molecular biology began, however, there initially was widely voiced concern that recombinant DNA procedures might generate dangerous and pathogenic new organisms. It was not until after much deliberation that in 1977 the cloning of the genes of higher organisms began in earnest. The third edition could barely mention the potential of recombinant DNA, and of necessity its brief discussions of how genes function in eucaryotic organisms were tentative, and sometimes quite speculative.

It is only in this fourth edition that we see the extraordinary fruits of the recombinant DNA revolution. Hardly any contemporary experiment on gene structure or function is done today without recourse to ever more powerful methods for cloning and sequencing genes. As a result, we are barraged daily by arresting new facts of such importance that we seldom can relax long enough to take comfort in the accomplishments of the immediate past. The science described in this edition is by any measure an extraordinary example of human achievement.

Because of the immense breadth of today's research on the gene, none of us can speak with real authority except in those areas where our own research efforts are concentrated. Thus it was clear from the first discussions about the fourth edition that writing it would be beyond the capability of any one scientist who also had other major responsibilities. So the task of preparing this edition has required several authors. We also realized that it would be a formidable undertaking to keep the book within a manageable length; even by adopting a larger page format, we saw no way not to exceed a thousand pages. DNA can no longer be portrayed with the grandeur it deserves in a handy volume that would be pleasant to carry across a campus. Although this edition could have been shortened by eliminating the introductory material found in the first eight chapters, we never seriously considered this alternative. To do so would remove the background material that so many readers of previous editions have found valuable, and which has let many novices in molecular biology use this book as their first real introduction to gene structure and function.

Now that we are at last finished, we find that the book is even longer than we had planned. In part this happened because we are two years behind schedule, and

150 additional pages were needed to accommodate the immediate past. We also seriously underestimated how many words and illustrations would be required to describe the extraordinary variety of gene structures and functions that underlie the complexity of eucaryotic cells. We therefore have made the decision to split the fourth edition into two volumes. In the first volume we cover the general principles that govern the structure and function of both procaryotic and eucaryotic genes. It can be used as the sole text for a one-term course in molecular biology at the undergraduate level. The second volume concentrates on those specialized aspects of the gene that underlie multicellular existence, and it concludes with a chapter on the evolution of DNA. In this edition the second volume is appreciably smaller than the first. This will not be true of subsequent editions. Now that it is at last possible to study differentiation at the DNA level, we can easily foresee the time when, in fact, more than one volume will be required for even an introductory description of how genes are organized and expressed in the specialized cells of multicellular organisms.

We hope that this new edition, like its predecessors, will be found to be a highly suitable text for teaching at the undergraduate level, and that it also will provide all molecular biologists with an easy reference to the basic facts about genes. We have shown sections of the manuscripts to a variety of colleagues who are listed as reviewers. Their comments have been taken seriously, and we hope that the final manuscript faithfully reflects their expertise. Any mistakes that remain are, of course, our responsibility. Those who have made major contributions by writing or rewriting large sections of the text are Thomas Steitz (Chapter 6), Ira Herskowitz (Chapters 18 and 19), John Coffin (Chapter 24), and Brent Cochran (Chapter 25). Their generous contributions of specialized knowledge has vastly upgraded those portions of the book. In addition, John Coffin, Scott Powers, Haruo Saito, Lisa Steiner, and Parmjit Jat helped with the references for various chapters in Volume II. The excellent index was prepared by Maija Hinkle.

Equally important have been the efforts at Cold Spring Harbor of Andrea Stephenson, whose competent secretarial assistance helped coordinate our diverse labors, and Susan Scheib, whose intelligent attention to detail kept the manuscript and the galleys moving on a forward course. We also wish to acknowledge the pleasure of working with the staff of The Benjamin/Cummings Publishing Company, including Editor-in-Chief Jim Behnke and Production Supervisors Karen Gulliver and Betsy Dilernia. In particular we wish to thank Jane Gillen, who has functioned as the responsible editor during the entire writing and production of the book. An especially satisfying aspect of the process has been seeing rough drawings come alive through the efforts of the talented illustrator Georg Klatt, who has been responsible for the vast majority of the hundreds of new drawings prepared for this edition, and whose commitment and interest have greatly improved the book. And finally we gratefully acknowledge the strong support of our families throughout this endeavor, which was of course far more difficult and protracted than we ever foresaw.

James D. Watson

Nancy H. Hopkins

Jeffrey W. Roberts

Joan Argetsinger Steitz

Alan M. Weiner

Reviewers

John Abelson, California Institute of Technology
Bruce Alberts, University of California-San Francisco
Manny Ares, Yale University
Spyros Artanvanis-Tsakonas, Yale University
Piet Borst, The Netherlands Cancer Institute
Clifford Brunk, University of California-Los Angeles
Tom Cech, University of Colorado
Brent Cochran, Massachusetts Institute of Technology
John Coffin, Tufts University
Nick Cozzarelli, University of California-Berkeley
Don Crothers, Yale University
Steve Dellaporta, Cold Spring Harbor Laboratory
Ashley Dunn, Ludwig Institute for Cancer Research
Gary Felsenfeld, National Institutes of Health
John Fessler, University of California-Los Angeles
Wally Gilbert, Harvard University
Nigel Godson, New York University
Howard Green, Harvard Medical School
Nigel Grindley, Yale University
Carol Gross, University of Wisconsin
Ron Guggenheimer, Cold Spring Harbor Laboratory
Gary Gussin, University of Iowa
David Helfman, Cold Spring Harbor Laboratory
Roger Hendrix, University of Pittsburgh
Ira Herskowitz, University of California-San Francisco
Andy Hiatt, Cold Spring Harbor Laboratory
Bob Horvitz, Massachusetts Institute of Technology
David Housman, Massachusetts Institute of Technology
Martha Howe, University of Wisconsin
Richard Hynes, Massachusetts Institute of Technology
Amar Klar, Cold Spring Harbor Laboratory
Nancy Kleckner, Harvard University
Larry Klobutcher, University of Connecticut
Marilyn Kozak, University of Pittsburgh
Chuck Kurland, Uppsala University
Peter Laird, The Netherlands Cancer Institute
Richard Losick, Harvard University
Brenda Lowe, Cold Spring Harbor Laboratory
Tony Mahowald, Case Western Reserve University
Nancy Maizels, Yale University
Will McClure, Carnegie Mellon University
Bill McGinnis, Yale University
Jeffrey Miller, University of California-Los Angeles
Paul Modrich, Duke University

Peter Moore, Yale University
Steve Mount, University of California-Berkeley
Kim Mowry, Yale University
Tim Nelson, Yale University
Harry Noller, University of California-Santa Cruz
Bill Nunn, University of California-Irvine
Paul Nurse, Imperial Cancer Research Fund
Kim Nasmyth, Laboratory of Molecular Biology, Cambridge
Norm Pace, Indiana University
Carl Parker, California Institute of Technology
Sheldon Penman, Massachusetts Institute of Technology
Scott Powers, Cold Spring Harbor Laboratory
Mark Ptashne, Harvard University
Charles Radding, Yale University
David Raulet, Massachusetts Institute of Technology
Phil Robbins, Massachusetts Institute of Technology
Hugh Robertson, Rockefeller University
Lucia Rothman-Denes, University of Chicago
Earl Ruley, Massachusetts Institute of Technology
Haruo Saito, Harvard University
Robert Schleif, Brandeis University
Thomas Shenk, Princeton University
Gerry Smith, Hutchinson Cancer Center, University of Washington
Deborah Steege, Duke University
Lisa Steiner, Massachusetts Institute of Technology
Tom Steitz, Yale University
Bruce Stillman, Cold Spring Harbor Laboratory
Bill Studier, Brookhaven National Laboratory
Bob Symons, University of Adelaide
Susumu Tonegawa, Massachusetts Institute of Technology
Olke Uhlenbeck, University of Colorado
Axel Ullrich, Genentech
Chris Walsh, Massachusetts Institute of Technology
Jim Wang, Harvard University
Bob Webster, Duke University
Allan Wilson, University of California-Berkeley
Barbara Wold, California Institute of Technology
Sandra Wolin, University of California-San Francisco
Keith Yamamoto, University of California-San Francisco
Jorge Yunis, University of Minnesota

Brief Contents

Detailed Contents

VOLUME II   SPECIALIZED ASPECTS

Part X
Cancer at the Genetic Level 961

Note to the Reader

The following features are intended to add to the usefulness of this book as both a text and a reference.

- **Pagination of Volumes I and II** of this text is consecutive, with the first chapter of Volume II (Chapter 22) beginning on page 745. (Cross-references refer to chapter or page numbers only.)
- **The index** at the end of this volume covers both Volumes I and II.
- **Key terms** within the text are highlighted by boldface type at the point in a chapter where the first full definition and major discussion of each term occur. Boldface type is also used in the index to identify the page where the full definition appears.
- **The concept headings,** which originated in the first edition of this text, have been retained. In addition, the longer chapters of Volume II have been subdivided into several major sections set off by briefer headings, to help organize the material for the reader. A complete list of all the headings in this volume may be found in the Detailed Contents beginning on page viii.
- **Summaries** follow the main text for each chapter.
- **Bibliographies** at the ends of the chapters provide a bridge to the scientific literature. Included are a relatively short list of recommended **General References,** which are mainly books and review articles, and a longer list of **Cited References.** The Cited References include the original papers in which important discoveries were first reported as well as a selection of more recent papers. The citations of these references within the text appear as **superscript numbers** that accompany text headings. Thus the Cited References provide a convenient way of finding more detailed information on specific topics.
- **Color plates** showing computer graphics are included in both volumes. Volume II (following p. 952) includes Immunoglobulin and Viral Hemagglutinin (Plate 7) and Capsid Structure of Icosahedral Viruses (Plate 8).

IX

GENE FUNCTION IN SPECIALIZED EUCARYOTIC SYSTEMS

CHAPTER 22

The Molecular Biology of Development

Embryology began as a descriptive science, growing out of the fascination that biologists experienced as they watched the remarkable events of early development in organisms as diverse as the fruit fly, the sea urchin, and the mouse. For example, the female frog deposits her eggs on the bottom of a pond, and the male frog fertilizes them. Shortly thereafter, tiny tadpoles can be seen freely swimming about, searching for food and shelter, completely on their own in a very large (and often dangerous) world. Viewed under the microscope, both the huge, nearly spherical frog egg and the much smaller, flagellated sperm have a characteristic morphology, but neither of these germ cells resembles the tadpole or the adult frog in any obvious way. How, then, do these two highly specialized germ cells fuse and develop so quickly into an independent organism with a functional digestive tract and a nervous system capable of coordinating rapid swimming motions? We know that the instructions for how the egg develops into an adult are written in the linear sequence of bases along the DNA of the germ cells. However, this genetic information would be useless if the fertilized egg could not express the information in an orderly fashion. During development, gene expression must therefore be regulated both in space (an adult fly must not have a leg in the middle of its forehead) and in time (the larval fly must not prematurely develop wings). In addition, it is not sufficient for the DNA within each cell to be properly expressed; the cells must also interact with one another so as to build complicated multicellular structures such as wings and legs. Embryologists search for the general principles of development by concentrating on the early developmental stages in the life of the organism when cellular differentiation and interaction are most apparent.

The Heart of Embryology Is the Problem of Cell Differentiation

All higher plants and animals are constructed from a large variety of cell types (e.g., nerve cells, muscle cells, thyroid cells, and blood cells) that must arise in an exquisitely coordinated way. In some organisms, specialization begins with the first few cell divisions after fertilization. In other organisms, a large number of divisions occur before any progeny cell is fixed in its fate. Regardless of the exact time that **differentiation** occurs, however, it always results in the transformation of the parental cell into a large number of morphologically different progeny cell types.

Classical developmental biologists examined differentiation from three viewpoints. First, what are the external (and internal) influences acting on the original undifferentiated cell that might initiate a chain of events resulting in two progeny cells of different constitution? Sometimes, asymmetrically acting external forces are easy to perceive. For example, gravity forces the yolk of an amphibian egg to the bottom. Thus, after the first few cell divisions subdivide the fertilized egg, some of the embryonic cells have more yolk than others.

Second, are the molecular differences between differentiated cells extreme, or does morphological differentiation arise from the presence of only a few unique proteins in especially large numbers? We now know that each type of differentiated cell contains many molecular species peculiar to that cell type. Thus, a complete description of differentiation at the molecular level would necessarily be a most formidable task, and would not necessarily help us to understand the essential mechanisms responsible for differentiation.

Third, are the various changes that bring about differentiation irreversible, and if so, how they are perpetuated in a heritable fashion from one cell to the next? As we shall see, these are not easy questions to answer. Whether differentiation is reversible or irreversible depends not only on which cell type we are talking about, and in which organism, but also to a surprising extent on the precise details of the experiment. Moreover, we are only just now beginning to have vague hints of the mechanisms by which cells can maintain or change their state of differentiation.

Until quite recently, embryology was largely studied as an isolated subject, apart from modern genetic or biochemical ideas. Now, however, it is clear that the morphological tools of the classical embryologist cannot give satisfying answers by themselves. Instead, as in genetics, fundamental answers require analysis at the molecular level. Thus, just as recent methodological advances have made certain aspects of biochemistry and genetics indistinguishable, so embryology is being transformed by progress in biochemistry and genetics brought about by the recombinant DNA revolution (Chapter 19).

A Hierarchy of Genes Controls Development[1, 2, 3]

How, then, can we understand development at the molecular level when differentiation involves so many changes in the protein composition of the cell (together with parallel changes in key metabolic pathways and intermediates, as well as in RNA, lipid, polysaccharide, and perhaps even ionic composition)? If each of the many hundreds of gene products that distinguish a liver cell from a muscle cell had to be studied in minute detail before we could begin to understand the molecular causes of development, the situation would be virtually hopeless. However, we shall see that for many, and perhaps all, developing organisms (both procaryotic and eucaryotic), the gene products responsible for development can be arranged in a hierarchy, with some genes controlling the expression of other genes. This conclusion should not really surprise us. For example, we have already seen that the λ CI protein (repressor) controls the developmental decision to grow lytically or to lysogenize (see Figure 17-16). Similarly, the *MAT* locus in yeast controls the expression of a large array of genes responsible for the differences between the two mating types and the diploid (see Figure 18-31). These are only a few obvious examples. Thus, the many hundreds of gene products that distinguish one mammalian cell from another are not all equal; some are more important than others, because they control the initial decisions to differentiate along one developmental pathway or another. The goal of molecular biologists studying development is therefore to discover which genes control the expression of other genes.

Traditionally, one of the problems that has confounded and confused the molecular study of development is that cells usually become "committed" or "determined" to differentiate long before any actual morphological differentiation is apparent. For example, stem cells in the bone marrow of mammals give rise to at least two different kinds of progenitor cells—those that proliferate and differentiate into oxygen-carrying red blood cells and those that proliferate and differentiate into antibody-producing lymphocytes. Yet, in the very early stages of proliferation, the two kinds of progenitor cells (although committed to different fates) are difficult to distinguish. Once we realize that genes are arranged in a hierarchy, it becomes clear that in most cases, **commitment** or **determination** corresponds to the expression of a controlling gene, while differentiation reflects the myriad molecular consequences of that initial developmental decision.

Necessity of Finding Good Model Systems for Studying Differentiation

We have already noted (Chapter 20) that the relative amount of genomic DNA increases by a factor of approximately 800 from *E. coli* to a mammalian cell, but not all of this additional DNA represents a real increase in genetic complexity. Satellite sequences, moderately repeated DNA sequence families, and introns account for much of the genome in higher cells. Nonetheless, any particular mammalian cell type (say, liver) does contain a much larger number of distinct protein species than does *E. coli*. In addition, although all of the several hundred different cell types in the body share a core of "housekeeping" proteins that are necessary for metabolism and replication, different mammalian cell types (say, liver and muscle) express very different subsets of the total protein-coding capacity of the genome.

Faced with such biological complexity, how are we to choose the best organism for studying selected aspects of differentiation? Historically, rapid progress in understanding *E. coli* was due primarily to the invention of powerful genetic techniques for dissecting complicated biochemical events; these experimental advantages, in turn, attracted the concentrated efforts of many scientists. Similarly, with only a few important exceptions, molecular biologists have tended to concentrate on organisms where relatively powerful genetic techniques can be brought to bear on otherwise intractable developmental problems. Thus, even if we are ultimately curious about human biology, common sense may direct us to work on a simpler or more convenient organism such as bakers' yeast *(Saccharomyces cerevisiae)*, the fruit fly *(Drosophila melanogaster)*, the soil nematode worm *(Caenorhabditis elegans)*, or the laboratory mouse *(Mus musculus)*. However, it is important to keep in mind that biologists have always been experimental opportunists, studying whatever systems promised to yield interesting results most readily, and today's molecular biologists are no different. Some organisms, such as the African clawed toad *(Xenopus laevis)* and the cellular slime mold *(Dictyostelium discoideum)*, have

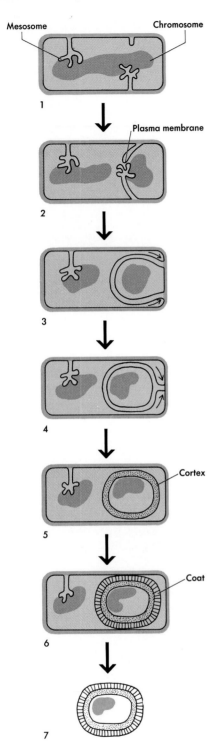

Figure 22-1

Diagrammatic representation of the stages in sporulation. (1) Preseptation: cell with axial chromosome. (2) Septation: plasma membrane folds in and cuts off an area containing intact chromosome. (3, 4) Stages of protoplast envelopment. (5) Cortex formation: material laid down between two membranes. (6) Coat formation and maturation. (7) Eventually the spore is released.

not been as extensively characterized genetically, but, as we shall see, offer unique experimental advantages for analyzing particular developmental phenomena.

To illustrate the advantage of using a model system, consider the problem of studying the mechanism of hormone action in a human (or even a mouse), where few (if any) mutants are available and where both the hormone and the hormone-responsive cells are often difficult to obtain in sufficient quantities. In contrast, work on the premier model system—induction of mating competence in yeast **a** cells by the mating hormone (pheromone) α factor (see Figure 18-32)—is moving swiftly because yeast is at least two orders of magnitude easier and cheaper to work with than higher cells.

THE MEASUREMENT OF TIME DURING DEVELOPMENT

Bacterial Sporulation Is the Simplest of All Model Systems[4]

The colossal magnitude of the task of attempting to understand the molecular basis of a complex differentiation process (e.g., the origin of a nerve cell) has led many people to look for cell systems much, much simpler than those studied by the classical embryologists. Some biochemists have gone so far as to focus their attention on the formation of bacterial **spores,** highly dehydrated cells that form inside certain bacteria (e.g., the genus *Bacillus*) as a response to a suboptimal environment. As spores possess virtually no metabolic activity, they resemble seeds of higher plants. Compared to vegetative bacteria, they are highly resistant to adverse conditions such as extreme heat or dryness. Thus, those bacteria that can produce spores have a much higher capacity to survive the extreme environmental conditions under which many forms of bacteria must live.

Some steps in the development of a spore are shown in Figures 22-1 and 22-2, which indicate that an early stage involves invagination of a portion of the bacterial membrane to enclose the chromosome and a small amount of cytoplasm. Afterward, a very thick and tough surface layer of protein is laid down on the outside. This coat has a completely different composition from the surface of the corresponding vegetative forms. Accompanying these morphological transformations are important changes in enzyme composition. The various cytochrome molecules responsible for aerobic metabolism completely disappear, and a new electron transport system appears. Ribosome content is also reduced, with an even greater decrease in the number of messenger RNA molecules. Small numbers of the enzymes necessary for protein synthesis remain; these are necessary for the resumption of protein synthesis when the spore germinates under the stimulus of more favorable nutritional conditions.

Sporulation thus requires cessation of the synthesis of almost all the proteins necessary for vegetative existence; conversely, germination converts spores to a condition in which they can make the vegetative proteins. So the making of a spore must entail an inhibition of most, if not all, vegetative mRNA synthesis, replacing it with the synthesis of mRNA specific for sporulation-specific proteins.

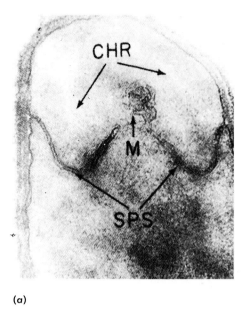

(a)

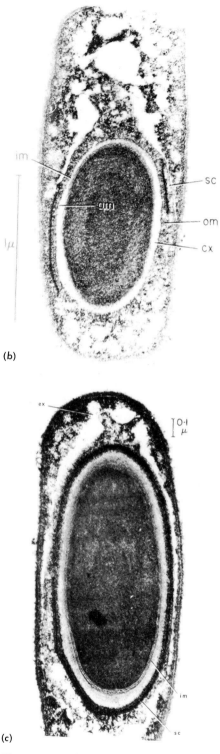

(b)

(c)

A Cascade of *σ* Factors for RNA Polymerase[5,6]

Until recently, the molecular basis of the changeover from vegetative to sporulation-specific transcription was a total mystery. Now, a combination of genetic and biochemical experiments tells us that sporulation is accompanied by the appearance of new RNA polymerase specificity factors that read the genes necessary for sporulation. The overall structure of RNA polymerase in *B. subtilis* is essentially identical to that in *E. coli* ($\alpha_2\beta\beta'\sigma$), although the precise molecular weights of the various subunits differ between the Gram-positive and Gram-negative bacteria. However, vegetatively growing *B. subtilis* contains at least four different *σ* factors, called σ^{55}, σ^{37}, σ^{32}, and σ^{28} (with molecular weights of 55, 37, 32, and 28 kdal). Each of these different *σ* factors can bind interchangeably to the core RNA polymerase ($\alpha_2\beta\beta'$) and confer upon it the ability to recognize a different set of promoter sequences. Factors σ^{37} and σ^{32} seem to be involved in switching on genes that are expressed at the very onset of sporulation. When sporulation in *B. subtilis* is initiated by starvation for carbon, nitrogen, or phosphorus, at least two new sporulation-specific factors are synthesized: one is a *σ* factor known as σ^{29}, and the other is a factor that dissociates the vegetative σ^{55} (and probably σ^{37}) from the core RNA polymerase. In this way, σ^{29} replaces σ^{55} on core polymerase and allows the transcription of sporulation-specific promoters to begin. (Surprisingly, the factor that dissociates σ^{55} from core RNA polymerase does not inactivate the σ^{55} and σ^{37}, since these can be recovered in active form from extracts of sporulating bacteria.)

Interestingly, the existence of multiple *σ* factors that determine the expression of specialized sets of genes is emerging as a general feature of bacteria as diverse as *E. coli*, *Bacillus*, and *Streptomyces*. For example, the expression of heat-shock genes and nitrogen-regulated genes in *E. coli* is now known to be dictated by specific *σ* factors encoded by genes *htpR* and *ntrA* (Chapter 16). Also, multiple *σ* factors are a feature of most complex bacteriophages such as T4, λ, T7, *μ*, and *B. subtilis* phage SPO1 (where novel *σ* species were actually first discovered) and are responsible for the temporal cascade from early to late phage gene expression (Chapter 17).

Figure 22-2
Electron micrographs of thin sections of sporulating *Bacillus cereus* cells showing (a) septation, (b) cortex development, and (c) coat formation. Note that CHR designates the chromosome; m, the mesosome; sps, the transverse spore septum; cx, the cortex; sc, the spore coat; ex, the exosporium; im and om, the inner and outer membranes. (Courtesy of Dr. W. G. Morrell.)

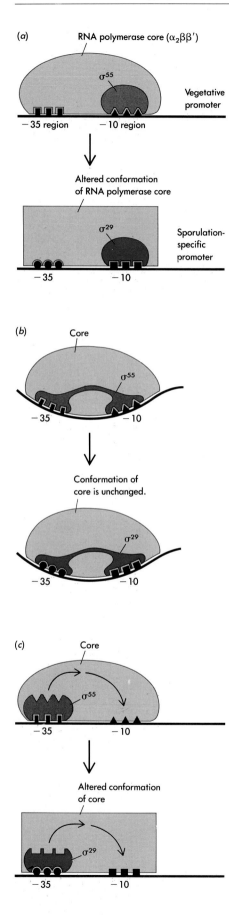

Figure 22-3

How do σ factors change the specificity of RNA polymerase? (a) A "core-conformation" model. σ^{29} displaces σ^{55} from RNA polymerase core. Binding of σ^{29} could change the conformation of core polymerase in such a way that it recognizes a different −35 region on DNA. This model is difficult to envision because core enzyme can interact with at least six different σ factors. (b) A simultaneous "direct-contact" model. σ^{55} and σ^{29} could be elongated molecules that touch the −10 and −35 regions simultaneously. Experiments supporting this model show that *E. coli* σ can be chemically crosslinked to both the −35 and −10 regions. (c) A sequential "direct-contact" model. σ factors could bind first to the −35 region (closed promoter complex) and then to the −10 region (open complex). This model is also consistent with the chemical crosslinking experiments, but would require that RNA polymerase core undergo a significant conformational change. [After R. Losick and J. Pero, *Cell* 25 (1981):582–584.]

The σ factors are truly remarkable, because the binding of a single additional polypeptide chain can confer a new promoter specificity on the general RNA-synthesizing machinery of core RNA polymerase. How can a σ factor do this? Current evidence suggests that σ factors are bifunctional proteins that can simultaneously bind to core RNA polymerase and recognize specific promoter sequences in the DNA. When a comparison was made between many different promoters, all recognized by *E. coli* core RNA polymerase complexed with σ^{70} (a complex called a holoenzyme), it was possible to deduce an average (or consensus) sequence for the σ^{70}-specific promoter (see Figure 13-8). Similarly, when the consensus promoters for *B. subtilis* RNA polymerase bearing the σ^{55}, σ^{37}, and σ^{28} bacterial factors or phage SPO1 σ factors were aligned, it immediately became apparent that each of these different σ factors enables core polymerase to recognize a different combination of sequences centered approximately around positions −35 and −10 upstream from the start site of transcription. These are, of course, the same regions recognized by *E. coli* RNA polymerase holoenzyme. Since it is difficult to imagine that σ factors can alter the conformation of core RNA polymerase in such specific ways as to change the precise sequence of nucleotides recognized as a promoter, it seems more likely that σ factors are themselves very highly elongated molecules that touch DNA sequences as far apart as 16 to 19 base pairs (−35 and −10 regions) when the factor is bound to RNA polymerase (Figure 22-3). However, binding to the −35 region may be relatively weak or less essential, since the bacteriophage T4-encoded σ factor required for reading the late phage genes appears to bind only to the −10 region (Chapter 13).

Multicellular Organisms Must Have Devices to Control When Genes Act

In multicellular organisms, as in bacteria, not all genes in a cell function at the same time. Something must dictate that muscle cells, for example, selectively synthesize the various proteins used to construct muscle fibers. The understanding of embryology will thus, in one sense, be the understanding of how genes function selectively. Moreover, we must ask not only what causes two daughter cells to synthesize different proteins (differentiation), but also what enables their daughters to continue synthesizing the same particular group of proteins (stable inheritance of the differentiated state). With the problem phrased in this way, it is clear that no one will ever be able to work out all the chemical details that accompany the development of any higher plant or animal. For even a modest approach to a comprehensive understanding, we would have to look at the behavior of many hundreds of different proteins. Still, we should be able to discover certain general principles governing the selective expression of specific proteins. In fact, there is increasingly good evidence that in multicellular organisms, as in bacteria, differentiation is often controlled at the chromosomal level by devices that regulate the transcription of specific genes.

Embryologists are now asking whether the fundamental control mechanisms act negatively, like the lactose repressor and the $MAT\alpha2$ gene product in yeast (see Figure 18-31), or positively, like the bacteriophage-specific SPO1 and T4 σ factors and the steroid receptors of higher cells (Figure 21-48). The dilemma, however, is that most of the systems that embryologists study are hopelessly complex from the biochemical viewpoint and cannot be correlated with the occurrence of a few well-defined enzymatic reactions. For example, although nerve cells are easy to identify morphologically, we are just beginning to understand their molecular structure, and only a few of the hundreds of protein molecules in nervous tissue have been well characterized. Although we are much more familiar with the muscle proteins, again large gaps in our knowledge are likely to make precise analysis of the function of each protein a most tricky endeavor.

The Measurement of Biological Time[7]

To control when genes act, cells must be able to measure time. In fact, the regularity of the cell cycle in both eucaryotic microorganisms (Chapter 18) and higher cells (Chapter 25) suggests that eucaryotes as well as procaryotes have accurate clock mechanisms at the cellular level. We are accustomed to thinking about time as a quantity that can be measured by a simple physical device, like the escapement of a mechanical clock, or a simple electronic device, like the crystal oscillator in a solid-state chip. But we have comparatively few clues as to how biological systems are able to measure and mark elapsed time.

In principle, biological systems could measure time in two very different ways. One possibility is that cells have an independent (autonomous) internal clock, together with devices to ensure that key reactions occur at certain times (much like an alarm clock). In this case, cellular events *a*, *b*, and *c* are *independent* of each other. Event *b* is scheduled for time *b*, and event *c* for time *c*; but if event *b* fails to occur for whatever reason, event *c* can still take place as scheduled. Alternatively, key reactions may be designed so that event *c* cannot take place unless event *b* has already occurred. For example, event *c* might be a chemical reaction that requires a product of event *b*. In this case, each cellular event is *dependent* on the previous events, and all events are linked together in a fixed order. The very sketchy evidence now available suggests that both independent and dependent clocks exist in eucaryotic development.

In Chapter 17, we saw that most complex bacteriophages use a cascade of positive and negative transcriptional regulators to control the order of gene expression during phage infection. The λ antitermination proteins N and Q function positively, while the T7 gene *2* protein (and possibly the gene *0.7* protein) functions negatively by

Figure 22-4
A mechanism by which sequential synthesis of different specificity factors might be used to count time.

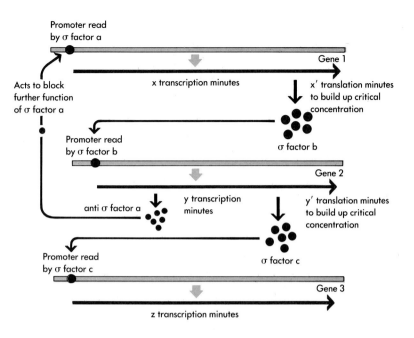

inactivating the bacterial RNA polymerase. In contrast, σ factors (encoded by the SPO1 genes *28*, *33*, and *34* and the T4 genes *33* and *55*) function both positively (to turn on expression of later genes) and negatively (to turn off expression of the earlier genes by displacing the old σ factor). Thus, even the very small genome of a bacteriophage can use a sequence of dependent events to count time. However, as a consequence of the relatively slow rate at which RNA polymerase molecules move along DNA templates (DNA replication occurs some 100 times faster!), bacteriophages can also use transcription as an independent measure of biological time. For example, at 37°C, the rate of movement of *E. coli* RNA polymerase is only 40 nucleotides per second, and so complete transcription of the λ genome by a single polymerase molecule would require over 25 minutes. Since several operons are transcribed simultaneously, this much time is not required for production of infectious phage. Nevertheless, transcription of the very long "late" operon encoding the capsid proteins counts out some 15 minutes, with the head proteins being made substantially before the tail proteins, as expected from the gene order (Figure 17-15). (Why the lysis genes *S* and *R precede* all of the capsid genes in the late transcription unit remains a mystery.)

Similar analysis applied to the *E. coli* chromosome gives a total possible transcription time of 33 hours. Thus, only 1 percent of its total genome would have to be used to code for periodic events occurring once every 20-minute division cycle. And even the 3 to 4 hours required for a *Bacillus* spore to germinate could easily be directly measured out on a DNA tape. There is, of course, no reason why such relatively long intervals must be coded along one continuous DNA segment. Several genes, each coding for a unique specificity factor necessary for reading the subsequently transcribed gene, would also do the job (Figure 22-4).

Although the division cycles of higher cells are much longer than those of bacteria or eucaryotic microorganisms, their DNA contents are correspondingly much greater; one RNA polymerase molecule would require a thousand days at 37°C to completely transcribe the haploid complement of human genes. Thus, much of the timing required for a 24-hour human cell cycle can easily be imagined to occur at the transcriptional level. Such speculation will probably receive its first experimental test from analysis of yeast *CDC* mutants (Chapter 18), which become blocked at specific stages of the cell cycle when they are subjected to slightly elevated temperature. Although yeast is a very simple eucaryote, it will often be the model system of choice for genetic and biochemical characterization of basic metabolic processes that it shares with higher eucaryotes.

The timing of certain crucial steps in the embryological development of higher organisms may also be controlled by transcription. We will see shortly that some of the homeotic genes responsible for morphogenesis in the fruit fly *Drosophila* have huge introns, and it is possible that the function of these very long transcription units (approximately 100 kilobases) is to delay expression of the gene product until a particular time in early embryogenesis.

Concentration on Organisms with Easily Observable Cleavage Divisions[8]

The mechanisms by which fertilized eggs develop into multicellular organisms have been a continuous source of mystery to biologists for almost a hundred years. During this time, experimental studies have been concentrated on echinoderms (especially the sea urchin) and amphibians such as frogs, salamanders, and toads. The reason for such emphasis lay in the ease with which echinoderm and amphibian eggs could be procured, fertilized, observed, and experimentally manipulated as they passed through the normal stages of development. In contrast, birds and mammals—where all the embryological stages take place within a thick-shelled egg or female parent—are much more tricky to work with. Only recently has real progress been made in opening up the early stages of mouse embryogenesis to experimental manipulation.

A very important feature of both urchin and frog embryology is that all the nutrients (mostly stored in the form of yolk) required for development through the blastula and gastrula stages must be present in the unfertilized egg. Only when the developing embryo becomes able to feed itself—as the frog does, for example, at the tadpole stage—does real growth in mass occur. Consequently, echinoderm and amphibian eggs are very, very large in comparison with their respective adult cells, and the cell divisions that occur after fertilization progressively produce smaller and smaller cells: hence the term **cleavage divisions** (Figure 22-5).

How the Frog Egg Counts Twelve Cell Divisions[9, 10, 11]

Although we have firm proof that transcription can measure biological time in bacteriophage, biologists have had only one major success in explaining the measurement of time during embryogenesis in a complex higher eucaryote. The crucial experiments took advantage of the extraordinarily large size of the frog egg (about 1 mm, or $\frac{1}{25}$ inch, in diameter): This egg is so large that foreign substances can easily be microinjected through a hollow glass needle without inflicting any damage on the developing embryo.

Immediately after fertilization, the frog egg undergoes 12 very rapid and almost synchronous cleavage divisions, characterized by

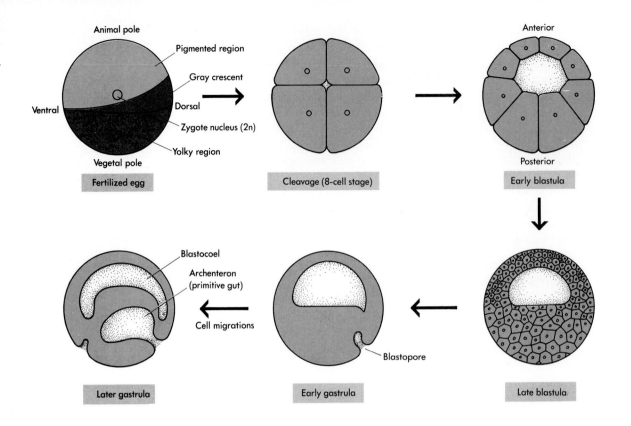

Figure 22-5
Early stages in the development of an amphibian egg.

short interphase periods when DNA synthesis occurs at a very accelerated rate. There is no measurable transcription of newly synthesized DNA during this time, perhaps because the presence of RNA polymerase molecules on the DNA would interfere with rapid DNA replication. After the twelfth division, when the fertilized egg has transformed itself into a blastula containing approximately 2^{12}, or 4096, cells, the rate of division slows down to a pace more typical of normal somatic cells, and transcription begins by RNA polymerases II and III (RNA polymerase I begins slightly later). This abrupt physiological change is called the **midblastula transition,** because it happens midway through **blastulation** (formation of a solid sphere of cells) but before **gastrulation** (when the embryonic cells become motile and the blastula invaginates to form various tissue layers such as endoderm, mesoderm, and ectoderm).

How do cells within the blastula know when they have undergone exactly 12 divisions? There are two obvious possibilities. The blastula could count each round of cell division directly (for example, by sequential modification of a site on the DNA once per division cycle), or it could count a product of the cell division (for example, the total amount of DNA in the entire blastula). We now believe the second possibility.

Experimentally, it has been found that the midblastula transition occurs earlier than usual in fertilized eggs that have been microinjected with foreign DNA. *Any* foreign DNA will work (plasmid, bacteriophage, or bacterial DNA), and the greater the quantity of DNA microinjected, the earlier the transition will occur. These observations strongly suggest that the mature frog egg contains a preformed factor (presumably a protein) that binds *stoichiometrically* to DNA: Twice as much DNA binds twice as much factor. The unfertilized egg contains exactly the right amount of this substance to bind the total mass of DNA in approximately 4096 cells, and there is no additional synthesis

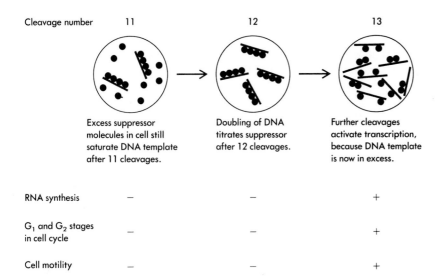

Cleavage number	11	12	13
	Excess suppressor molecules in cell still saturate DNA template after 11 cleavages.	Doubling of DNA titrates suppressor after 12 cleavages.	Further cleavages activate transcription, because DNA template is now in excess.
RNA synthesis	−	−	+
G_1 and G_2 stages in cell cycle	−	−	+
Cell motility	−	−	+

Figure 22-6
A titration model for the onset of transcription at the midblastula transition. [After M. Kirschner and J. Newport, *Cell* 30 (1982):687–696.]

of the factor between fertilization and the midblastula transition. Each cleavage division increases the amount of DNA in the developing blastula, and when the last trace of preformed factor is bound by newly synthesized DNA, the disappearance of free (unbound) factor triggers both transcription and the onset of cell motility required for gastrulation (Figure 22-6). We might speculate that the factor is a DNA binding protein that binds loosely and nonspecifically to any DNA, thereby preventing transcription. Initially, a high concentration of free factor in the egg would saturate all potential transcription units with factor. Only when free factor had been fully titrated by 4096 cells worth of new DNA could transcription resume. Mechanistically, regulation of the midblastula transition resembles regulation of the amount of oocyte-specific 5S RNA by titration of preformed transcription factor TFIIIA with the 5S gene product (Chapter 21).

Titration of preformed factors in the egg cytoplasm is an elegant, satisfying, and potentially general mechanism for timing many different developmental events. Nevertheless, it is extremely difficult to imagine how the maturing frog oocyte (egg) manages to synthesize exactly the right amount of the preformed factor. This dilemma brings us face to face with the central philosophical problem of all developmental biology. How can we claim to have explained the behavior of the developing frog blastula in terms of the properties of the unfertilized egg, when we do not know how the egg itself was made? Thus, the molecular study of development appears to be an endless hall of mirrors. Because all cells come from preexisting cells, and all organisms from preexisting organisms, almost every answer gives rise to a prior question. Our fundamental faith as molecular biologists is that by amassing a great many such partial answers, we will eventually be able to come full circle and escape from the paradox of the chicken and the egg.

THE NATURE OF DIFFERENTIATION

Cellular Differentiation Is Often Irreversible

At present, it is possible to isolate a variety of differentiated cells and grow them (like bacteria) outside living organisms under well-defined nutrient conditions. This technique of "cell culture" (Chapter 25) allows us to ask, for example, whether a nerve cell continues to look like a nerve cell when growing outside its normal cellular environment: The answer in this case is yes. Something has happened that appears to have permanently destroyed this cell's capacity to synthesize proteins other than those found in nerve cells. Similar results are obtained when many other cell types from higher animals are studied. But with higher plants, the opposite answer is more often found. As we shall see, a complete plant can often be regenerated, starting from differentiated cells of the mature plant.

Differentiation Is Usually Not Due to Chromosome Gain or Loss

Perhaps the most obvious hypothesis to explain irreversible differentiation is that only a fraction of the genes of the fertilized egg are passed on to a nerve cell, a muscle cell, and so on. This scheme, however, appears to be completely wrong. As far as we can tell, most cells of most organisms, with the obvious exception of the haploid sex cells, contain roughly the same chromosomal complement. Virtually all cell divisions are preceded by a regular mitosis, so that daughter cells receive an identical set of chromosomes. We cannot say, however, that no permanent changes have occurred at the level of individual genes. Even though the mass of evidence gathered by detailed characterization of many developmentally regulated genes in many organisms overwhelmingly supports the idea that differentiation rarely involves an irreversible change in genomic DNA, important exceptions are already known. For example, the mammalian immune response depends on both irreversible genomic rearrangements and selective mutation of lymphocyte genes responsible for antibody production and regulation (Chapter 23).

Differentiation Is Usually Not Irreversible at the Nuclear Level[12]

Nuclear transplantation experiments provide conclusive evidence against the notion that the selective synthesis of a restricted set of gene products in a differentiated cell requires irreversible changes in genomic DNA. In these experiments, diploid nuclei from differentiated cells were transplanted into unfertilized eggs whose haploid nuclei had been previously removed. The resulting genetically complete diploid eggs were then artificially induced to divide and grow, occasionally forming adult organisms whose chromosomal makeup was derived entirely from clonal reproduction of donor nuclei (Figure 22-7). Although nuclei derived from larval intestinal cells support clonal reproduction, nuclei from an adult frog do not work. This suggests that the pluripotency of some larval intestinal cells (the ability to *de*differentiate completely) is lost upon maturation. So far, the frog is the most complex organism for which nuclear transplantation has allowed clonal reproduction. Success in this case owes much to the very large size of the amphibian egg, which allows conventional microsurgical removal of the maternal nuclei.

Cellular Memory and the Replication Fork[13]

Just as bacteria divide to produce more bacteria and elephants give birth to elephants, so most differentiated cells in our own bodies breed true. Liver cells give rise to more hepatocytes, and fat cells

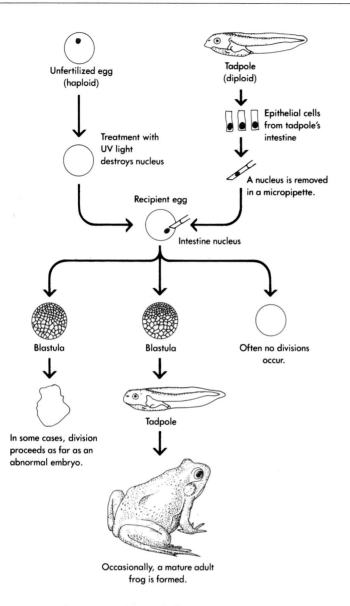

Figure 22-7
Steps in the creation of a clonal frog.
[Redrawn from J. R. Gurdon, *Sci. Amer.*
219 (1968):24–35.]

Unfertilized egg
(haploid)

Tadpole
(diploid)

Treatment with
UV light
destroys nucleus

Epithelial cells
from tadpole's
intestine

A nucleus is removed
in a micropipette.

Recipient egg

Intestine nucleus

Blastula

Blastula

Often no divisions
occur.

In some cases, division
proceeds as far as an
abnormal embryo.

Tadpole

Occasionally, a mature adult
frog is formed.

produce more adipocytes. Since different cell types express very different (but overlapping) sets of proteins, maintenance of the differentiated state must reflect mechanisms that allow both daughter cells to inherit the same pattern of active and inactive genes found in the parent.

Semiconservative DNA replication provides both daughter cells with identical copies of each gene, but the DNA sequences by themselves cannot specify whether a gene is to be turned on or off in the daughter cells. Instead, regulation of individual genes is accomplished by a variety of mechanisms (Chapter 21): diffusable regulatory factors, nondiffusable covalent modifications of the DNA (methylation of the dinucleotide CpG), and local changes in histone and nonhistone chromosomal proteins leading to active and inactive chromosomal domains. Each of these mechanisms could in principle contribute to the inheritance by each daughter cell of a characteristic pattern of active and inactive genes.

Unfortunately, a crucial gap in our knowledge of the enzymology of eucaryotic DNA replication prevents us from becoming too specific regarding the mechanisms that might allow the inheritance of active and inactive genes: We do not know what happens to chromatin (the complex of DNA with protein) as it passes through the replication fork. In other words, can proteins bound to the parental duplex remain associated with the gene while it is being replicated? Since passage through the fork occurs quickly, and since most sequence-specific DNA binding proteins also bind nonspecifically to DNA, a reasonable guess might be that some (but not all) proteins will remain loosely bound to the sugar-phosphate backbone of the single-stranded DNA while the bases are used as a template by DNA polymerase for synthesis of the complementary strand; these proteins could then resume tight binding to the newly reformed daughter duplex.

Our ability to make firm statements about the mechanisms for perpetuation of active and inactive genes is further qualified by lack of precise quantitative knowledge regarding the speed with which various proteins are able to find their binding site on DNA (Chapter 16). For example, consider a gene that is activated in the parental cell by a positive regulatory protein (e.g., the steroid receptor protein). Whether this gene remains active in the daughter duplexes after it passes through the replication fork may in some cases reflect a competition: Can the regulatory protein bind (or rebind) to the gene specifically before other proteins (perhaps histones) bind nonspecifically and shut the gene down? Thus, if the parental cell has a sufficient concentration of the regulatory protein, both gene copies could be activated after replication and the active state of the gene would be inherited by both daughter cells.

Similar arguments can be advanced regarding the perpetuation of active and inactive chromosomal domains during DNA replication; however, such speculation must be tempered by our lack of knowledge about the structure of active chromatin, the behavior of the nucleosomes at the replication fork, and the rate of nucleosome assembly onto naked DNA. The one fact that seems clear is that nucleosomes segregate conservatively at the fork, one daughter duplex receiving the old nucleosome intact while the other daughter duplex acquires an entirely new nucleosome. Since active nucleosomes appear to differ structurally from inactive nucleosomes (Chapter 21), a very important question is whether the deposition of old nucleosomes can occur preferentially on the leading or lagging strand.

DNA Methylation and Inheritance of the Differentiated State[14, 15, 16]

The DNA of most higher eucaryotes is modified by methylation at the 5 position of cytosine in the dinucleotide sequence ^mCpG (where ^mC denotes 5-methylcytosine) (Chapters 9 and 21). We have seen that many active or potentially active genes are undermethylated at certain CpG sequences, but examples are also known where decreases in methylation accompany rather than precede gene activity. Although the exact molecular basis for these effects is unknown, our current knowledge suggests that the presence of a methyl group in the major groove of the DNA can regulate gene expression by either increasing or decreasing the affinity of the DNA for specific regulatory proteins (repressors and activators). Because certain organisms, such as insects, lack detectable DNA methylation, methylation cannot be indispensable for regulation of eucaryotic genes. Nonetheless, the experimental evidence correlating DNA methylation with changes in gene expression in other organisms, and particularly in mammalian cells, is very impressive. Thus, we are justified in assuming that DNA methylation does play a role in regulation of gene expression, but alternative mechanisms must exist that can accomplish the same tasks.

During the S phase that precedes cell division, when a DNA replication fork passes through an ^{m}CpG doublet, each of the daughter helices is transiently hemimethylated (methylated only on the old DNA strand). If there were no way of restoring the symmetrical methyl group, methylation at this particular genomic locus would quickly be diluted out of the population through successive generations. However, we know that methylation is heritable, and thus it is not surprising that cells contain a maintenance methylase that is capable of methylating the cytosine opposite an ^{m}CpG doublet soon after it emerges from the replication fork.

When we ponder the existence of maintenance methylases in eucaryotes, together with the ability of methylation to regulate gene activity, it is very tempting to speculate that methylation at CpG doublets may be one of the mechanisms by which eucaryotic cells maintain the differentiated state. If genomic DNA in germ cells or in some very early stage of embryogenesis were automatically restored to a kind of methylation ground state corresponding to no differentiation, methylation and demethylation at particular chromosomal sites could play a role in programming expression of genes throughout the course of development (Figure 22-8). The main objection to such a model is that it is difficult to imagine how either methylation or demethylation would itself be regulated (the hall of mirrors effect, again). All the known eucaryotic maintenance methylases appear to be nonspecific in their activity: *Any* accessible hemimethylated CpG doublet will be methylated. As a result, we are forced to invoke other factors, such as the binding of specific regulatory proteins or changes in chromatin structure (Chapter 21), that could affect the accessibility of critical cytosine residues to methylation. However, once methylation or demethylation has taken place at a particular regulatory site, maintenance of the differentiated state could then be entrusted to a *non*specific methylase activity.

Irreversible Cytoplasmic Differentiation Concomitant with Loss of Ability to Divide

Many cells contain nuclei that, under normal conditions, will never divide again. Often, they represent cases of extreme differentiation, where a very large fraction of the total synthesis is devoted to making just a few different proteins. Immature red blood cells, or reticulocytes, are perhaps the most striking example. These cells never divide, having as their only function the production of hemoglobin molecules able to combine reversibly with oxygen. Moreover, in some species, including all mammals, the nuclei of these cells eventually disintegrate, so that the mature erythrocytes (red blood cells) are unable to make any more hemoglobin. Behaving somewhat similarly are the adult plasma cells, which produce specific antibodies. They likewise never divide, but secrete their antibodies for several days and then die. During this time, their nuclei remain fully functional, synthesizing the mRNA molecules coding for their antibody products. In contrast, many other specialized cells (e.g., nerve cells), though never dividing, may live many, many years. The functioning of such long-

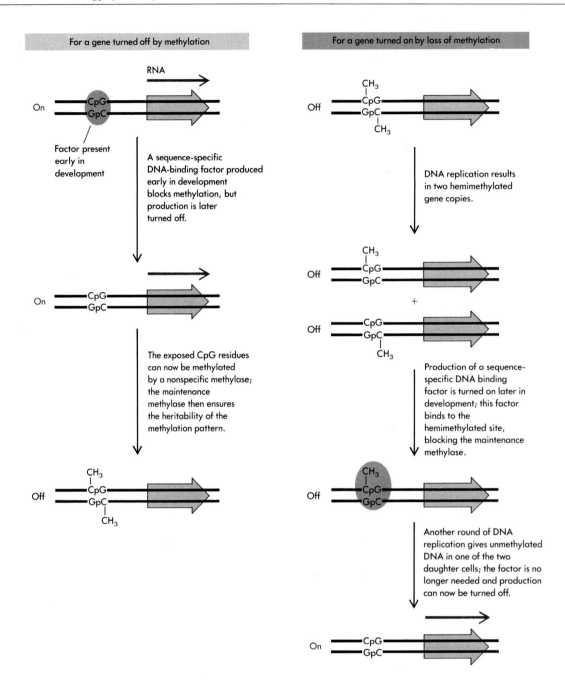

Figure 22-8
Methylation (or lack of it) is heritable and could maintain the differentiated state.

lived cells always depends on the presence of a nucleus able to make mRNA. Some protein synthesis, even at a very low level, seems to be necessary for the continued functioning of all cells. This synthesis may replace aging proteins targeted for destruction by the ATP- and ubiquitin-dependent proteolytic system (Chapter 21).

Stable Transcription Complexes Could Also Maintain the Differentiated State[17]

Experimental evidence suggests that genes transcribed by all three eucaryotic RNA polymerases are capable of forming stable transcription complexes with their respective transcription factors. In the case of RNA polymerase III, it is an essential transcription factor (*not* the

RNA polymerase III itself) that forms a stable complex with the DNA (Chapter 21). Once the transcription factor binds stably to the gene, that particular locus is activated and can be repeatedly transcribed by RNA polymerase III at least until DNA replication occurs. Recent experiments suggest that much the same situation holds true for both RNA polymerases I and II. Thus, stable transcription complexes provide an attractive possible mechanism for maintenance of the differentiated state during interphase (i.e., between cell divisions).

Stem Cells Divide Asymmetrically[18]

Stable transcription complexes can also help to explain the heritability of gene activity during DNA replication and subsequent cell division if the complexes can pass unharmed through the DNA replication fork. This possibility suggests an explanation for the behavior of stem cells. Not all cells give rise to two identical daughters. Stem cells divide to give one daughter identical to the parent, thus perpetuating the stem cell line itself, and another daughter committed to differentiate. For example, the highly differentiated erythrocytes circulating in the human body have a lifetime of only 120 days, after which they wear out and are destroyed in the spleen. Red blood cells are continuously replaced by repeated divisions of hematopoietic stem cells in the bone marrow, which give rise to an identical daughter stem cell and an immature erythrocyte (reticulocyte).

Recent experiments have shown that the RNA polymerase III transcription factor TFIIIA (Figure 21-34) binds preferentially to the noncoding strand of the 5S ribosomal RNA gene; other stably bound transcription factors for RNA polymerases I and II may also have a strong strand preference. We can speculate that when the DNA replication fork passes through such an activated transcription unit, the transcription factor(s) may remain bound to the parental noncoding strand. The daughter chromosome that inherits this parental DNA strand will therefore inherit the factor, and along with it the gene in a transcriptionally activated state. But the gene on the daughter chromosome that inherits the parental coding strand may simply be packaged into nucleosomes like any unremarkable stretch of DNA and thus be repressed (Figure 22-9). The conservative segregation of parental nucleosomes at the replication fork provides yet another possible mechanism for producing two daughter cells, only one of which retains the gene in an active (or inactive) state.

DNA REARRANGEMENTS AND
AMPLIFICATION DURING DEVELOPMENT

Extensive DNA Rearrangements
Occur During the Life Cycle of Ciliates[19–22]

In certain rare cases, differentiation does require rearrangement of DNA. For example, the bacterium *Salmonella* alternates between two antigenically distinct flagellar proteins (flagellins) by inversion of a specific promoter-containing DNA segment (Figure 11-26). Bakers' yeast switches between the **a** and α mating types by unidirectional gene conversion of the *MAT* locus, using one of the two silent

Figure 22-9
Stem cell lineages could be maintained by conservative segregation of DNA-binding factors at the replication fork.

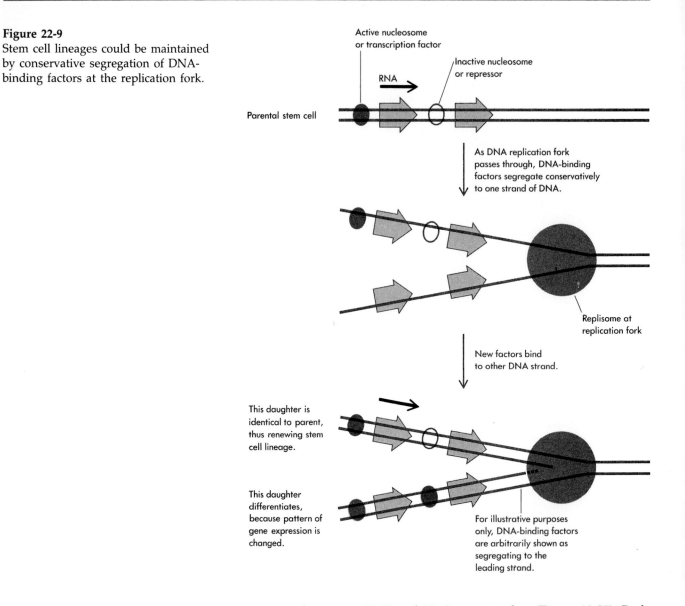

cassettes known as *HMR* and *HML* as a template (Figure 18-25). Both antigen alternation in *Salmonella* and mating-type switching in yeast are reversible DNA rearrangements that do not affect the total genetic information inherited by daughter cells. Similarly, as we shall see in Chapter 23, the production of immunoglobins and certain antigen-specific cell-surface proteins during the mammalian immune response involve complex rearrangements of the corresponding genes. However, these rearrangements occur exclusively in certain somatic cells called B and T lymphocytes, never in the germ line. Thus, these DNA rearrangements, although irreversible, cannot be passed on to the next generation.

Here we will consider the curious predicament of the ciliates, whose somatic and germ line nuclei share a common cytoplasm. These unicellular aquatic organisms provide by far the most dramatic exception to the general rule that differentiation occurs without rearrangement of DNA. The two nuclei within each ciliate cell serve very different functions. The larger **macronucleus** is the workaday nucleus responsible for all transcription during ordinary vegetative growth and asexual reproduction by binary fission. This somatic nucleus is polyploid and contains many copies of the genes required for vegetative growth and binary fission. In contrast, the smaller germ line **micronucleus** is diploid and transcriptionally silent during the asexual life cycle; this nucleus functions only when opposite mating types come together for sexual conjugation. Then, the diploid micronucleus undergoes meiosis, and the resulting haploid micronuclei serve as germ cells. After the haploid micronuclei of opposite mating type fuse to form a diploid zygotic nucleus, the old macronuclei degenerate. Finally, through a complex succession of programmed nuclear divisions and disintegrations, the zygotic nucleus gives rise to a new haploid micronucleus and an immature macronucleus (Figure 22-10).

In many, and perhaps most, species of ciliates, polyploidization of the macronucleus occurs in two stages. First, repeated rounds of DNA replication without subsequent nuclear division produce many daughter chromatids in perfect lateral register. In some species (especially the hypotrichs such as *Oxytricha*), the number of rounds of DNA replication is sufficient to produce visible polytene chromosomes that bear a striking resemblance to those found in the salivary gland and fat body of the fruit fly *Drosophila* (see Figure 21-12). However, unlike the *Drosophila* polytene chromosomes, which remain intact, the newly polytenized chromosomes in the immature ciliate macronucleus are immediately transected into fragments. Fragmentation is more extreme in some species than in others; in *Oxytricha nova*, the fragments range in size from 1000 to 20,000 base pairs. Transection is followed by massive degradation of as much as 95 percent of the polytenized DNA in certain ciliate species. Those pieces that survive degradation seem to contain the genes required for vegetative growth and asexual reproduction; DNA that is not necessary for the asexual phase of the life cycle may simply be discarded.

In the second stage of polyploidization, the surviving linear DNA fragments in the maturing macronucleus replicate to still higher copy number (about 45 copies per macronucleus in the case of *Tetrahymena*). Thus, each linear fragment is capable of autonomous replication and must be equipped with an origin of DNA replication and two telomeres (see Figures 18-14 and 18-15). As expected, the origins of DNA replication appear intact in germ line micronuclear DNA, but interestingly, the telomeres do not. Instead, macronuclear telomeres (composed of tandem repeats of the simple sequence C_4A_2 or C_4A_4, depending on the species of ciliate) appear to be polymerized de novo onto the ends of the fragments by a T_2G_4- or T_4G_4-adding enzyme, which is primer- but not template-dependent. In other words, depending on the species from which the enzyme is isolated, the telomere-adding enzyme will polymerize additional tandem repeats of T_2G_4 or T_4G_4 onto the free 3' hydroxyl group of a single-stranded primer with the sequence $(T_2G_4)_n$ or $(T_4G_4)_n$, where n is greater than or equal to 1. Presumably, the complementary strand containing C_4A_2 or C_4A_4 repeats is then synthesized by a template-dependent DNA polymerase, which copies the new T_2G_4 or T_4G_4 repeats. (The CCA-adding enzyme that polymerizes *ribo*nucleotides onto the 3' end of tRNA is another example of a primer-dependent but template-independent reaction; see Chapter 14.)

The linear fragments of DNA in the macronucleus are not true chromosomes, because they lack the centromere required for attachment of spindle fibers to the chromosome (Chapter 18). Without a spindle apparatus to ensure that each daughter cell receives an equivalent number of macronuclear DNA fragments, the macronucleus must divide by uneven budding rather than by conventional mitosis.

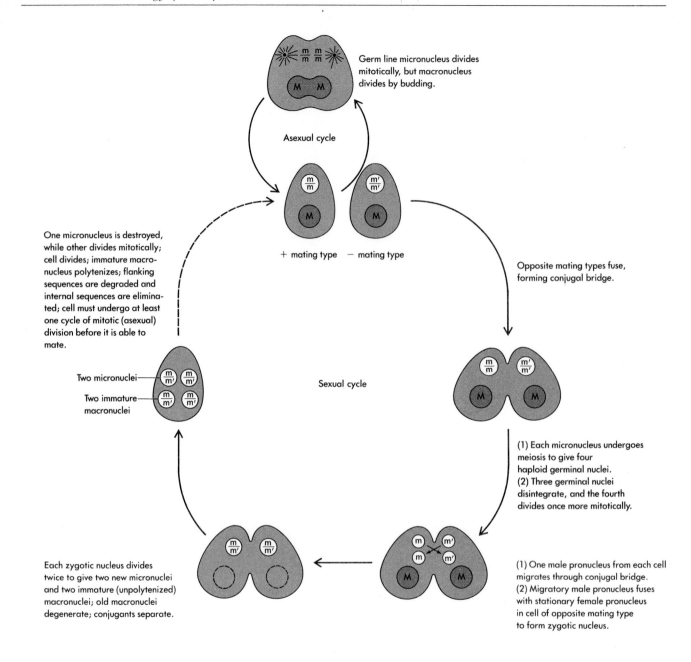

Germ line micronucleus divides mitotically, but macronucleus divides by budding.

Asexual cycle

+ mating type   − mating type

One micronucleus is destroyed, while other divides mitotically; cell divides; immature macronucleus polytenizes; flanking sequences are degraded and internal sequences are eliminated; cell must undergo at least one cycle of mitotic (asexual) division before it is able to mate.

Opposite mating types fuse, forming conjugal bridge.

Sexual cycle

Two micronuclei
Two immature macronuclei

(1) Each micronucleus undergoes meiosis to give four haploid germinal nuclei.
(2) Three germinal nuclei disintegrate, and the fourth divides once more mitotically.

Each zygotic nucleus divides twice to give two new micronuclei and two immature (unpolytenized) macronuclei; old macronuclei degenerate; conjugants separate.

(1) One male pronucleus from each cell migrates through conjugal bridge.
(2) Migratory male pronucleus fuses with stationary female pronucleus in cell of opposite mating type to form zygotic nucleus.

Figure 22-10
The complex nuclear life cycle of the ciliate *Tetrahymena*. The polyploid macronucleus is denoted by M; the diploid micronucleus, by m/m. As many as seven different mating types are known in this organism, but for simplicity, only two mating types are shown here (arbitrarily designated + and −). Because the genetic determination of mating type is complex, cells derived from a mating (exconjugants) may have a different mating type from either parent (the conjugants). Thus, the dotted arrow in the last stage of the sexual life cycle is meant to indicate that the exconjugants may differ from the parental strains in mating type as well as in genetic constitution. Cells of any mating type can divide mitotically, but for clarity, the asexual life cycle is illustrated for the + mating type alone.

Perhaps the initiation of DNA replication on the linear DNA fragments is subject to feedback inhibition, so that each macronuclear fragment can be maintained at constant copy number despite unequal distribution of macronuclear DNA fragments to the daughter macronuclei during budding.

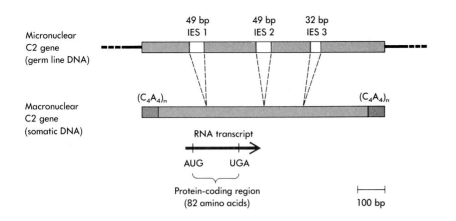

Figure 22-11
Internal sequences are eliminated from genes during macronuclear development in the ciliated protozoan *Oxytricha nova*. Gray segments show micronuclear sequences that are preserved in the macronucleus; heavy hollow line, internal sequences eliminated during regeneration of the macronucleus; thin line, flanking sequences degraded during macronuclear regeneration; colored segments, telomeres presumably added after site-specific cleavage of micronuclear DNA. [After L. A. Klobutcher, C. L. Jahn, and D. M. Prescott, *Cell* 36 (1984):1045–1055.]

Why is the macronucleus polyploid? Unicellular ciliates are complex and gigantic cells by normal standards; such size and complexity have many advantages, perhaps the most obvious of which is that ciliates are able to engulf and consume smaller organisms. Many gene copies are required to synthesize enough structural RNA (rRNA, tRNA, etc.) and mRNA to stock the cytoplasm with sufficient quantities of each gene product. Thus, polyploidization enables the ciliate to achieve large size and complexity without solving the many new developmental problems associated with multicellularity.

Do Ciliates Splice DNA?[23]

Very recently, it has become clear that the macronuclear DNA fragments are not always colinear with the micronuclear DNA sequences from which they are derived. In addition to elimination of DNA sequences *between* genes during transection and selective degradation of the polytene chromosomes, internal DNA sequences *within* protein-coding regions are also precisely excised (Figure 22-11). From the limited data now available, such **internally eliminated sequences (IESs)** can be as short as 32 base pairs or as long as several kilobase pairs. Since some IESs interrupt protein-coding regions, they must be eliminated precisely lest nonsense proteins be synthesized. IESs are flanked by short, direct repeats and contain internal inverted repeats that could potentially function as signals for excision at either the RNA or DNA level; however, IESs do not resemble introns found in mRNA, tRNA, or rRNA, all of which are removed by RNA splicing. IESs might be removed by direct DNA splicing; by reverse transcription of a spliced RNA, as has been postulated for the generation of processed genes in mammalian genomes (Figures 20-30 and 20-31); or by gene conversion of the DNA using a spliced RNA (or its reverse transcript) as template. Only further experiments can decide between these unprecedented possibilities.

Trypanosomes Evade the Host Immune Response by Switching Surface Antigens

Trypanosomes are flagellated unicellular protozoan parasites that live in the bloodstream of infected mammals and (like the malarial parasite) are spread from one animal to another by an insect vector, in this case the tsetse fly. The most common human disease caused by trypanosomes is African sleeping sickness, a name that graphically describes

Figure 22-12
Activation and replacement of trypanosome VSG genes at the telomeric expression sites. (a) Early after infection, only one particular VSG antigen is expressed; three or more chromosomes have telomeric VSG genes, but only the expression site is actively transcribed. (b) Later during infection, directed gene conversions replace one or more of the telomeric VSG genes with new variant genes drawn from the transcriptionally silent basic copies. (c) During first relapse, a new variant VSG is expressed when the old telomeric expression site is turned off and a new site is activated.

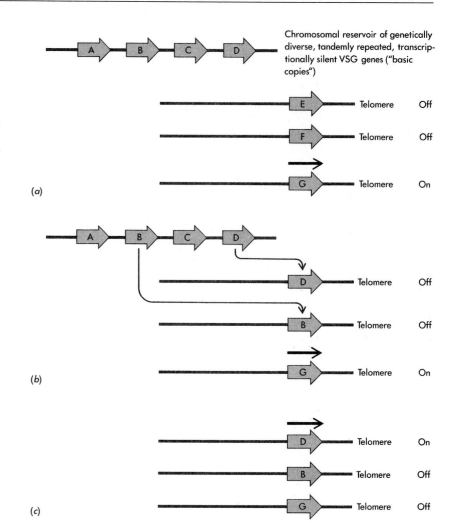

the complete fatigue accompanying persistent infection with an aggressive parasite. In Africa and South America, trypanosomes frequently infect cattle, severely lowering the yield of milk and meat and increasing susceptibility of the animals to other diseases. The lives of millions of people in underdeveloped countries are crippled by trypanosomiasis, and the development of practical treatments to combat this dread endemic disease is now the focus of a major worldwide effort.

The surface of each bloodstream trypanosome is concealed beneath a dense coat of a **variant surface glycoprotein (VSG)**. The host immune system recognizes this VSG as a foreign antigen and makes antibodies against it in an effort to target the parasite for destruction. The host immune response would succeed in eliminating the trypanosome infection were it not for the amazing ability of the parasite to replace the first VSG with a second antigenically unrelated VSG. Such **antigen switching** apparently occurs quickly enough to keep the host on the defensive, and thereby effectively outwit the remarkable resources of the mammalian immune system (Chapter 23). Moreover, the trypanosome's genome contains a repertoire of perhaps a thousand different VSG genes. Thus, it is impossible to manufacture a conventional vaccine to guard against trypanosome infection, because any particular vaccine would react with only one of the many potential VSGs. The hope is that a better understanding of the basic molecular biology of trypanosome infection will suggest new approaches for eradicating the disease.

In Chapter 11, we saw how the bacterium *Salmonella* can switch from one surface antigen to another (a process called phase variation for historical reasons) by inverting a small promoter-containing DNA segment that can drive transcription of either of two divergently transcribed surface antigen genes (Figure 11-26). The trypanosome, however, can switch between at least 100 different VSGs. Clearly, a simple promoter inversion cannot explain this switching behavior.

In fact, trypanosome antigen switching is a much more difficult logistical operation than bacterial phase variation for two main reasons. First, the protozoan cell is far larger than the tiny bacterium, and the particular VSG being expressed at any one time constitutes at least 10 percent of the total trypanosome protein. To make so much VSG protein, either the VSG must be encoded by multiple genes, or transcription must take place at an extraordinary rate, or translation of the VSG messenger RNA must be unusually efficient, or the VSG mRNA must be unusually stable. Second, the trypanosome must be able to switch VSG disguises at a relatively high frequency (estimated to be about 10^{-4} to 10^{-5} per generation). In this way, in any given population of bloodstream parasites expressing one VSG antigen, at least one individual parasite will switch to a new VSG antigen before the host immune response is able to kill all members of the population. Moreover, trypanosomes expressing the new VSG antigen must not continue to express the old VSG, or they will be the target for attack by antibodies directed against the previous dominant VSG antigen. Such rapid antigenic switching requires that the VSG mRNA be relatively *un*stable. The necessary instability of the VSG mRNA only emphasizes the importance of producing huge amounts of translationally active VSG mRNA.

Multiple Mechanisms of Trypanosome Antigen Switching[24]

Antigen switching in the parasite *Trypanosome brucei* can be studied in the laboratory using infected rabbits. After initial innoculation, the rabbits become sick as *T. brucei* bearing a particular VSG multiply within the bloodstream. The rabbits recover as their immune system mounts a response against the first VSG, but soon relapse as trypanosomes bearing a new VSG multiply and eventually become the dominant VSG type in the bloodstream. By cloning mRNA from the bloodstream parasites after the first, second, and third cycles of parasitemia, it is possible to assemble a collection of VSG mRNAs corresponding to successive switching from one VSG gene to the next within a single cell lineage. When these cDNA clones are used as probes for the genomic organization of the VSG genes in the three successive parasite populations, a truly remarkable pattern emerges (Figure 22-12).

VSG genes are found in two places in the trypanosome genome: as tandemly repeated copies at internal positions within the chromosomes, and as individual copies at the chromosomal ends (telomeres). Telomeric VSG genes can be found on at least three different chromosomes, and quite possibly on more. (The situation is complicated because trypanosomes have eight to ten relatively large chromosomes of 200 to 2000 kilobase pairs and about a hundred minichromosomes ranging in size from about 50 to 150 kilobase pairs.) The expressed VSG gene is always located at one of the telomeric sites, so these are called **expression sites.** Only one telomeric VSG gene is activated at a time, but the mechanism responsible for such mutually exclusive activation remains a mystery. The several thousand VSG gene copies located internally within the chromosome are transcriptionally silent and can only be expressed after they are duplicated and moved from an internal tandem array to a telomeric expression site (presumably by a directed gene conversion). Quite possibly, basic copies of the VSG genes are continually moved from the internal tandem arrays to telomeric sites to replenish the variety of VSG genes awaiting activation at a telomeric expression site.

The similarity between switching of yeast mating types (Chapter 18) and trypanosome variant surface antigens should be obvious. The silent mating type cassettes at *HMR* and *HML* are analogous to the tandem repeats of transcriptionally silent basic VSG gene copies. The directed gene conversion that moves silent genetic information from *HMR* or *HML* to *MAT* (the expression site) is analogous to antigen switching. And the analogy may extend still further. In the yeast mating switch, not only must the genetic information at *MAT* be expressed, but the silent gene copies at *HMR* and *HML* must also be repressed. The selective repression of the *HMR* and *HML* loci (but not *MAT*) is the job of the *SIR* (silent information repression) gene products, which appear to alter the chromatin structure surrounding the silent cassettes. Perhaps a similar mechanism, based on the difference between active and inactive chromatin (Chapter 21), may allow trypanosomes to express only one telomeric VSG gene at a time.

Interestingly, the first 35 base pairs of all expressed VSG mRNAs are identical, regardless of antigenic type. However, the DNA encoding this 35-base-pair "mini-exon" sequence is not part of the basic VSG gene copy that is moved to the telomeric expression site, nor does the expression site contain a 35-base-pair mini-exon-coding sequence that can be used by whichever basic VSG gene copy happens to occupy the expression site. Instead, about 200 copies of the 35-base-pair sequence are present elsewhere in the trypanosome genome, largely in the form of a 1.35-kilobase-pair tandemly repeated DNA segment (Figure 22-13). In particular, at least one telomeric expression site is located on a 550-kilobase-pair chromosome completely lacking any 35-base-pair mini-exon sequences. Thus, transcription of a complete VSG mRNA must be discontinuous; the 35-base-pair mini-exon sequence is transcribed independently of the body of the VSG mRNA and subsequently added to it.

How is the 35-base-pair mini-exon sequence assembled onto the body of the expressed VSG mRNA? We now know that the 35-base-pair mini-exon-coding regions are transcribed to give 140-base RNAs known as "mini-exon-derived RNAs" or medRNAs. We also know that both the 140-base medRNAs and the mature VSG mRNA possess the 5' cap structure typical of all eucaryotic mRNAs. Thus, there are two obvious possible mechanisms for the joining reaction (see Figure 22-13). Either the medRNAs could serve as primers for transcription of the body of the VSG mRNA, or the medRNAs could be joined to the body of the VSG mRNA by an unusual *trans* (intermolecular) RNA splicing reaction. Experiments now in progress will soon distinguish between these possibilities.

Although first discovered in VSG mRNAs, the evidence now suggests that most (and perhaps all) trypanosome mRNAs share the same 5' mini-exon sequence. Why trypanosomes have resorted to such an exotic procedure for initiating mRNA transcription remains a mystery, but recent work suggests that *trans* splicing may not be confined to trypanosomes. In extracts of human cells, *trans* splicing can

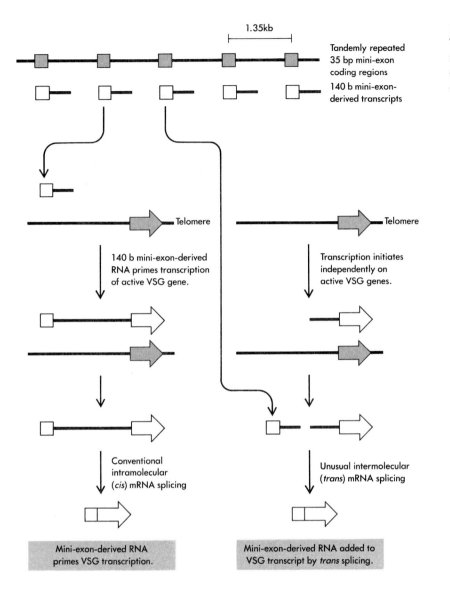

Figure 22-13
Two models for assembly of the common 35-nucleotide mini-exon sequence onto the body of the expressed trypanosome VSG gene (not drawn to scale).

proceed efficiently if the two separate RNA molecules carrying the 5' and 3' exons are constructed in such a way that base pairing between complementary sequences in the intron can hold the two molecules together. No natural instance of *trans* splicing has yet been documented in higher eucaryotes, but it seems unlikely that evolution has failed to take advantage of a reaction that can proceed so easily in vitro.

Unusual Excision and Amplification of the Gene Encoding *Tetrahymena* Ribosomal RNA[29, 30]

In Chapter 20, we discussed the ability of the ribosomal RNA precursor (pre-rRNA) in the ciliate *Tetrahymena* to function as an enzyme (ribozyme). However, autocatalytic splicing is not the only extraordinary aspect of the rRNA in this ciliate. In most eucaryotes ranging from yeast to mammals, the genes encoding ribosomal RNA constitute a large multigene family with several hundred germ line copies per haploid genome (Chapter 20). Remarkably, the germ line micronucleus of *Tetrahymena* contains only a single copy of the rDNA gene.

Figure 22-14

Multiple copies of the 21-kilobase linear extrachromosomal palindromic rDNA in the *Tetrahymena* macronucleus are derived by excision and amplification from a single 10.5-kilobase half palindrome in the germinal micronucleus. A plausible mechanism for this process is shown here. To distinguish between the two strands of the rDNA, one of them is drawn as a thick line. The flanking chromosomal sequences in the micronucleus are represented by wavy lines. M represents the 42-nucleotide inverted repeats on either side of the future center of palindromic symmetry. A′ represents the 20-nucleotide inverted repeats flanking the rDNA. Excision involves breakage of the chromosome at or near the two ends of the rDNA. The A′ sequence might serve as the recognition sequence for breakage. Some flanking sequences are lost from both breakage sites, probably by exonuclease activity. These events lead to the formation of four free ends, at least three of which are subsequently modified by addition of the telomeric T_2G_4 repeat sequences to one strand (black boxes) and C_4A_2 repeats to the other strand (white boxes). The 5′ end of the rDNA, which is not modified by the addition of telomeres, folds back to allow pairing between the two inverted M sequences. Folding back is possible because the micronuclear half palindrome is actually a little larger than half and includes the future center of palindromic symmetry. Intramolecular homologous recombination then joins the free end of the thick line to an internal site on the thin line, producing one short and one long hairpin molecule. The short hairpin molecule primes replication of the long one, and after one complete round of replication, the first mature extrachromosomal 21-kilobase palindrome is formed. The rRNA transcription units are shown only on this first mature rDNA molecule (large shaded arrows). The figure is not drawn to scale. [After M.-C. Yao, S.-G. Zhu, and C.-H. Yao, *Mol. Cell. Biol.* 5 (1985):1260.]

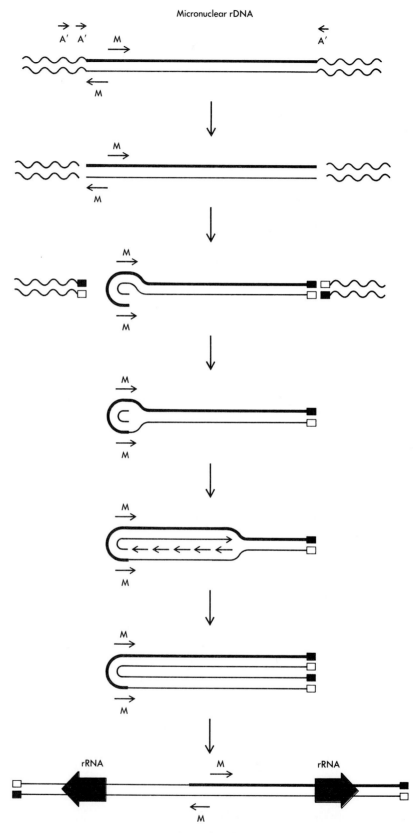

During regeneration of the new macronucleus after mating, this single rDNA copy is excised from micronuclear DNA as a 10.5-kilobase-pair linear molecule, transformed into a nearly perfect 21-kilobase-pair palindromic dimer, and subsequently amplified to higher copy number much like any other macronuclear DNA fragment (Figure 22-14).

Early studies had established that the palindromic rDNA dimer, like other linear DNA molecules in the macronucleus, has both an origin of DNA replication and telomeres. The origin is located very near (or even at) the center of symmetry, as judged by the distribution of replication bubbles visualized in the electron microscope (Chapter 21). Telomeres (which also function in yeast; see Figure 18-14) consisting of 40 to 70 tandem repeats of the simple sequence C_4A_2 were found at either end of the palindrome. In principle, then, we might have guessed that the single micronuclear (germ line) copy of the gene encoding rRNA would be excised as a simple half palindrome, where one endonuclease cutting site would be defined by the center of rDNA symmetry and the other site by tandem germ line repeats of C_4A_2. However, these expectations proved to be naive.

The first surprising result was that the DNA sequence surrounding the origin of replication near the center of the palindrome was not perfectly symmetrical. This was a clue that the extrachromosomal palindromic rDNA could not be generated by simple excision and dimerization of an exact germ line half palindrome (monomer). In fact, after conjugation, the earliest detectable rDNA excision product was not cut at the center of (imperfect) symmetry but rather 42 base pairs away (see Figure 22-14). Apparently, after excision of the germ line rDNA monomer, this 42-base-pair sequence can fold back to form an imperfect hairpin that primes DNA replication, thereby converting the newly excised linear monomers into mature palindromic dimers.

The second surprising result was that the germ line rDNA monomer contained only a single C_4A_2 where the telomere should be. Thus, the mature telomeres are probably synthesized de novo by the T_2G_4-adding enzyme together with DNA polymerase. (For historical reasons, telomere sequences are conventionally given as the C_nA_m-containing strand, although the properties of the T_mG_n-adding enzyme appear to determine the exact sequence of the telomeres in each species.) In principle, the solitary germ line C_4A_2 could serve as the recognition sequence for excision by a C_4A_2-specific endonuclease. Alternatively, excision could occur beyond the germ line C_4A_2 sequence, and an exonuclease would then nibble back the free end until the exposed germ line T_2G_4 could be recognized by the T_2G_4-adding enzyme. Recently, the germ line rDNA monomer was found to be flanked on either side by inverted repeats of a 20-base-pair sequence. This suggests that the 20-base-pair sequence (which is also found at many other places in the *Tetrahymena* genome) is the recognition site for an endonuclease that fragments micronuclear DNA at specific sites during regeneration of the macronucleus.

Excision and Selective Amplification of the rRNA Genes Within Amphibian Oocytes[31]

The number of rRNA genes within an amphibian oocyte is about a hundred- to a thousandfold greater than within somatic cells of the same organism. But this extra DNA does not reflect an increase in chromosome number from the $4N$ characteristic of oocytes prior to

Figure 22-15
Electron micrograph of an extrachromosomal nucleolar core from one oocyte of the African toad *Xenopus laevis*. About ten rRNA genes are tandemly arranged along the circular DNA molecule. The arrows point to individual genes. (Courtesy of O. L. Miller and B. R. Beatty, Oak Ridge National Laboratory.)

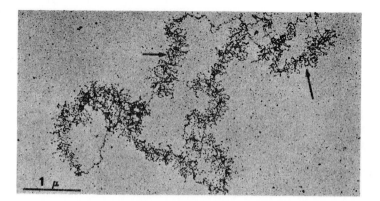

the first meiotic division. Instead, almost all the increase is due to large numbers of extrachromosomal nucleoli. Each nucleolus contains one or more circular DNA molecules with length varying from 12 to about 210 kilobase pairs. This corresponds to 1 to 20 tandemly arranged rRNA genes, each coding for the 45S rRNA precursor. Figure 22-15 shows a most elegant electron micrograph of one of these extrachromosomal nucleoli to which are attached over a thousand growing rRNA precursor molecules.

Very large numbers of extrachromosomal nucleolar circles are needed to produce sufficient RNA components (18S, 5.8S, and 28S rRNA) for the massive stock of ribosomes found in a mature oocyte, the largest of all cells. This is especially true because structural RNA species such as ribosomal RNA are the final product of synthesis, and unlike messenger RNA, the product cannot be further amplified by multiple rounds of translation. All the ribosomes present before gastrulation are synthesized before fertilization in the oocyte. During growth of the frog oocyte, for example, some 10^{12} ribosomes must be made in the course of only a few weeks. The number of rRNA genes within the oocyte chromosomes, however, reflects the requirements of the ordinary frog somatic cell, which is some 10^{5} times smaller than the oocyte. Only by extensive rDNA amplification can oocyte growth proceed at a biologically acceptable rate.

Most of the nucleolar circles in mature oocytes arise by a rolling circle mechanism (Chapter 10) from previously made circles (Figure 22-16). Still very unclear, however, is the origin of the first extrachromosomal gene copies. Genetic evidence obtained from hybrid frogs shows that the oocyte's extrachromosomal rRNA genes must be derived from the tandemly repeated chromosomal rRNA genes. Conceivably, the first extrachromosomal gene copies could originate by a crossing-over mechanism similar to that which releases λ prophages from the *E. coli* chromosome. But this scheme requires the later reinsertion of rDNA so that the number of chromosomal rRNA genes will remain relatively constant. Another possibility is the enzymatic creation of specific single-strand cuts in chromosomal rDNA, which lead to the displacement of long, single-stranded sections. Such displaced chains could then be converted into double-stranded circles by a process of the rolling circle type. This matter will most likely be settled only when such circles can be made in vitro by oocyte extracts.

The Frog Oocyte 5S RNA Genes Are "Preamplified"

In all eukaryotes, the 18S, 5.8S, and 28S rRNA components of the ribosome are derived by RNA processing from a single primary transcript synthesized by RNA polymerase I; however, the 5S RNA component of ribosomes is transcribed from an entirely separate family of genes by RNA polymerase III (Chapter 20). Some provision must therefore be made during oogenesis for producing the same amount of 5S RNA as there is 18S plus 5.8S plus 28S rRNA synthesized by the thousandfold amplified extrachromosomal rDNA. As we saw in Chapter 20, the frog accomplishes this by having two different sets of 5S RNA genes, one coding for the 5S RNA found in oocyte ribosomes, the other for the 5S RNA of somatic cells. While there are only 500 somatic 5S genes, over 10,000 copies of the oocyte variety are present per haploid genome; both types of genes are organized into separate tandemly repeated arrays. Thus, no extrachromosomal amplification of 5S DNA need occur during oogenesis; the genes are in effect "preamplified." However, some control mechanism must selectively turn on the much larger gene set needed for oocyte ribosome synthesis.

The several hundred germ line copies of the 12,000-base-pair rDNA repeat unit normally account for only 0.2 percent of frog DNA, but after thousandfold amplification, the extrachromosomal rDNA accounts for 75 percent of the total oocyte DNA. This explains why developmentally regulated amplification of the rDNA is a better biological strategy than carrying a thousandfold excess of rDNA genes in every somatic cell. In contrast to the 12,000-base-pair rDNA repeat unit, the 5S gene repeat unit is only about 500 base pairs long. Consequently, even 20,000 germ line copies of the oocyte 5S genes still account for only 0.01 percent of the frog DNA, a tolerable amount.

In contrast to the tandemly repeated germ line rDNA genes, which are located on only one or a few frog chromosomes, the tandemly repeated 5S genes are located near (or at) one of the telomeres of almost every chromosome (Figure 22-17). We might attempt to rationalize the telomeric location of both oocyte and somatic 5S genes by noting that the major steps in ribosome assembly are completed *within* the nucleolus using newly synthesized ribosomal proteins imported from the cytoplasm (Chapter 21). Moreover, the many extrachromosomal nucleoli in an oocyte are physically located just inside the nuclear membrane, perhaps to minimize the distance that incoming ribosomal proteins and outgoing ribosomes must traverse during the massive accumulation of ribosomal subunits. Since telomeres are often associated with the nuclear membrane, the telomeric location of oocyte 5S genes might facilitate the efficient incorporation of newly synthesized 5S RNA into ribosomes. Unfortunately for this hypothesis, 5S genes in the fruit fly *Drosophila melanogaster* are located internally on the same chromosome that carries the rDNA genes. This discrepancy suggests that it may be a long time before we understand why these (and other) genes are located where they are on chromosomes.

Lampbrush Chromosomes Fulfill the Massive Transcriptional Demands of Large Oocytes[32]

To make 10^{12} ribosomes for use in early embryogenesis, the *Xenopus* oocyte must synthesize more than just rRNAs. A correspondingly high rate of transcription (and translation) of the mRNAs coding for ribosomal proteins is required. Also provisioned for the egg during the few months of oogenesis are huge stores of many other proteins (e.g., enough histones to package 4096 genomes worth of DNA), as well as most of the mRNAs that are translated only after fertilization during the early stages of development (maternal mRNAs).

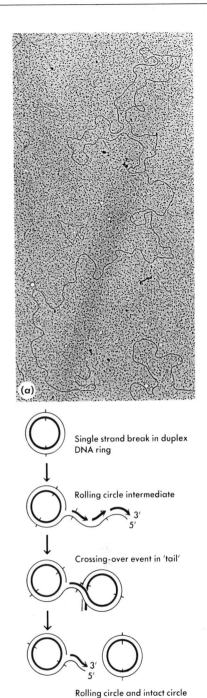

(a)

Single strand break in duplex DNA ring

Rolling circle intermediate

3'
5'

Crossing-over event in 'tail'

3'
5'

Rolling circle and intact circle

(b)

Figure 22-16

(a) An electron micrograph of a ribosomal DNA rolling circle from a *Xenopus laevis* oocyte. The circumference of the template circle corresponds to three ribosomal RNA genes and the attached tail to 3.4 ribosomal RNA genes. [Reproduced with permission from D. Hourcade, D. Dressler, and J. Wolfson, *Cold Spring Harbor Symp. Quant. Biol.* 38 (1973):537–550.]

(b) Ribosomal DNA amplification by a rolling circle mechanism from an extrachromosomal nucleolar circle. Compare with Figure 10-21.

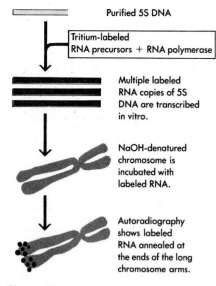

Purified 5S DNA

Tritium-labeled RNA precursors + RNA polymerase

Multiple labeled RNA copies of 5S DNA are transcribed in vitro.

NaOH-denatured chromosome is incubated with labeled RNA.

Autoradiography shows labeled RNA annealed at the ends of the long chromosome arms.

Figure 22-17

The demonstration that the multiple copies of 5S RNA genes are located at or near the telomere region on the long arms of most chromosomes in the toad *Xenopus*. Labeled RNA copies of the 5S gene (shown), or in vivo labeled 5S RNA molecules, anneal with denatured DNA in the chromosomes and are revealed by autoradiography.

Figure 22-18

A partial view of a pair of highly extended lampbrush chromosomes from an oocyte of the newt *Triturus viridescens*. Two chiasmata are seen in this micrograph, taken in the diplotene phase of meiosis. The arrow points to a section where the double-stranded character of each premeiotic chromosome is revealed. (Courtesy of J. Gall, Carnegie Institution of Washington.)

To meet such incredible demands for RNA synthesis, the chromosomes of large oocytes assume a bizarre extended configuration. As shown in Figure 22-18, much of the DNA is compacted into tight masses called **chromomeres** that contain the main chromosomal axes. But a significant fraction of the genome (from 5 to 30 percent, depending on the species) is extended into very long lateral loops that contain up to 100 kilobase pairs; hence the name **lampbrush chromosomes.** (Lampbrushes were used to clean soot out of the globe of an oil or gas lamp before the advent of the incandescent electric bulb.) While the condensed chromomeric regions are largely inactive in transcription, the majority of loops are transcribed at nearly maximal rates. That is, a nascent RNA chain emanates about every 100 base pairs (Figure 22-19) from the looped-out DNA helix. This can be compared to the average, highly transcribed gene region in somatic cells, where one RNA transcript is seen only every 10,000 base pairs. The presence of lampbrush chromosomes is nicely correlated with large oocyte size, which can occur in amphibians, reptiles, birds, and insects, but not in mammals. Among echinoderms, starfish have lampbrush chromosomes, whereas sea urchins do not; urchin eggs are several hundredfold smaller than those of starfish.

The most curious aspect of lampbrush chromosomes, which distinguishes them from somatic chromosomes, is that many of the lampbrush transcripts are gigantic. Even huge loops are often transcribed

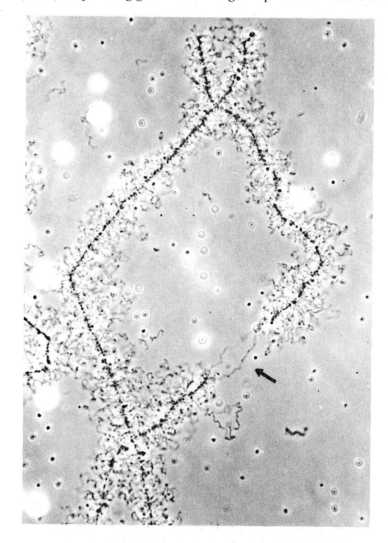

into a single RNA species. Such loops have a distinctively polar appearance caused by the association of proteins to form RNP particles on the nascent RNA chains (see Figure 22-19). Alternatively, two or more transcription units can be distinguished on some loops. Accounting for the generation of the unusually long transcripts is a failure in transcription termination and/or RNA processing. While the transcripts appear to initiate at bona fide promoters, they proceed far past the normal 3' ends and thus read into repetitive sequences of various sorts. Late in oogenesis, readthrough increases, giving rise to a higher fraction of such transcripts. Many of these RNAs are exported to the cytoplasm, and it is not known whether they find some eventual use. Meanwhile, however, vast quantities of functional mRNAs have also been produced both for translation during oogenesis and for storage in the egg.

Figure 22-19
A section along a loop of a *Triturus* lampbrush chromosome, showing a gradient of increasingly long RNA chains attached to their DNA template. The black dots at the sites where the RNA chains attach are RNA polymerase molecules. (Courtesy of O. L. Miller and B. Beatty, Oak Ridge National Laboratory.)

The *Drosophila* Chorion Genes Can Be Amplified Without Excision[33]

We have seen how ciliates achieve selective gene amplification in the somatic macronucleus by a curious and roundabout strategy: The entire chromosome complement is first polytenized, and then DNA segments that are unnecessary for vegetative growth are degraded. Amphibian oocytes achieve selective gene amplification more directly by excision of chromosomal rDNA and subsequent extrachromosomal amplification. However, as we shall see for the chorion genes of the fruit fly, developmentally regulated gene amplification need not involve chromosome fragmentation or extrachromosomal replication.

The *Drosophila* egg is such a large cell that a single haploid nucleus cannot possibly be transcribed at a sufficiently high rate to stock the entire egg cytoplasm in the brief 5- to 7-day time span available for oogenesis. Instead, thousands of follicle cells surround and actively nourish the maturing egg. In addition, 15 nurse cells located at the anterior end of the egg build cytoplasmic bridges to secrete various factors (including mRNA!) into the egg. Thus, we can think of the

Figure 22-20
Scanning electron micrograph of the surface of a mature eggshell (chorion) from another insect, the silk moth *Antheraea polyphemus* (1400×). The single micropyle for sperm entry is visible in the foreground; numerous breathing tubes (aeropyles) occupy the background. Note the extraordinary degree of structural organization in this tiny egg. (Courtesy of G. D. Mazur, Harvard University.)

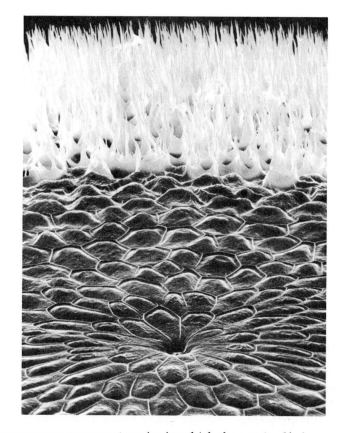

mature egg as a group enterprise in which the egg itself plays a major role but the nurse and follicle cells do much of the hard work. As the egg matures, the surrounding follicle cells construct an eggshell, or **chorion,** to protect the egg from physical stress and drying out. During the 5 hours of choriogenesis, the follicle cells secrete a carefully orchestrated succession of some 20 cysteine-rich chorion proteins that assemble and crosslink to form a complex, multilayered eggshell. The only holes in the eggshell are the breathing tubes that allow respiration and the single anterior micropyle that permits sperm entry during fertilization (Figure 22-20). The mature eggshell (chorion) has a recognizable ventral and dorsal side, as well as an anterior and posterior end (shown schematically in Figures 22-24 and 22-27).

In *Drosophila*, the chorion proteins must be synthesized so quickly, and in such large quantities, that single gene copies are not enough; instead, the follicle cells are developmentally programmed to selectively amplify their chorion genes at the appropriate time in oogenesis. The genes encoding the chorionic proteins are located in two clusters, one at cytological band 7F1-2 on the *Drosophila* X chromosome and the other at band 66D12-15 on the third chromosome. Each gene is present as only a single copy. However, in follicle cells that are midway through oogenesis, an origin of DNA replication located within each chorion gene cluster is programmed to fire three to six times, eventually leading to a 32- to 64-fold amplification (2^6) of the adjacent chorion genes. (Note that these unusual developmentally regulated origins fire during interphase in follicle cells. We do not know whether the same origins also function during S phase in other cell types.) Replication proceeds bidirectionally, but each successive round of replication is less extensive than the one before: Perhaps topological constraints limit the number of replication forks that can lie close together on the DNA. The final structure resembles an onion-skin and remains an integral part of the chromosome. The gradient of

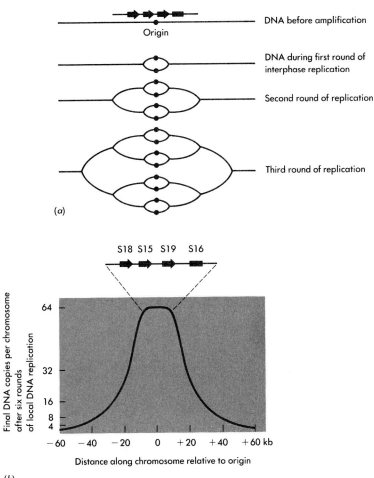

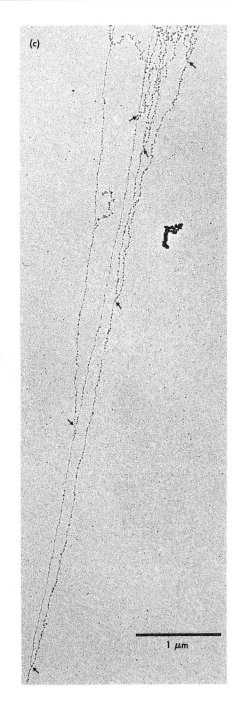

Figure 22-21

Developmentally regulated amplification of the *Drosophila* chorion genes in follicle cells. (a) Diagram showing first three rounds of interphase DNA replication at the 66D locus on chromosome 3. The three small arrows, shown only for the unamplified DNA, represent three well-characterized chorionic genes in this cluster. The polarity of a fourth chorionic gene, and the precise location of the origin relative to the genes, is not yet known. (b) A graphical representation of the final levels of gene amplification as a function of position along the chromosome. The genes are shown above the symmetrical amplification gradient to emphasize that the boundaries of the amplified DNA are much larger than the chorionic protein transcription units within it. (c) Chromatin strands from the "onion-skin" of an early-stage 11-follicle cell, showing multiple replication forks (arrows). [(b) After Figure 5 and (c) reproduced from Figure 8 of Y. N. Osheim and O. L. Miller, Jr., *Cell* 33 (1983):543–553, with permission.]

amplification on the third chromosome extends for some 50 kilobase pairs in either direction and has been visualized directly in the electron microscope (Figure 22-21).

Developmental Problems May Have Multiple Solutions[34]

Why does *Drosophila* selectively amplify the chorion genes just before they are expressed, rather than creating a germ line multigene family that is selectively turned on during choriogenesis? We might have guessed that selective amplification is part of the mechanism by which the genes are turned on, but the data tend to rule out this attractive possibility: In the normal course of development, chorion genes are amplified *before* they are turned on. Still worse for any simple rationalization of chorion gene amplification, in two silk moths (*Bombyx mori* and *Antheraea polyphemus*), the chorion genes are in fact present in multiple germ line copies; their expression, but not their copy number, is developmentally regulated.

Both the fruit fly and the silk moth regulate expression of the chorion genes, but the moth maintains amplified genes in the germ line, while the fly subjects the genes to somatic amplification just prior to use. This discrepancy may illustrate a general developmental truth: Organisms solve their developmental problems in any way that works, whether sloppy or elegant. And since evolution proceeds by sequential modification of preexisting developmental patterns, even a sloppy or awkward developmental mechanism may become so intricately entangled in the overall developmental program of an organism that it cannot be changed thereafter. Viewed in this light, developmental programs are partly logical, and partly mere historical expedients that have been frozen into place by subsequent evolution. A corollary of this truth is that there may be no universal developmental mechanism, but rather a multiplicity of mechanisms that can be combined and adapted in many surprising ways.

Stored mRNAs in the Egg May Be Regulated by Translational Control[35–42]

A mature oocyte (egg) may be huge, but it is still only a single cell. The fertilized egg must therefore undergo many cleavage divisions before there are enough cells for morphogenesis (blastulation and gastrulation; see Figure 22-5). In organisms where the life cycle puts a premium on rapid development (a toad egg may hatch in a mud puddle that is rapidly drying out in the hot sun), these cleavage divisions take place very fast. Such rapid cell divisions leave no time for an increase in embryo mass (growth) or for transcription, and protein synthesis during this time must use preformed mRNAs present in the oocyte at the moment of fertilization (stored "maternal" mRNAs). (Sperm cells are very small, have little cytoplasm, and carry essentially no mRNA into the egg.) Only after the midblastula transition, when RNA polymerase II and III are activated (see Figure 22-6) and mRNA synthesis commences, is the paternal contribution to the embryonic genome transcribed and expressed for the first time.

Not surprisingly, the transition from the mature oocyte to the newly fertilized egg is accompanied by large changes in protein synthesis. Many proteins specifically required for building the oocyte are no longer synthesized, while many stored maternal mRNAs that encode proteins required for early embryogenesis must be activated. How this switch from oocyte-specific to embryo-specific translation occurs is only just now beginning to be understood. In the surf clam *Spisula solidissima*, the switch is not accompanied by any detectable changes in the population of cellular mRNAs. Instead, utilization of different subsets of the total mRNA population at the two developmental stages is accomplished by translational control. Experimentally, it was found that in vitro translation of deproteinized mRNA from the egg and embryo yielded the same pattern of protein synthesis. However, when the mRNA was translated in vitro without prior deproteinization, egg mRNA ribonucleoprotein particles (mRNPs; Chapter 21) were translated into egg proteins, and embryo mRNPs were translated into embryonic proteins. This suggests that translation of specific mRNA species can be negatively regulated (masked) by some of the proteins that package mRNA. Other evidence suggests that positive translational control may be responsible for preferential expression of hsp70 protein (Chapter 21) during the heat-shock response in frog oocytes and in *Drosophila* tissue culture cells. Presumably, both positive and negative translational control reflect the sequence-specific recognition of mRNA by proteins, but recognition of mRNA by complementary ("antisense") RNA species remains a possibility (see the model for regulation of Tn10 expression in Figure 16-26). In addition, the level of posttranscriptional modifications such as methylation of 5' cap structures and polymerization of 3' poly A tails may selectively influence the translation of different subpopulations of mRNA.

What proteins might the early embryo need in large amounts? Rapid DNA replication requires both the deoxyribonucleotide triphosphate building blocks for DNA and histone proteins to package the newly made DNA into chromatin. Thus, one of the most abundant stored maternal mRNAs that is activated during early embryogenesis of the surf clam encodes the small subunit of ribonucleotide reductase, a key metabolic enzyme required for synthesizing the deoxyribonucleotide precursors of DNA. Similarly, some of the most abundant stored maternal mRNAs in frog and sea urchin eggs encode histones. Although the egg does contain a stock of histone proteins, this stock is quickly exhausted by the synthesis of new DNA. Translation of the stored mRNA population must then sustain the oocyte until mRNA synthesis resumes after the midblastula transition. In contrast to the situation in normal somatic cells (where histone mRNA is synthesized only during DNA replication and afterward is rapidly degraded in the cytoplasm), histone mRNA is stored in the nucleus of the oocyte before fertilization and is stable in the absence of DNA replication. The relative stability of histone mRNA in the cytoplasm after fertilization makes sense, since DNA replication in the early embryo is so fast that the cell is effectively always in S phase. However, the maternal stock of histone mRNA is ultimately degraded when DNA synthesis slows down after the midblastula transition, and the intervals between successive S phases increase.

Although the early embryo synthesizes many proteins just prior to use, other macromolecular components must be stored in massive amounts in the oocyte cytoplasm because there is simply no time to make them during embryogenesis. We have already seen that amplification of the genes encoding the large rRNAs and "preamplification" of the 5S rRNA genes allow the oocyte to accumulate a huge stock of ribosomes during oogenesis. In fact, from the study of mutant frogs lacking rDNA ("anucleolate," or *nu*⁻, frogs), we know that this ribosome stock is sufficient to carry the developing embryo all the way through to the tadpole stage! In contrast to ribosomes, which are stored as mature ribonucleoprotein complexes (ribosomal subunits), the oocyte cytoplasm stores snRNP proteins rather than mature snRNPs. When snRNAs (U1, U2, etc.) and mRNA precursors begin to be synthesized by RNA polymerase II after the midblastula transition, newly synthesized U1 and U2 snRNA assemble with the stored snRNP proteins to make mature U1 and U2 snRNPs that are capable of splicing the newly made mRNA precursors (Figures 20-18 and 20-19). Both ribosomes and snRNPs are ribonucleoprotein complexes, and ribosomes outnumber snRNPs by a factor of about 5 in a typical somatic cell. Why, then, does the oocyte store complete ribosomal subunits, but only snRNP proteins? Perhaps this illustrates once again that similar developmental problems often have very different solutions.

A Second Independent Clock in Fertilized Amphibian Eggs[43]

We have seen previously that titration of a preformed (protein?) factor tells the cleaving frog embryo when it has synthesized about 4000 cell equivalents of DNA; titration then triggers the many changes known collectively as the midblastula transition. This titration mechanism is not the only clock in the early embryo. Among the stored maternal mRNAs in the frog oocyte is the mRNA encoding fibronectin, a major protein of the extracellular matrix that plays a role in cell attachment and migration. Like many other changes in the embryo, translation of the stored fibronectin mRNA is normally scheduled to begin at the midblastula transition. Taking advantage of an experimental trick discovered by the pioneering embryologists of the last century, it was possible to demonstrate recently that translation of fibronectin is controlled by a second *independent* clock. The trick is that merely pricking an amphibian egg with a needle will activate it to begin the early developmental program. Apparently, a small hole in the egg can mimic the major physiological changes accompanying normal fertilization, such as the rapid increase in cytoplasmic pH and the release of bound calcium ions. These pricked eggs synthesize little or no DNA and consequently can never proceed beyond the midblastula transition; however, fibronectin synthesis begins on schedule, just as though the egg had been fertilized naturally. Thus, the egg contains a second independent clock that does not require DNA synthesis and counts time from the moment of activation, whether it occurs by natural means or by experimental trickery.

GENETIC CONTROL OF MORPHOGENESIS

Cyclic AMP Is a Morphogen in Slime Mold Development[44-47]

Clearly, the most interesting microorganisms to the embryologist are those in which cell differentiation is particularly prominent. Thus, much attention has been devoted to the cellular slime mold, *Dictyostelium discoideum,* whose life cycle is illustrated in Figure 22-22. The cycle begins with spores germinating in damp earth to yield amoeboid cells called myxamoebas, which feed by engulfing bacteria and reproduce by binary fission. When the multiplying amoebas exhaust their local food supply, they aggregate to form a multicellular slug, which migrates in a search of more food and better environmental conditions. Eventually, the slug undergoes a complex morphogenesis to construct a fruiting body that elevates the spores above soil level for most efficient dispersal to new locations.

Both aggregation and subsequent construction of the fruiting body

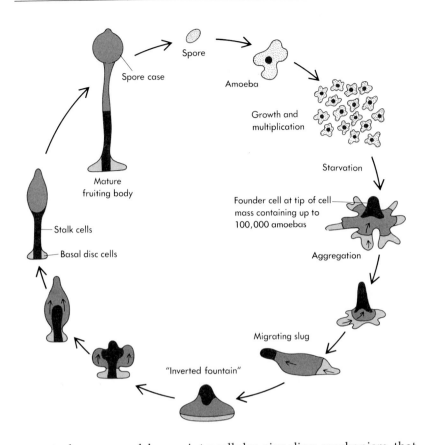

Figure 22-22
The life cycle of the cellular slime mold *Dictyostelium discoideum*. Small arrows show morphogenetic movement of cells. The first amoebas to aggregate are arbitrarily shown in the darkest color.

seem to be governed by an intracellular signaling mechanism that uses cAMP as a chemical attractant. Within 6 to 8 hours after the onset of starvation, each amoeba has become competent to sense a transient increase in local cAMP concentration (reception) and to secrete a short pulse of cAMP in response (transmission). This is accomplished by a membrane-bound cAMP receptor protein that measures cAMP concentration and then triggers a burst of activity by a membrane-bound adenylate cyclase. In addition, the cells possess a cAMP phosphodiesterase to destroy excess extracellular cAMP in the medium and thereby prevent the signaling from being jammed by accumulation of chemical "noise."

Note that the center of each morphogenetic field is defined by the first cell that spontaneously secretes a burst of cAMP: All the cells become competent to signal at about the same time, but whichever cell signals first (the **founder cell**) effectively summons all the others. The evidence for this mechanism comes from a simple experiment in which a glass micropipette filled with a solution of cAMP is used to deliver small drops of cAMP to a random position within a field of competent amoebas that have not yet begun to aggregate. The amoebas immediately begin to migrate toward the pipette, as though it were a founder cell!

Mathematical models have been proposed to show that the ability of each individual amoeba to secrete a pulse of cAMP shortly after sensing one should lead to aggregation, but perhaps the clearest evidence for signaling comes from an elegant experiment that directly visualizes concentric waves of cAMP surrounding the founder cell at early stages of aggregation (Figure 22-23). Later, the formation of intercellular contacts between cell surface glycoproteins causes the amoebas to adhere to one another and to stream toward the center.

Most important is the realization that both the size of an aggregation field and the speed with which it forms rule out the possibility

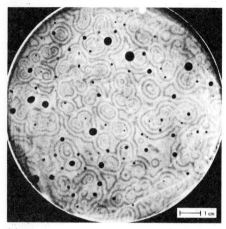

(a)

Figure 22-23

Visualizing a gradient of the slime mold morphogen, cyclic AMP, by an isotope dilution method. (a) An autoradiograph. (b) The experimental procedure. [Photograph reproduced with permission from K. J. Tomchik and P. N. Devreotes, *Science* 212 (1981):443–446.]

(b)

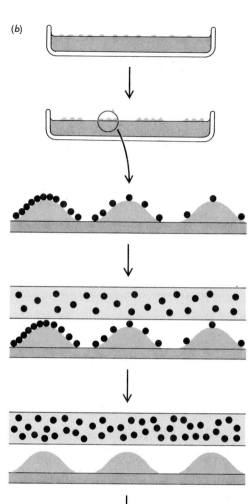

Amoebas growing on nutrient agar (side view)

Nutrients exhausted

Fields of aggregating amoebas

Greatly enlarged side view of aggregating amoebas, with black dots representing a concentration gradient of nonradioactive cAMP secreted by the cells.

A dry filter paper that had previously been soaked in a dilute solution of radioactive cAMP (color dots) is gently placed on top of the aggregating cells.

Nonradioactive cAMP synthesized by the cells dilutes the radioactive cAMP in the presoaked filter paper.

A second filter paper bearing cAMP binding protein (CAP factor) is gently placed on top of the first filter. Because the nonradioactive cAMP is present in great excess over the binding sites on CAP, the amount of radioactive cAMP bound by the second filter is inversely proportional to the local cAMP concentration synthesized by the cells.

The second filter is dried and exposed to X-ray film for autoradiography. After development, dark bands appear on the film where low concentrations of cellular cAMP have failed to dilute the radioactive cAMP in the first filter.

that cAMP secreted by the founder cell is merely diffusing outward and telling amoebas at the edge of the field to migrate inward toward the source of the signal. Such passive diffusion would be much too slow, and the cAMP would be destroyed much too quickly by the phosphodiesterase. Like a bucket brigade, the morphogenetic signal must be actively relayed from one amoeba to the next, until the entire field behaves as a single unit.

Cyclic AMP does more than simply bring together a mass of cells with radial symmetry; it also allows them to organize into a multicellular organism with a distinct front and back end (an anterior-posterior axis). The first cells to aggregate are at the center and form the front tip of the slug when the conical mass topples over on its side. Similarly, the last cells to join the aggregate form the tail of the slug. This cellular organization is preserved while the slug migrates, but then undergoes a curious inversion during morphogenesis to form a fruiting body. When the slug comes to rest, the front tip (organizer region) travels down through the rest of the slug to form an "inverted fountain." These migrating tip cells differentiate into a tough stalk, where the cells then suffer a programmed cell death (see the section Invariant Cell Lineages for more on this fascinating subject). Cells from the remainder of the slug are elevated by stalk formation and differentiate into spores that can begin a new life cycle. Except for the final steps of sporulation and stalk cell death, differentiation is completely reversible. For example, when stalk cells are isolated from the fruiting body before cell death begins, they may transform back into the ameoboid phase and proceed to divide by fission.

Strictly speaking, we do not know whether the behavior of slime mold amoebas is a good model for, say, cell movement during gastrulation of an amphibian or human embryo. However, it is an article of faith for experimental biologists that radically different organisms may solve similar problems in similar ways. The basis for this fundamental faith is that natural selection tends to (but need not) conserve basic developmental mechanisms in the course of evolution. Paradoxically, this faith does not conflict with the knowledge that similar developmental problems have often been solved in multiple (and surprisingly different) ways. Thus, we may hope that gradients of small chemical morphogens like cAMP or small polypeptides may direct cellular movements during the development and differentiation of tissues in higher organisms.

Embryogenesis Establishes the Basic Body Plan[48]

Perhaps the greatest challenge for modern embryologists is to explain in molecular terms how the simple, one-dimensional array of genetic information in DNA can be transformed into the complex, three-dimensional organization of the adult. Classical embryologists had already realized that the egg and the sperm could not be entirely "shapeless." Somehow, the germ cells had to contain either a miniaturized version of the adult body plan or a biochemical mechanism for regenerating that body plan afresh with each new generation. In the 1700s, this led to the theory of preformation, which held that the egg and the sperm are actually tiny adults (a homunculus in the case of *Homo sapiens*). The main function of development was then to make this hidden little creature bigger. We can look back on such speculation with a comfortable sense of superiority, but there was truth in the basic idea. We now know that certain developmental patterns *are* preformed within the egg, and other shape-determining mechanisms are in fact actuated by fertilization.

For insects, as for amphibians, early embryogenesis takes place so fast that there is no time for cell growth. Instead, the task of early embryogenesis is to subdivide the very large egg into many smaller cells so that **morphogenesis** can begin. Drosophila and other insects go about this task in a curious way (Figure 22-24). Immediately after fertilization in *Drosophila*, the zygotic nucleus undergoes roughly 9 very rapid divisions to yield approximately 512 nuclei without accom-

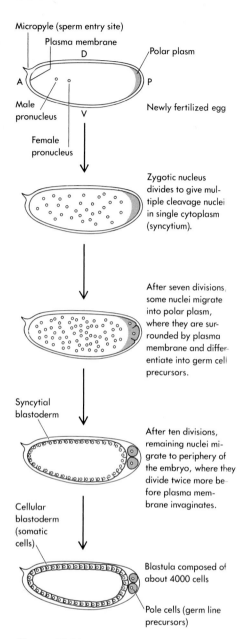

Figure 22-24
Early embryogenesis of the fruit fly *Drosophila*. A stands for anterior; P, posterior; D, dorsal; V, ventral. [Modified from W. J. Gehring, *Sym. Soc. Dev. Biol.* 31 (1973):103–128; and V. E. Foe and B. M. Alberts, *J. Cell. Sci.* 61 (1983):31–70.]

panying cellularization: The egg becomes a giant syncytium (literally, "one cell") filled with nuclei. After the ninth division, the nuclei migrate to the surface of the egg and divide four more times in synchrony. The plasma membrane of the egg invaginates to enclose each nucleus in a separate plasma membrane, thus forming the cellular **blastoderm.** Complex folding of this hollow ellipsoidal ball of cells then gives rise to a segmented larva known as the first **instar,** which has a head, three thoracic segments (T1–T3), and eight abdominal segments (A1–A8; Figure 22-25).

This instar is the first of three larval stages, separated by molts, during which the wormlike creature eats voraciously to accumulate sufficient biomass to construct an adult fly. After the third instar, the larva pupates to provide a protected environment for metamorphosis. During metamorphosis, the larva digests many of its own tissues and replaces them with adult structures; however, the body of the adult fly retains the segmental boundaries of the larval instar forms. Surprisingly, structures on the surface of the adult fly (such as legs, wings, and antennas) are constructed *not* from differentiated larval cells but rather from small groups of proliferating but relatively undifferentiated cells called **imaginal discs** (Figure 22-26). (An imago is an adult fly.) The 19 imaginal discs (9 paired and 1 fused genital disc) are set aside very early in embryogenesis, and each is committed to a particular developmental pathway (e.g., left leg, right leg, left eye, or right eye). The imaginal discs remain in a relatively undifferentiated state near the surface of the larva until changing hormonal levels during pupation cause them to differentiate and undergo morphogenesis. Thus, much of the basic body plan of the fly is established in embryogenesis, long before overt differentiation of the imaginal discs creates the adult cuticular structures that clothe the internal organs of the animal.

Genes That Encode Shape-Determining Factors in the Fruit Fly

To understand morphogenesis, modern biologists almost always start by identifying genes and gene products that affect the visible appearance of the organism. This is the strategy that was used to analyze morphogenesis of the bacteriophage λ and T4 capsids (Chapter 17). The dorsalizing genes of the fruit fly *Drosophila* have recently become a very promising focus for understanding the mechanisms that govern morphogenesis in higher eucaryotes.

We have seen that the unfertilized *Drosophila* egg is surrounded by a highly structured eggshell (chorion) whose head and tail (anterior and posterior) as well as front and back sides (ventral and dorsal) are readily distinguishable. Such external differentiation suggests that the egg within already possesses anterior-posterior and dorsal-ventral polarity, and this was confirmed by genetic and biochemical analysis.

The products of at least ten different genes are now known to be required for constructing an egg with proper dorsoventral polarity. If the mother lacks a wild-type copy of any of these genes, the dorsalized embryo cannot be restored to health by a wild-type gene copy contributed by the sperm at the time of fertilization. Evidently, it is too late to reconstruct the egg using the wild-type gene product. For this reason, such genes are said to exhibit a strict **maternal effect:** The genotype of the father is completely irrelevant.

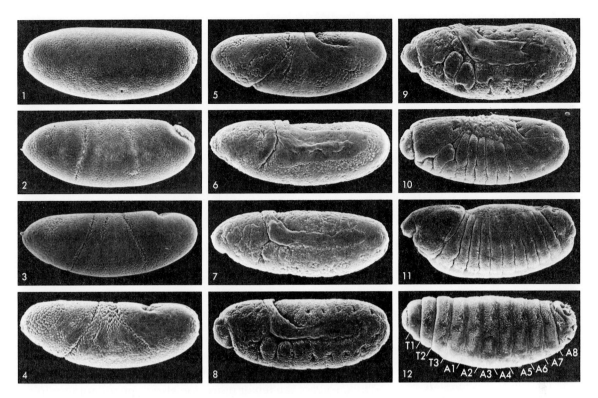

For recessive mutations in the ten dorsalizing genes, maternal effects are only observed in the eggs of a homozygous female where both gene copies are defective. Eggs of a wild-type or heterozygous female, where a wild-type gene copy can supply the normal gene product, are completely normal. The reason for this is that essentially all of the mRNA and factors present in the mature egg are secreted into the egg cytoplasm by the 15 diploid nurse cells; *dorsal* mRNA, for example, can be identified in nurse cells before it appears in the oocyte. Transcription from the haploid egg nucleus itself makes little, if any, contribution toward stocking the huge egg cytoplasm.

A Morphogenetic Gradient
May Establish Dorsoventral Polarity[49]

The appearance of a fully dorsalized embryo is stunning: It seems to be all back (dorsal) and no front (ventral)! The embryo cannot survive, because it lacks a variety of dorsolateral structures such as Filzkörper (breathing tubes) and ventral structures such as denticle bands (rows of small bumps that help the larva to crawl). Instead, dorsal structures extend all around the embryo. Furthermore, geneticists have been able to identify weaker mutations in some of these genes, which lead to loss of some, but not all, ventral structures. The existence of such intermediate (partially dorsalized) embryos suggests that dorsoventral polarity in the unfertilized egg is defined by a gradient of some sort. The collective job of the various dorsalizing genes may be to generate this gradient or to enable cells to interpret their position within the gradient.

Surprisingly, mutations in some of the ten dorsalizing genes can be partially or completely "rescued" by microinjection of cytoplasm taken from an early wild-type embryo before cellularization of the blastoderm (Figure 22-27). The rescuing activity for at least five of the mutants appears to be mRNA. Moreover, RNA from a mutant

Figure 22-25

A sequence of twelve scanning electron micrographs showing early embryogenesis in *Drosophila* from 3.5 to 14 hours after fertilization. The eggshell, or chorion, has been chemically removed to reveal the details of morphogenesis. The embryo is about 0.4 mm long, and individual cells are visible as small bumps. Note that all photographs are reproduced at the same scale: The first larval instar is no bigger than the egg, but morphologically it appears to be immensely more complex. (1) The syncytial blastoderm stage, with the plasma membrane invaginating to enclose individual cells. (2) Pole cells are clearly visible at the posterior of the embryo, and the remains of the micropyle can be seen at the anterior. The first furrows appear in the cellular blastoderm, indicating that morphogenesis has begun. (8) Intricate folding of the blastoderm is well under way, and segmentation has become apparent. (12) The late embryo, about to hatch as a first larval instar. Only at this late stage does it become difficult to discern individual cells in the embryo. (Courtesy of R. Turner and A. Mahowald, Case Western Reserve Medical Center.)

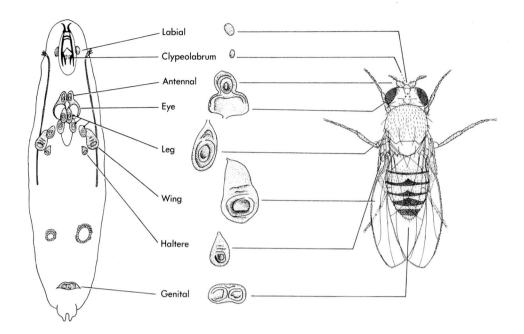

Figure 22-26
The imaginal discs of *Drosophila*. The adult eye and antennal structures are both derived from a single pair of large, complex discs. The larva has three pairs of leg discs and also a pair of dorsal thoracic discs that are not shown here. Segmental boundaries have been omitted for clarity. [After J. W. Fristrom, R. Raikow, W. Petri, and D. Stewart, in *Problems in Biology: RNA in Development,* ed. E. W. Hanly. (Salt Lake City: University of Utah Press, 1969), p. 382.]

embryo with a defect in one dorsalizing gene will rescue a mutant embryo with a defect in a different dorsalizing gene, and vice versa. Thus, each of the ten or more dorsalizing genes appears to contribute a different gene product to the complex molecular machinery responsible for generating dorsoventral polarity.

Especially striking is the observation that rescue of ventral structures in embryos with a homozygous mutation in the *dorsal* gene occurs only when wild-type cytoplasm is injected into the ventral side of the mutant embryo (see Figure 22-27). Thus, the *dorsal* embryo has already established a gradient of positional information along the dorsoventral axis, but in the absence of the *dorsal* gene product, the cells are unable to interpret their position within this gradient and to differentiate accordingly. Equally interesting is the observation that injection of wild-type cytoplasm into the posterior ventral region of a *dorsal* embryo (where breathing tubes form in the wild type) rescues

Figure 22-27
Much of the information specifying the dorsoventral polarity of the *Drosophila* embryo is stored as maternal mRNA. Note that some mutations can be rescued by injections of wild-type cytoplasm but not wild-type mRNA, and that mutations in some of the ten dorsalizing genes are more completely rescued by this protocol than others. Note also that the table does not apply to homozygous *Toll* embryos. As described in the text, no matter where wild-type cytoplasm is injected into a *Toll* embryo, the site of injection defines the ventral side of the rescued embryo. This suggests that the *Toll* gene product is itself a morphogen or provides the source for a gradient of a morphogen.

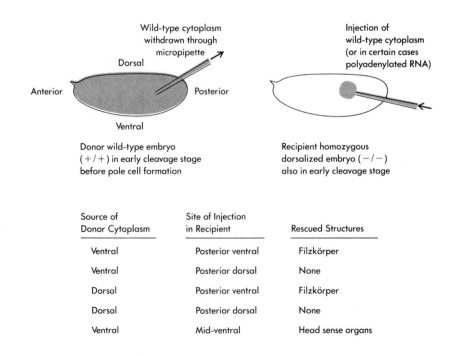

Source of Donor Cytoplasm	Site of Injection in Recipient	Rescued Structures
Ventral	Posterior ventral	Filzkörper
Ventral	Posterior dorsal	None
Dorsal	Posterior ventral	Filzkörper
Dorsal	Posterior dorsal	None
Ventral	Mid-ventral	Head sense organs

Filzkörper, whereas injection into the mid-ventral region (where head structures form in the wild type) rescues head sense organs. Thus, the mature *Drosophila* egg has also established a gradient of positional information along the anterior-posterior axis. Moreover, since defects in dorsoventral polarity do not seem to affect the anterior-posterior gradient of information, these two gradients appear to be independent of each other.

The *Toll* Gene Product Determines the Polarity of a Morphogenetic Gradient[50]

The notion that the dorsoventral polarity of the egg is determined by a gradient of some sort is reinforced by the behavior of mutations in the dorsalizing gene known as *Toll*. Recessive mutations in *Toll*, like those in the other nine dorsalizing genes, produce embryos lacking ventral and lateral structures. However, the *Toll* gene product must behave differently from the products of the other dorsalizing genes, because it has been possible to isolate dominant mutations in *Toll*. Amazingly, these dominant *Toll* mutations cause ventral structures to extend all the way to the dorsal side of the embryo! This suggests that the *Toll* gene product may actually be the morphogen (or generate the morphogen) that causes ventral structures to form. Dominant (ventralizing) *Toll* mutations would result in overproduction of the *Toll* gene product at the normal ventral site or inappropriate expression of the *Toll* gene product throughout the egg, so that cells are told to make ventral structures in the wrong place.

Experiments using wild-type cytoplasm to rescue mutant *Toll* embryos support the interpretation that the *Toll* gene product is (or produces) a morphogen. As we have seen, the *Drosophila* embryo normally displays the same polarity as the eggshell (chorion) around it. When mutations in any of the other nine dorsalizing genes are rescued by microinjection of wild-type cytoplasm, the rescued (or partially rescued) embryo always retains the same polarity as the eggshell. But for *Toll* mutations, the polarity of the rescued embryo is determined by the site of injection of the wild-type cytoplasm: Dorsal injection reverses the normal polarity of the egg with respect to the shell. Thus, unlike the products of the other nine dorsalizing genes, which simply help cells of the blastoderm to interpret their position within a preexisting dorsoventral gradient, the *Toll* gene product appears to determine the polarity of the gradient itself.

We do not know whether this positional information might be encoded by a gradient of a morphogen (like cAMP in the slime mold); by an electrical potential akin to the proton motive force in bacteria (Chapter 6) or the propagation of a neural impulse in higher eucaryotes; by polar interactions between cell surface glycoproteins on adjacent cells; by large cytoskeletal structures that might span the entire embryo; or by mechanisms that are currently unimaginable.

Using recombinant DNA technology, the various dorsalizing genes can now be cloned and mutagenized in vitro. The ability of such mutagenized gene products to rescue dorsalized embryos can then be tested either by microinjecting the embryo with mutant mRNA synthesized in vitro, or by putting the mutagenized gene back into the *Drosophila* germ line using P element–mediated transformation. These studies may reveal the physical nature of positional information and the way cells interpret their position within a developmental gradient.

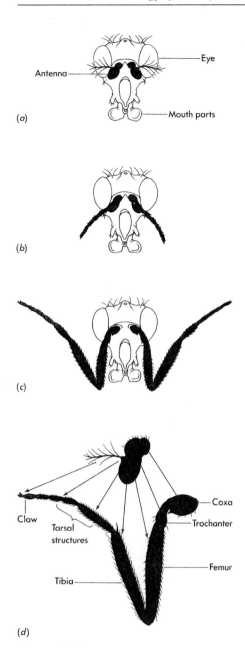

Figure 22-28
Homeotic mutations in the fruit fly
Drosophila can convert one body part
into another. (a) Head of a wild-type
fly. (b) A relatively weak *Antennapedia*
mutation called *Aristapedia* transforms
the distal part of the antenna (the
arista) into the corresponding distal
part of a leg. (c) A strong *Antennapedia*
mutation completely transforms the antenna into a leg. (d) Diagram showing
the precise correspondence between
parts of antenna and leg. [After J. H.
Postlethwait and H. A. Schneiderman,
Dev. Biol. 25 (1971):606–640; and B.
Alberts, D. Bray, J. Lewis, M. Raff, K.
Roberts, and J. D. Watson, *Molecular
Biology of the Cell* (New York: Garland,
1983), p. 840.]

Homeotic Mutations in the Fruit Fly Change One Body Part into Another

The appearance of a dorsalized embryo may be curious, but strangest
of all developmental derangements in the fly are **homeotic mutations,**
which transform one part of the adult body into another. For example, the effect of the dominant homeotic mutation called *Antennapedia*
(literally, "antenna-foot") is to transform the antenna on the head of
the fly into a perfectly formed but absurdly misplaced leg (Figure
22-28). A less extreme mutation known as *Aristapedia* transforms only
the outermost part of the antenna (the arista) into the corresponding
parts of the leg (tarsus and claw). Because many *Antennapedia* mutations have visible chromosomal abnormalities (translocations, inversions, deletions) at a single locus, we know that the DNA responsible
for *Antennapedia* is located at band 84B on the right arm of chromosome 3. Thus, the existence of the *Antennapedia* mutation immediately
implies that the product of a single gene (or small group of genes) can
work like a simple molecular switch to select one developmental fate
or another for a related group of cells in the eye-antennal imaginal
disc (see Figure 22-26). Weaker mutations, such as *Aristapedia*, further
suggest that the switch can be thrown independently in different sub-
groups of these imaginal disc cells. Thus, the differentiated state of a
cell, or group of cells, may simply reflect which combination of molecular switches has been thrown.

Molecular Clones of the Homeotic Genes Have Been Identified[51, 52]

Even the body plan of so small and inconsequential a creature as the
fruit fly is immensely complicated, and it should come as no surprise
that *Antennapedia* is not the only homeotic locus in *Drosophila*. There
are, in fact, two large clusters of homeotic genes known as the **Antennapedia complex** (ANT-C) and the **bithorax complex** (BX-C). As
shown in Figure 22-29, genes of the Antennapedia complex regulate
the development of the head and the three thoracic segments (T1, T2,
and T3) of the fly, while the bithorax complex controls development
of two thoracic segments (T2 and T3) as well as all eight abdominal
segments (A1–A8). BX-C contains three complementation groups
(*Ultrabithorax, abdominal-A,* and *Abdominal-B*; abbreviated *Ubx, abd-A,*
and *Abd-B*), each containing more than one gene. ANT-C contains at
least four genes (*Deformed, Sex combs reduced, fushi tarazu,* and *Antennapedia*; abbreviated *Dfd, Scr, ftz,* and *Antp*). Mutations in these
homeotic genes usually cause such severe defects that the fly dies in
late embryogenesis. Occasionally, however, nonlethal mutations
allow survival to adulthood, and these flies display visible abnormalities. In Figure 22-30, three separate nonlethal homeotic mutations in
the *Ubx* complementation group (*abx, bx³,* and *pbx*) have been combined by genetic crosses to produce a four-winged fly where T3 is
completely transformed into T2. (T3 normally bears two diminutive
winglike balancing organs called halteres, and these are transformed
into a nearly normal pair of wings in the triple mutant.)

Advances in recombinant DNA technology have made it possible
to obtain molecular clones for several different (and perhaps all)
homeotic genes in the Antennapedia and bithorax complexes. Although many subtle genetic tricks were required to assemble a com-

plete array of recombinant clones spanning each locus, the basic experimental strategy for cloning the homeotic genes was simple (Chapter 19). Salivary gland polytene chromosomes were probed by in situ hybridization with randomly cloned fragments of *Drosophila* DNA, and in a few lucky cases, a fragment was found to be "near" the cytological location of a known homeotic gene (within a few hundred thousand base pairs!). It then became thinkable for scientists of sufficiently strong constitution to "walk along the chromosome" from the initial site of hybridization to the homeotic gene by generating a series of as many as 100 overlapping genomic clones. The first such chromosomal walk of 150 kilobase pairs took place within the bithorax complex of *Drosophila,* and it represented both a mental and a technical breakthrough. Steady progress in recombinant DNA technology has now made it an increasingly straightforward matter to obtain molecular clones for almost any chromosomal region of interest in almost any organism. Inevitably, the power of the recombinant DNA revolution has driven molecular geneticists to conceive of ever more ambitious projects. The 170-kilobase-pair genome of Epstein-Barr Virus (EBV) was recently sequenced in its entirety, and total DNA sequence analysis of the 3000-kilobase-pair circular chromosome of *E. coli* is no longer completely unthinkable. Although complete DNA sequence analysis of even the smallest human chromosome (about 50,000 kilobase pairs) must still be considered a dream, it should be possible to construct a complete restriction map for some human chromosomes. In fact, a complete restriction map for the 80,000-kilobase-pair genome of the small nematode worm *Caenorhabditis elegans* is now nearing completion.

The Genetic Control of Segmentation[53, 54]

Although the *ftz* gene is located within the Antennapedia complex between homeotic genes that affect development of the head and thorax (see Figure 22-29), mutations in the *ftz* gene are not strictly homeotic. Instead, *ftz* belongs to a fascinating group of at least 15 genes responsible for proper segmentation of the embryo. These segmentation genes fall into three classes, *ftz* itself belonging to the pair-rule class (Figure 22-31). In a homozygous *ftz* embryo, the posterior half of every "even" segment (T2, A1, A3, A5, and A7) and the anterior half of every "odd" segment (T1, T3, A2, A4, A6, and A8) is

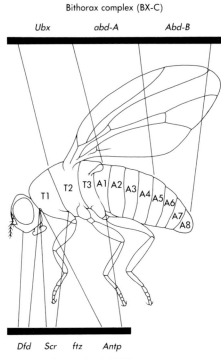

Figure 22-29
Primary domains of homeotic gene function. After metamorphosis, the body of the adult fly retains the segmental boundaries of the larval instar forms. The colored lines show the primary domain of function for each of the three complementation groups in the bithorax complex (BX-C) and each of the genes in the Antennapedia complex (ANT-C). The genetic maps are not drawn to scale. Note the wings on the second thoracic segment (T2) and the halteres, or balancing organs, on the third thoracic segment (T3). [After K. Harding, C. Wedeen, W. McGinnis, and M. Levine, *Science* 229 (1985):1236.]

Figure 22-30
A four-winged fruit fly resulting from mutations in three separate genetic loci (*abx, bx³,* and *pbx*) of the *Ultrabithorax* complementation group. This fly has two essentially normal second thoracic segments (T2) and no third thoracic segment (T3), because the combined effect of the three mutations is to transform T3 into T2 without affecting any other parts of the fly. (Courtesy of E. B. Lewis, California Institute of Technology.)

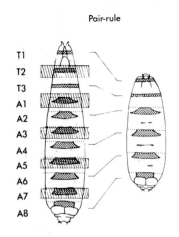

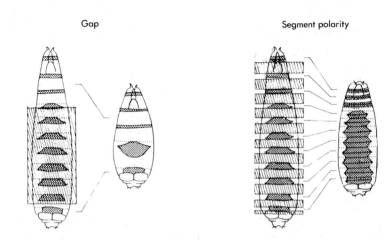

Figure 22-31

Three classes of genes govern embryonic segmentation in *Drosophila*. For each class, the schematic drawing shows a normal larva on the left and a mutant on the right. Dotted regions indicate denticle bands; dotted lines, the segmental boundaries; hatched boxes, the regions missing from the mutant larva; and transverse lines, corresponding parts of the normal and mutant larvas. [Reproduced from C. Nüsslein-Volhard and E. Wieschaus, *Nature* 287 (1980):795.]

deleted. The remaining half-segments appear to be fused so that the resulting embryo has half the normal number of segments (*fushi tarazu* is Japanese for "not enough segments"). Mutations in genes of the second class delete several adjacent segments (gap mutations), whereas mutations in genes of the third class delete half of each segment and replace it with a mirror image of the remaining half-segment (segment polarity mutations).

The behavior of many segmentation mutants was initially very puzzling, because defects frequently did not affect a single segment or group of segments, but rather half of one segment and the adjacent half of the next segment. Now, however, it has become apparent that the basic units of organization in the body of the larval and adult fly may not be the obvious morphological segments (visually defined by constrictions of the body) but rather parasegments (which extend from the middle of one morphological segment to the middle of the next, with the constrictions falling within each parasegment). This illustrates a common source of confusion in the molecular study of development: It is very easy to mistake a visible boundary (such as a segmental constriction, or the edge of a wing) for a domain of gene function.

The observation that only a few subtle mutations in controlling genes are sufficient to transform one body segment of the fly into another (see Figure 22-30) or to reduce the total number of body segments in the larva (see Figure 22-31) has tantalizing evolutionary implications. Higher insects have comparatively few body segments and either two or four wings, but are known from the fossil record to have evolved from annelids such as the earthworm, with many identical body segments. Thus, the evolution of the insects appears to reflect naturally occurring mutations no different from those identified experimentally by the geneticist in the laboratory. The ability of single mutations to dramatically change the body plan of an organism helps us to understand how the accumulation of many independent mutations has led, over the course of evolution, to the incredible variety of contemporary life forms.

Some Homeotic Genes Are Very Long[51, 52, 55]

Several exciting generalizations have already emerged from the preliminary analysis of the DNA sequence, transcription patterns, and developmental regulation of homeotic genes from the Antennapedia

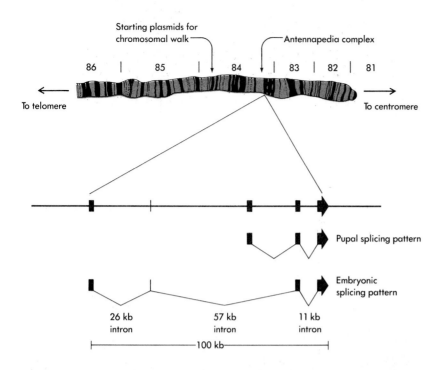

Figure 22-32
The alternative splicing of *Antennapedia* transcripts is developmentally regulated. (Upper) Right arm of salivary gland polytene chromosome 3, showing Antennapedia complex at band 84B and the initial site of hybridization of the starting plasmids for the chromosomal walk at band 84F more than 2000 kilobase pairs away. (Lower) Some alternative splicing patterns. The most downstream exon contains the homeobox homology. [After R. L. Garber, A. Kuroiwa and W. J. Gehring, *EMBO J.* 2 (1983):2027–2036.]

and bithorax complexes. First, several of these homeotic genes appear to be huge in comparison with most other genes in the fruit fly. Of course, all eucaryotes have introns in some of their protein-coding genes, but in general, lower eucaryotes have fewer and smaller introns (Chapter 20). Surprisingly, at least one embryonic mRNA precursor encoded by the Antennapedia complex is nearly all intron, so that mRNAs of only a few kilobases are derived by splicing from a primary transcript of approximately 100 kilobases! Similarly, the primary transcript of the ultrabithorax region of the bithorax complex is about 75 kilobases, although the mature mRNAs derived from it are not unusual in size.

Why are some homeotic transcripts so long? One idea, discussed at the beginning of this chapter, is that both procaryotes and eucaryotes may use long transcription units to control when genes act. This possibility is especially attractive for eucaryotes, since the mature mRNA cannot be completely spliced out of the primary transcript until the most downstream exon has been transcribed. Another possibility, shown in Figure 22-32, is that multiple mRNA species can be spliced out of a single primary transcript in a developmentally regulated fashion. For example, one embryonic *Antennapedia* mRNA includes at least four exons, while an mRNA species synthesized later during the pupal stage includes two of these embryonic exons together with an additional pupal-specific exon. Since many exons appear to encode protein domains with distinct structural or enzymatic functions (Chapter 20), alternative splicing patterns like those seen in the Antennapedia and bithorax complexes may represent a mechanism for organisms to adapt one basic protein structure (encoded by core exons) to different but related developmental purposes. By adding or subtracting functional protein domains encoded by optional developmentally regulated exons, the structural and enzymatic properties of the homeotic gene product could be subtly modified, and the ability of the protein to interact with other cellular components could be changed as development progresses.

Extraordinary Conservation of a Protein-Coding Domain in Homeotic Genes[56–59]

The most startling insight to emerge from DNA sequence analysis of genes controlling pattern formation in *Drosophila* is the existence of a short, 180-base-pair sequence called the **homeobox,** which is found within the 3' exon of many homeotic and segmentation genes, including the *Ubx, abd-A,* and *Abd-B* regions of the bithorax complex and the *Dfd, Scr, ftz,* and *Antp* genes of the Antennapedia complex. The sequences of the individual homeoboxes are nearly identical and evidently code for part of a larger polypeptide: When the homeobox DNA sequences are translated in the correct reading phase, all the minor sequence divergence can be accounted for by conservative substitutions of one amino acid for another in a way that would presumably not affect the enzymatic activity or structural properties of the gene product. But even more amazing is the recent observation that when the *Antennapedia* homeobox is used as a radioactive probe to look for comparable DNA sequences in other eucaryotes, the African clawed toad *Xenopus,* the mouse, and humans were all found to have multiple related sequences. In fact, one of the *Xenopus* homeoboxes is so similar to the *Antennapedia* homeobox in *Drosophila* that only 1 amino acid out of 60 has changed despite millions of years of evolutionary separation between the two species. This raises the possibility that some of the same molecular mechanisms that play a role in morphogenesis of the fruit fly are at work in vertebrates as well, and that the gene products (or parts of them) have been subject to the most extreme selective pressure imaginable during evolution. There is even a preliminary indication that the *MAT*a1 and *MAT*α2 transcripts in yeast (Chapter 18) share slight homology with a part of the fly and vertebrate homeoboxes. Since the *MAT* locus is a well-characterized binary switch determining whether the yeast mating type is **a** or α, this finding fuels speculation that homeotic genes may be turned on and off in groups of cells in order to generate a code capable of specifying the developmental fate of major body parts.

Localization of Homeotic Transcripts in the Early Embryo[60]

The simplest interpretation of the available evidence is that a homeotic gene like *Antennapedia* affects the initial commitment of all or part of one imaginal disc to become all or part of a particular structure on the surface (cuticle) of the fly. An alternative but somewhat more complicated interpretation is that the homeotic gene product might govern not the initial commitment of the disc cells in embryogenesis, but rather the ability of the embryonic cells to retain that commitment through the three larval instars that precede metamorphosis. In either case, as long as the homeotic gene product is turned on and off at the level of transcription or RNA processing (Chapter 21), we might expect that particular homeotic gene transcripts will be found in certain regions of the embryo and not in others. That this is true can be shown dramatically by probing a cross section of a developing embryo with a radioactively labeled DNA fragment from a homeotic gene (Figure 22-33). This in situ hybridization technique is essentially the same as that used to map the location of genes on a chromosome, except here the radioactively labeled probe anneals

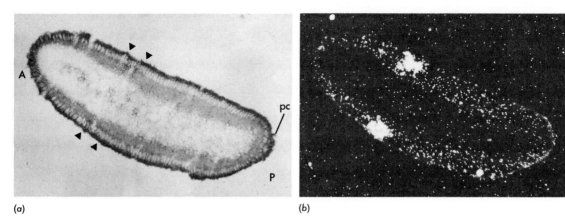

(a)   (b)

with single-stranded mRNA in the thin section of tissue, rather than with denatured DNA in the partially disrupted chromosome (see Figure 22-17). We can see that a DNA fragment from the *Dfd* gene of the Antennapedia complex detects a concentration of transcripts just after formation of the cellular blastoderm. The mRNA is distributed in a belt about one-third of the way back from the anterior tip of the embryo, in a region known to form the posterior head and first thoracic segments of the fly. This is the same region identified by geneticists as the target affected by mutations in the *Dfd* gene.

Transdetermination: Differentiation as a Hierarchy of Developmental Switches

An ingenious set of serial transplantation experiments provides strong support for the notion that commitment of an imaginal disc to become a specific cuticular structure depends on the particular combination of developmental switches thrown in that disc (Figure 22-34). A specific imaginal disc (say, an eye disc) is surgically excised from the larva and transplanted into the abdomen of an adult fly. There, bathed in hemolymph (insect blood), the disc cells survive and proliferate, outliving their host, so that for long-term culture the discs must be cut into smaller pieces and serially transplanted from one adult fly to another. The stable hormonal environment of the adult fly (lacking the molting hormone, ecdysterone) effectively mimics that of the larva, so the cultured disc cells, although committed to a particular fate, do not differentiate. Only when a cultured disc is transplanted back into a third instar larva on the verge of pupation can its developmental potential be tested. When this is done, the usual and not very surprising result is that after metamorphosis, the transplanted discs have retained their commitment to become a specific adult body part. Commitment and subsequent differentiation appear irreversible.

Occasionally, however, an imaginal disc will partially or completely transdetermine to a different body part as though the original commitment of the disc to its particular fate was not completely stable over the long term in culture. By analyzing many such **transdetermination** events, transdeterminations were found to obey strict rules (Figure 22-35). For example, an antenna disc can become a leg disc, and vice versa, but an antenna disc cannot become an eye disc except by two successive transdeterminations. Occasionally, a disc will transdetermine twice or more during serial transplantation—say, from genital to antennal disc, and then from antennal to wing disc. However, a genital disc is never observed to transdetermine *directly* to

Figure 22-33
Homeotic transcripts are localized in the *Drosophila* embryo. Serial sections of a wild-type embryo at the cellular blastoderm stage were hybridized with a radioactively labeled DNA probe from the *Dfd* gene in the Antennapedia complex. To do this, the tissue sections were spread out on a microscope slide and then overlaid with liquid photographic emulsion. After a sufficient amount of time had elapsed for radioactive disintegrations from the hybridized probe to expose the emulsion, the slide was developed. (a) Bright-field photomicrograph of the embryo after the in situ hybridization. A denotes anterior; P, posterior; pc, pole cells. (b) Dark-field autoradiograph of the same embryo, with silver grains showing the location of mRNA which hybridized with the labeled *Dfd* probe. Arrows in panel (a) show the boundaries of hybridization in panel (b). [Reproduced with permission from W. McGinnis, M. S. Levine, E. Hafen, A. Kuroiwa, and W. J. Gehring, *Nature* 308 (1984):428–433.]

Figure 22-34
When cultured in the abdomen of an adult fly, imaginal discs generally remain committed to their original fate, but occasionally transdetermine partially or completely to a disc of a different developmental fate. [After B. Alberts, D. Bray, J. Lewis, M. Raff, K. Roberts, and J. D. Watson, *Molecular Biology of the Cell* (New York: Garland, 1983, p. 839.]

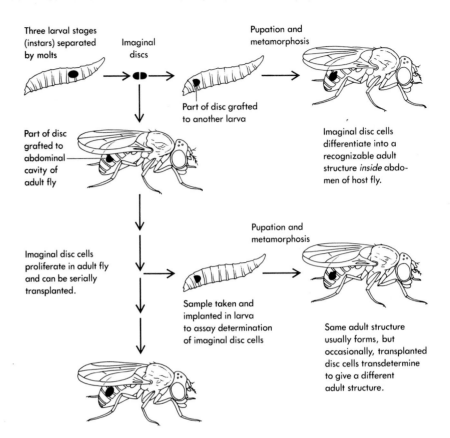

Figure 22-35
Only certain transdeterminations are allowed, and many of these resemble known homeotic mutations such as *Antennapedia*. [After E. Hadorn, *Sci. Amer.* 219 (1968):110–120. © 1968 by Scientific American, Inc. All rights reserved.]

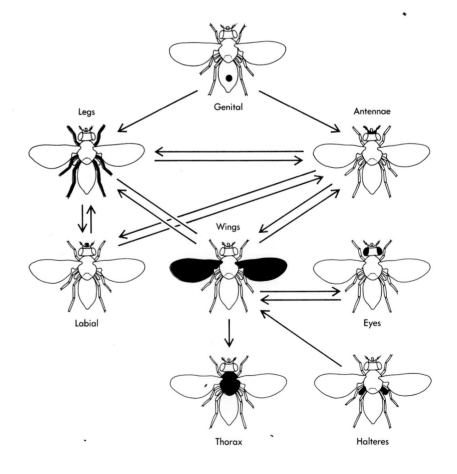

a wing disc. This suggests that each transdetermination corresponds to throwing a single molecular switch, and the ultimate developmental fate of each disc is determined by the combination of molecular switches thrown in it. Transdetermination bears a suspicious resemblance to the behavior of homeotic mutations such as *Antennapedia*, as though the consequence of certain homeotic mutations was to increase the frequency of experimentally induced transdetermination events from very low to essentially 100 percent. Taken together, these observations suggest that the products of homeotic genes are responsible for the initial commitment of imaginal discs, or the maintenance of that commitment, in the normal development of the fly.

Polyclones: Cells Can Communicate and Throw the Same Developmental Switch Together[61, 62]

We have seen that an imaginal disc consisting of many thousands of cells can partially or completely transdetermine from one body part to another (say, an antenna to a wing). In principle, this rather startling phenomenon could be explained by either of two contrasting mechanisms—clonal or polyclonal transdetermination—and these have radically different implications for development in general. In clonal transdetermination, a single cell in the antenna disc would first transdetermine in a heritable fashion into a wing cell and then rapidly outgrow the neighboring antenna cells. The transdetermined wing tissue would constitute a **clone**, the progeny of a single initial transdetermined cell. Alternatively, many cells within the antenna disc could transdetermine simultaneously and then proliferate so that the wing tissue would be a **polyclone**, the progeny of many different initial cells.

Clonal and polyclonal transdeterminations can be distinguished experimentally when a very small number of cells in the initial antenna discs are genetically marked with a special visible mutation that makes the tissue yellow (Figure 22-36). If the transdetermined wing tissue is always found to be exclusively wild-type or yellow, then a single transdetermined cell gives rise to the entire mass of transdetermined tissue. But if the transdetermined tissue is sometimes a mixture of wild-type and yellow cells, then a group of cells (both wild-type and mutant) must have transdetermined together. In fact, transdetermined wing tissue is often observed to be genetically mixed, so many different cells in the initial disc must have some way of communicating with one another in order to throw the same molecular switch (antenna to wing).

How do cells within a polyclone communicate with one another? We might speculate that ionic currents like a nerve impulse, or chemical morphogens like those in *Dictyostelium*, are able to pass through extracellular spaces or even through intracellular cytoplasmic bridges. Or perhaps cells recognize other cells within the same compartment by means of compartment-specific cell surface glycoproteins. (We shall see in Chapter 23 that cells of the mammalian immune system make extensive use of such cell surface recognition elements.) We also do not know the nature of the molecular switches that enable cells within a polyclone to remember their state of determination in a heritable fashion. The memory might correspond to a physical change in the DNA such as a chromosomal inversion (as in *Salmonella*

Antennal disc in larva

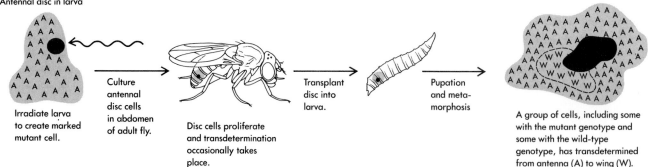

Culture antennal disc cells in abdomen of adult fly.

Irradiate larva to create marked mutant cell.

Disc cells proliferate and transdetermination occasionally takes place.

Transplant disc into larva.

Pupation and meta-morphosis

A group of cells, including some with the mutant genotype and some with the wild-type genotype, has transdetermined from antenna (A) to wing (W).

Figure 22-36
Groups of unrelated cells (polyclones) transdetermine together. [After B. Alberts, D. Bray, J. Lewis, M. Raff, K. Roberts, and J. D. Watson, *Molecular Biology of the Cell* (New York: Garland, 1983, p. 839.]

phase variation), directed gene conversion (as in the yeast mating-type switch), or methylation/demethylation of specific DNA sequences. Another possibility is that the memory reflects a heritable change in the chromosomal proteins that package and regulate expression of DNA. A still more remote possibility is that the memory is retained within the nucleoplasm or cytoplasm by metabolic feedback circuits or self-replicating cytoskeletal structures, with no direct chromosomal record even being made.

Development of the Fruit Fly Is Modular[63, 64]

An immense wealth of genetic evidence has now been amassed to show that pattern formation in the developing *Drosophila* embryo is determined by a hierarchy of controlling genes. Maternal genes expressed in nurse cells during oogenesis are responsible for establishing gradients of positional information along the major developmental axes of the egg (e.g., the 10 or more dorsalizing genes). Genes expressed in the early embryo itself determine the initial folding of the cellular blastoderm (e.g., the 15 or more "segmentation" genes). And genes expressed still later in embryogenesis can partially or completely transform one adult body part into another by transdetermining the fate of a polyclone within an imaginal disc.

A hierarchy of developmental decisions may also help to explain the behavior of a homeotic mutation like *Aristapedia*, where only the most distal part of the antenna is transformed into the most distal part of the leg (see Figure 22-28). Why is the effect of the *Aristapedia* mutation confined to the distal part of the antenna, while the *Antennapedia* mutation transforms the entire antenna? One attractive (but hypothetical) explanation is that just as a set of molecular switches determines whether the cells within an imaginal disc are fated to become one adult cuticular structure or another, so another set of switches may determine whether particular subsets of cells within the imaginal disc (polyclones) are fated to be nearer the body of the adult fly (proximal) or farther away (distal). Seen in this way, "antenna" and "leg" would be alternative positions for one switch, "proximal" and "distal" alternative positions for a second switch, and the switching mechanism would cause these two switches to interact with each other. The *Aristapedia* mutation might prevent the first switch from being on "antenna" whenever the second switch was on "distal," or alternatively, *Aristapedia* might cause the first switch to reverse from "antenna" to "leg" in cells that threw the second switch to "distal." In contrast, *Antennapedia* might lock the first switch on "leg" regardless of the position of the second switch.

Thus, we see that the fruit fly *Drosophila* is likely to be put together from many modules, each differentiating in response to a hierarchy of developmental switches thrown in succession. Lest the word "switch" itself imply any particular molecular mechanism, we should remember that biological switches can be as different as the competition between RNA polymerase and CI repressor during the early stages of infection by bacteriophage λ (see Figure 17-16) and the displacement of one σ factor by another during sporulation in *B. subtilis*. We currently have no idea what kind of molecular switches will be found to govern the hierarchy of genes controlling modular morphogenesis in the fruit fly.

We shall see shortly that this kind of modular behavior is carried to even greater extremes in the development of a small nematode worm named *Caenorhabditis elegans*. Although comparable modules are not readily apparent in the development of higher organisms (except for segmentation of the vertebrate spinal column and rib cage), it would be most surprising if modular construction did not turn out to be a universal developmental mechanism, albeit in modified or somewhat disguised form.

Positive Control of Gene Function by Steroids[65, 66]

At the genetic level, the best-understood steroid hormone is ecdysterone, the growth hormone responsible for the successful passage of developing insects from their larval to their pupal stages. In the early larval period, the puffing pattern of the polytene *Drosophila* salivary chromosomes is stable. But as soon as ecdysterone is released into the insect hemolymph, the preexisting puffs begin to regress, and over 100 new puffs quickly develop during the transition from the larval to the pupal stage. Each band always puffs at a precise time and remains active for a well-defined interval (Figure 22-37). The first such puffs (the "early" puffs) begin appearing within 5 minutes of hormone release. They appear to be direct responses to the hormone, since their appearance is unaffected by inhibitors of protein synthesis. In contrast, those "late" puffs that only appear some hours after ecdysterone is added do not appear when protein synthesis is blocked. Presumably, a protein product(s) encoded by one or more of the early puffs induces the formation of the late puffs.

The ability of ecdysterone to induce very rapid expansion of the early puffs initially suggested that the hormone binds directly to specific sites on chromatin. This tends to be confirmed by experiments showing that ecdysterone can be photochemically crosslinked to the early puff sites. Thus, ecdysterone may function in insects just as other steroids do in higher cells. The hormone first binds to a specific receptor protein, and this hormone-receptor complex then binds directly to specific DNA sequences, where it modulates transcription. Although steroids usually function as positive activators of transcription (Chapter 21), this may not always be the case. In addition to inducing the expansion of early puffs, ecdysterone also causes a different group of preexisting puffs to regress (recondense). Moreover, ecdysterone appears to bind directly to these regressing puffs, just as it does to the expanding hormone-inducible puffs. These interesting observations suggest that at least in insects, and perhaps in higher eucaryotes, steroid hormones may also function as negative regulators of transcription. This possibility should not really surprise us. In bacteria, repressors work by physically blocking the sites on DNA

Figure 22-37
Schematic representation of the action of ecdysterone in the sequential production of early and late chromosome puffs.

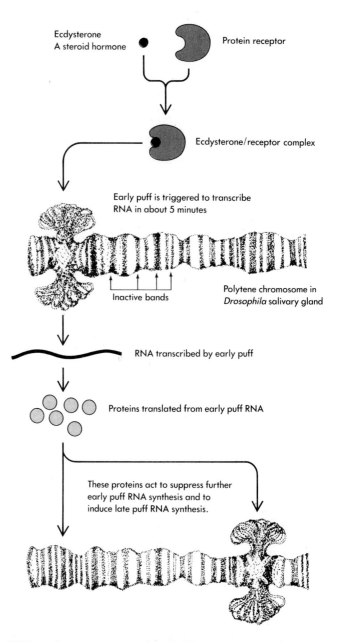

Ecdysterone
A steroid hormone

Protein receptor

Ecdysterone/receptor complex

Early puff is triggered to transcribe RNA in about 5 minutes

Inactive bands

Polytene chromosome in *Drosophila* salivary gland

RNA transcribed by early puff

Proteins translated from early puff RNA

These proteins act to suppress further early puff RNA synthesis and to induce late puff RNA synthesis.

where RNA polymerase must bind. Thus, in principle, any protein that functions as a positive activator of transcription by recognizing a specific DNA sequence in one gene can also function as a repressor of another gene. For the activator protein to function as a repressor, the protein binding site in the second gene need only be located in a region that is essential for transcription. Thus, the position of a protein binding site relative to the promoter may determine whether binding of a particular protein modulates transcription positively or negatively.

Bigger Is Not Always Better— Or, Why Study a Worm?

The nematode *Caenorhabditis elegans* is a very small worm that has recently come to occupy a very large place in the molecular biology of development. There are several reasons for the sudden scientific pop-

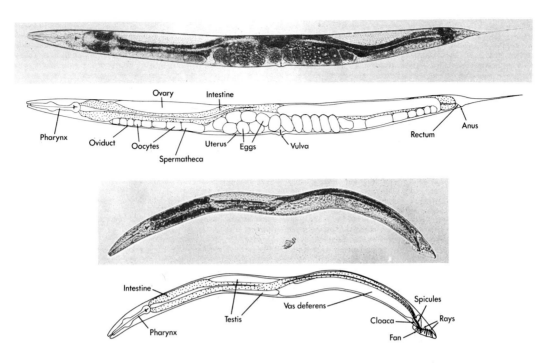

Figure 22-38

The sexes of *Caenorhabditis elegans* are the self-fertilizing hermaphrodite (above) and the male (below). The reproductive system of the adult hermaphrodite produces both sperm and eggs. Eggs from the ovary pass through the oviduct and are fertilized in a specialized sperm-producing region of the oviduct called the spermatheca. The fertilized eggs then enter the uterus, where embryogenesis takes place. The juvenile worms leave the parent through the vulva. Eggs that escape fertilization within the hermaphrodite can be fertilized by sperm that the male deposits in the vulva of the hermaphrodite. [Reprinted with permission from J. E. Sulston and H. R. Horvitz, *Dev. Biol.* 56 (1977):110–156.]

ularity of this organism. The worm is small (about 1 mm long) and has a very simple body plan (Figure 22-38). Nematodes can be grown in large numbers on Petri dishes seeded with bacteria (their favorite food), and they reproduce with a remarkable generation time of a little over 3 days from conception to maturity. There are two sexes (hermaphrodites and males) and a large number of progeny (300 per parent) so that the organism can be analyzed genetically with great ease. Perhaps most important of all, the worm is transparent at all stages of the life cycle from egg to adult. As a result, the cellular anatomy of the developing worm can be examined and recorded with absolute precision in the living animal by using a special light microscope equipped with Nomarksi interference optics. Interference optics display differences in refractive index between various cellular organelles (Figure 22-39). By following both the movements and divisions of the highly refractile cell nuclei starting immediately after fertilization, it has been possible to reconstruct the entire anatomical history of each cell as the fertilized egg develops into a multicellular adult.

Invariant Cell Lineages[67, 68]

Two amazing conclusions emerged when the information gathered from detailed anatomical studies of living worms was combined with more classical anatomical studies obtained by electron microscopy of serial thin sections of the worm at different developmental stages. First, each cell in the adult worm is derived from the zygote by a virtually invariant series of cell divisions, cell migrations, and occasional programmed cell deaths, collectively called a **cell lineage** (Figure 22-40). Second, as a direct consequence of the invariant cell lineages, individual nematodes are anatomically invariant carbon copies of each other. The mature adult hermaphrodite is always put together from exactly 945 cells (the actual number of nuclei is slightly larger, because some tissues such as the intestine are partially syncytial). If

Figure 22-39
Early development of the nematode *Caenorhabditis elegans* as viewed by Nomarski interference optics. (a) Fertilized egg just before the first cleavage. The male and female pronuclei have not yet fused completely. (b) The beginning of gastrulation, marked by cell movements and the formation of primitive tissues. For example, Ea and Ep are the anterior and posterior daughters of the first intestinal precursor cell called the E cell, D is a mesodermal precursor, and P4 gives rise to germ cells. (c) By 7 hours, the embryo starts to move, and individual cells are more difficult to discern. (d) The late embryo is clearly a miniature worm. The white arrow points to the mouth, and the two black arrows to germ cells. [Reprinted with permission from J. E. Sulston, E. Schierenberg, J. G. White, and J. N. Thomson, *Dev. Biol.* 100 (1983):64–119.]

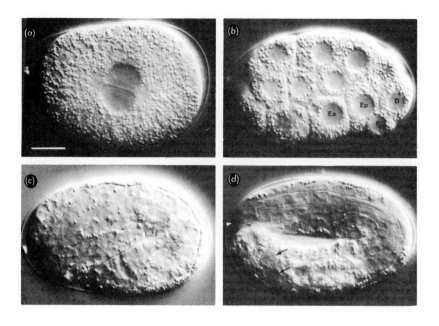

we focus on the nervous system for the purpose of illustration, 302 of the 945 cells are always neurons. These 302 neural cells are always derived from 407 neural precursor cells, of which 105 always suffer a programmed cell death. Of the 302 surviving neurons, 8 are always dopaminergic (responding to the neurotransmitter dopamine) and 2 are always serotonergic (responding to serotonin). The remaining 643 nonneural cells in the adult hermaphrodite can be described with the same precision as those of the nervous system. No other eucaryotic organism can currently rival this degree of anatomical detail.

The 302-element neural circuitry of the adult worm may be as simple (and is certainly as reproducible) as the printed circuits in a mass-produced portable calculator. Such small size and simplicity will speed molecular analysis. Knowing how a worm coordinates the wiggling movements of its body musculature or how neural events trigger the mating of a male with a hermaphrodite (there is no female) should give us general insights into neural function. With luck, these insights will lay the groundwork for understanding the vastly more complex wiring pattern in our own brains, where literally billions of cells are hooked up in a variable and somewhat sloppy fashion.

Laser Microsurgery Complements Classical Genetic Analysis[69]

Because *C. elegans* grows readily on Petri dishes, tens of thousands of individual animals can be visually surveyed through a low-power microscope. This makes it exceptionally easy to identify and isolate mutants with obvious behavioral abnormalities in movement, thermotaxis, chemotaxis, avoidance of high osmotic strength, touch sensitivity, egg laying, and synthesis and degradation of neurotransmitters. For example, *unc* mutants wiggle in an *unc*oordinated fashion or fail to wiggle at all; these worms have a defect in either the nerves, muscles, or neuromuscular junctions required for smooth and efficient gliding motions. Characterization of *unc* mutants has already led to the identification of many genes, including those for actin and myosin.

Intensive genetic analysis can reveal the developmental consequences of a deficiency or alteration in a single gene product, or a mutation in a regulatory DNA sequence. But for studying the molecular mechanisms of development, it is sometimes experimentally advantageous simply to eliminate a particular cell from the embryo. There is no systematic way to isolate such mutants genetically, so biologists have traditionally resorted to physical techniques. Classical biologists surgically transplanted (or excised) relatively large multicellular chunks of embryo. In this way, for example, the dorsal lip of the amphibian blastopore (see Figure 22-5) was shown to be largely responsible for organizing the initial stages of blastulation: After the grafting of the dorsal lip from one embryo into the blastocoel of another, the extra lip region would instruct the surrounding cells to form a secondary embryo alongside the first. Modern developmental biologists can use a sharply focused laser beam to destroy (ablate) single cells in the embryonic nematode without any apparent damage to neighboring cells.

Mosaic Versus Regulative Development

The results of extensive laser ablation experiments, together with genetic analysis, are consistent with the notion that differentiation can proceed in at least two profoundly different styles—mosaic and regulatory—which presumably reflect profoundly different molecular mechanisms. In **mosaic development** (so called because the organism is assembled from independent parts), the differentiation of a cell does not depend on the behavior or even the existence of neighboring cells: Laser ablation eliminates its target cell but does not affect the behavior or differentiation of any other cells. Apparently, in mosaic development, internal events within each cell instruct that cell what to do; such events could be triggered by cell division itself or by the ticking of an internal biological clock that is set in motion by fertilization. In **regulatory development,** differentiation is partially or completely dependent on the interaction between neighboring cells. For example, during embryogenesis, the lone gonadal anchor cell induces a certain subset of ventral hypodermal cells to form a vulva (the opening through which eggs are laid in the hermaphrodite). If the anchor cell is destroyed by laser ablation, no vulva develops; but if almost all the gonadal cells other than the anchor cell are ablated, a normal vulva can still be formed.

Mosaic development governed by strict cell lineages is the overwhelming rule in *C. elegans,* and regulative development the exception (occurring only in vulval morphogenesis and differentiation of germ cells within the gonad). In most other organisms (ourselves included), the reverse is almost certainly true. Mosaic development may dominate in the nematode because survival of the species requires the life cycle to unfold with the utmost possible speed. Thus, there may not be sufficient time for complex intercellular signaling and interaction (many hours are required to establish a slime mold aggregation field; see Figure 22-23). Instead, each cell is on automatic pilot, with programmed instructions so precise that the independently developing cells can all fit together neatly into a functional multicellular organism. The price of such mosaicism may be very small size (perhaps only a limited number of cell divisions can be so rigidly programmed) and only modest complexity (cell-cell interactions may be required to construct a more elaborate anatomy).

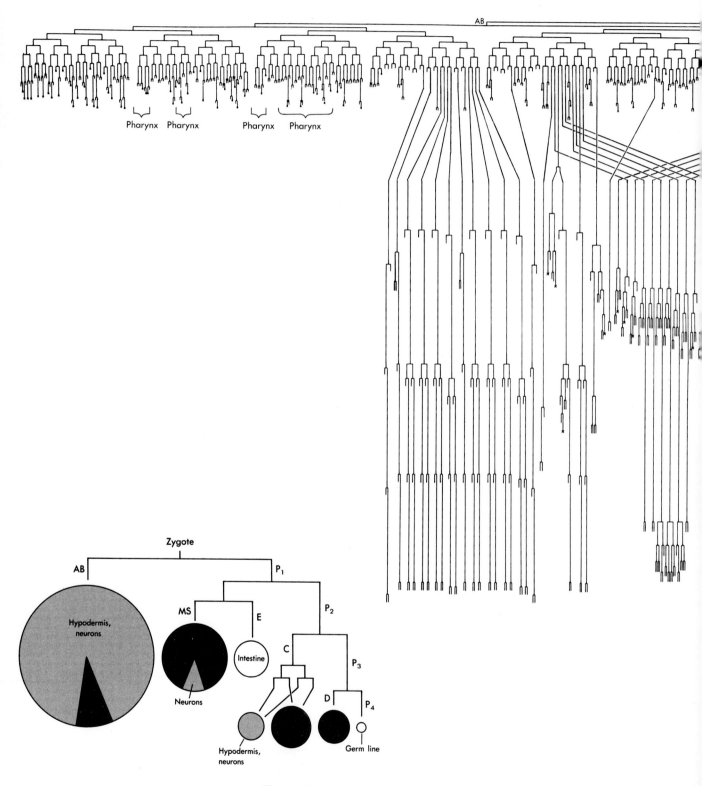

Pharynx Pharynx   Pharynx Pharynx

Zygote

AB   P₁

Hypodermis,
neurons

MS   E

Neurons

C   P₂

Intestine

P₃

D   P₄

Hypodermis,
neurons

Germ line

Figure 22-40
The complete cell lineage tree of the adult hermaphrodite sex of the nematode *Caenorhabditis elegans*. Vertical axis represents time, increasing downward (about 50 hours at 20°C). Horizontal lines represent cell divisions; vertical lines ending in "X" represent cell deaths (131 total); other vertical lines represent surviving cells (945 total). As summarized diagrammatically in the inset, the fertilized egg (zygote) initially divides into the AB, MS, E, C, D, and P₄ founder cells from which the major cell lineages are

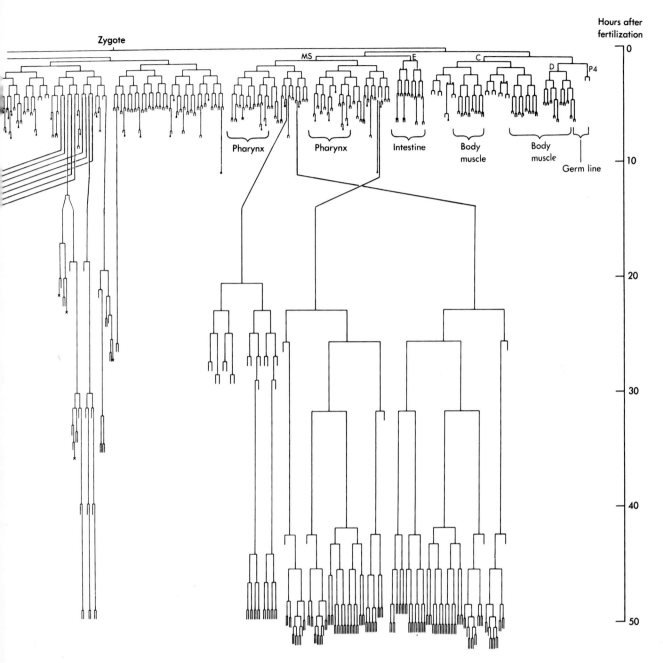

derived. For illustrative purposes, only the lineages that generate the pharynx (mouth), intestine, germ line, and parts of the body muscle are labeled in the complete lineage tree. The germ line is derived from the P_4 cell, and the intestine from the E cell. Only the first embryonic division of the P_4 cell is shown, since its subsequent lineage is indeterminate (the number and pattern of cell divisions are not fixed). The pharynx is composed of cells from both the AB and MS lineages. The body musculature is derived from four different founder cells (AB, MS, C, and D), but only the D lineage gives rise exclusively to muscle cells. Most of the unlabeled cells in the complete lineage represent neurons or hypodermal cells. Hypodermal cells (i.e., cells lying just below the skin) secrete the protective cuticle that covers the surface of the worm and serves as an attachment site for the body muscles. Note that left and right in the figure do *not* correspond to anterior and posterior; for example, the pharynx at the front of the worm is composed of cells from both sides of the figure.

[After a figure by Dr. J. E. Sulston, MRC Laboratory of Molecular Biology, Cambridge; and J. E. Sulston, E. Schierenberg, J. G. White, and J. N. Thomson, *Dev. Biol.* 100 (1983):64–119.]

What Can Cell Lineages Tell Us About Molecular Mechanisms of Development?

Most organisms with the anatomical complexity of the nematode have perhaps 10 to 1000 times as many cells. For the nematode to achieve the same level of biological complexity with as few as 945 cells means that a very high percentage of the cell divisions in the invariant cell lineages must produce daughter cells committed to different developmental fates. Consequently, cell lineages alone appear to determine cell fate (with the aforementioned exception of regulative development in certain vulval and germ line cells). Consider, for example, the posterior daughter of the anterior daughter of the posterior daughter of the zygotic nucleus. This particular cell (the E cell) will be the precursor of all the intestinal cells in the worm, and only cells descended from this cell can contribute to formation of the gut (see Figure 22-40).

How might cell fate be progressively determined by an unfolding cell lineage? Since each lineage is defined by a succession of cell divisions, we might speculate that it is DNA replication or cell division itself that results in commitment of the daughter cells. Current evidence suggests that there are at least four general mechanisms by which cell division might seal cell fate:

- Physical modification of the DNA (e.g., 5-methylation of cytosine) might be gained or lost during replication.

- Chromosomal proteins might preferentially associate with one strand of the parental duplex during semiconservative DNA replication.

- Passage of chromatin through the replication fork might strip old regulatory proteins from the DNA, and freshly replicated DNA emerging from the fork might be able to bind a new set of regulatory proteins.

- Cytoplasmic factors might segregate asymmetrically during cell division.

We have already discussed the first three possibilities as conceivable mechanisms for generating stem cell lineages; next we examine some of the evidence for cytoplasmic determination.

Insect Polar Plasm: Cytoplasmic Determinants of Nuclear Differentiation

Classical embryology abounds with instances of **cytoplasmic determinants**—cytoplasmic substances that tell a nucleus what to do. We have already seen that the female egg is not simply an undifferentiated, spherically symmetrical, haploid cell awaiting fertilization. Unfertilized eggs already possess complex internal architecture, reflecting extensive regional differentiation of the egg cytoplasm. Cytoplasmic determinants, which are distributed in a precise, three-dimensional pattern, are partitioned into different blastula cells by cleavage divisions following fertilization. These substances then control the subsequent differentiation of the cells (blastomeres) that receive them.

Among the best-studied cytoplasmic determinants are the **polar**

granules found in the posterior cytoplasm of insect eggs, which cause cleavage nuclei in their vicinity to differentiate into primordial germ cells (see Figure 22-24). Immediately after fertilization, the zygotic nucleus of a *Drosophila* egg undergoes eight nuclear divisions without cellularization. We already saw how most of the nuclei migrate to the outside of the egg after nine divisions, undergo four more synchronous divisions, and are then enclosed by invagination of the egg membrane to form a cellular blastoderm. These cells will give rise to the somatic (non–germ line) tissues of the fly. However, we did not discuss the remarkable observation that a few nuclei—destined to become the primordial germ cells—are set aside after only eight nuclear divisions to follow a distinct developmental pathway. At that time, nuclei that happen to be near small cytoplasmic structures called polar granules in the posterior of the egg are induced to differentiate into primordial germ cells. These **pole cells** are larger and rounder than the small cells of the blastoderm. The pole cells initially lie outside the sheet of cellular blastoderm, but later migrate to the gonads (a somatic tissue), where they give rise to thousands of progeny germ cells.

Taking advantage of the wealth of genetic mutations in *Drosophila*, an elegant experiment was devised to prove that the polar cytoplasm (containing polar granules) determines the differentiation of embryonic nuclei (Figure 22-41). The experimental design took advantage of a recognizable but genetically recessive mutation called yellow. The wild-type gene y^+ is dominant over y^- mutations, since one copy of the normal gene product is sufficient to confer normal body color on the fly. Thus, homozygous y^+/y^+ flies and heterozygous y^+/y^- flies have normal color, but homozygous y^-/y^- flies are a lighter yellow. Polar cytoplasm was withdrawn from the posterior end of a wild-type embryo with a tiny pipette and transplanted to the anterior end of another y^+/y^+ embryo (normal color). The transplanted polar cytoplasm induced formation of large, round cells resembling pole cells in the anterior of the second embryo, where no pole cells usually form. To test whether the presumptive pole cells were genetically functional, they were transplanted to yet a third embryo with the y^-/y^- constitution (yellow). These embryos developed into yellow flies because all their somatic tissues had the y^-/y^- genetic constitution, but some of their progeny were wild-type in appearance. Such wild-type flies could only have obtained the wild-type y^+ gene from y^+/y^+ nuclei in the second embryo. This proves that polar cytoplasm induced functional pole cells in the second embryo.

Cytoplasmic Determinants in the Amphibian Egg[70–73]

Polar plasm also exists in vertebrate eggs, although the experimental proof does not quite have the genetic elegance that is possible in *Drosophila*. Frog eggs contain a special **germinal cytoplasm** at the vegetal pole where primordial germ cells eventually form (see Figure 22-5). One of the active factors in this cytoplasm appears to be nucleic acid (presumably RNA, but conceivably DNA), because selective irradiation of the germinal cytoplasm with ultraviolet light prior to fertilization results in sterile but otherwise normal frogs. However, the irradiated eggs can be rescued by microinjecting unirradiated germinal cytoplasm from a second egg; tadpoles with fully formed gonads and germ cells are then obtained. Similar experiments have been

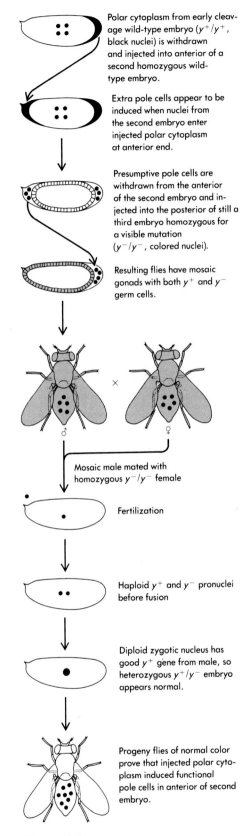

Polar cytoplasm from early cleavage wild-type embryo (y^+/y^+, black nuclei) is withdrawn and injected into anterior of a second homozygous wild-type embryo.

Extra pole cells appear to be induced when nuclei from the second embryo enter injected polar cytoplasm at anterior end.

Presumptive pole cells are withdrawn from the anterior of the second embryo and injected into the posterior of still a third embryo homozygous for a visible mutation (y^-/y^-, colored nuclei).

Resulting flies have mosaic gonads with both y^+ and y^- germ cells.

Mosaic male mated with homozygous y^-/y^- female

Fertilization

Haploid y^+ and y^- pronuclei before fusion

Diploid zygotic nucleus has good y^+ gene from male, so heterozygous y^+/y^- embryo appears normal.

Progeny flies of normal color prove that injected polar cytoplasm induced functional pole cells in anterior of second embryo.

Figure 22-41
Polar cytoplasm in the egg induces adjacent nuclei to differentiate into primordial germ cells. [After K. Illmensee and A. P. Mahowald, *Proc. Nat. Acad. Sci.* 71 (1974):1016–1020.]

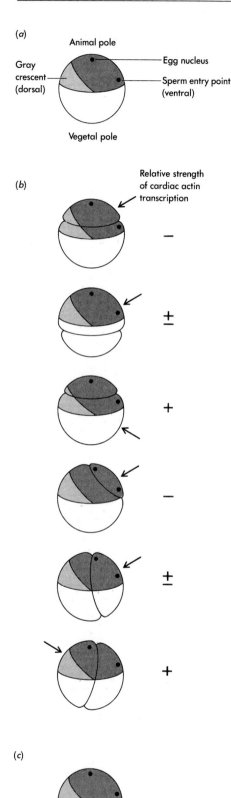

(a)

Animal pole

Gray crescent (dorsal)

Egg nucleus

Sperm entry point (ventral)

Vegetal pole

(b)

Relative strength of cardiac actin transcription

—

±

+

—

±

+

(c)

Figure 22-42

Cytoplasmic determinants are localized in the amphibian egg. (a) Axes of the fertilized frog egg. (b) Plane of ligation determines the level of cardiac actin transcription in the part of the embryo (denoted by arrow) that was grown and subsequently analyzed. (c) Schematic interpretation of results. Dots represent distribution of cytoplasmic substances that are responsible for activating cardiac actin genes in the developing embryo. [After J. B. Gurdon, T. J. Mohun, S. Fairman, and S. Brennan, *Proc. Nat. Acad. Sci.* 82 (1985):139–143.]

performed on eggs of the midge *Smittia* (an insect), with analogous results. The strong resemblance between germ cell determination in an insect and an amphibian suggests that there are highly conserved mechanisms to protect primordial germ cells from the events of somatic differentiation. We do not know why such protection is necessary, but we might speculate that somatic differentiation sometimes involves physical modification of the DNA, somatic rearrangement of the DNA, or changes in DNA packaging (chromatin structure) that cannot easily be reversed.

A related series of experiments has been performed to demonstrate that cytoplasmic determinants contained in the gray crescent of the amphibian egg are required for transcription of muscle-specific actin genes at the neurula stage (following gastrulation). Instead of using ultraviolet light to inactivate nucleic acid components in the egg, a tiny loop composed of a single human hair was carefully tied into a tight knot around the newly fertilized amphibian egg (Figure 22-42). Depending on the orientation of this ligature with respect to the natural axes of the egg (the animal and vegetal poles, and the gray crescent opposite the point of sperm entry), the embryo that develops from the nucleated half of the ligated egg will be deprived of different parts of the original egg cytoplasm. When the ligated eggs reached the equivalent of the neurula stage, the expression of muscle-specific actin genes was examined by measuring the ability of neurula RNA to form RNA-DNA hybrids with a cloned actin gene. The muscle actin genes were expressed only when ligated embryos retained a subequatorial region of the egg near the gray crescent area. Since the gray crescent normally becomes the dorsal side of the animal, where embryonic muscle first appears, we can see that the cytoplasm of the frog egg contains molecular determinants. These determinants (whose chemical nature is still unknown) instruct cleavage nuclei entering different regions of the cytoplasm to express different genes.

Cytoplasmic determinants and morphogenetic gradients are fundamentally different ways of determining the differentiation of embryonic cells. Cytoplasmic determinants are only able to instruct nearby nuclei to differentiate into a particular cell type. In contrast, because cells respond differently depending on where they lie within a gradient, gradients of positional information are able to influence a much larger field of cells. The advantage of cytoplasmic determinants may be that embryogenesis can proceed fastest when cells do not have to spend time creating and interpreting gradients of positional information. In the nematode, for example, where embryogenesis is extraordinarily rapid, cytoplasmic determinants direct one blastomere to differentiate into the E cell, which subsequently gives rise to the entire intestinal cell lineage (see Figure 22-40). The disadvantage of cytoplasmic determinants may be that only certain kinds of differentiation are simple enough to be determined solely by the influence of a single factor (or small group of factors) on an embryonic nucleus.

PUTTING CLONED GENES BACK INTO THE GERM LINE

The Triumph of Modern Embryology: Mutating Genes at Will

Recombinant DNA techniques enable us to mutate cloned genes in any way we wish, but we cannot learn everything about gene regulation just by studying the transcription of these mutant genes in vitro or by putting the mutant genes back into immortal somatic cell lines (Chapters 19 and 25). The real test of a mutant gene is to put it back into the organism and see how it functions in its natural environment. An additional (but more subtle) advantage of putting new genes into the germ line is that the genes will then experience the entire developmental history of the organism. This is important because cell lineage plays a role in determining cell fate (as we have seen in the nematode *C. elegans*), and this implies that sequential modification of gene or chromatin structure in the course of development may be necessary for proper developmental regulation. For each gene, then, development begins during spermatogenesis and oogenesis, when germ cell DNA first acquires its characteristic pattern of DNA modification and protein packaging. A newly introduced gene may have to begin at the beginning in order to be in its "natural" state at the moment when it should be turned on or off.

Classical Genetic Techniques Are Still Indispensable

We want to study the developmental consequences of replacing a normal gene in an organism with one that has been mutated in vitro in carefully engineered ways. This ambitious experimental program requires a subtle marriage of classical genetics and contemporary recombinant DNA technology. The power of classical genetics is that it asks a very general question: How many different kinds of mutations in how many different genes can produce the same or a related defect? In this way, we find out which genes are important for what, and we accumulate the knowledge necessary to clone each of these genes from the organism. For example, by looking for nematodes that are paralyzed or that move in an uncoordinated fashion *(unc)*, we assemble a catalog of all those genes involved in proper construction of the nervous system and body musculature. Similarly, by identifying fruit flies whose eggs fail to complete embryogenesis, we can gain access to a dazzling variety of genes, some of which determine the shape of the organism (e.g., *dorsal* embryos are all back and no front).

The major limitation of classical genetics is that the nature of the mutations (single base changes, insertions, deletions, inversions, and translocations) cannot be controlled. This limits our ability to design the most informative experiments. For example, one naturally arising single-base change in the homeobox of the *fushi tarazu* gene of the

Antennapedia complex is known to be lethal for development of the fly at slightly elevated temperature. However, there is no way to use classical genetic analysis to isolate additional mutations in this particular homeobox. Thus, it would be extremely helpful to be able to systematically vary the amino acid sequence of the homeoboxes in the *Drosophila* Antennapedia or bithorax gene complexes. Fortunately, the power of recombinant DNA technology is that it can make exactly such mutations in cloned DNA fragments. The trick, then, is not so much to clone the genes in the first place, nor to make interesting mutations in them in vitro, but to get the mutant genes back into the organism in a functional form.

Precise Gene Substitution Is Now Possible Only in Bacteria and Yeast[74a, b]

In yeast alone has it proved possible to use homologous recombination to precisely replace a resident wild-type gene with a mutant gene copy that has been genetically engineered in vitro (Chapters 18 and 19). We should expect no less from a eucaryotic microorganism that is as easy to work with and to analyze genetically as *E. coli*. So far, it has not proved possible to achieve gene replacement in any higher eucaryote than yeast, but this impasse may soon prove temporary. In any event, cloned genes have been successfully reintroduced into the germ line of both the fruit fly *Drosophila* and the laboratory mouse. Although integration occurs at apparently random chromosomal positions, it is particularly encouraging that these "misplaced" genes are in general subject to correct developmental regulation. Random integration is not as nice as precise gene replacement, because the resident wild-type (or mutant) gene is still present and may confuse the analysis; but at least a start has been made.

Putting Genes Back into *Drosophila* with P Element Vectors[75, 76]

We have already seen that the P elements in *Drosophila* are similar to many procaryotic and eucaryotic transposable elements (Figure 20-32). A complete P factor consists of 31-base-pair perfect terminal inverted repeats surrounding a 3-kilobase protein-coding region that encodes two functions: a transposase (which catalyzes movement of the element to new sites) and a repressor of transposition (which accumulates until further transpositions are blocked). During a dysgenic cross, sperm from a male fly bearing genomic P elements (P male) fertilize eggs from a female fly lacking P elements (M female). Genes in the tightly packed sperm head are transcriptionally silent, but P elements on the paternal chromosomes begin to be transcribed soon after entering the egg. Although P elements appear to be transcribed in all cells, transposition occurs exclusively in the germ line as a result of tissue-specific alternative mRNA splicing: The major P element transcript (spanning nearly the entire element) is spliced in somatic cells to produce an mRNA encoding the repressor function, but the same primary transcript can be alternatively spliced in germ line cells to produce an mRNA encoding the transposase. A variety of genetic evidence suggests that P element transposition occurs only in embryonic germ cells (pole cells); this could mean that the tissue-specific splice is restricted to pole cells, or that sufficient repressor accumulates in maturing cells of the germ line to block further transposition.

Tissue-specific alternative splicing of the major P element transcript raises the intriguing possibility that the transposase and repressor are related proteins that share one or more common exons. For example, a transposase-like protein that recognized the terminal inverted repeats of the P element, but lacked the active catalytic site of the transposase, could function as a repressor of transposition by competing with functional transposase for binding to the terminal repeats.

When insertion of a paternal P element happens to disrupt an essential gene (or chromosomal region), heritable new mutations are observed in the offspring of a P male and an M female. (The parental flies display no mutations, since transposition takes place only in the germ line.) The cumulative effect of such mutations (as well as accompanying chromosomal rearrangements) is called **hybrid dysgenesis,** because the progeny of the dysgenic cross are genetically less fit than the parents. The reciprocal cross of an M male with a P female does not lead to dysgenesis, since P elements in the haploid egg genome are repressed by high levels of accumulated repressor in the egg cytoplasm.

Interestingly, most P elements in the genome of a *Drosophila* P strain are internally deleted: Part or all of the transposase- and repressor-coding region is missing. However, as long as the 31-base-pair terminal inverted repeats are intact, these incomplete P elements can be mobilized when transposase is supplied by an intact P element. Apparently, the transposase need only recognize the inverted repeats of the P element. This observation suggested a strategy for putting genes back into *Drosophila* by using recombinant P elements as vectors to insert new genes into germ line DNA (Figure 22-43). Recombinant DNA technology was used to replace the internal protein-coding region of a complete P element with a foreign gene of interest. The resulting recombinant P element (carrying the foreign gene), together with an intact P element (to supply transposase activity), was then microinjected through a thin glass needle into a freshly fertilized nondysgenic egg (produced by crossing an M male with an M female). These injected P elements then behaved exactly like P elements that had been naturally introduced by sperm at the moment of fertilization. The transposase gene on the complete P element was transcribed in the embryo, the mRNA translated, and the transposase catalyzed transposition of both the complete P element and the recombinant P element into embryonic chromosomes.

To maximize the chance that the recombinant P element would integrate into germ line chromosomes, the DNA was microinjected near the posterior end of the fly, where cytoplasmic polar granules determine which blastula nuclei will become germ (pole) cells (see Figure 22-24). The microinjected flies (G_0 generation) did not generally express the new gene, since it was present in only a small percentage of their cells. However, some of the G_0 generation flies carried the new gene in germ line cells; thus, some of the progeny of these flies (G_1 generation) had the gene in every cell, so that the effects of the new gene could be properly studied. Flies that have successfully taken up the new gene are not difficult to recognize. For example, flies with a defect in the enzyme xanthine dehydrogenase (*rosy*$^-$) are fatally poisoned by excess purines in their food. After microinjection of *rosy*$^-$ embryos with a wild-type *rosy*$^+$ gene, transformed flies expressing the new gene can be selected by growth on purine-enriched medium.

Figure 22-43
Use of recombinant P elements to introduce new germ line genes into *Drosophila*. (a) Structure of injected plasmids. (b) Protocol for introduction of a *rosy*⁺ gene into the germ line of a *rosy*⁻ strain.

(a)

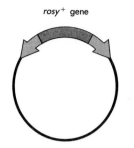

rosy⁺ gene

Intact P element

Plasmid with *rosy*⁺ transposon consisting of a *Drosophila* xanthine dehydrogenase gene (*rosy*⁺) inserted into an internally deleted P element

Helper plasmid with intact P element to supply transposase activity

(b)

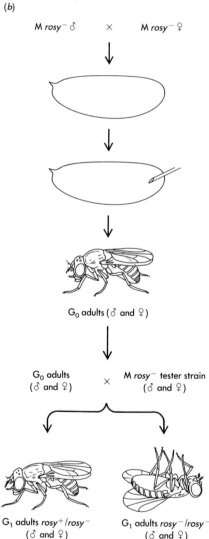

M *rosy*⁻ ♂   ×   M *rosy*⁻ ♀

rosy⁻ M flies (lacking P elements) are mated to produce *rosy*⁻ embryos.

Plasmid containing *rosy*⁺ transposon and helper plasmid are coinjected into site of germ cell formation at posterior pole of *rosy*⁻ embryo lacking P elements.

G₀ adults (♂ and ♀)

First-generation flies (G₀) are genetic mosaics, with *rosy*⁺ transposon inserted into the DNA of some germ cells, but not all.

G₀ adults   M *rosy*⁻ tester strain
(♂ and ♀)   ×   (♂ and ♀)

G₀ flies are mated with tester strain of *rosy*⁻ M flies.

G₁ adults *rosy*⁺/*rosy*⁻   G₁ adults *rosy*⁻/*rosy*⁻
(♂ and ♀)   (♂ and ♀)

Second-generation flies (G₁) derived from germ cells containing the *rosy*⁺ transposon are heterozygous in all somatic cells (*rosy*⁺/*rosy*⁻) and can be selected on purine-rich medium.

Many Microinjected Genes Are Not Very Sensitive to Chromosomal Environment[33, 76–80]

Unlike targeted gene substitution in yeast (Chapter 18), microinjected P elements seem to insert randomly into new chromosomal sites. Initially, biologists had feared that correct expression of the newly introduced gene might be so strongly dependent on its immediate chromosomal environment that only genes that had integrated at or near the site of the resident wild-type gene would function properly. This worry was based on the existence of position effects in classical *Drosophila* genetics: Occasionally, genes were poorly expressed or even turned off completely after being moved by transposition or inversion to a new chromosomal location. Fortunately, the expression of new germ line genes introduced by recombinant P elements is generally not subject to strong position effects. For example, a recombinant P element carrying a cloned DNA fragment from the 66D12-15 chorion gene cluster will program correct developmentally regulated amplification of flanking DNA (see Figure 22-21) after insertion at a variety of different sites. Also almost independent of the site of chromosomal integration was the tissue-specific expression of metabolic enzymes such as alcohol dehydrogenase and dopa decarboxylase, as well as the inducibility of heat-shock genes by high temperature (Chapter 21).

Circadian Rhythms[79, 80, 81]

The accurate measurement of biological time is not restricted to the rapidly unfolding cell movements (morphogenesis) of early development. Many (and perhaps most) higher eucaryotes, from *Drosophila* to humans, exhibit daily **circadian rhythms** (from the Latin *circa die*, meaning "about one day long"). Psychologists have shown that when human subjects are isolated in a clockless environment lacking all visual and auditory clues about the passage of day and night, most individuals establish a "free-running clock" with a period of about 23 hours. This clock governs cyclic patterns of waking, sleeping, activity, and hunger. Evidently, our bodies "know" the length of the day, even without environmental clues; the alternation of day and night simply modifies or "entrains" our natural biological rhythm to a slightly longer period of 24 hours.

Circadian rhythms in the fruit fly *Drosophila* are not conspicuously different from our own. Behavioral mutants of the fly can be isolated from mutagenized populations, just as biochemical mutants are, so long as the investigator is patient enough to observe the behavior of individual flies or clever enough to devise a selection procedure. One particularly obvious selection for flightless mutants is to study those flies that do not fly away when the top is taken off the culture bottle. Similarly, flies can be isolated whose free-running clock is slower or faster than usual, or even completely irregular. These mutations map to a single genetic region called the *per* (for *period*) locus, and DNA from the locus has been cloned. Recently, a 7-kilobase-pair fragment of wild-type *per* DNA from *Drosophila* was introduced by P element transformation into an arhythmic fly, and the mutation was corrected to a normal 24-hour cycle (Figure 22-44). The 7-kilobase-pair fragment encodes a 4.5-kilobase mRNA, and parts of this mRNA are homologous to DNA in the chicken, mouse, and human genomes. Studies

Figure 22-44
Introduction of a wild-type *per* locus into the germ line of an aperiodic mutant fly using a recombinant P element restores normal circadian rhythms. Wild-type flies (*per*⁺), an aperiodic mutant (*per*⁰), and the *per*⁰ mutant transformed with a wild-type *per*⁺ locus, were grown in culture bottles on solid food. The flies reproduce under these conditions, and all stages of the life cycle—egg, larva, pupa, and adult—are represented. For 3 days before the start of the experiment, the flies were entrained to a 24-hour cycle with 12 hours of alternating light and dark. Beginning on day 1, the adults that had enclosed (emerged from the pupal cases after metamorphosis) were collected every 2 hours and counted; the graphs show the number of adults emerging in each 2-hour interval. Note that the alternating cycle of light and dark was discontinued after the fifth day, and the flies maintained thereafter in total darkness. Interestingly, the 24-hour circadian rhythm persists in darkness, showing that the fly's internal clock, once entrained, no longer requires daily stimulation. [After R. A. Bargiello, F. R. Jackson, and M. W. Young, *Nature* 312 (1984):752–754.]

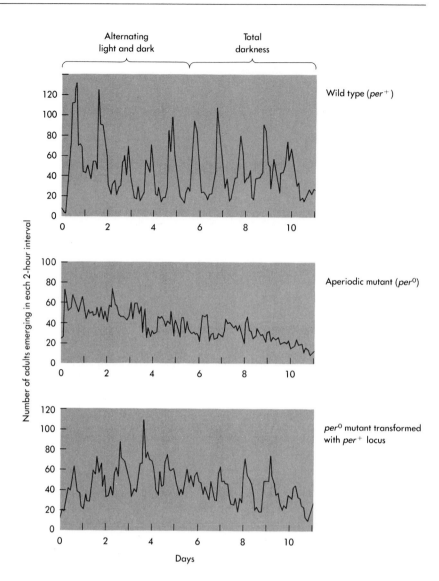

such as these form the basis for a new science: the molecular genetics of behavior. The inescapable conclusion is that we human beings, proud possessors of sophisticated intelligence, will find that our behavior is governed to some extent by elementary biochemical reactions.

Transgenic Mice: Mice with New Germ Line Genes[82–85]

The production (or construction, as molecular biologists sometimes like to say) of a transgenic mouse carrying a new germ line gene is superficially similar to transformation of *Drosophila* embryos with recombinant P elements; however, there is one important difference. No mammalian transposable element has yet been fully adapted to function as a recombinant vector analogous to P elements, although some very promising vectors derived from RNA tumor viruses (Chapter 24) are now being tested. Instead, transformation of mouse germ line DNA currently depends on the recombinational abilities of the fertilized mouse egg itself.

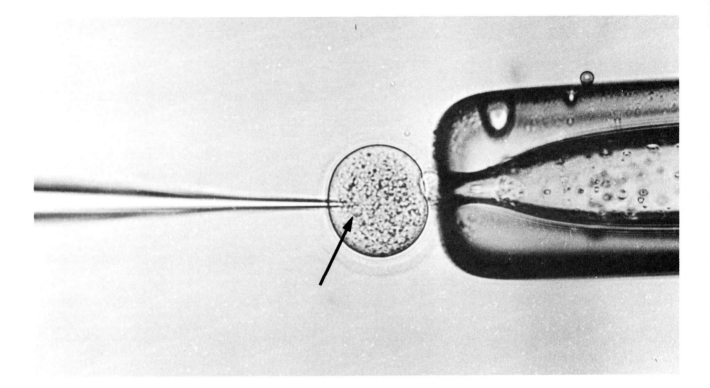

To make a transgenic mouse, eggs are first surgically removed from a mature female and then fertilized with sperm in the test tube. A recombinant plasmid carrying a recognizable new gene (one that can be distinguished from the mouse's own gene) is microinjected with a thin glass needle into the male pronucleus of the freshly fertilized egg before the haploid sperm and egg nuclei have fused to form a diploid zygotic nucleus (Figure 22-45). The circular plasmids generally recombine homologously with each other within the egg, forming a long, tandemly repeated concatemer; this concatemer then integrates randomly to give tandemly repeated genes at a single chromosomal site. The microinjected eggs are surgically implanted in a foster mother and allowed to come to term. The resulting mice are called **transgenic,** since part of their genome comes from another genetically unrelated organism (be it animal, vegetable, or bacterial). Because the foreign gene is introduced even before fusion of the egg and sperm nuclei, chromosomal integration takes place early, and the progeny transgenic mice often carry the new gene in some germ line and somatic cells. These mice (the G_0 generation) must then be bred to produce mouse lines in which all cells (both germ line and somatic) are genetically identical.

The initial experiments with transgenic mice have been extraordinarily rewarding. For example, elastase is a pancreatic protease that is normally secreted into the upper intestine to aid digestion. When the rat elastase gene was cloned into a plasmid and microinjected into fertilized mouse eggs, the transgenic mice expressed the rat elastase mRNA and protein only in pancreatic secretory cells, although the gene was present in the cells of all other tissues. Evidently, these two rodents are so closely related that control signals in rat DNA are correctly recognized by the corresponding mouse regulatory factors. Nonetheless, rats and mice are sufficiently different that the rat elastase mRNA and protein can be easily distinguished from the mouse products.

Figure 22-45
Injection of foreign DNA into a fertilized mouse egg before the male and female pronuclei have fused. The egg is first immobilized by applying mild suction to the large, blunt holding pipette (right). Several hundred copies of the gene of interest are then injected through the sharp end of the narrow microneedle (left). A large nucleolus can be seen near the tip of the microneedle (arrow). [Courtesy of J. Gordon and F. H. Ruddle, *Methods in Enzymology* 101 (1982):411–433.]

Figure 22-46
A "gigantic" male mouse grown from an egg injected with multiple copies of a rat growth hormone gene under the control of a mouse metallothionein promoter. A normal male sibling is shown for comparison. The transgenic mouse weighs 44 grams, the normal mouse 29 grams. The new gene is passed on to offspring, which grow two to three times faster than normal and reach a size up to twice normal. Note that both mice have short tails. This is not an inherited trait; rather, a small piece of tail has been chopped off to extract DNA for an analysis of the number and organization of the growth hormone gene ("transgene") in the normal and transgenic animals. (Courtesy of Dr. R. L. Brinster, University of Pennsylvania.)

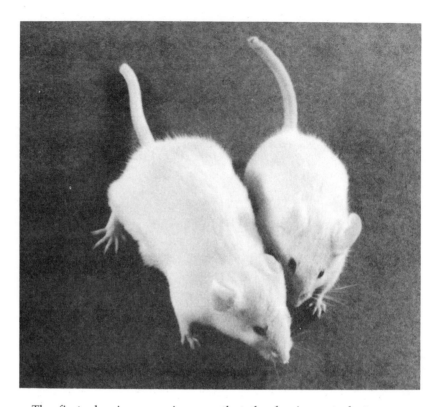

The first pleasing surprise was that the foreign rat elastase gene could be correctly regulated despite the fact that its site of chromosomal integration differed in each of the transgenic mouse lines. Equally pleasing was the discovery that at least some regulatory sequences in mammalian DNA are quite small. When a series of transgenic mice was made where less and less upstream flanking sequence preceded the rat elastase gene, it was found that merely 204 base pairs were sufficient for correct tissue-specific expression. This suggests that even in as complicated a eucaryote as the mouse, control regions may generally lie upstream of the gene and need not be much (if at all) larger than those found in bacteria. Similarly encouraging results have been obtained showing tissue-specific expression of transgenic light and heavy immunoglobulin chains (Chapter 23). Perhaps the most dramatic experiment done so far with transgenic mice involved the construction of a fusion gene in which the coding region for rat growth hormone was attached to a powerful promoter derived from the mouse metallothionein gene (metallothioneins are small cysteine-rich proteins that transport essential heavy metals such as zinc and copper throughout the body). Transgenic mice carrying multiple copies of the highly expressed rat growth hormone gene grew very fast. The mouse with the most copies (approximately 35) grew fastest, to almost twice the size of a normal mouse (Figure 22-46); but sadly, this gigantic animal (by mouse standards) died a premature death. Although large mice are merely a scientific curiosity, the same kinds of experiments can also be performed on farm animals of obvious economic importance, such as sheep and goats.

Working with a Mammal Takes Time

Putting genes back into yeast is easy. One hundred million (10^8) cells can be transformed and plated on selective media in less than an hour's time, and the resulting transformants grow with a doubling time of 2 hours. The fruit fly *Drosophila* is not difficult to work with. Perhaps 100 successful microinjections can be done in a working day, and tens of thousands (10^4) of progeny flies can be grown in small bottles with a generation time of a few weeks and visually inspected for the desired mutants. Mice are the most tedious of all to work with. Only 5 to 10 successful microinjections can be done in a day, but not all of these necessarily express the foreign gene. Moreover, a standard mouse cage has room for at most a dozen mice, and their generation time is nearly 2 months. Fortunately, natural curiosity about our own species (humans are not so different genetically from mice) gives us the drive to carry out the laborious experiments of making many transgenic mice.

PLANT MOLECULAR GENETICS COMES OF AGE

The Large and Frequently Polyploid Genomes of Most Plants

Until recently, incisive work on the structure and expression of plant genomes lagged far behind similar work with animals. Perhaps in part this sorry state of affairs was caused by our natural preference (as animals) for studying other animals. But the main problem was that genetic analysis of higher plants is really quite difficult. Plants grow relatively slowly, and only a few generations of genetic crosses can be performed each year. In addition, most economically important plants are large and require expensive greenhouse facilities for growth in temperate climates. Even today, no plant has been as thoroughly analyzed at the genetic level as *Drosophila*. Genetic analysis of plants is further complicated because many important plants not only have large numbers of chromosomes but are highly **polyploid** and contain at least four copies of any particular chromosome. Multiple copies of the wild-type gene in a polyploid organism make recessive mutations (which are far more common than dominant mutations) exceedingly laborious to detect. Consequently, botanists and agronomists have usually depended on traditional plant-breeding techniques for crop improvement. These techniques involve the repeated crossing of large numbers of plants with desirable characteristics (disease resistance, large fruit, drought tolerance, etc.) in the hope that such characteristics can be stably combined in one germ plasm. Unfortunately, such breeding procedures are tedious; moreover, the results are often uncertain, because desirable traits may in fact be the product of several different genes, and genes may function differently in different genetic backgrounds.

The very large genomes of most plants represent yet another obstacle to understanding plant genetics at the molecular level. For example, the genome of corn is actually several-fold larger than the human genome, despite the obvious difference between the two organisms in biological complexity. Thus there is no doubt that most DNA in the corn genome does not encode gene products. Instead, as is also true for the genomes of higher animals (Chapter 20), plant genomes are full of functionally obscure DNA. Much of this "extra" DNA seems to arise from the insertion of repetitive DNA elements whose widespread presence often greatly complicates the cloning and sequencing of plant genes.

Chloroplasts Possess Their Own Unique DNA

The chloroplasts of green plants are cytoplasmic organelles that house the various pigments and enzymes of the light-harvesting photosynthetic apparatus. Even before the turn of the century, it was clear that green pigmentation was one of the easiest traits to observe in plant-breeding experiments. Although some pigmentation traits obeyed Mendel's laws, other color traits were only transmitted through the female parent that provided the cytoplasm of the zygote. These observations of **cytoplasmic** or **maternal inheritance** eventually led to the hypothesis that chloroplasts must carry genes.

We now know that chloroplasts contain a unique circular DNA genome that is completely different from the nuclear genome. Electron micrographs indicate that the chloroplast DNA is some ten to twenty times smaller than the *E. coli* chromosome. For example, the chloroplast genome of maize (corn) contains 140,000 base pairs of DNA. Such genomes are much too small to encode the approximately 1000 different proteins found in chloroplasts. Instead, biosynthesis of the chloroplast involves an intimate collaboration between the nuclear and chloroplast genomes. In fact, every known multimeric protein component of chloroplasts is a mixture of the products of both nuclear and chloroplast genes.

Most chloroplast proteins are encoded by nuclear DNA, translated in the cytoplasm, and imported into the chloroplast by a specific transport mechanism that enables polypeptides to cross the outer membrane of the organelle. However, some 100 chloroplast-specific proteins are synthesized within the chloroplast itself. These proteins are encoded by chloroplast DNA, transcribed by the chloroplast-specific RNA polymerase, and translated by the chloroplast-specific protein synthesizing machinery. Since RNA cannot cross the outer membrane of the chloroplast, chloroplast ribosomal RNAs and tRNAs must be encoded in chloroplast DNA.

Chloroplasts are not static organelles, but can adapt to different physiological conditions such as high or low levels of light. For example, when grown entirely in the dark, chloroplasts lack chlorophyll but retain carotenoid pigments. Thus many chloroplast genes are light-regulated, in certain cases by light-sensitive promoters!

Maize May Be the Best Understood Plant Genetically

Partly because of its economic importance, and partly because of many favorable biological characteristics, maize (corn) has become one of the best understood plants at the genetic level. Only an acre of land is needed to grow some 10,000 maize plants. Each plant bears several ears containing 500 or more kernels, and each kernel is the product of a separate genetic cross. A maize geneticist working with only 1 to 2 acres of corn can thus investigate the progeny of over a million genetic crosses in the course of a year. Most importantly, the male and female parts of maize are found on separate flowers. This simplifies controlled pollination, and it is the principle reason why maize genetics has advanced so far compared with the genetics of cereals such as wheat, rice, or barley.

Over 170 different genetic loci have been mapped onto the ten chromosomes of the maize genome. Gene mapping in maize is often facilitated by the existence of knobbed protuberances on meiotic chromosomes. These protuberances are frequently unique to each strain of maize, and make it possible to identify a chromosome (bearing a particular combination of wild-type and mutant genes) by rapid cytological observation rather than by slow genetic analysis. Another key to steady progress in the molecular genetics of maize is the behavior of the well-characterized transposable elements *Ac* and *Ds*. Many spontaneous mutations in maize involve insertion of these elements near or within genes; moreover, both *Ac* and *Ds* have been cloned and sequenced, so that they can be used as probes for the genes into which they have inserted (see Figure 20-34).

Using the Ti Plasmid to Introduce New Genes into Plants[86–93]

We have seen that the diploidy (and usually polyploidy) of most crop plants poses a substantial problem for classical genetic analysis, so that new plant varieties are often more easily made by classical plant-breeding procedures than by a molecular approach that first requires the identification and isolation of particular genes. A much simpler and more straightforward approach to crop improvement would be to introduce known genes into the chromosomes of a healthy plant in such a way that the genes could be strongly expressed at all times in all tissues, or only in certain tissues (say, leaves), or only under certain environmental conditions (say, strong light or high temperature). We shall see that this goal can now be achieved by using the Ti plasmid to introduce foreign or genetically engineered genes into healthy plants in a heritable fashion (Figure 22-47).

Plants can generally be divided into **monocots,** such as wheat and corn (where the sprouting seed has a single leaf), and **dicots,** such as tomatoes or apples (where the sprout has two leaves). Most dicotyledonous plants are naturally susceptible to crown gall disease, in which tumors (uncontrolled cell growths) occur at the site of a wound (often at the crown, separating the stem from the roots of the plant). These tumors are caused by infection with the Gram-negative soil bacterium *Agrobacterium tumefaciens*, which transforms plant cells at the wound site so that they have the capacity for autonomous growth in the absence of any further stimulation by the infecting bacteria. In addition, depending on the type of infecting *Agrobacterium*, the transformed plant cells acquire the ability to synthesize one of two unusual arginine derivatives known as octopine [*N*-alpha-

Figure 22-47
Recombinant Ti plasmids can introduce new genes into plants. Note that only the T-DNA is transferred to the plant genome, while a different region of the Ti plasmid known as *vir* encodes the enzymes required to mobilize the T-DNA. Small arrows above the T-DNA segment represent the 25-base-pair direct repeats that define the ends of the transposable element and are required for transposition. A large fragment of the Ti plasmid encompassing both the T-DNA and *vir* regions is first cloned into the common bacterial plasmid pBR322 (Figure 19-1). The new gene is then cloned into a nonessential region of the T-DNA and introduced into *A. tumefaciens* carrying an intact Ti plasmid. When plants are infected with these bacteria, the gene products of the *vir* region on the intact Ti plasmid mobilize the recombinant T-DNA, and the recombinant T-DNA integrates into the plant genome. [After J. D. Watson, J. Tooze, and D. T. Kurtz, *Recombinant DNA* (New York: Freeman, 1983), p. 171.]

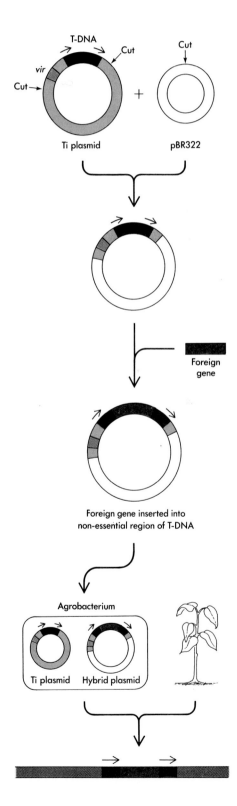

(D-1-carboxyethyl)-L-arginine] or nopaline [*N*-alpha-(1,3-dicarboxy-propyl)-L-arginine]. These **opines** serve both as food and as an energy source for the infecting bacteria. Interestingly, the octopine-inducing bacteria can grow on octopine but not on nopaline, and the nopaline-inducing bacteria can grow on nopaline but not octopine. Thus, tumor formation can be seen as an evolutionary strategy devised by *Agrobacterium* to divert the metabolic resources of the plant into protecting, feeding, and perpetuating a bacterium from which it derives no obvious benefit.

In principle, the genes encoding the opine synthetases could have been present (but silent) in the uninfected plant cell genome and activated by infection with *Agrobacterium*. Alternatively, the opine synthetase genes could be carried by the bacterium and transferred to the plant cell genome during infection. We now know that the genes for both octopine and nopaline synthetase come from the infecting bacterium, where they are initially present on a giant plasmid known as the **Ti** (*tumor-inducing*) **plasmid.** Interaction between the infecting bacterium and the host plant cell stimulates the bacterium to excise a 30-kilobase-pair region of the Ti plasmid known as **T-DNA** (for transforming DNA) that is flanked by two directly repeated 25-base-pair sequences. Excision appears to occur by site-specific recombination between the two 25-base-pair terminal direct repeats and results in an extrachromosomal circle of T-DNA containing a single copy of the 25-base-pair sequence. Curiously, these T-DNA circles do not appear to be covalently closed and may be held together by proteins bound to the terminal repeats. These circular T-DNA molecules (or perhaps copies derived from them) are then transferred from the bacterium to the nucleus of the host plant cell, where the T-DNA integrates into the host nuclear genome, occasionally in the form of tandem repeats. In this way, the host acquires a variety of new genes carried on the T-DNA, including a number of genes responsible for host cell transformation, as well as the gene for an opine synthetase.

The gene products required for excision, transfer, and integration of the T-DNA into the host plant cell are not encoded by the T-DNA itself, but by a different part of the Ti plasmid known as the *vir* region (for virulence). *Vir* genes are induced by specific plant products, ensuring that the soil bacterium activates T-DNA transfer only in the presence of a potential plant host. Successful mobilization and integration of the T-DNA requires only the 25-base-pair terminal direct repeats; all internal T-DNA sequences are dispensable. Thus, the Ti plasmid (and the T-DNA within it) constitutes a natural vector for introducing new DNA sequences into the nuclear genome of somatic cells from susceptible plant species. The internal transforming genes of a cloned T-DNA can be removed by recombinant DNA techniques and replaced by any gene (whether of plant, animal, fungal, or bacterial origin) that can be expressed in plant tissues. Most commonly, the coding sequence of the foreign gene is substituted for the coding region of the opine synthetase gene: In this way, the natural promoter and polyadenylation signals of the opine synthetase gene confer high-level expression of the foreign protein. More recently, it has proved possible to replace the constitutive opine synthetase promoter with a light-inducible promoter derived from a gene encoding the small subunit of the chloroplast enzyme ribulose 1,5-bisphosphate carboxylase. The occurrence of the heat-shock response in plants as well as all other kingdoms (Chapter 21) promises to permit heat-inducible expression of foreign genes as well.

Fertile Plants Can Be Regenerated from Somatic Cells[94]

Introduction of new genes into plant somatic cells would be useful for basic studies of plant gene expression, but not for crop improvement, had not classical botanists discovered earlier that normal plants can often be regenerated from somatic tissues. Many plants (including such common house plants as the rex begonia) can be readily regenerated from a small disk of somatic tissue punched out of a leaf and placed on moist nutrient medium (Figure 22-48). To the naked eye, roots and shoots appear to sprout directly from the leaf disk; but on a microscopic level, differentiated cells within the leaf must first dedifferentiate, proliferate, and finally redifferentiate into the many tissues required to construct a complete plant. More technically challenging, but useful in many ways, is the regeneration of a complete plant from a **callus** (a disorganized, undifferentiated cell mass originally excised from a mature plant and then grown indefinitely in tissue culture). Treatment with the proper plant hormones induces differentiation of the callus into roots and shoots, and eventually a complete plant with fertile flowers can be regenerated (see Figure 22-48). No animal of apparently comparable biological complexity can yet be regenerated from mature somatic cells. (The clonal frog in Figure 22-7 was derived from intestinal nuclei of a tadpole.)

In principle, it should be possible to transform both leaf disks and callus cultures with T-DNA vectors, and then to regenerate complete transgenic plants (see Figure 22-48). (Although the top and bottom of a leaf disk are covered by a waxy cuticle, the open sides of the disk serve as wounds for T-DNA transformation.) However, several problems currently complicate the use of T-DNA vectors. Only some plant species can be persuaded to grow in callus culture, and not all of these can be regenerated from the callus; other plant species will not regenerate from a leaf disk; and lastly, the T-DNA vector itself does not infect leaf disks or callus cells of all plant species. Given the current pace of research in plant molecular genetics, such technical obstacles may soon be surmounted. Interestingly, transformation of somatic plant cells with wild-type T-DNA blocks regeneration, because the transforming genes subvert the normal mechanisms of growth control and development. Thus, it is essential that the transforming genes be altered or deleted in the T-DNA vector carrying the new genetic information into the plant genome.

The one major problem remaining in the use of genetically engineered Ti plasmids for crop improvement is that many important crop plants are monocots originally thought to be resistant to infection by *Agrobacteria*. Fortunately, recent evidence suggests that *Agrobacteria* can infect some monocots such as lily and asparagus (but not yet corn or wheat). Opine synthetase genes are expressed, but the transforming genes of the T-DNA are unable to work effectively in the monocot environment. Presumably, tricks will be devised in the near future to infect economically important monocots.

Crop Improvement Through Genetic Engineering [86, 95]

What new genes would we want to introduce to improve important crop plants? The seed proteins of such staple crops as wheat, beans,

(a)

(b)

(c)

Figure 22-48
Plants can be regenerated from mature somatic cells. (a) Regeneration of shoots from an undifferentiated mass of callus cells derived from the common garden flower *Petunia hybrida*. The callus on the left was maintained on normal plant tissue culture medium, whereas the callus on the right was grown on a special regeneration medium that allows organogenesis. (b) Shoots regenerating directly from the edge of a leaf disk on medium that supports both callus growth and organogenesis. (c) Transformed, kanamycin-resistant petunia shoots regenerating from a leaf disk grown on regeneration medium containing kanamycin. Only the disk on the left was infected with a genetically engineered strain of *Agrobacterium tumefaciens* that expresses the bacterial gene for kanamycin resistance in the transformed plant cells. (Courtesy R. B. Horsch, Monsanto Company.)

and corn are often deficient in particular amino acids; corn is especially poor in lysine, threonine and methionine. Genes encoding the major seed storage proteins have already been identified and cloned, and it should not prove difficult to use recombinant DNA technology to insert nucleotide sequences encoding the deficient amino acids into the correct reading frames. The structure of seed proteins (which are broken down to feed the germinating plant embryo) are presumably not so critical that insertion of a few new amino acids would interfere with function. Since abundant seed proteins are generally encoded by large multigene families, the genetically engineered seed protein genes will have to be introduced into the plant genome in relatively high copy number in order to have a significant effect on the nutritional characteristics of the seed. This presents no technical problem, because T-DNA can integrate in the form of long tandem repeats. However, because T-DNA integrates randomly into the plant genome (as do P elements in *Drosophila* and microinjected DNA in fertilized mouse eggs), one initial worry was that position effects might prevent genes on T-DNA vectors from being properly regulated in their new chromosomal environment. Fortunately, as is the case for transgenic flies and mice, the immediate chromosomal environment usually has little effect on regulation of the transgene. For example, soybean seed storage proteins are subject to proper developmental regulation when introduced into tobacco plants: The transgenes are expressed in seeds, but not in roots, shoots, or leaves.

One potential use for T-DNA vectors is in the development of insect-resistant plant lines. Certain bacteria such as *Bacillus thuringiensis* are known to produce peptides that are highly toxic to many insects. The genes encoding such natural insecticides could be incorporated directly into plant genomes, rendering repeated spraying with chemical insecticides unnecessary. However, in a pilot experiment where the gene encoding the *Bacillus thuringiensis* toxin was incorporated into a tobacco plant, the toxin produced by the newly introduced gene was inactivated by the ultraviolet component of sunlight! This unexpected result warns us that biological systems can be very subtle, and it will often be easier to dream up revolutionary applications of recombinant DNA technology than to put these ideas into practice.

Another potential use for recombinant T-DNA vectors stems from the observation that many soil bacteria are able to degrade various herbicides enzymatically. In principle, these bacterial genes could be cloned and expressed in new plant lines, so that herbicides could replace mechanical cultivation as a method for weed control. Still other bacteria may be found to produce antifungal agents, the biosynthetic gene(s) for which could be incorporated into crop plants. However, ecosystems are often delicately balanced, and the environmental impact of introducing transgenic plants will sometimes require serious scrutiny.

Agrobacterium tumefaciens Can Also Transform Chloroplasts[96]

Chloroplasts are thought to be derived from symbiotic photosynthetic procaryotes that colonized the cytoplasm of primitive eucaryotes. As predicted by this **endosymbiont hypothesis,** chloroplasts resemble bacteria in many ways. For example, ribosomes found in the cytoplasm of eucaryotes are resistant to the antibiotic chloramphenicol, but both procaryotic and chloroplast ribosomes are sensitive.

The differing antibiotic sensitivity of cytoplasmic and chloroplast ribosomes was used in an ingenious experiment designed to test whether T-DNA can insert into chloroplast as well as nuclear genomes. A gene fusion was constructed between the promoter of the nopaline synthetase gene and the coding region of the bacterial chloramphenicol resistance gene. Because the nopaline synthetase promoter was already known to function in bacteria, and the chloramphenicol resistance gene was itself derived from bacteria, it seemed possible that the gene fusion might function in chloroplasts. When the gene fusion was introduced into chloramphenicol-sensitive tobacco cells using a recombinant T-DNA vector, mature tobacco plants could be regenerated in the presence of chloramphenicol. The enzyme chloramphenicol transacetylase (which degrades chloramphenicol and is the product of the chloramphenicol resistance gene) was found exclusively in chloroplasts of the resistant plants; moreover, the chloramphenicol resistance trait was maternally inherited. These observations confirmed that the T-DNA had integrated directly into the chloroplast genome.

The ability of T-DNA to insert into the chloroplast genome has important agricultural implications. Many herbicides work by uncoupling light-driven electron transfer from production of high-energy chemical bonds during photosynthesis, and resistance to these herbicides often resides in chloroplast DNA. (Remember that the macromolecular complexes responsible for photosynthesis are generally put together from gene products encoded by both the nuclear and chloroplast genomes.) Thus it should prove technically possible to transfer resistance to various kinds of herbicides from one species of plant to another, using T-DNA as the vector to move the appropriate region of chloroplast DNA from the resistant to the sensitive chloroplasts. Thus herbicide-resistant crop plants may soon be a reality.

Plant Molecular Geneticists
Need Better Model Organisms[97]

Recently, there has been renewed interest in a small flowering plant called *Arabidopsis thaliana*, which has been studied by classical genetic techniques for over 40 years. This harmless weed of the mustard family has no economic or nutritional value, but it has many other qualities that commend it to the plant molecular biologist. *Arabidopsis* has a generation time of only 5 weeks, produces 10^4 seeds per plant, and is so small that tens of thousands of plants can be grown in a small laboratory room. The plants are self-fertile (so that new mutations are naturally made homozygous) and can also be cross-fertilized by hand (so that genes can be mapped and multiple mutations can be introduced into a single plant stock). Over 75 mutations have already been mapped on the *Arabidopsis* genome. In addition, *Arabidopsis* is susceptible to infection with *Agrobacterium tumifaciens*, so it should eventually be possible to introduce foreign genes into the chromosomes of the organism.

Perhaps most important for the molecular biologist are the small genome size and comparative lack of repetitive DNA in *Arabidopsis*. The smaller the genome of an organism, the easier it is to clone genes, because there is less DNA to sift through. Also, DNA·sequences that are repeated many times in the genome are an impediment to many recombinant DNA manipulations because these sequences tend to hybridize with each other. The genome of *Arabidopsis* is only 10 percent as large as that of cotton, 5 percent as large as that of tobacco, and 1 percent as large as that of wheat. Nonetheless, this "small" plant genome (7×10^7 base pairs) is still about 20 times larger than that of *E. coli.* Since distantly related organisms often share similar mechanisms of gene regulation, studies of *Arabidopsis* may prove useful or even directly applicable to work on economically important but genetically less tractable plants.

Symbiotic Development: Nitrogen-Fixing Bacteria in the Roots of Leguminous Plants[98]

Until now, we have discussed the development of individual organisms in isolation. The nitrogen-fixing bacteria of the genus *Rhizobium* that live in the roots of legumes such as beans, clover, and alfalfa provide a striking example of complex developmental interactions between two different organisms. Most plants can use only fixed nitrogen in the form of nitrates (NO_3^-). Although molecular nitrogen (N_2) constitutes about 70 percent of the atmosphere, this potential food stuff would be unavailable to either plants or animals if it were not for the nitrogen-fixing bacteria that can convert atmospheric molecular nitrogen into fixed nitrates. Some nitrogen-fixing bacteria are free-living, but most of the economically important species of *Rhizobium* live within specialized root nodules of legumes. The formation of such a nodule must require an intimate genetic collaboration between the bacteria and their plant host, since many mutations are known in both the bacterial and plant genomes that can block development of a functional nitrogen-fixing nodule.

Over the past fifteen years, there has been enormous progress in understanding the precise details of the genetic collaboration between rhizobia and legumes. In addition to the intrinsic biological fascination of the subject, these studies have also been motivated by the realization that many high-yielding cereal crops (such as corn) quickly exhaust the soil they grow in. The fixed nitrogen content of the soil can be rebuilt by time-consuming crop rotation with leguminous plants (alfalfa, clover, soybeans), or the soil can be artificially renewed by heavy applications of chemical fertilizers rich in phosphates and nitrates. But the cost of such fertilizer is almost prohibitive for many underdeveloped countries, and the ecological consequences of chemical runoff from croplands into local streams and lakes can be serious, even in a technologically advanced nation. These problems might be overcome if cereal and other crops could be naturally bred, or genetically engineered, to become functional hosts for rhizobia.

Symbiosis begins with attachment of rhizobia in the soil to the root hair of a germinating legume (Figure 22-49). Because each species of *Rhizobium* has a narrow **host range** (for example, *Rhizobium melliloti* will grow on alfalfa but not on clover), this initial attachment is thought to depend on a cell recognition event between surface molecules (possibly glycoproteins) on the bacteria and the plant. Following attachment, the bacterium is encapsulated within a pocket of the plant cell wall and carried deep inside the root by formation of an infection thread consisting primarily of cellulose. Cortical cells on the outside of the root dedifferentiate to form a growing meristem that bulges out into a nodule. Proliferating infection threads then release

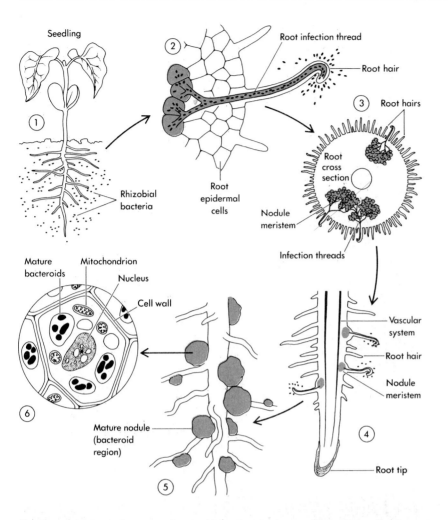

Figure 22-49
Development of a legume nodule.

the bacteria into the cytoplasm of the host plant cells, where the bacteria—after they are covered with a special membrane synthesized by the host—are known as **bacteroids.** Within this extremely protective environment, the bacteria finally differentiate into a nitrogen-fixing state by derepressing the genes for nitrogenase and bacteroid-specific cytochromes.

One of the most cogent examples of the degree of cooperation between rhizobia and the host plant is the biosynthesis of leghemoglobin, the plant's oxygen binding protein. The molybdenum-containing active site of the nitrogenase enzyme complex is specifically designed to reduce molecular nitrogen (N_2), but it can easily be poisoned by free molecular oxygen (O_2). Yet, the bacteroids also need a steady supply of oxygen for their general metabolism. The host plant meets these apparently contradictory demands by synthesizing large quantities of leghemoglobin (Chapter 20), which carries oxygen to the bacteroids while maintaining the overall oxygen tension at a sufficiently low level so as not to harm the nitrogenase. Remarkably, although the plant synthesizes the leghemoglobin polypeptide, the heme cofactor is supplied by the bacteria themselves. Coordinating the relative amounts of heme and leghemoglobin under these circumstances may not be a simple matter. In vertebrate erythrocytes, the synthesis of heme and hemoglobin is coordinated at the translational level; heme is required for formation of an initiation complex between the mRNA and the ribosome (Chapter 21). Perhaps a similar mechanism is at work within the legume cytoplasm.

Interestingly, bacteroids are approximately the same size as mitochondria, and in this sense they can be considered to be a nitrogen-fixing, plant cell organelle. In fact, the analogy with mitochondria is not only morphological but also metabolic. Mitochondria reduce atmospheric oxygen, and bacteroids reduce atmospheric nitrogen. Mitochondria donate the bulk of the ATP they generate to the general cellular metabolism, although a small fraction of the ATP is consumed by the mitochondrial DNA, RNA, and protein-synthesizing systems. Rhizobia, curiously, are completely unable to assimilate the ammonia they make by reducing molecular nitrogen. Instead, the ammonia is exported to the surrounding plant cytoplasm, where it is incorporated into larger metabolites such as amino acids, which can be imported and assimilated by the bacteria.

Both mitochondria and chloroplasts are thought to have descended from initially free-living photosynthetic bacteria that were engulfed by, or developed a symbiotic relationship with, a primitive nonphotosynthetic plant cell (endosymbiont hypothesis). Once the bacterial cell was lodged within the eucaryotic host, mutations within the bacterial genome could have deprived it of the ability to live outside its host. Subsequent mutations in both the nuclear and algal genomes could have slowly stripped the proto-mitochondrion and proto-chloroplast of all but the most indispensable functions. Conceivably, the symbiotic relationship between today's rhizobia and legumes may be giving us a glimpse of a similar kind of evolution in progress. Perhaps in the distant future, some species of rhizobia may lose the ability to live outside their plant hosts and thus become permanent nitrogen-fixing organelles. The same result might also be achieved deliberately (and hopefully sooner) by genetic engineering, if no unforeseen problems are encountered. Perhaps the most exciting challenge would be to use T-DNA transformation to extend the host range of *Rhizobium* species to include nonleguminous plants such as corn and wheat. The intimate and complex nature of the genetic collaboration between the Rhizobium bacteria and the host plant cell (see Figure 22-49) suggests that this task may be very difficult; but the potential benefits to humankind are so overwhelming that there is no shortage of effort directed toward accomplishing this goal.

The Genetics of Nitrogen Fixation

An investigation of the genes responsible for development is difficult, and genetic characterization of the cooperation between two different organisms will be doubly so. A second problem is that free-living rhizobia do not differentiate into bacteroids outside the plant host, nor does the host plant construct nodules in the absence of infecting bacteria. Thus, mutations in either the bacterial or the plant genes that are required for functional nodulation can only be identified by the laborious procedure of inoculating individual plants with individual bacterial strains. A third problem is that the genetic analysis of plants has not been easy. Plants grow slowly; techniques had to be developed for the rapid and convenient growth of large numbers of plants on a small laboratory scale; and plants are frequently polyploid. Ironically, the very difficulty of plant genetics has simplified the prevailing experimental approach to nitrogen fixation. Molecular biologists have chosen to take advantage of the power of bacterial genetics and to concentrate on the bacterial genes required for functional nodule formation. A deeper understanding of these genes should provide future leverage for investigating the genetic contribution of the host plant.

Cloning the Symbiosis Genes of Rhizobia[98, 99]

The bacterial genes required for formation of functional *sym*biotic nitrogen-fixing nodules are known as *sym* genes. The *sym* genes have been provisionally divided into three classes, depending on which step in the nodulation process appears to be blocked by mutation. Additional analysis will surely lead to further subdivision of these classes. The *nod* genes are required for initiation and the early steps of *nod*ule formation. The *fix* genes are required for later steps in the construction of a nodule capable of nitrogen *fix*ation. And finally, the *nif* genes encode the three highly conserved subunits of the nitrogenase (which reduces molecular nitrogen to ammonia) as well as auxiliary proteins required for the nitrogenase to function. No good estimate can yet be made for the total number of *sym* genes, except for the *nif* genes, which number at least 17 in the enteric bacterium *Klebsiella pneumoniae*. The *sym* genes are thought to be located on a very large autonomously replicating plasmid, but a chromosomal location has not been strictly excluded.

The bacterial *sym* genes can be identified and cloned using the technique of **transposon tagging** (Chapter 11). The bacterial transposon Tn5 is a mobile genetic element carrying a gene that confers resistance to the antibiotic kanamycin. The kanamycin resistance acts as a molecular tag that allows geneticists to follow the movement of the transposon from one chromosomal site to another within a bacterium, or from one bacterium to another during conjugation. Tn5 inserts randomly during transposition. If the insertion site lies within the coding or controlling regions of a gene, the gene will be inactivated. To identify *sym* genes in *Rhizobium*, a strain of *E. coli* carrying Tn5 in its genome (and therefore resistant to kanamycin) is conjugated with a wild-type strain of *Rhizobium* lacking Tn5 (and therefore sensitive to kanamycin). After mating, rhizobia with a newly inserted Tn5 can be selected by growth on kanamycin; rhizobia lacking Tn5 will be killed by the antibiotic. The surviving bacterial colonies are then tested for mutations in the *sym* genes by inoculation onto wild-type seedlings of the appropriate plant species. Because the *sym* genes constitute only a small percentage of the total number of genes in the rhizobial chromosome, most of the Tn5 insertions will not affect the nitrogen-fixing genes. However, between 1 in 100 and 1 in 1000 seedlings tested is blocked at some stage preceding formation of a fully functional nodule. Although growth and examination of 1000 little plants is not a trivial exercise, the procedure has proved to be manageable.

Transposon mutagenesis yields a collection of mutant rhizobial strains, each carrying a Tn5 insertion in one of the *sym* genes. Molecular clones of these *sym* genes can easily be obtained by using the Tn5 DNA sequences as a molecular tag. First, radioactive Tn5 DNA is used as a probe to fish out from the mutant rhizobial DNA a large DNA fragment spanning the Tn5 insertion. Since Tn5 is an insertional mutagen, the rhizobial DNA sequences flanking the Tn5 insertion belong to the inactivated *sym* gene. Fragments of the inactivated *sym* genes are then used as radioactive probes to fish out an undamaged copy of the *sym* from the unmutated parental strain of *Rhizobium*.

Recently, expression of the *nod* genes was found to be induced by soluble plant metabolites that are excreted into the medium when plants are grown in culture. This important discovery helps to explain why the bacteria normally do not differentiate outside the plant host. Moreover, since the *nod* genes can now be turned on simply by treating bacteria with culture medium from growing host plants, this discovery also gives geneticists a way to study certain aspects of nodulation using the rapid techniques of classical microbial genetics, rather than the laborious inoculation of thousands of plants.

Summary

The expression of many genes must be exquisitely coordinated during differentiation and development. These developmentally regulated genes appear to be arranged in a hierarchy, with genes at the top of the hierarchy controlling the expression of those below them. Thus, the best hope for understanding complex developmental processes lies in identifying the controlling genes.

Genes are turned on and off during development by a variety of mechanisms. Displacement of one σ factor by another determines the order of gene expression during sporulation of Gram-positive bacteria like *B. subtilis*. Titration of a preformed factor in the frog egg by DNA synthesized during embryogenesis determines the timing of the midblastula transition. And the extraordinarily long transcription units of some homeotic genes in *Drosophila* may be a mechanism for delaying expression of the gene product until a specified time after transcription resumes in the newly fertilized egg.

We are just beginning to understand what causes cells to differentiate and how the differentiated state is maintained and stably inherited. Differentiation can be reversible or irreversible, depending on the cell type and the organism. Differentiation usually does not involve DNA rearrangement or loss, but rather may reflect changes in chromatin structure, the binding of specific regulatory proteins (repressors and activators), the formation of stable transcription complexes, and perhaps methylation and demethylation at the dinucleotide CpG. In addition, conservative segregation of nucleosomes at the replication fork provides an attractive (albeit hypothetical) explanation for how a stem cell can divide asymmetrically, giving rise to a differentiated daughter and another stem cell.

Sometimes, DNA rearrangements do accompany differentiation. In ciliates, extensive tracts of germ line DNA are eliminated from micronuclear DNA during regeneration of a new macronucleus, and many of the remaining DNA fragments are then precisely joined (perhaps by DNA splicing). Trypanosomes vary their surface antigens, probably by a directed gene conversion that moves a transcriptionally silent copy of the antigen-coding region on one chromosome to a transcriptionally active expression site on another chromosome.

Developmentally regulated gene amplifications are a way to provide extra gene copies in specific cell types where the demand for the gene product is high, but without imposing on every cell of the organism the genomic burden of carrying a large multigene family. During amphibian oogenesis, the ribosomal RNA genes are amplified extrachromosomally. In contrast, repeated rounds of localized DNA replication lead to chromosomal amplification of the chorion genes in follicle cells during *Drosophila* oogenesis.

Rapid progress is being made in understanding the genetic control of pattern formation during embryogenesis. The various dorsalizing genes appear to establish a gradient of dorsoventral polarity in the *Drosophila* egg, and the *Toll* gene product may actually be (or produce) a morphogen. The *Aristapedia* mutation in the fly causes only the distal part of the antenna to differentiate into a leg, as though the cells were differentiating in response to their position in some kind of gradient. Yet, except for the role of cyclic AMP as the morphogen that directs aggregation of starved *Dictyostelium* amoebas, we have no idea what the molecular nature of such a gradient might be, or how the cells might perceive their position within it.

Both internal and external forces play a role in determining whether a cell will differentiate along one developmental pathway or another. In a few simple cases, cytoplasmic determinants present in the unfertilized egg cause the embryonic cells containing them to differentiate into a specific cell type. Cytoplasmic determinants cause the formation of germ cells in insect and amphibian embryos, intestinal cells in the nematode, and muscle cells in the frog. In most cases, however, determination seems to reflect a much more complex interaction between many different gene products. Nonetheless, we suspect that the early stages of cellular differentiation correspond to expression of a few master genes high in the controlling hierarchy. Subsequent differentiation may simply reflect these initial decisions. Thus, the *Drosophila* segmentation genes determine whether the pattern of embryonic segments is correctly formed, but subsequent differentiation of these segments (e.g., into a thoracic segment bearing wings or an abdominal segment bearing reproductive organs) is the responsibility of other subordinate genes.

In mosaic development, interactions between cells play little or no role in determining cell fate. Instead, cytoplasmic determinants and cell lineage determine the nature and course of cellular differentiation. Although embryogenesis in the nematode worm *Caenorhabditis elegans* is almost exclusively mosaic, proper development of the gonad is

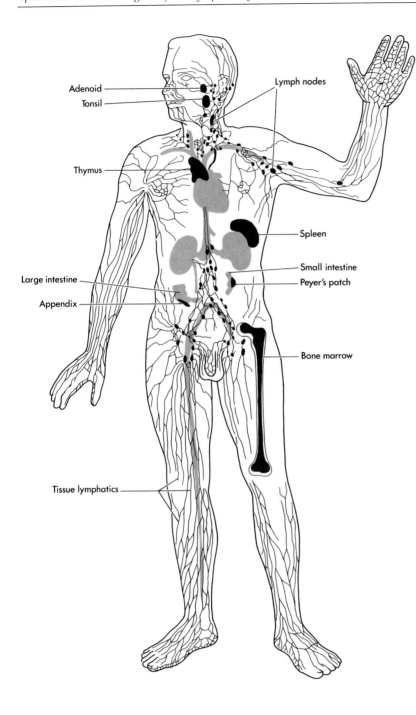

Figure 23-2
Diagram of the human lymphoid system. The system consists of the lymphoid organs shown here and of millions of lymphocytes that circulate between the lymphoid system and the bloodstream. Lymphocytes originate in the bone marrow, which is present in the long bones (one of which is indicated in the picture). After maturing in the bone marrow (B cells) or thymus (T cells), the lymphocytes migrate out and congregate in the other (secondary) lymphoid organs. From "The Immune System" by N. K. Jerne. 1973, *Sci. Amer.* All rights reserved.

called Peyer's patches (see Figure 23-2). In these sites, lymphocytes may be activated by exposure to the antigens that have been gathered up by the lymph and concentrated in the secondary lymphoid organs. From here the lymphocytes, some of them now primed for attack, begin their circulation to the blood vessels and back again to the lymph system, seeking out and responding to antigens as they go.

In addition to lymphocytes, there are other types of accessory cells involved in immune responses. These include macrophages, monocytes, and Langerhans cells. Many accessory cells are located in the lymph nodes, but others are well distributed throughout the body, including in the skin. Some of these cells perform the critically important function of accumulating antigens. Macrophages, for example,

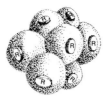

Figure 23-3
Diagrammatic view of an antigen. The symbol R represents a single determinant of immunological specificity, the actual group that combines with an antibody molecule.

are scavengers that gobble up foreign substances and cell debris. They do not simply destroy the things they ingest. Rather, they reposition fragments of the ingested material at their surface. As we shall learn later, many lymphocytes can only recognize antigens and so respond to them when the antigens are displayed in this way. Accessory cells that gather up antigens and help lymphocytes to interact with them are called antigen-presenting cells.

Certain Properties Make Substances Antigenic[3, 8, 9, 10]

While the primary function of the immune system may be to save us from disease-causing microorganisms, the system itself cannot distinguish potentially pathogenic from nonpathogenic invaders. Rather, it is because they are foreign, or nonself, that invading microorganisms are antigenic. An object is potentially antigenic when it possesses an arrangement of atoms at its surface that differs from the surface configuration of any normal host component.

Being foreign is not the only property that makes an object antigenic, however. Another major requirement is that an antigen must either be a macromolecule or be built up from macromolecules (e.g., a virus particle). Most proteins and some polysaccharides and nucleic acids are antigens. Small molecules by themselves can seldom induce specific circulating antibodies. This is not due to a lack of specificity, since many small molecules, nonantigenic by themselves, change the antigenic properties of a larger molecule (e.g., a protein) when covalently coupled to it. The resulting antibodies then become partially directed against the small molecule. Small molecules that become antigenic when coupled to protein carriers are called **haptens.**

It is unlikely that the entire surface of a large molecule is necessary for its antigenicity. The immunological system probably responds to specific groups of atoms, called **antigenic determinants** or **epitopes,** located at a number of sites about a molecular surface (Figure 23-3). In proteins, an epitope is composed of five to eight amino acids. So a given protein molecule usually possesses many antigenic determinants and induces the formation of many types of antibodies. Of course, objects the size of bacteria possess a very large number of different antigenic determinants.

Although we are uncertain as to exactly how many unique antigenic determinants exist, the number is certainly large—larger than 10,000 and probably on the order of millions. The number 10,000 is based on experiments that test whether antibodies induced by a given protein ever accidentally combine with a completely unrelated protein. If there were only a limited number of antigenic determinants, we would expect the same determinant to be found on many different proteins, leading to unexpected antigenic homologies between unrelated proteins. But, in fact, such cross-reactions almost never occur.

Antibody Specificity (T Cell Receptor Specificity) Resides in Amino Acid Sequence[11]

The most striking property of the vertebrate immune system is its ability to specifically recognize thousands, perhaps even millions, of foreign antigens. This recognition is performed by the specialized proteins—antibody molecules and T cell receptors—that specifically bind to antigenic substances in much the same way that enzyme molecules bind specifically to their substrates. For a long time, various theories vied to explain how an organism could possibly be born with the ability to produce proteins capable of specifically recognizing the many different antigenic determinants that may be encountered in its lifetime. According to one long-dominant theory, only a small number of different antibody molecules were required, since the antigen itself would determine which antibody should be formed by combining with a newly synthesized antibody chain before it had folded into the final three-dimensional form. The interaction would allow the antibody molecule to fold around the antigen, automatically creating a complementary shape between the two molecules. Because, according to this scheme, the antigen directly determines the shape of the antibody, this model was called the *instructive theory* of antibody formation.

However, the instructive theory was ruled out many years ago. Convincing evidence against it came first from experiments in which the three-dimensional structure of an antibody with known specificity was temporarily destroyed (by denaturation) and then allowed to re-form in the absence of antigen: The antibody molecules resumed their original specificity! Now we know that the vast diversity and remarkable specificity of immune responses are due to the fact that vertebrates synthesize a vast number of different antibody molecules, each with a different amino acid sequence. Among this vast repertoire of molecules there are invariably some that can interact with any of the vast array of antigens the organism will meet with in its lifetime.

Far less is known about the structure of T cell receptor proteins than about antibody molecules, since the former were only detected biochemically a few years ago, whereas antibody molecules became available for study several decades ago. However, there is evidence that different T cell receptors have different amino acid sequences, and no doubt their antigen binding specificity is also determined by their amino acid sequence.

Clonal Selection Explains Specific Immune Responses and Immunological Memory[12, 13]

Fundamental to the way immune responses work is the fact that each B lymphocyte produces antibody molecules of a single specificity, and each T cell carries a single type of T cell receptor molecule. Rather than having each lymphocyte produce antibodies or T cell receptors that can recognize many different antigens, diversity is achieved by having a very large number of B and T cells, each synthesizing a unique antigen recognition molecule. Before B cells differentiate to mature antibody-secreting cells, called **plasma cells,** they carry their antibodies as membrane-bound surface molecules. In this form, the antibody serves as a receptor for antigen, much like the receptors on T cells. When an antigen collides with an appropriate antibody receptor on an immature B cell, the specific binding, along with other cellular interactions, triggers the proliferation and maturation of that particular B cell. The end result is to generate plasma cells that secrete antibodies of the same specificity as the original membrane-bound

Figure 23-4

The theory of clonal selection as it applies to B cells and T cells. Among the millions of B and T lymphocytes present in the vertebrate body, there are always some that can interact with any antigen that enters the body. These particular cells, whose antibody receptors or T cell receptors can bind the incoming antigen, are triggered by the binding to undergo proliferation and maturation. The result is expanded clones of activated T cells that can attack antigen-bearing cells, and expanded clones of B cells that secrete antibodies.

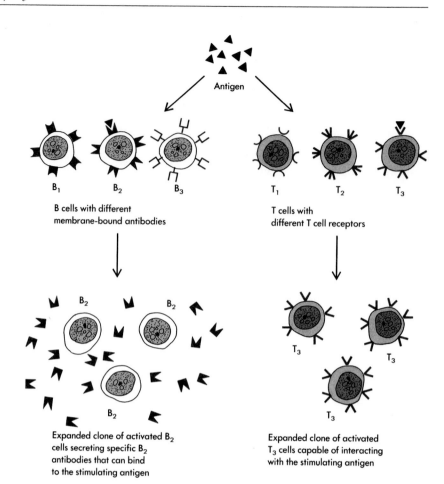

B cells with different
membrane-bound antibodies

T cells with
different T cell receptors

Expanded clone of activated B₂
cells secreting specific B₂
antibodies that can bind
to the stimulating antigen

Expanded clone of activated
T₃ cells capable of interacting
with the stimulating antigen

antibody carried by the immature B cell (Figure 23-4). The idea that antigens stimulate division and maturation of the B cells whose antibodies they can bind to was first predicted in the 1950s and is known as the **theory of clonal selection.** (A clone is a group of cells all descended from a common ancestor.) No longer regarded as just a theory, clonal selection is one of the cornerstones of immunology.

Antigen binding to a T cell receptor also triggers cellular proliferation (clonal selection) and maturation of the immature cell to cells capable of carrying out T cell functions, such as the destruction of virally infected cells. Thus, the theory of clonal selection applies equally well to both T cells and B cells. Mature, functional T cells and plasma cells are called **effector cells.** The maturation of a B cell from a resting small lymphocyte to an activated plasma cell involves a considerable increase in size, and the endoplasmic reticulum is amplified enormously to cope with the massive synthesis of antibody molecules (Figure 23-5). Activated T cells are also enlarged relative to resting T cells, but they lack the extensive endoplasmic reticulum found in plasma cells.

Clonal selection can explain immunological memory. When an animal first encounters an antigen, it produces a relatively slow **primary immune response,** because only a small number of its cells have membrane-bound antibody molecules or T cell receptors that can respond to the antigen. However, the next encounter with the same antigen leads to a far more rapid and intense **secondary response,** because cells that can bind that particular antigen were stimulated to divide by the previous encounter (Figure 23-6). The proliferation and

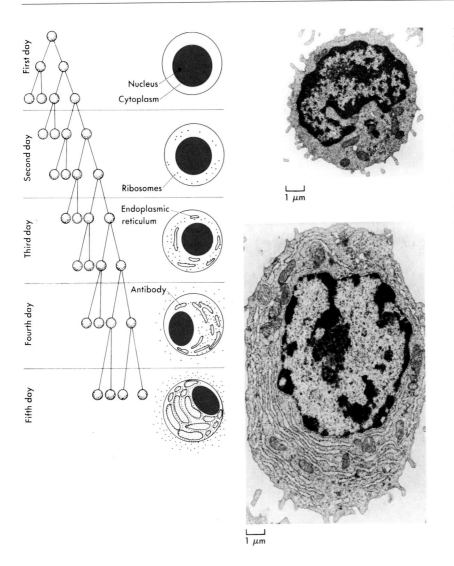

Figure 23-5
Successive stages in the development of mature plasma cells. Immature B cells are small, round cells that carry their antibodies as membrane-anchored receptors. Following antigen stimulation, at least eight cell generations and five days of growth are required before the appearance of cells producing a great deal of antibody. The most noticeable feature of the mature cell is the extensive endoplasmic reticulum, whose internal cavity is filled with antibody molecules. The electron micrographs show the cell as it would appear at the beginning and end of the process. Note that the two cells are shown at the same magnification. [Diagram after G. J. V. Nossal. 1964, *Sci. Amer.* All rights reserved. Electron micrographs courtesy of Dr. Dorthea Zucker-Franklin, NYU Medical Center.]

maturation of particular B or T lymphocytes induced by the first exposure to antigen lead not only to mature antibody-producing plasma cells or functional T cells, but also to expanded clones of long-lived, not yet mature B and T cells called **memory cells.** Memory cells can respond to the specific antigen the next time it is encountered, again differentiating to produce functional B and T effector cells (Figure 23-7). The human body has about 10^{12} lymphocytes. These cells are thought to arise from about 10^6 to 10^8 lymphocytes with different receptor specificities. Among the 10^{12} cells, lymphocytes carrying any particular receptor may be represented only once, or a particular cell may have been stimulated to divide by an encounter with an antigen and so have become a clone of as many as 10^6 cells.

The Immune System Actively Acquires and Maintains Tolerance to Self Antigens[14, 15, 16]

Just as remarkable as the immune system's ability to respond specifically to thousands of different antigens is its failure to respond to the host's own cells and proteins, the phenomenon known as tolerance. Tolerance is not due to any intrinsic lack of antigenicity in our own cells and proteins, since if they are transplanted to another

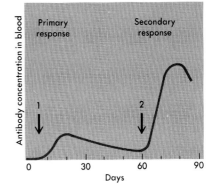

Figure 23-6
A graph showing the primary and secondary responses of antibody production in a rabbit. Arrows 1 and 2 represent the times of injection with the antigen. The secondary response is both more rapid and more intense than the primary response.

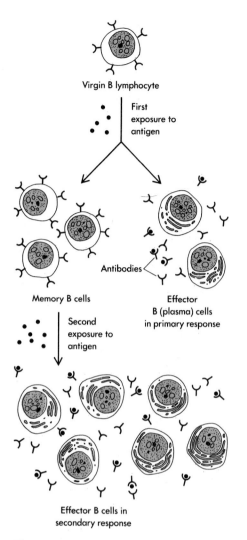

Virgin B lymphocyte

First exposure to antigen

Memory B cells

Antibodies

Effector B (plasma) cells in primary response

Second exposure to antigen

Effector B cells in secondary response

Figure 23-7
The cellular basis for the greater speed and intensity of a secondary antibody response. The first encounter of a virgin B lymphocyte with its cognate antigen causes the cell to proliferate and to differentiate into two types of cells: effector B plasma cells that synthesize antibodies for the primary response, and memory B cells that can differentiate into effector cells the next time the same antigen is encountered.

individual, that individual's immune system soon mounts a response to them and rejects them. Tolerance is not inherited, but rather is acquired and then actively maintained by the immune system throughout an individual's life. This maintenance requires the continued presence of the substances to which the individual is tolerant.

Our understanding of tolerance first came from experiments with cattle performed in the 1940s. Whereas tissue transplants between most individuals were generally unsuccessful owing to the immune response, transplants between dizygotic cattle twins succeeded. Dizygotic twins are genetically related as siblings, but unlike fraternal twins, they share a common blood supply and hence exchange cells in utero. This sharing apparently results in their being tolerant to one another's cells, proving that tolerance is acquired during embryonic development. Later experiments showed that almost any antigen can be made to induce tolerance experimentally. Whether a foreign substance acts as an antigen and induces an immune response or acts as a **toleragen** and induces tolerance depends on the amount of the substance and the route by which the substance enters an animal's body. Also, it is easier to induce tolerance in younger animals, rather than in adults.

One can imagine two plausible mechanisms by which tolerance may be achieved. Either those specific clones of T and B cells that can interact with self antigens are destroyed before they mature, or they mature but are then prevented from responding to self. Apparently, both mechanisms are used to achieve tolerance, but *how* these mechanisms work is still a mystery.

THE B CELL RESPONSE

Typical Antibody Molecules Are Tetramers Composed of Two Identical Light Chains and Two Identical Heavy Chains[17, 18, 19]

Typical antibody molecules are proteins of about 150,000 daltons that consist of four polypeptide chains. There are two identical **light chains,** containing about 220 amino acids each, and two identical **heavy chains,** containing about 440 amino acids each. Each light chain is linked to a heavy chain by one covalent S–S (disulfide) bond, and S–S bonds also run between the two heavy chains to give the schematic picture shown in Figure 23-8. Each chain also has regularly spaced intrachain disulfide bridges. Carbohydrate moieties are attached to the heavy chains, making antibody molecules glycoproteins.

The four polypeptide chains are arranged so that each antibody molecule contains two identical, widely separated antigen binding sites, each of which can combine with an antigen, using secondary bonds to hold their complementary surfaces together. Visualization of antibodies in the electron microscope reveals Y-shaped or T-shaped molecules with one such active site at the end of each arm. Each arm is effectively hinged to the central axis with the distance between active sites sometimes increasing when they are bound to antigen.

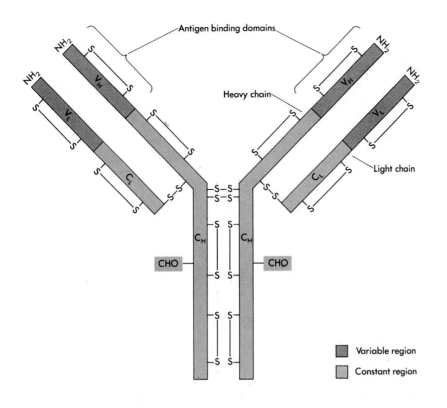

Figure 23-8
A model of the structure of a typical human antibody molecule, composed of two light and two heavy polypeptide chains. Interchain and intrachain disulfide bonds are indicated, and CHO marks the approximate position in the heavy chain of the carbohydrate moiety. Two identical active sites for antigen binding are located in the arms of the molecule.

The existence of two identical binding sites permits a single antibody molecule to link together two similar antigens. This feature is of great advantage in allowing antibodies to defeat an infection by a microorganism, since all microorganisms contain a large number of identical antigenic determinants. In the presence of specific antibodies, a microorganism thus becomes linked to a large number of similar microorganisms through antibody bridges (Figure 23-9), and these aggregates then tend to be taken up and destroyed by macrophages.

Some of the earliest information about antibody structure and its relationship to function came from breaking the antibody molecule apart with the proteolytic enzymes papain and pepsin. As shown in

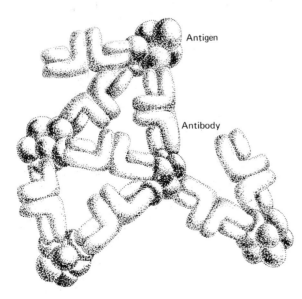

Figure 23-9
A view of how antigens and antibodies combine to form large aggregates.

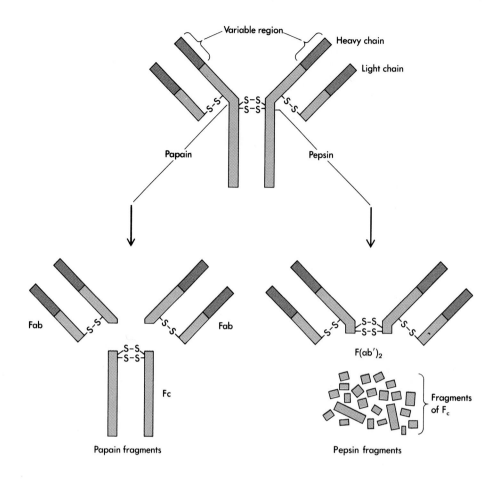

Figure 23-10

Creation of Fab, Fc, and F(ab')₂ fragments by specific proteolytic digestion of intact antibody molecules.

Figure 23-10, papain produces two identical antigen binding fragments, called **Fab fragments,** each with a single antigen binding site, and an **Fc fragment,** whose name reflects its ability to crystallize readily. Pepsin treatment yields an F(ab')₂ fragment that has two antigen-combining sites and is still capable of crosslinking antigen.

The stem portion of the Y-shaped antibody molecule, the Fc region, is designed to carry out the so-called effector functions of the molecule. **Effector functions** are the various activities that antibodies must perform in addition to antigen binding. For example, antigen-antibody complexes must interact with phagocytic cells that will ingest the complexes, and this interaction occurs between the Fc regions and the ingesting cells. Another effector function of these regions is interaction with a set of blood proteins collectively called the **complement system.** When complement proteins bind to antigen-antibody complexes, they can bring about destruction (lysis) of the antigen (Figure 23-11).

Myelomas and Hybridomas Provide Pure Antibodies for Amino Acid Sequencing[20, 21, 22]

Major insights into the molecular basis of immunological diversity and specificity first came from determining the amino acid sequences of the light and heavy polypeptide chains of many different antibody molecules. These studies require preparations of pure antibodies. Although antibodies account for about 20 percent of the protein in blood

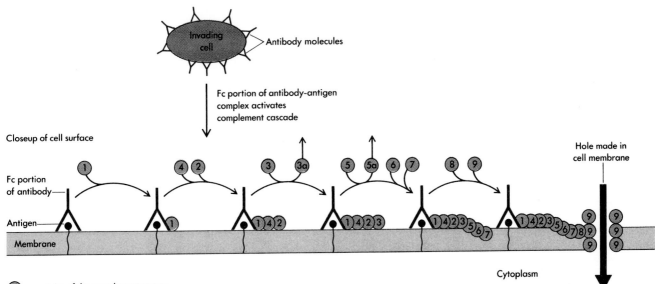

Figure 23-11

Antibody bound to antigen can activate the complement cascade and bring about lysis of the invading cell (microorganism). Complement is a system of about 20 proteins that are present in the blood. In the presence of antibody molecules that are bound to antigen, particularly those on the surface of a cell, the first component of the cascade, C1, becomes activated. In this form, it both binds to the antigen-antibody complex on the cell surface and also activates the next component of the complement pathway. The cascade continues until a complex called C9 is generated. This complex can make a hole in a cell membrane. The hole allows the influx of ions and water into the cell, causing it to swell and then burst.

plasma, at any given time the antibodies present in an individual are a collection of many, many different species directed against a large variety of different antigens. Even after continued immunization with a single strong antigen, no single antibody species is present in pure enough form to allow its isolation. Traces remain of the many different antibodies previously induced in that individual's history. Even more important, a given single antigen always seems to promote synthesis of several different antibodies, each with a different amino acid sequence, yet able to specifically bind the same antigen.

Single antibody species can, however, be obtained in two ways. They were first obtained from the blood of certain humans (and mice) afflicted with the bone marrow disease multiple myeloma. (In mice, the disease is called myeloma or plasmacytoma.) A form of cancer, this disease involves the uncontrolled multiplication of antibody-producing cells. Most important, like all cancers, it arises from a single cell; thus, a given specific tumor produces only one antibody type characterized by a specific amino acid sequence. Because the tumor cells proliferate wildly and continue producing their unique antibody, a myeloma patient becomes a source of a highly pure, single type of antibody. As expected, the antibodies produced by different myeloma tumors each have different amino acid sequences, permitting the comparative study of many different antibody sequences. In the case of multiple myelomas, it is usually impossible to know what antigen could potentially bind to the antibody produced by any particular tumor. Such an antibody is usually called an **immunoglobulin (Ig)**, a term that refers to any antibody-like molecule regardless of its antigen binding specificity.

The amount of a given myeloma protein made in a myeloma patient is very large, easily permitting protein sequence studies. Moreover, many myeloma patients excrete large amounts of specific proteins, called **Bence-Jones proteins** after their discoverer. A given Bence-Jones protein is identical to the light chain of the corresponding myeloma immunoglobulin. Its excretion into the urine is the result of overproduction of immunoglobulin light chains relative to heavy chains by the myeloma cells. The presence of very large amounts of Bence-Jones proteins makes their isolation extremely

Figure 23-12
Procedure for making a hybridoma. A mouse is immunized to stimulate proliferation of desired B cells. The animal's B cells are isolated from the spleen and fused to myeloma cells using polyethylene glycol (PEG) to promote fusion. The myeloma cell line used in the procedure is unusual in two ways: It has stopped synthesizing antibodies, and it is a mutant, called HGPRT⁻, that cannot synthesize the enzyme hypoxanthine-guanine phosphoribosyl transferase (HGPRT). When these myeloma cells are fused to normal B cells, the resulting hybrid cells, but neither parental cell alone, can grow in selective hypoxanthine aminopterin thymidine (HAT) medium. HAT medium contains a drug, aminopterin, that blocks one pathway for nucleotide synthesis, making the cell dependent on a salvage pathway that uses the HGPRT enzyme. Since the myeloma cells lack this enzyme, all cells that fail to fuse to a normal B cell die in HAT medium. Normal B cells that fail to fuse to a myeloma cell die because they lack the tumorigenic property of immortal growth. Surviving hybridomas, which inherit the HGPRT gene from the normal B cell and the cancerous property of immortal growth from the myeloma cell, are screened to find those that inherited the antibody genes of the normal B cell and are secreting monoclonal antibodies that can bind to the immunizing antigen.

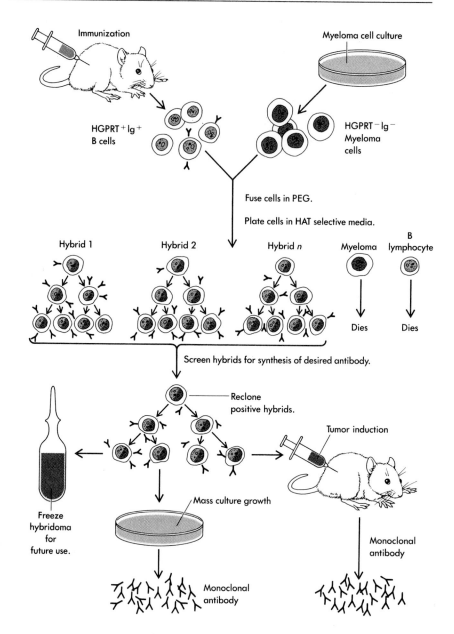

simple, and the first important results with antibody amino acid sequences arose from the study of these proteins.

Our general inability to know which antigen a myeloma protein recognizes is a major drawback for many types of studies. In contrast, a recent and enormously powerful technique has made it possible to produce pure antibodies, called **monoclonal antibodies,** with specificity to bind to any desired antigen. In this method, a mouse is repeatedly immunized with an antigen of choice. As a result, B cells making antibodies specific for that antigen proliferate. Then all the animal's B lymphocytes, now highly enriched for B cell clones that recognize the immunizing antigen, are removed and fused to a mutant myeloma cell whose own antibody synthesis has shut off. When cells fuse, their nuclei can also fuse to yield a hybrid cell that contains chromosomes from both parents. Hybrids between the myeloma and normal B cells are selected and screened for the synthesis of antibodies to the immunizing antigen (Figure 23-12). Usually, about one

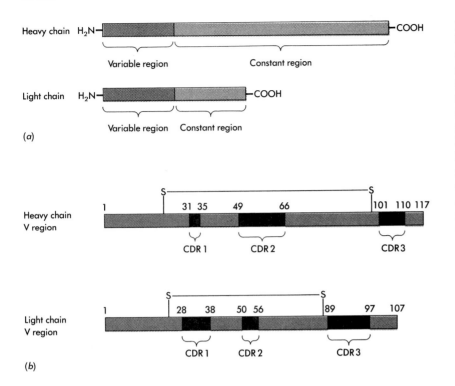

Figure 23-13
Location of variable and hypervariable (CDR) regions in light and heavy chains. (a) Variable (V) regions of light and heavy chains are about 110 amino acids long and are located at the amino-terminal ends of the polypeptides. (b) Diagram of V regions of light and heavy chains, showing positions of hypervariable (complementarity-determining) regions. Numbers indicate typical amino acid position, and S–S represents a disulfide bridge. CDRs are flanked by the framework portions of the V regions.

in several hundred cell hybrids produces an antibody of the desired specificity, and this cell is purified as the desired **hybridoma.** The role of the myeloma cell in this process is to provide the normal B lymphocytes with the cancerous property of uncontrolled growth, thereby allowing hybridomas to grow easily both in tissue culture and after injection into mice. Normal noncancerous B cells would quickly die in either situation.

Light and Heavy Chains Have Variable Regions for Antigen Recognition and Constant Regions to Perform Effector Functions[23, 24, 25]

The first experiments to examine the differences in amino acid sequence between closely related light chains or heavy chains isolated from different antibody molecules were done by the method of tryptic peptide mapping. These experiments showed that similar types of chains from different molecules have many peptides in common and differ in others. As soon as amino acid sequence data were obtained for two very similar light chains, it was clear that the variability was all at the amino-terminal end of the chains, while the carboxyl halves were identical. Further sequencing soon showed the same situation in heavy chains: **variable (V) regions** at the amino terminus and **constant (C) regions** in the carboxyl portion (Figure 23-13).

As more and more amino acid sequences of antibody molecules were determined, it became clear that variability is not evenly distributed through the V regions. It is concentrated in three segments, called **complementarity determining regions (CDRs)** or **hypervariable regions,** and this is true in both the light- and heavy-chain variable segments (Figures 23-13 and 23-14). It is these hypervariable regions that line the walls of the antigen binding cavities in the native

Figure 23-14

The amino acid sequences of the variable regions of nine different human heavy chains. Unbroken lines mean identity with the first protein, called Tie. Dashes in the brackets mean that the corresponding protein has no amino acid at this position. These gaps were introduced to maximize the homology. Sequence variations (shaded area) are clustered in the three complementarity-determining regions, CDR1, CDR2, and CDR3. A fourth region of considerable variation, between amino acid residues 84 and 88, is not located in the antigen binding site, and its function is unknown. [After J. D. Capra and J. M. Kehoe, *Proc. Nat. Acad. Sci.* 71 (1974):4032.]

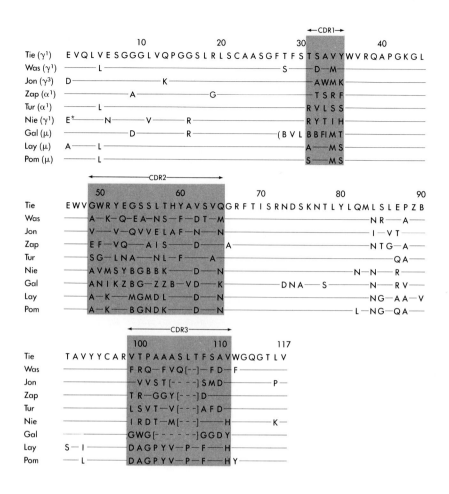

antibody molecule. The more highly conserved portions of variable regions are called the **framework.**

The first complete amino acid sequence of an antibody molecule revealed important sequence relationships between the variable regions of light and heavy chains and also between their constant regions. The sequences also revealed the presence of repeated domains within heavy-chain constant regions. Examination of the heavy-chain sequence strongly suggests that the gene encoding the constant portion of this polypeptide arose by successive duplication of a primitive antibody gene. This can be seen clearly by dividing the sequence of a heavy-chain constant region into four parts. The first, the variable portion, shows similarity to the variable part of the light chain. In turn, the three regions of the constant portion, C_H1, C_H2, and C_H3, show strong homologies both to each other and to the constant region of a light chain (Figure 23-15). Thus, the variable regions of light and heavy chains consist of one type of **homology unit,** while the constant regions of light chains and the three C_H segments consist of a second type of homology unit. Each homology unit is about 110 amino acids long, and each has one centrally located S—S bond, with the two sulfur atoms separated by some 65 or so amino acids. Even though V and C homology units have no apparent amino acid sequence homology to one another, their similar length and the comparable location of the S—S bonds suggest that they, too, arose from a common ancestor way back in evolutionary time.

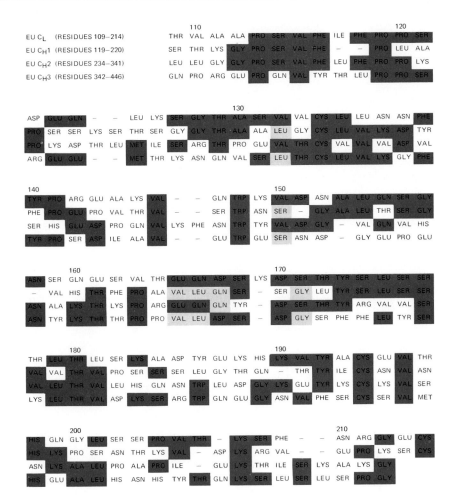

X-Ray Diffraction Reveals the Three-Dimensional Structure of Antibody Molecules[26–29]

To determine the three-dimensional arrangement of a protein, the protein must first be crystallized and then subjected to X-ray diffraction analysis. For antibody molecules, this was done first with Fab and Fc fragments (see Figure 23-10). Later, it was achieved with a complete immunoglobulin. From this analysis, it is clear that V and C homology units are folded in similar configurations, called the **immunoglobulin fold;** an important difference is that the V regions have extra polypeptide loops, composed of the hypervariable segments, for antigen binding (Figure 23-16). Each V or C homology unit contains two layers of extended β sheets that are roughly parallel to each other and that surround an interior portion in which are packed the side groups of hydrophobic amino acids. The V_L and V_H units then pack together to form the variable domains of the completed antibody molecule, while C_L and C_{H1} units lie together, as do the remaining pairs of C_H homology units (Figure 23-17). Each pair of homology units thus forms a discrete globular structural domain. This organizational tactic explains why V regions must have conserved framework portions and carefully localized hypervariable segments: If their structure were allowed to diverge too much, they would be unable to associate with another V homology unit and thereby take up the proper three-dimensional configuration within the molecule.

Figure 23-15
Sequence homology in the constant regions of the light and heavy chains of an antibody molecule, called EU, the first antibody molecule whose complete sequence was determined. Deletions indicated by dashes have been introduced to maximize the homology. Identical residues are shaded. Both dark and light shadings indicate identities that occur in pairs in the same positions. The figure emphasizes the presence of repeating domains within the heavy-chain constant region and their similarity to the light-chain constant region. [After G. M. Edelman et al., *Proc. Nat. Acad. Sci.* 63 (1969):78.]

Figure 23-16
The three-dimensional configuration of a single light chain. Both the variable and constant regions exhibit the characteristic immunoglobulin fold consisting of two β-pleated sheets, one composed of three strands and one of four strands. The strands within the two sheets are indicated in different colors. The characteristic intrachain disulfide (S–S) bonds, which are present in both the V and C regions of the chain, bridge the two β sheets in each domain. The hypervariable regions within the V region form loops that lie near similarly placed loops in the heavy-chain V region. Together these loops form the antigen binding site. [After M. Schiffer, R. L. Girling, K. R. Ely, and A. B Edmundson, *Biochemistry* 12 (1970):4620.]

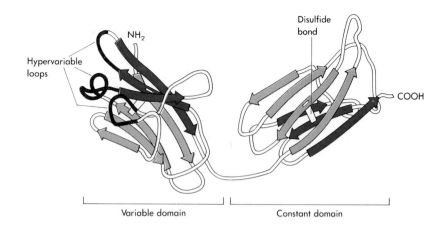

The antigen binding cavities (active sites) formed by the hypervariable region loops of V_L and V_H segments vary in size and shape from antibody to antibody, depending on their precise amino acid sequence. Examples of two different cavities and their interactions with their respective antigens are shown in Figure 23-18. An antigen that binds to a specific antibody is called a **cognate antigen.**

There Are Two Major Types of Light Chains, λ and κ[19]

After the amino acid sequences of many different light chains were determined, it was apparent that in each vertebrate species, they can be assigned to one of two clearly distinct types, called **κ** (kappa) and **λ** (lambda), based on the amino acid sequences of their constant regions. κ and λ constant regions are about 30 to 40 percent homologous (Figure 23-19). There is only one type of κ light-chain constant region ($C_κ$) in an animal. However, there can be several types of λ

Figure 23-17
The three-dimensional structure of an antibody molecule. Each small sphere represents one amino acid. The carbohydrate chain attached to the C_H2 portion of the heavy chains is shown as large spheres. [After E. W. Silverton, M. A. Navia, and D. R. Davies, *Proc. Nat. Acad. Sci.* 74 (1977):5140.]

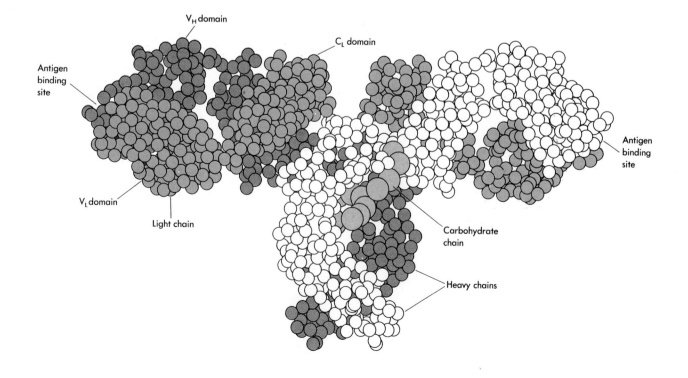

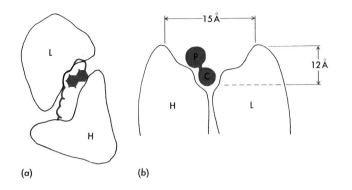

(a)　　　　　(b)

Figure 23-18
Schematic drawings of two antigen binding cavities interacting with their cognate antigens. (a) Top view of the shallow cleft (16 x 7 x 6 Å) between the heavy and light variable chains of the human myeloma protein NEW, which specifically binds the hapten vitamin K_1OH (shown in color). (b) Phosphoryl-choline (PC, in color) binding in the larger (15 x 20 x 12 Å) cavity formed by the hypervariable regions of the heavy and light variable chains of a mouse myeloma protein McPC 603. This side view reveals that most antibody-antigen interaction involves the heavy chain in this case. [(a) after F. F. Richards et al., in *The Immune System*, Third ICN-UCLA Symposium on Molecular Biology (New York: Academic Press, 1974). (b) after E. A. Padlan et al., in *The Immune System*, Third ICN-UCLA Symposium on Molecular Biology (New York: Academic Press, 1974).]

chains. For example, in the mouse, there are three types, λ_1, λ_2, and λ_3, whose constant regions are nearly identical in amino acid sequence. Any particular B cell synthesizes only one type of light chain, either κ or λ (and if λ, then only λ_1, λ_2, or λ_3). The variable regions of λ and κ chains can also usually be distinguished by characteristic amino acids in their framework portions (see Figure 23-19).

There Are Five Major Types of Heavy Chains (α, δ, ε, γ, μ) That Correspond to the Five Classes of Antibodies (IgA, IgD, IgE, IgG, and IgM)[19]

Antibody molecules can be assigned to different classes, depending on the amino acid sequence of the constant region of their heavy chains. As a result of these differences in amino acid sequence, antibody molecules belonging to some classes are composed of more than one light-heavy chain tetramer, and molecules belonging to different classes can also differ in the positions of the disulfide bridges that link their two heavy chains. The most important consequence of the differences in amino acid sequence is that distinct biological functions are conferred to different antibody classes. There are five major classes of antibodies, named **IgA, IgD, IgE, IgG, and IgM**, and several of these can be further divided into subclasses (e.g., IgG_1). The heavy-chain constant regions that correspond to the different classes of antibody molecules are called α, δ, ε, γ, and μ. Their amino acid sequences are about 40 percent homologous. The structures of the different classes of antibody molecules are shown in Figure 23-20, and their properties are summarized in Table 23-1. When more than one

Figure 23-19
Selected amino acid sequences from the V and C regions of several human λ (top) and κ (bottom) light chains. Not all chains have exactly the same number of amino acids, and the sequences are aligned for maximum homology. [After H. Eisen, *Immunology* (New York: Harper & Row, 1974).]

Variable	Hypervariable	Variable	Constant

```
 9                                                                     1               2             3               4
 1 2 3 4 5 6 7 8 9 0 1 2 3 4 5 6 7 8 9 0 1 2 3 4 5 6 7 8 9 0 1 2 3 4 5 6 7 8 9 0 1 2 3 4 5 6 7 8 9 0
 A D Y Y C N S R D S S G K H V L F G G G T K L T V L G Q P K A A P S V T L F P P S S E E L Q A N K A
 A D Y Y C S S Y V D N N N F X V F G G G T K L T V L R Q P K A A P S V T L F P P S S E E L Q A N K A
 A H Y H C A A W D Y R L S A V V F G G G T Q L T V L R Q P K A A P S V T L F P P S S E E L Q A N K A
 A D Y Y C Q A W D S S L N A V V F G G G T K V T V L G Q P K A A P S V T L F P P S S E E L Q A N K A
 A D Y Y C Q A W D - - S M S V V F G G G T R L T V L S Q P K A A P S V T L F P P S S E E L Q A N K A
 A T Y Y C Q Q Y D - - T L P R T F G G G T K L E I K R T Y - A A P S V F I F P P S N E Q L K S G T A
 A T Y Y C Q Q F D - - N L P L T F G G G T K V D F K R T Y - A A P S V F I F P P S D E Q L K S G T A
 G V Y Y C Q M R L - - E I P Y T F G Q G T K L E I R R T Y - A A P S V F I F P P S D E Q L K S G T A
 G V Y Y C M Q A L - - Q T P L T F G G G T N V E I K R T Y - A A P S V F I F P P S B Z Z L K S G T A
 A V Y Y C Q Q Y G - - S S P S T F G Q G T K V E L K R T Y - A A P S V F I F P P S D E Q L K S G T A
```

A = ala	E = glu	I = ile	N = asn	S = ser	Y = tyr
B = asx	F = phe	K = lys	P = pro	T = thr	Z = glx
C = cys	G = gly	L = leu	Q = gln	V = val	
D = asp	H = his	M = met	R = arg	W = trp	

Figure 23-20

The subunit structures of different classes of human immunoglobulins. Disulfide bridges are indicated by –S–S–. IgG1, IgG2, IgG3, and IgG4 are subclasses of IgG. IgA1 and IgA2 are monomeric subclasses of IgA. The J polypeptide chain joins the polymeric form of IgA and may initiate the joining of the five subunits that comprise extracellular IgM molecules. [Adapted from J. Gally in *The Antigens*, M. Sela (Ed.), Academic Press, New York, 1973, p. 209.]

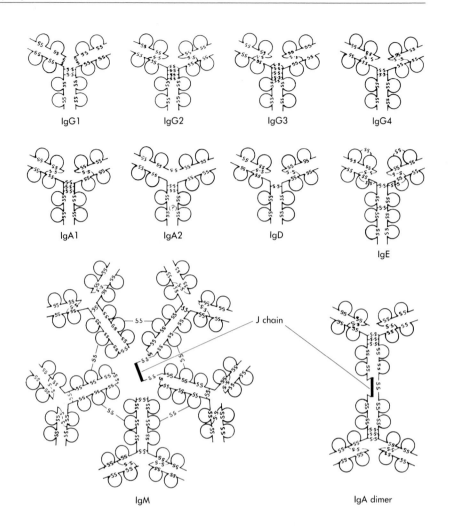

tetramer is present in a molecule (e.g., the IgM pentamer), the tetramers are held together by disulfide bridges, and some of these are joined by a special polypeptide called the **J (joining) chain.**

Different antibody classes have evolved to perform specialized functions. For example, IgM pentamers appear early during primary immune responses; then, during secondary responses, they are largely replaced by the more common IgG antibodies. IgG antibodies are the only ones that can cross the placenta to travel from mother to fetus during pregnancy. IgA molecules are most common in the secretions (tears, saliva, milk, etc.). The different behavior of the molecules is due to their heavy-chain constant regions. For example, specific receptors on placental cells recognize the constant region of IgG molecules, causing them to be ingested by endocytosis, shuttled across the cell, and released into the fetal blood on the other side of the cell. As we noted before, these diverse functions that antibodies perform are called effector functions.

Random Association of Heavy and Light Chains Can Help Explain Antibody Diversity

An antibody molecule of a particular class can have either a λ or a κ light chain. In other words, any type of light chain can combine with any type of heavy chain, a phenomenon known as **combinatorial**

Table 23-1   Properties of the Major Classes of Antibodies in Humans

Property	Antibody Class				
	IgM	*IgD*	*IgG*	*IgA*	*IgE*
Heavy chain	μ	δ	γ	α	ε
Light chain	κ or λ	κ or λ	κ or λ	κ or λ	κ or λ
Number of four-chain units	5	1	1	1 or 2	1
Structure	$(\mu_2\kappa_2)_5$ or $(\mu_2\lambda_2)_5$	$\delta_2\kappa_2$ or $\delta_2\lambda_2$	$\gamma_2\kappa_2$ or $\gamma_2\lambda_2$	$(\alpha_2\kappa_2)_{1-2}$ or $(\alpha_2\lambda_2)_{1-2}$	$\varepsilon_2\kappa_2$ or $\varepsilon_2\lambda_2$
Molecular weight (daltons)	900,000	185,000	150,000	160,000 320,000	200,000
% total serum immunoglobulin	6	1	80	13	0.002
% carbohydrate content	12	13	3	8	12
Biological function:					
Activates complement	+	–	+	–	–
Crosses placenta	–	–	+	–	–
First antibody to appear after immunization	+	–	–	–	–
High in secretions	–	–	–	+	–
Allergic reactions due to interaction with other cell types	–	–	–	–	+

+ = possesses the property. – = lacks the property.

association. This simple fact has extremely important implications. Because antigen binding cavities are formed by the variable regions of both the light and heavy chains, if any light chain can associate with any heavy chain, then the number of different antigen binding cavities that can be formed by 1000 different light chains and 1000 different heavy chains is $10^3 \times 10^3 = 10^6$. Of course, we have no way of knowing whether all light- and heavy-chain combinations can form functional antigen binding sites. Nonetheless, it seems possible that many combinations might result in molecules that can recognize some antigen or other. Thus, it is believed that the random association of light and heavy chains probably contributes substantially to antibody diversity.

The probable existence of combinatorial association provided the first major explanation of how an organism might produce a vast number of different antibody molecules while using only a relatively small number of genes.

Germ Line Versus Somatic Theories to Explain the Genetic Basis of Antibody Diversity[30, 31]

As soon as the instructive theory of antibody diversity was laid to rest and it was clear that antibody specificity and diversity are achieved by the synthesis of thousands of different antibody molecules, the central puzzle of immunology was to understand the genetic basis for this ability. The accumulation of amino acid sequence data only served to highlight the problem, since virtually every light- or heavy-chain variable region had a different sequence. Two theories vied to explain the genetic basis of antibody diversity. According to the *germ line theory*, a different gene for each antibody light or heavy chain would be inherited in the DNA. This theory was repugnant to those who realized the vast number of possible antibody molecules an animal can make and who believed that far too much of the genome would have to be devoted to the immune system. According to rival *somatic theories,* a smaller number of light and heavy chain genes would be inherited in the germ line. Then in B cells, these genes would undergo some sort of genetic alteration in their V-coding portions to generate novel variable regions and hence novel antigen binding specificities.

As we shall see in the following sections, the somatic theory turned out to be correct, although aspects of the germ line theory also help explain diversity. In all the vertebrates that have been studied so far, not a single intact antibody gene is inherited in the germ line. Rather, antibody genes are inherited as fragments that must be assembled in somatic (B) cells to generate functional genes. Because the numerous DNA fragments that encode variable regions can be assembled in many combinations, this mechanism of modular gene construction contributes greatly to V-region diversity. In addition, fully assembled antibody genes undergo further somatic mutation in their V-region-coding portions to increase diversity even more.

Antibody Structure Implies That Separate Genes Code for Variable and Constant Regions[32]

Before direct proof was possible, two considerations led some insightful immunologists to postulate that separate genes must encode the variable and constant regions of antibody polypeptide chains. The first problem was trying to imagine a mechanism that could generate so much diversity in one half of a protein while conserving the sequences of the remainder. Although elaborate schemes could be concocted, some researchers thought that the simplest hypothesis involved a few genes for the different constant regions of the heavy and light chains (e.g., one gene for C_μ, one gene for $C_{\gamma 1}$, and one gene for $C_{\lambda 2}$), many genes for the variable regions, and a mechanism that could join the two types of genes together.

A second observation that pointed to a two genes–one polypeptide model for antibodies came from the phenomenon known as **class switching,** first observed in a patient with multiple myeloma. This individual's tumor secreted antibodies with a single type of light chain but two types of heavy chains. The heavy chains had identical variable regions, but they differed in their constant regions. Later, normal B cells were found to undergo a transition from IgM production to the synthesis of other heavy-chain classes as part of their differentiation program. Once again, the same heavy-chain variable region was present before and after the class switch (Figure 23-21). At the time these observations were made, RNA splicing, which could have provided a plausible mechanism for the observations, had not yet been discovered; so the best hypothesis was that a gene coding for a particular heavy-chain variable region could associate first with one type of C_H (e.g., C_μ) and later with a different type. Remarkably, this explanation for class switching turned out to be essentially correct, although some class switches occur by changes in RNA processing rather than as a result of DNA rearrangements.

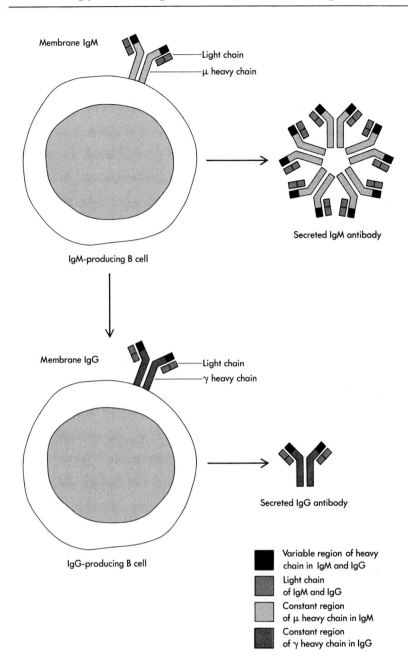

Membrane IgM

Light chain

μ heavy chain

IgM-producing B cell

Secreted IgM antibody

Membrane IgG

Light chain

γ heavy chain

IgG-producing B cell

Secreted IgG antibody

Variable region of heavy chain in IgM and IgG

Light chain of IgM and IgG

Constant region of μ heavy chain in IgM

Constant region of γ heavy chain in IgG

Figure 23-21
Attachment of the same V_H region to a different C_H region during an IgM to IgG class switch.

DNA Coding for Variable Regions Moves Closer to Constant-Region DNA to Create Functional Antibody Genes in B Cells[33]

In 1976, direct experimental evidence showed that antibody genes are assembled by recombination events that move DNA coding for variable regions closer to DNA encoding the constant regions. The idea behind the experiment was to compare the structure of a κ light-chain gene in two types of cells: embryonic mouse cells that do not synthesize antibodies, and a mouse myeloma that synthesizes a κ light chain. A change in the gene structure would imply that a DNA rearrangement had occurred to produce the functional gene. First, DNAs isolated from the two types of cells were digested with a restriction endonuclease and the resulting fragments separated by electrophoresis.

Figure 23-22
An experiment showing that antibody genes undergo DNA rearrangement in B cells that synthesize antibodies.

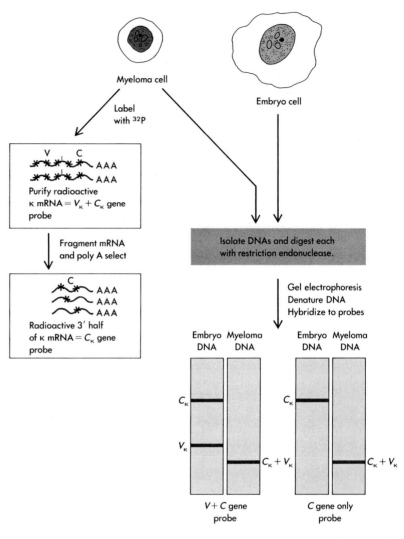

Then two nucleic acid probes for a κ gene were prepared from radioactively labeled κ mRNA isolated from the myeloma (Figure 23-22). One probe, the full-length mRNA, could hybridize to DNA encoding both the variable and constant regions, while the other, consisting of just the 3' half of the mRNA, could only hybridize to DNA coding for the constant region. When each probe was hybridized to the digested embryonic DNA, the full-length mRNA probe detected two different-size fragments (one tentatively the C gene, one the V gene), while the smaller probe detected only one of these (presumably the C gene). In contrast, in the myeloma DNA, the two probes hybridized to the same-size DNA fragment, indicating that in this DNA, the V and C genes were close together, in fact, close enough to lie on the same DNA fragment.

   Similar experiments soon showed that DNA rearrangements are also involved in generating functional genes for λ light chains and for heavy chains. More recently, the same mechanism has been found for T cell receptor genes. So far, these are the only genes we know of in vertebrates that are assembled in somatic cells. All the rest are inherited intact, although, as we have seen, most genes of higher eucaryotes undergo editing at the RNA level to remove their intervening sequences. Assembling genes at the DNA level in specific cell types is a novel form of gene regulation, and, as we might expect, the process is closely tied in with B and T cell differentiation.

Three Unlinked Multigene Families Encode the λ and κ Light Chains and Heavy Chains[34]

If DNA coding for V regions moves near to C region genes in B cells, the question immediately arises as to the physical arrangement of the genes before they are moved. Are they near one another? Do they lie on separate chromosomes? If they lie together, what becomes of the DNA between them when they are rearranged? The first information about the arrangement of antibody genes came from genetic studies. The studies showed that genes coding for λ, κ, and heavy chains lie on separate chromosomes, but that in each case the *V* genes are near to their respective *C* genes.

The genetic studies that first mapped antibody genes were possible because antibody molecules, being proteins, can themselves be antigenic. As a result, it is possible to raise antisera to antibodies, and the antisera can then be used to follow the inheritance of particular antibody types. Antigenic determinants on variable regions are called **idiotypes.** Antigenic determinants that distinguish different types of constant regions (e.g., C_μ from C_γ, $C_{\lambda 1}$ from $C_{\lambda 2}$) are called **isotypes.** And antigenic determinants that distinguish allelic genes are called **allotypes.** By following the inheritance of idiotypic, isotypic, and allotypic determinants on antibody molecules, it was shown that the genes for λ and κ chains and heavy chains are genetically unlinked, but at each locus, *V* genes are linked to their respective *C* genes. Because there are several hundred variable-region genes and because there are vast distances between coding segments, the heavy-chain locus alone probably occupies at least 2000 kilobases of DNA. The λ-, κ-, and heavy-chain loci have been mapped to specific chromosomes in different vertebrate species (Table 23-2).

Light-Chain Genes Are Assembled by *V-J* Joining[35, 36, 37]

The discovery that antibody genes are assembled by DNA rearrangements still did not answer the question of whether antibody V-region diversity is explained by the germ line or somatic theories. This issue was resolved only when molecular cloning techniques made it possible to study the detailed structure of germ line and rearranged antibody genes. Then it was clear that the mechanism of antibody gene assembly generates a great deal of amino acid sequence diversity, while mutation of rearranged genes generates the rest: The somatic theory was the correct one.

Extensive data now exist for the structure of the antibody genes of humans and mice. As we shall see in this and following sections, the general principles for how λ-, κ-, and heavy-chain genes are arranged and how they are assembled by DNA recombination are quite similar. However, only a single DNA recombination event is needed to make a light-chain gene, while two, and sometimes three, DNA recombination events are involved in assembling functional heavy-chain genes. We can begin by describing the structure and assembly of the λ light-chain genes of the mouse, since they are the simplest.

Figure 23-23 shows the structure of a mouse λ light-chain gene as it appears in most cells of the body, as well as the rearrangement it undergoes as it becomes a functional gene in a B cell. The gene is inherited in four segments. Two of these segments are exons, and

Table 23–2 Chromosome Locations of the Heavy-Chain Genes and the λ and κ Light-Chain Genes in Mouse and Human

Polypeptide chain	Chromosome	
	Human	*Mouse*
H	14	12
κ	2	6
λ	22	16

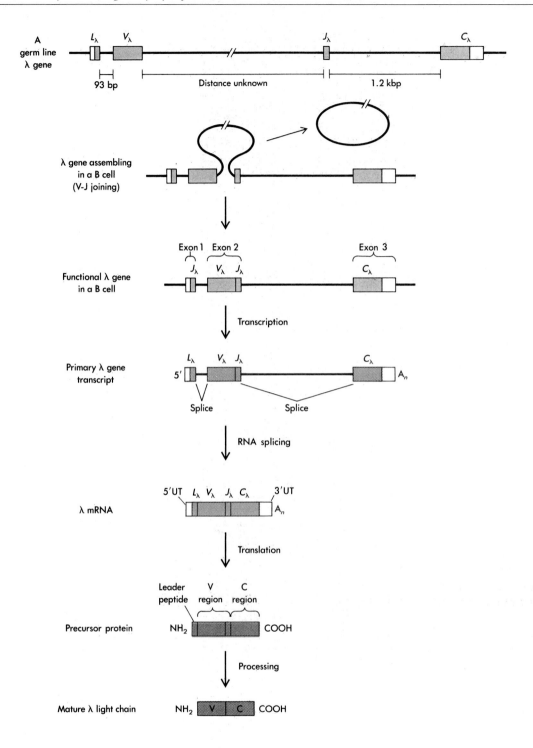

Figure 23-23
Diagram of a λ light-chain gene as it appears in germ line DNA, and the DNA rearrangement that could occur to produce a functional gene in a B cell. Except for the gap between *V* and *J*, the gene is drawn to scale. Boxes in DNA and RNA molecules represent exons or parts of exons. White portions of exons encode 5′ or 3′ untranslated regions of the final mRNA. These are designated 5′ UT and 3′ UT, respectively. L_λ encodes amino acids -19 to -5 of the leader peptide; V_λ encodes amino acids -4 to 0 of the leader peptide and 1 to about 98 of the variable region; and J_λ encodes amino acids 99 to 110 of the variable region.

two are incomplete fragments that must be brought together in the DNA to complete a third exon. The C_λ segment is an exon that encodes the complete constant region of the chain and also includes sequences 3' to the coding region that are transcribed to yield a 3' untranslated region (of unknown function) on the λ mRNA. The other three segments, L_λ, V_λ, and J_λ, are involved in specifying the variable region. Most important are the V_λ and J_λ segments. These two fragments must be brought together at the DNA level, a process known as **V-J joining,** to create a functional exon and thus a functional antibody gene. During this rearrangement, the J_λ DNA segment and the C_λ exon retain their relative configurations. The additional exon, L_λ, which lies just 93 bases upstream from the V_λ segment, also maintains its relationship to V_λ during the rearrangement. This is important, since a transcriptional promoter for the gene lies just in front of the L_λ segment. L_λ specifies a 5' untranslated region of the λ mRNA followed by sequences that encode most of a 15–20 amino acid–long hydrophobic leader peptide that guides the nascent light chain through the membrane of the rough endoplasmic reticulum but is then promptly cleaved from the chain. V_λ encodes the last four amino acids of the leader peptide and the N-terminal 98 amino acids of the variable region. J_λ encodes the last 12 or so amino acids of the approximately 110–amino acid–long V_λ region.

It is apparent from Figure 23-23 that the newly assembled V-J exon is still separated from the single C_λ exon by about 1.2 kilobase pairs. This intervening sequence is like any other conventional intron and is removed from the primary λ gene transcript by RNA splicing. The 93-base gap between L_λ and the V-J exon is also removed at the RNA level by splicing.

In the description just given (and in Figure 23-23), we have implied that the L_λ and V_λ segments lie upstream from the J_λ and C_λ segments. While this is almost certainly the case, the distance between the two pairs of segments is so large that it has not been possible to clone a fragment of DNA from mouse (or human) embryonic cells carrying all four segments. (This has been achieved for a λ gene of the chicken, however, where the distance from V_λ to J_λ turns out to be much shorter than it is in the mouse or human.)

So far, we have described only a single mouse λ light-chain gene. The complete mouse λ light-chain locus has the structure shown in Figure 23-24. There are two clusters of J and C segments, although the distances between them are not known. In agreement with amino acid sequences of mouse λ light chains, there are three functional C_λ exons. A fourth C_λ exon is present in the DNA, but it has suffered mutations that make it nonfunctional.

What happens to the DNA between a V and J segment when the two are brought together by recombination? In many cases, it can be shown that the DNA is lost as if it had been excised during the recombination process. How this might occur is indicated in Figure 23-23.

We have already noted that an antibody-producing plasma cell synthesizes only one light chain. Once a DNA rearrangement generates a functional light-chain gene (e.g., a λ_1 gene), further rearrangements of light chain genes are apparently inhibited.

The strategy for constructing functional κ light-chain genes is just the same as for λ genes, but the overall structure of the mouse κ multigene family is quite different from that of the mouse λ locus. Figure 23-25 shows the κ gene family. Most important, there are at

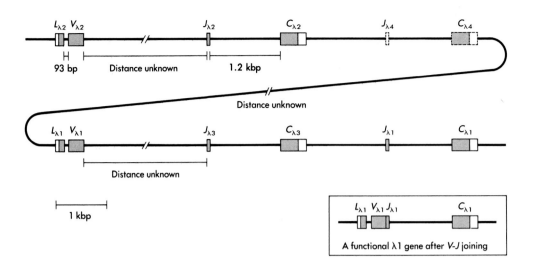

Figure 23-24

The probable structure of the mouse λ light-chain locus in its germ line configuration. The relative positions of the two *V-J-C* clusters are not known with certainty. Except for the unknown distances, the figure is drawn to scale. The structure of one possible light-chain gene after DNA rearrangement is shown. Note that $V_{\lambda 1}$ can also join with $J_{\lambda 3}$ to make a functional λ_3 light-chain gene. Boxes represent exons or parts of exons. White portions of some exons encode 5' or 3' untranslated regions of mRNA.

least several hundred V_κ segments, each with its adjacent L_κ exon. Furthermore, there are five J_κ segments that lie between the V_κ segments and the single C_κ exon (although one of the five J_κ segments is defective and cannot be used to make a functional gene). Clearly, the possibility for making different κ light chains is very great. This is because (as far as we know) within a particular cell, any one of the V_κ DNA segments may be joined to any J_κ segment. If we estimate that there are 300 V_κ segments, then 300 $V \times 4 J$ = 1200 possible V_κ exons that can be generated by V-J joining. The random joining of any V to any J is called **combinatorial joining.** It is an important source of diversity in the immune system, both in the κ genes and, in a more limited way, in the mouse λ light-chain genes. In humans, both the λ and κ loci resemble the mouse κ locus with its multiple V segments. It seems that inbred mice may have suffered a genetic catastrophe in which they simply lost many of their λ V-coding fragments.

The hundreds of V_κ segments can be divided into subfamilies according to their nucleotide sequence homology. These relationships are shown by preparing radioactively labeled copies (probes) of different V_κ segments and hybridizing them one at a time to mouse DNA that has been digested with a restriction endonuclease and separated by gel electrophoresis (the technique known as Southern blotting). Each V_κ probe will usually hybridize to five to ten different-size fragments of cellular DNA.

Heavy-Chain Variable-Region Exons Are Assembled by Two DNA Recombination Events: *D-J* Joining and *V-(D-J)* Joining[38, 39, 40]

Not surprisingly perhaps, the basic strategy for constructing functional heavy-chain genes is similar to that used at the light-chain loci. But there are additional complexities that give heavy-chain polypeptides their special characteristics. The structure of the mouse heavy-chain multigene family is shown in Figure 23-26. Three types of segments, V_H, D, and J_H, are joined at the DNA level to generate a *V-D-J* exon that encodes the mature heavy-chain variable region. *D* stands for diversity, since this segment encodes the third hypervariable re-

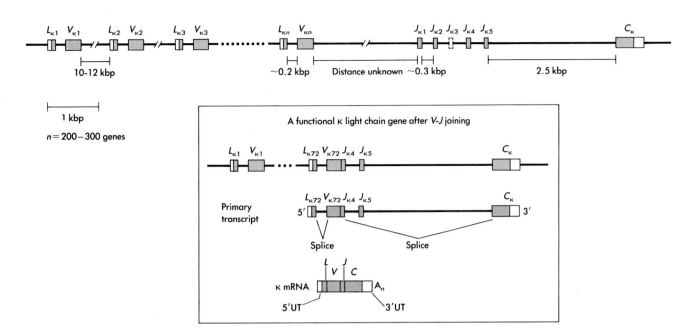

gion, long known to be particularly variable in amino acid sequence and in length from one heavy chain to another (see Figure 23-14). There are several hundred V_H segments. As with the light-chain genes, each has an attendant L_H exon that encodes the hydrophobic leader peptide. These L_H segments do not contribute to diversity, of course, since the leader is removed during maturation of the polypeptide. There are 10 to 12 D segments that encode between 3 and 14 amino acids, and there are 4 to 6 J segments that each encode about 15 amino acids. To create a functional heavy-chain gene, one D segment, presumably selected at random, is first joined to a J_H segment. Then a second DNA recombination event joins a V_H segment to the D-J_H piece. If we take the current, and perhaps conservative, estimate of 250 V_H segments, then assuming that combinatorial joining will allow all combinations of V_H-D-J_H, we can calculate that random joining of these segments can generate approximately 250 $V \times 10\ D \times 4\ J =$ 10,000 different V-D-J exons for heavy-chain variable regions!

The heavy-chain V_H-coding segments probably lie upstream of the constant region exons, which in turn are quite close to the J_H segments (see Figure 23-26). As you will recall, there are five major types of heavy-chain constant regions—α, δ, ε, γ, and μ—and some of these types have additional subtypes (γ_1, γ_2, etc.). Immunologists had long suspected a separate germ line "gene" for each type, although as we now know, these "genes" are really exons. In contrast to the much smaller light-chain C regions encoded by single exons, the heavy-chain C regions are each coded for by 5 to 7 exons. How an assembled V-D-J exon chooses to associate with a particular heavy-chain constant region will be discussed later in the chapter. In the example shown in Figure 23-26, a V-D-J exon is using C_μ, whose exons lie closest to the V_H genes. In this case, a primary transcript containing the V-D-J and C_μ exons is spliced to produce a functional heavy-chain mRNA. As we shall see, association of a V-D-J exon with some C_H "genes" requires that DNA recombination first bring the two closer together. The heavy-chain constant-region "genes" occupy more than 150 kilobase pairs of DNA, and the distance between adjacent "genes" can be as great as 55 kilobase pairs.

Figure 23-25
The κ light-chain locus and a functional κ light-chain gene. There are 200 to 300 V_κ segments, each with an attendant L_κ exon, five J_κ segments (of which one is inactive in the mouse), and one C_κ exon. Boxes represent exons or parts of exons. White portions of boxes encode 5' or 3' untranslated regions of mRNA. Except for the gaps, the figure is drawn approximately to scale. Note that the $L + V$ segments may be spread over 2000 to 3000 kilobase pairs of DNA, since the distance between at least some of these segments is 10 to 12 kilobase pairs.

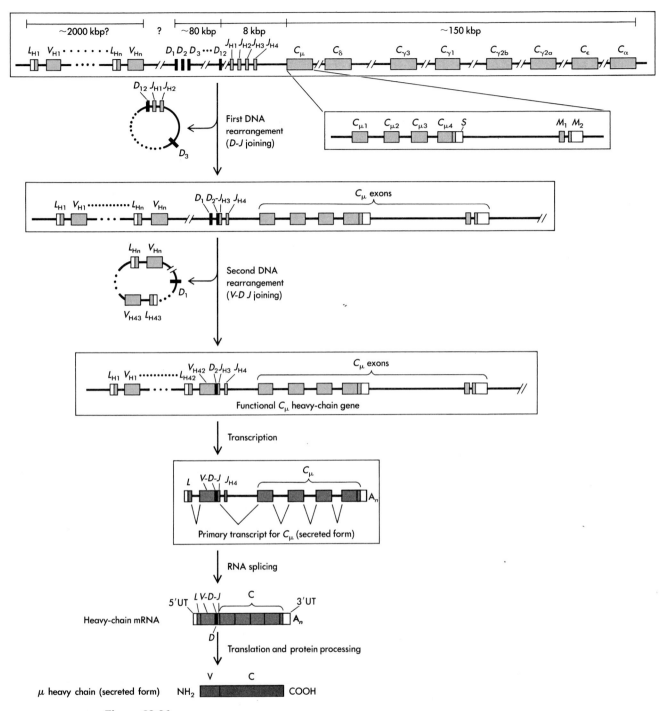

Figure 23-26

The heavy-chain locus as it appears in germ line DNA and as it appears after *D-J* and *V-D-J* DNA rearrangements that produce a functional heavy-chain gene in a B cell. Each different heavy-chain class has a constant region that is encoded by multiple exons, but for the sake of simplicity and space, only the exons of C_μ are shown in detail. In the figure, the DNA between *D* and *J* or between *V* and *D-J* segments that join is shown as being circularized and discarded as part of the recombination process, but this is only hypothetical. As will be discussed later, the carboxyl terminus of the membrane-bound and secreted forms of μ heavy chains are encoded by different exons, and different RNA-processing patterns determine which type of protein will be made. In the figure, a secreted μ chain is being made. Because of the vast size and complexity of the heavy-chain locus, the figure is not drawn to scale; but some distances are indicated.

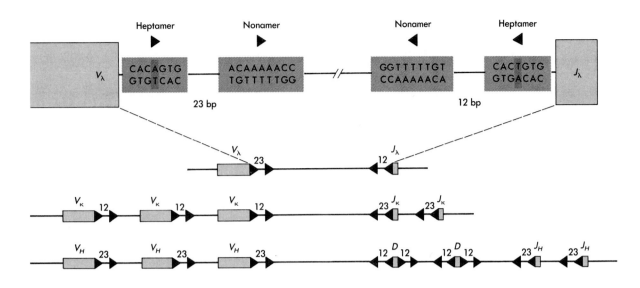

The heavy-chain V_H DNA segments, like the V_κ segments, can be divided into subfamilies based on their ability to hybridize to one another. As we discussed earlier in the chapter, antibody genes must arise by successive gene duplications. Presumably, V_H segments that belong to homologous families represent more recent duplications in evolutionary time, while families whose nucleotide sequences are less closely related diverged long ago.

Heptamer-Nonamers and the 12-23 Spacer Rule Govern *V-J* and *V-D-J* Joining[41, 42]

A specialized recombination system is probably needed for the unusual task of joining *V-J* and *V-D-J* DNA segments to make light- and heavy-chain genes. Based on other types of specialized recombination, we might expect to find unique nucleotide sequences serving as recognition signals in the system. Special sequences are indeed found at the borders of germ line *V*, *D*, and *J* segments, just near the sites where recombination events can take place. Figure 23-27 shows typical sequences that flank V_L and J_L segments and V_H, *D*, and J_H segments of light- and heavy-chain genes, respectively. What can be seen in this picture, and confirmed by analyzing many additional such sequences, is that each segment border that can participate in recombination has two highly conserved sequences, one seven bases long (a **heptamer** with the consensus sequence CACAGTG) and the other nine bases long (a **nonamer** with the consensus sequence ACAAAAACC), that are separated by a short nonconserved sequence. (Note that the heptamer sequence is a palindrome around the central AT base pair.) At two edges that can be joined (e.g., V_λ and J_λ in Figure 23-27), the conserved sequences are related to one another by a 180° rotation. Also very striking is the length of the nonconserved sequence that lies between the haptamer and the nonamer. When two segments can join, one always has a 12 ($\pm$1)-nucleotide–long spacer between its conserved heptamer and nonamer, while the other has a 23 ($\pm$2)-nucleotide–long spacer. Thus, three specific features characterize the sequences that specify *V-J* and *V-D-J* joining: the presence of the conserved heptamer and nonamer sequences, the orientation of the heptamer-nonamer to the segments

Figure 23-27
Nucleotide sequences that specify joining of V_L-J_L and V_H-D-J_H DNA segments in antibody genes. The top line of the figure is a close-up view of nucleotide sequences at the borders of a *V* and *J* segment that could be joined to make a functional λ light-chain gene. Note that the segments are bordered by a palindromic heptamer that is oriented in opposite directions at the two edges that will be joined. The heptamers are followed by either 23 or 12 nucleotides, whose sequence is not important, and then by a nonamer sequence. Like the heptamers, the nonamers are oriented in opposite directions. Color and black triangles represent the heptamer and nonamer, respectively, with the direction of the triangle giving the orientation of the sequence. Below the enlargement showing the sequence are abbreviated diagrams of the λ, κ, and heavy-chain V-region-coding DNA. Color and black triangles represent the positions of heptamers and nonamers that flank these sequences. The 12- or 23-base-pair spacing between them is indicated. These heptamer and nonamer sequences are consensus sequences, and the actual sequences can vary slightly from those given.

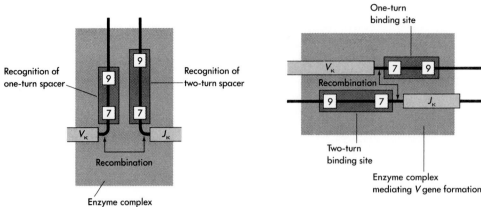

Figure 23-28
Two potential configurations of the heptamer-nonamer recognition sequences for the enzyme-mediated recombination of *V-J* and *V-D-J* DNA segments. [After *Immunology*, 2d edition by L. E. Hood, I. L. Weissman, W. B. Wood, and J. H. Wilson. 1984 by The Benjamin/Cummings Publishing Co.]

being joined, and the length of the spacer between the heptamer and nonamer.

According to the rules just described, we can see from the configuration of sequences that lie on the 3' sides of V_H segments, on both sides of D_H segments, and on the 5' sides of J_H segments why V_H segments never mistakenly skip over D_H sequences to join directly with J_H segments (see Figure 23-27). V_H-J_H joining would break the 12-23 spacer rule, while joining the 5' side of *D* to J_H would require recombination between incorrectly oriented heptamer-nonamers. However, it is also apparent that not all undesirable recombinations are prohibited by these rules alone. For example, recombinations could occur between segments belonging to different gene families (λ and κ, heavy and light). Perhaps this is why the three gene families lie on different chromosomes: to decrease the possibility of such a mistake.

The lengths of the spacers between the heptamers and nonamers, 12 and 23 base pairs, correspond approximately to one and two turns of a DNA double helix. This has led to the idea that perhaps it is a combination of specific sequences and three-dimensional arrangement that is recognized by the putative antibody gene recombinase. Figure 23-28 shows some models that have been drawn to help clarify the nature of the substrate for this still-hypothetical enzyme.

Variation in the Precise Positions of *V-J* and *V-D-J* Recombinations Contributes Further to Antibody Diversity (Junctional Diversity)[43, 44]

When amino acid sequences of the variable regions of, for example, different κ light chains are compared with those predicted from nucleotide sequences of V_κ and J_κ DNA, it is apparent that there is much more amino acid sequence variability at the position encoded by the *V-J* joint than we would expect. Further scrutiny of the nucleotide sequences reveals that the extra variability can be accounted for very easily if the recombination events that join *V* and *J* segments do not occur at a precise nucleotide relative to the heptamer-nonamer recognition sequence. An example of the flexibility of *V-J* joining is shown in Figure 23-29 for a V_κ-J_κ pair. This one joining can occur in four different ways to generate four different *V-J* nucleotide sequences that would encode three different amino acid sequences. When we

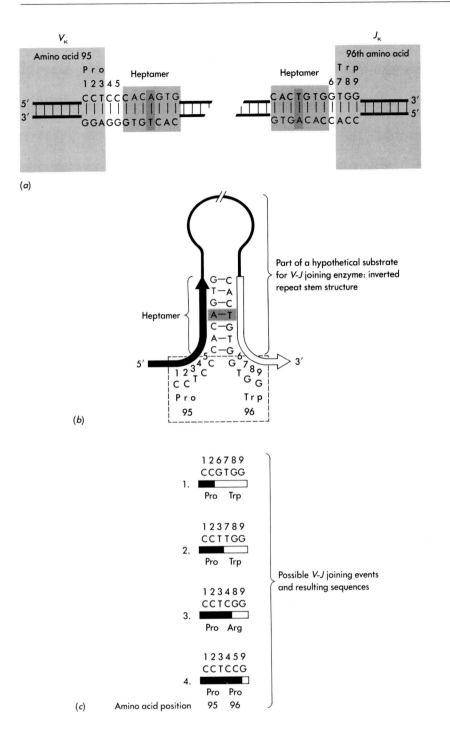

(a)

(b)

(c)

Figure 23-29

An example of how variation in the precise position of *V-J* joining generates amino acid sequence variation (junctional diversity) at the position of the joint. (a) Edges of a V_κ and a J_κ segment that are about to be joined together. (b) One DNA strand of a hypothetical substrate for the *V-J* joining enzyme. The heptamer sequence is being used to generate a stem structure. Nucleotides that can participate in the recombination are numbered. (c) The four possible nucleotide sequences that can be generated by recombination events occurring at slightly different positions result in three different amino acids at position 96 in the mature K light chain. [After Gottlieb, P. D. *Molec. Immunol.* 17 (1980): 1423; based on H. Sakano et al., *Nature* 280 (1970):288 and E. E. Max, et al., *Proc. Nat. Acad. Sci.* 76 (1979):3450.]

recall that there are five J_κ segments and that each of these can probably undergo joining with any of several hundred V_κ segments, it is immediately apparent that variability in *V-J* joining, called **junctional diversity,** can make an important contribution to antibody diversity. Junctional diversity is found at both the κ and λ light-chain loci. The DNA at *V-J* joints encodes a portion of the third hypervariable, or CDR, region of light chains, a portion long known from amino acid sequence data to be the most variable region of the chain and also known to form a part of the antigen binding cavity.

Junctional diversity also occurs in heavy-chain genes, both at V_H-D joints and at D-J_H joints. However, in the case of heavy chains, even

Figure 23-30

Model for *D-J* joining, showing how N nucleotides may be added at the joint in a template-free fashion. (1) Two segments that are to be joined are lined up with the help of the hypothetical DNA-joining enzymes. Nicks are introduced into all four DNA strands at the ends of the heptamers adjacent to the coding sequences (positions are indicated by arrows). (2) The two heptamers are joined end-to-end. The coding sequences do not join yet, but are held near to one another by the joining proteins. (3) Some nucleotides may be removed from the *D* and *J* ends by an exonuclease. (4) One or more nucleotides are added at the *D* and *J* ends in a template-free fashion. This reaction may be carried out by terminal deoxynucleotidyl transferase, an enzyme that can polymerize random deoxynucleotides at 3' ends. (5) Joining is completed by a DNA polymerase that copies the added nucleotides and by a ligase that seals the structure. [After F. Alt and D. Baltimore, *Proc. Nat. Acad. Sci.* 79 (1982):4118.]

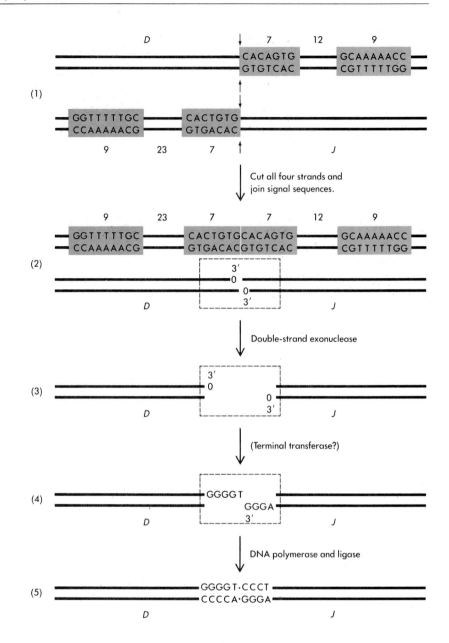

this mechanism of increasing variability can not account for all the diversity seen in V_H regions at the positions corresponding to the joints. There are cases where extra nucleotides seem to have been inserted between the V_H-D or D-J_H segments. These extra nucleotides are called **N regions.** Presently, it is thought that N nucleotides may be added in a template-free fashion as the gene is being assembled (Figure 23-30). An enzyme called terminal transferase, with the capacity to add nucleotides to the 3' end of a DNA strand, is known to exist, so perhaps it or some other enzyme with similar ability is involved. In heavy chains, the most variable part of the V region is the third hypervariable segment. Appropriately, this turns out to be encoded by the D segment and by the V_H-D and D-J_H joints.

The variability in the precise position of V-J and V-D-J joining leads to some recombination events that generate stop codons in the rearranged genes. Presumably, these genes are not functional. Such inactive rearrangements are seen frequently in antibody-producing cells. Apparently, the cell keeps on rearranging its DNA until a functional gene is formed. Presumably, it is worth this extra effort in order to have the extra variability that can be generated by junctional diversity.

Mutation of Fully Assembled Antibody Genes Further Increases Diversity[45-48]

From the preceding discussion, we might conclude that molecular biologists can now predict the complete amino acid sequence of every possible mouse λ light chain and, if they had the energy to clone and sequence enough DNA, every κ light chain as well. (Of course, even if they cloned and sequenced all the heavy-chain genes, they still could not predict all the heavy chains an animal can synthesize if indeed there is template-free addition of nucleotides into the genes.) However, when we compare the predicted amino acid sequences, for example, of mouse λ_1 light chains with the actual amino acid sequences of mouse λ_1 light chains, there are still more amino acid differences in the variable regions than can be accounted for by *V-J* joining and junctional diversity at the *V-J* joints. The additional diversity consists of single amino acid differences scattered here and there, and they can usually be explained by the introduction of single-base changes in the *V-J* exon. Additional experiments have shown that point mutations are also introduced into fully assembled V_H genes. This source of diversity is called **somatic hypermutation.** The mechanism is unknown, but remarkably, antibody genes that have been subjected to it only carry mutations in a limited region encompassing the variable-region exon, with their constant-region exons apparently not subjected to this mutational process.

There are biologically compelling reasons why somatic hypermutation may be advantageous. First, antibodies that appear during the later stages of an immune response frequently have a higher affinity for the immunizing antigen than do the antibodies found during the primary response. If the first B cells triggered by antigen binding undergo somatic hypermutation of their rearranged antibody genes as they divide, then starting from a set of antibodies that are *already* able to recognize the incoming antigen, one may more readily generate antibodies with an "improved fit" (higher affinity) for the antigen. Experimental evidence supports the idea that mutation of fully assembled antibody genes does indeed generate antibodies with higher affinity for an immunizing antigen.

A second reason why somatic hypermutation may be highly advantageous is that despite the vast number of different antibodies an animal can make by all the possible arrangements of its germ line information, this is still a fixed number, predetermined at birth. In contrast, the world of antigens, which consists of living microorganisms, is continually changing as the organisms grow and mutate. To keep up with the changing world of antigens, perhaps we need an immune system that can also change.

Mechanisms of Antibody Diversity Reviewed[49, 50]

For two decades, immunologists puzzled over how an organism could be genetically programmed to produce perhaps as many as a million different antibody molecules. Now that we know the answer,

Table 23-3   The Sources of Diversity in Mouse Antibody Molecules

	Heavy Chains	Light Chains	
		κ	λ
1. Multiple germ line gene segments for variable regions			
V	200–300 V_H	100–300 V_κ	2 V_λ
D	10 D	0	0
J	4 J_H	4 J_κ	3 J_λ
2. Combinatorial (random) joining of V-J or V-D-J segments	$250 \times 10 \times 4 = 10,000$	$250 \times 4 = 1000$	$2 \times 3 = 6$
3. Junctional diversity (variability in position) in V-J or V-D-J joining	$10,000 \times 3 \times 3 = 90,000$	$1000 \times 3 = 3000$	$6 \times 3 = 18$
4. N nucleotide addition at V-D-J joints	>90,000	—	—
5. Somatic hypermutation of rearranged genes	>>90,000	>3000	>18
6. Combinatorial joining of any light chain with any heavy chain	(>>90,000) $\times$ [>3000 + >18] =		
	>>>>2.7×10^8 possible antibodies		

Numbers of germ line V_H and V_κ segments are approximate. The calculations in the table are based on the assumption that any V, D (if present), or J at a particular locus can be joined, and also that any light chain can associate with any heavy chain to make a functional antibody molecule.

it is apparent that researchers had seriously underestimated the magnitude of the problem! Using the strategies described in the preceding sections, we can calculate that the average mouse or human can synthesize more than 200 million different antibody molecules, even before somatic hypermutation goes to work on the genes. Given somatic hypermutation as well, the number of possibilities becomes almost unlimited.

Combinatorial diversity provided the first simplification of the diversity problem. Since two polypeptide chains form the antigen binding cavity, if we assume that any light chain can associate with any heavy chain, the number of antibody molecules with different active sites is equal to the number of different light chains multiplied by the number of different heavy chains. So the question of antibody diversity becomes the number of possible light- and heavy-chain variable regions an animal can make. (Since constant regions do not contribute to the diversity of antibody specificities, they are excluded from this discussion.) For the mouse, the calculations are as follows and as shown in Table 23-3.

First let us consider the light chains. At the κ locus, there are estimated to be 100 to 300 V_κ segments and 4 functional J_κ segments. Using the number 250 for V_κ segments, the random association of V_κ segments and J_κ segments (combinatorial joining) would yield 1000 different V_κ regions. If we estimate that variability in the precise point of V-J joining (junctional diversity) can lead to three different amino acids at this position, then the number of V_κ genes rises to 3000.

There are only 2 V_λ segments and 3 functional J_λ segments in the mouse; so considering combinatorial joining and junctional diversity, there are 18 possible V_λ-coding sequences, too small a number to contribute significantly in our calculations. (It should be noted that most antibodies in a mouse have κ rather than λ light chains.)

For heavy chains, there are about 250 V_H segments, 10 D segments, and 4 functional J_H segments. Combinatorial joining of these segments would produce 10,000 V_H-region genes. If we add in the three amino acids for junctional diversity at both the V-D and D-J joints, we can calculate the possible number of V_H genes as $10,000 \times 3 \times 3 =$

90,000. This is probably an underestimate of the number of possible V_H sequences, since it does not account for the greater junction variability that can arise from template-free addition of N nucleotides at the *V-D* and *D-J* borders.

Multiplying the 90,000 heavy chains by the 3000 light chains, we get 2.7×10^8 possible antibody molecules. Since somatic hypermutation can introduce any of 20 different amino acids into many different sites in the V regions of these molecules, it is apparent that the number 2.7×10^8 is only a small fraction of the total number of antibodies an animal is likely to synthesize in its lifetime.

For many years, immunologists argued over whether antibody diversity is encoded in the germ line or generated by genetic events in somatic (B) cells. Now it is clear that aspects of both theories are correct, and the question has more or less been laid to rest. Still, it is significant that even though there are many germ line *V*-region segments, there are still only a few hundred for light chains and a few hundred for heavy chains. It is what happens to these segments in B cells that turns approximately 300 light-chain *V* and *J* segments into approximately 3000 V_L segments, and about 300 heavy-chain *V*, *D*, and *J* segments into about 90,000 V_H segments. Also, the fact that antibodies are made of two chains gives these 93,000 chains the potential to form more than 200 million different antibody molecules.

A Tissue-Specific Transcription Enhancer Element Is Located in the Intron Between *J* and *C* DNA[51, 52, 53]

In a plasma cell synthesizing antibodies, about 10 percent of the cell's mRNA is heavy-chain message. Unrearranged V_H segments are not transcribed; only the V_H segment that has been joined to *D-J* to make a functional gene is active. The DNA rearrangements that bring V_H segments and their adjacent L_H exons next to D-J_H segments do not alter the promoter elements that lie in front of the L_H exons, so the question arises as to why juxtaposition of *V-D-J* segments leads to such a high level of gene expression. An important part of the answer came from experiments that showed that this rearrangement brings the promoter of the chosen *V* segment near to a powerful enhancer element that is located between the *J* and *C* segments.

When a molecularly cloned rearranged heavy-chain gene was introduced by DNA transfection into a mouse myeloma cell, the introduced gene was transcribed very efficiently. This made it possible to experimentally delete portions of the gene to see which part was important for the high level of transcription. It turns out that an important set of sequences, about 200 nucleotides long, lies between *J* and *C* (Figure 23-31). This sequence has properties of the transcriptional enhancer elements that had already been discovered in the promoter regions of DNA and RNA tumor viruses: It can increase transcription when it is placed in either orientation relative to a gene, and it can influence transcription from a promoter located several kilobase pairs away. This would be a long enough range to enhance transcription from the promoter of a rearranged *V* gene, but not from the promoters lying next to *L* and *V* segments that are still in their germ line configuration. Moreover, the heavy-chain enhancer element is tissue specific: Although it enhances transcription dramatically in lymphoid cells of the B lineage, it does not seem to work in fibroblasts at all.

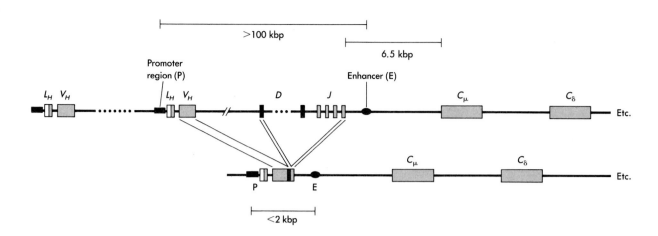

Figure 23-31
DNA rearrangements that join *V-D-J* to make a heavy-chain *V* gene bring the transcriptional promoter for the rearranged *V* gene under the influence of a powerful transcriptional enhancer element located in the intron between *J* and *C*. Note that before the DNA rearrangements, the nearest promoter to the enhancer element is at least 100 kilobase pairs away; but after *V-D-J* joining, the needed promoter is within about 2 kilobase pairs of the enhancer.

In addition to the enhancer, a promoter element lying 5′ of each *L* segment is also important for heavy-chain gene transcription. Recently, this element has also been shown to have tissue specificity, functioning more efficiently in B cells. A critical sequence in the element is a conserved octamer, with the consensus sequence ATGCAAAT, that lies just upstream of the TATA boxes in front of many L_H exons (see Figure 23-31). Interestingly, an octamer with the complementary sequence (ATTTGCAT) is present in the promoter regions of light-chain *V* genes. In addition, the κ light-chain locus, like the heavy-chain locus, has been found to contain an enhancer element in the intron separating the $J_κ$ cluster and the κ constant-region exon.

Antibody Gene Rearrangements Occur in a Developmentally Regulated Fashion[54, 55]

B cell differentiation culminates in a plasma cell that synthesizes a single type of light chain and a single type of heavy chain. This means that although vertebrates are diploid, only one of the two heavy-chain loci and one of the four light-chain loci are used in a given cell, a phenomenon known as **allelic exclusion,** which is unique to antibody genes (Figure 23-32). Antibody gene rearrangements follow a carefully ordered sequence that is intimately tied to, and in fact, helps to define, stages of B cell differentiation. V_H segments are rearranged first, then κ light-chain genes. At least in the mouse, it appears that λ light-chain genes are assembled only if rearrangements at the κ locus fail to produce a functional κ light chain. The first antibodies a B cell produces are membrane-bound IgM molecules. Later, the process of class switching will join the same heavy-chain *V-D-J* exon to other C_H-region exons. Researchers can study mouse cells at different stages of differentiation by using cancer viruses to transform the cells into permanently growing cell lines that are often frozen at a particular stage along the developmental pathway.

The developmentally regulated rearrangements and expression of antibody genes are diagrammed in Figure 23-33. In the **null cells,** the earliest cells to show evidence of antibody gene rearrangements, heavy-chain genes have begun to rearrange, but only *D-J* recombinations have occurred (see Figure 23-33). Although only one chromosome will ultimately carry a functional heavy-chain gene, both chro-

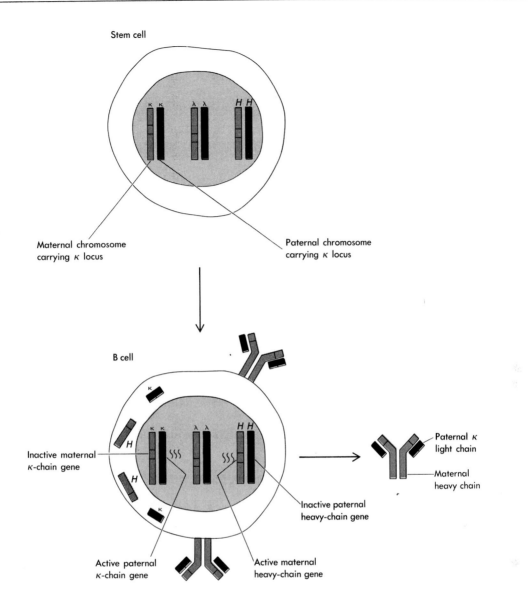

Stem cell

Maternal chromosome
carrying κ locus

Paternal chromosome
carrying κ locus

B cell

Inactive maternal
κ-chain gene

Active paternal
κ-chain gene

Active maternal
heavy-chain gene

Inactive paternal
heavy-chain gene

Paternal κ
light chain

Maternal
heavy chain

mosomes undergo D-J_H rearrangements. The next cell in the pathway is called a **pre–B cell.** Immature pre–B cells undergo V_H to D-J_H rearrangements. As soon as a functional V-D-J exon is generated, further heavy-chain gene rearrangements stop and the cell synthesizes heavy chains with a μ constant region. These chains remain in the cytoplasm until light chains are made. Once heavy-chain gene expression has begun, the μ-positive pre–B cell goes on to rearrange its κ light-chain genes. If a functional κ gene is made, apparently all V-region DNA rearrangements cease. Then the κ and μ chains assemble and move to the membrane, where they are anchored and serve as receptors for antigens. If rearrangements at the two κ loci fail to produce a functional gene, then λ-chain genes undergo rearrangement, and if successful, the B cell will have receptor molecules that are λ-μ tetramers. (Note that although secreted IgM antibodies are made up of five heavy-light-chain tetramers, membrane-bound IgM antibodies are simple tetramers.)

All the events described in the preceding paragraph occur independently of antigen stimulation. The end result is to generate a set of **virgin B$_\mu$ cells,** whose membrane-bound IgM receptors probably

Figure 23-32

Allelic exclusion at the antibody loci. A particular B cell synthesizes just one type of heavy chain and one type of light chain. Once a functional gene for a heavy chain (and then for a light chain) is assembled, further DNA rearrangements are inhibited, perhaps by the end product of the rearrangement. For example, synthesis of a μ heavy-chain polypeptide may inhibit further DNA rearrangement at the other heavy-chain locus.

Figure 23-33
A model for B cell differentiation. There is experimental evidence for the different steps in the picture, except for the progression from B_μ to $B_{\mu+\delta}$ and back to B_μ and for the possible role of membrane-bound IgD in signaling the elimination of self-reacting cells. [After *Immunology*, 2d edition by L. E. Hood, I. L. Weissman, W. B. Wood, and J. H. Wilson. 1984 by The Benjamin/Cummings Publishing Co.]

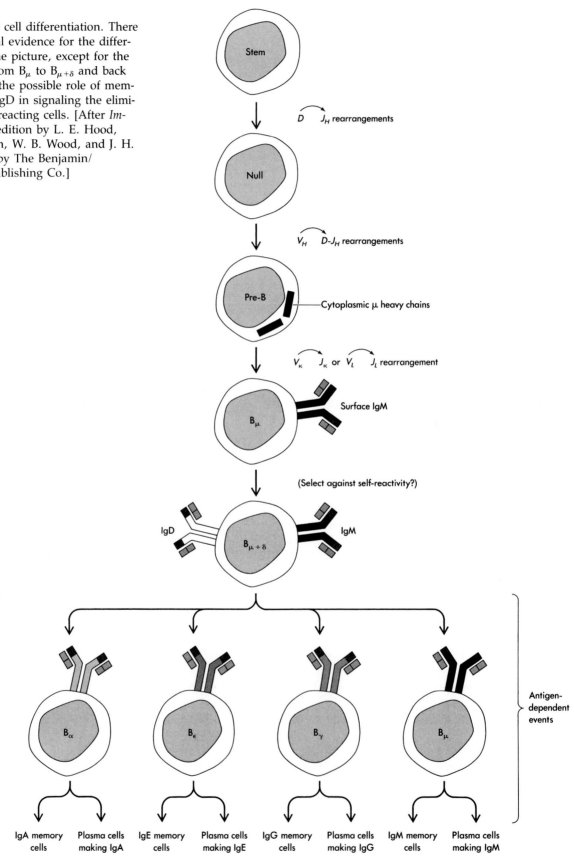

include the complete repertoire of antibody V regions an animal can generate before somatic hypermutation intervenes. The next step in B cell maturation is a $B_{\mu+\delta}$ cell, which produces both IgM and IgD antibody receptors. The two types of molecules have identical V regions, differing only in their heavy-chain constant regions. As we shall see, they arise by changes in RNA processing of the heavy-chain gene. Finally, in response to antigen binding, the $B_{\mu+\delta}$ cells differentiate into **blast cells,** which divide and secrete IgM and IgD antibodies. These cells also retain IgM and IgD as membrane-bound receptors. The activated blast cells then differentiate to yield short-lived plasma cells and long-lived memory cells.

Plasma cells secrete antibodies at the rate of about 10,000 molecules per cell per second. Because of class switching, which can join any heavy-chain V region exon to any heavy-chain constant region, a single B cell can give rise to plasma cells that synthesize antibodies of any class. Memory cells do not secrete antibodies, but rather keep them as membrane-bound receptors. Memory cells can also have antibodies belonging to any heavy-chain class. The memory cells will provide a prompt response during future encounters with their cognate antigens.

As discussed earlier, the differentiation of a resting B cell to a miniature antibody factory, the plasma cell, involves a massive expansion of the cell's cytoplasm and of the rough endoplasmic reticulum, where antibody synthesis occurs (see Figure 23-5).

Partially Rearranged Heavy-Chain Genes Are Transcribed and Translated to Yield D-J-C Products[56, 57, 58]

An exciting model to explain the ordered rearrangement of antibody genes and allelic exclusion holds that the protein product of the first rearrangement is involved in regulating the second rearrangement, that the product of the second rearrangement controls the third, and so on. For example, the synthesis of a complete μ heavy chain may signal the cell to stop further rearrangements of heavy-chain genes, while synthesis of a light chain that can combine with the cytoplasmic μ heavy chains to form a complete antibody molecule may put an end to light-chain gene rearrangements. This model could explain allelic exclusion, since gene rearrangements would be blocked as soon as one functional gene of each type was made. Support for this model comes from transgenic mice in which a functional κ light-chain gene was introduced into mouse embryo cells. In the B cells of adult animals that express the introduced gene, endogenous κ-chain genes do not undergo DNA rearrangement. Apparently, the cell is fooled into thinking it has already rearranged its light-chain genes, since a functional light chain is present.

How could the model explain the fact that *D-J* rearrangements at the heavy-chain locus must occur before V_H segments can join to *D-J* segments? In other words, could a gene product with a regulatory role be generated as a result of *D-J* joining? Recently, RNA species that contain D, J_H, and C_μ sequences but lack V_H have been found in immature B cells. Thus, there must be transcriptional promoters between V_H and D segments. Furthermore, examination of the nucleotide sequences of these transcripts shows that there are AUG (methionine initiation) codons near their 5' ends, and, at least in some cases, these are the start of open reading frames that extend through D, J_H, and C_μ. In fact, D-J-C polypeptides have been detected in some types of mouse B lymphoid tumors that have undergone D-J but not yet V-D-J recombination. Conceivably, once these D-J-C_μ products are synthesized, they signal to the cell that now it is time to begin joining V_H to D-J segments.

At first it was believed that V-region segments are transcriptionally dormant until they are joined to D-J segments and thereby brought near to enhancer elements. Now, however, transcription of unrearranged V_H segments has been detected, although the level of transcription is lower than that of complete heavy-chain genes. V_H transcription occurs in pre–B cells that have made D-J rearrangements and are at the stage of undergoing V_H to D-J_H rearrangements. Once a successful V-D-J rearrangement has occurred, however, the small germ line V_H transcripts are shut off. It is not known why transcription of the V_H segments is temporally related to V to D-J rearrangements. Perhaps the transcription simply reflects the presence of open (and hence active) chromatin, which could be an essential substrate for the putative recombinase involved in V-D-J DNA rearrangements. Alternatively, the small V_H transcripts could be translated to yield V_H polypeptides that play some part in the V to D-J DNA rearrangement.

Expression of D-J-C_μ transcripts does not show allelic exclusion; rather, transcripts from both chromosomes can be present in the same cell. Apparently, allelic exclusion occurs only after a V_H segment has successfully joined to the DNA of a rearranged D-J segment to generate a complete antibody gene.

Differential RNA Processing Explains the Progression from Membrane-Bound to Secreted Antibody[59, 60]

When B cells are first activated by interaction with their cognate antigens, they begin to secrete IgM antibodies, which, until then, were present only as membrane-bound receptors. This progression results from a change in RNA processing. Surface IgM is anchored to the cell membrane by the carboxyl end of its μ heavy chain. The 41 hydrophobic amino acids that make up the transmembrane anchor are encoded by two small exons that lie beyond four exons that encode the bulk of the C_μ chain. The switch to synthesis of secreted IgM results from a change in RNA processing that appears to activate an RNA termination site lying between the fourth C_μ exon and the two membrane-anchor-coding exons (Figure 23-34). As a result, RNA transcripts terminate before these two exons and a new carboxyl terminus, one lacking a transmembrane anchor, is specified by the fourth C_μ exon. This segment of the fourth C_μ exon is skipped over during RNA splicing to join the two membrane-anchor-coding mini exons.

How the binding of antigen to surface IgM brings about an alteration in RNA processing is a complete mystery at the moment. Furthermore, in activated B cells, both membrane-bound and secreted antibody are synthesized, while in the more mature plasma cells, only secreted antibody is made. Thus, this method of gene expression can be fine-tuned to match the needs of the cell.

All antibody classes can serve as membrane-bound receptors as

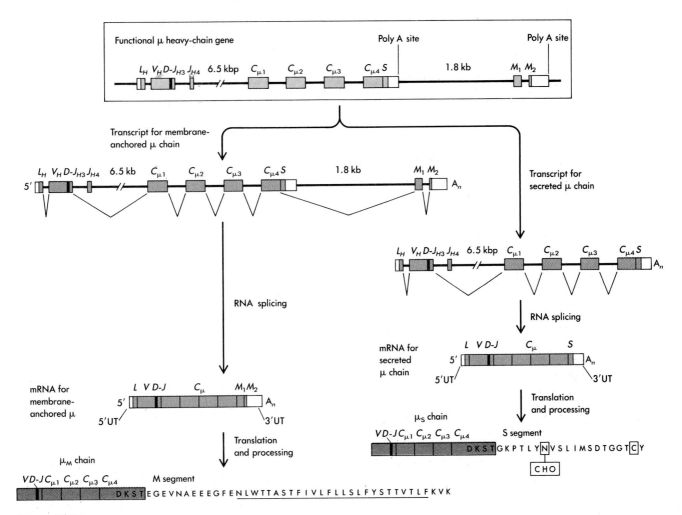

Figure 23-34
Synthesis of membrane-anchored or secreted μ heavy chains is a consequence of differential RNA processing. Membrane-anchored and secreted μ chains have identical amino acid sequences up to their carboxyl termini, where they diverge: The last 41 amino acids of the membrane form are replaced by 20 different amino acids in the secreted form. The membrane-anchored form is made from an RNA that splices out the end of the fourth C_μ exon. This portion of the fourth exon encodes the carboxyl terminus of the secreted form and the 3' untranslated region of the exon. To make secreted μ chains, either this splice must be suppressed or RNA synthesis must terminate at the first poly A addition site so that transcripts never reach the membrane-anchor-coding mini-exons. The amino acid sequences of the carboxyl termini of the two forms of μ heavy chain are shown at the bottom of the figure. The hydrophobic sequence in the membrane-bound form, which could serve to anchor the chain, is underlined. The secreted form has a cysteine residue (boxed) that can disulfide-bond to the J chain in secreted μ (see Figure 23-20), as well as an asparagine-linked carbohydrate chain (N–CHO). Boxes in DNA or RNA molecules represent exons. [After *Immunology*, 2d edition by L. E. Hood, I. L. Weissman, W. B. Wood, and J. H. Wilson. 1984 by The Benjamin/Cummings Publishing Co.]

well as being secreted; and as we might expect, all the heavy-chain constant-region genes have membrane-anchoring exons and alternative sequences that are used to encode the carboxyl-terminal ends of secreted chains.

IgM (μ)→IgD (δ) class switch by RNA processing

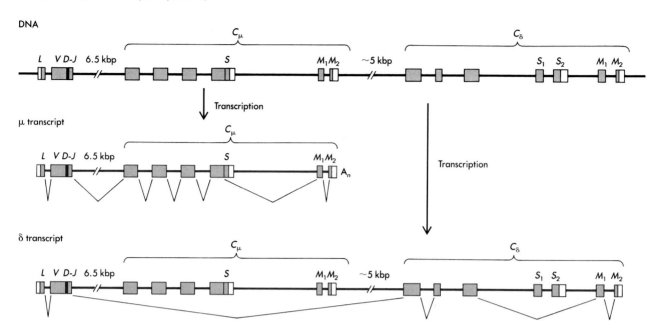

Figure 23-35

IgM to IgD class switching is accomplished by a change in RNA processing. Differential RNA processing is used to join the transcript of an assembled heavy-chain variable-region "gene" to either a μ or a δ constant-region transcript. Note that DNA that encodes both the μ and δ constant-regions consists of exons that encode both a membrane and a secreted form of the chain. In this figure, the splicing patterns indicated would result in synthesis of the membrane-anchored form of each chain. Boxes represent exons in DNA or RNA with white portions of boxes representing 5′ and 3′ untranslated regions.

Differential RNA Processing Allows the Synthesis of IgM and IgD Antibodies by a Single Plasma Cell[61, 62]

As we said earlier, heavy-chain constant regions perform the biologically important effector functions, such as localizing antibodies to certain sites in the body, interacting with complement components, and binding to macrophages. It is highly desirable for an organism to be able to combine its complete V-region repertoire with each of these different functions. Thus, antibody gene assembly includes mechanisms for joining fully assembled *V-D-J* exons to different C_H exons, the phenomenon of class switching. There are two mechanisms for class switching, and their use is developmentally regulated. One mechanism involves differential RNA splicing (for IgM to IgD), and the other involves DNA switch recombination (for IgM to any other class but IgD).

The first class switch that takes place in B cell differentiation allows a given *V-D-J* exon to acquire a δ constant region in addition to μ. Some unknown signal triggers virgin B_μ cells that express just surface IgM to progress to mature $B_{\mu+\delta}$ cells that have both membrane-bound IgM and IgD. The two types of molecules have heavy chains with identical variable regions, and of course, both are associated with the same light chain, since only one light chain is made in a given cell. Exons that encode C_δ lie downstream from C_μ coding exons (Figure 23-35), and differential RNA splicing of a long transcript allows the use of the downstream C_δ exons. This change in RNA splicing may result from a change in RNA termination. In a virgin B_μ cell, RNA transcripts may terminate before the C_δ sequences, while in $B_{\mu+\delta}$ cells, some primary transcripts may extend past C_μ and through C_δ. These transcripts can then be spliced to produce mRNAs using the C_δ exons. At present, it is not clear whether poly A addition or RNA

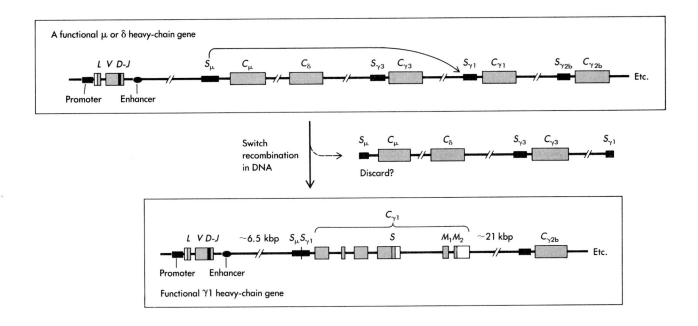

splicing is the means of regulating the μ to δ progression, or whether both processes are regulated to achieve the desired control.

The progression from a B_μ cell to $B_{\mu+\delta}$ usually must take place before B cells are competent to interact with foreign antigens. At some point after B_μ cells appear, they may need to be screened to remove or inactivate those cells that can react with self antigens, so that the animal will be tolerant to self antigens. Conceivably, the appearance of membrane-bound IgD antibodies in addition to IgM is a signal that this screening process has taken place and it is now safe for the cell to participate in immune responses.

Antibody Class Switching Beyond IgD Involves DNA Recombination in the Switch Regions of the Heavy-Chain Constant-Region Genes[63, 64, 65]

Once triggered by antigen recognition, $B_{\mu+\delta}$ cells have the potential to differentiate to produce memory cells and plasma cells that synthesize antibodies belonging to all possible heavy-chain classes. DNA rearrangements are involved in class switches that transfer *V-D-J* exons from C_μ to other heavy-chain types except, as we have just discussed, to C_δ. The DNA rearrangements involve recombination within special segments, called **switch regions,** that lie about 2 to 3 kilobase pairs upstream of all the heavy-chain constant-region genes, except C_δ (Figure 23-36). For example, a switch region in front of the C_μ exons is used to bring a *V-D-J* segment near to the $C_{\gamma1}$ exons. Notice in Figure 23-36 that this DNA rearrangement leaves the heavy-chain enhancer element in the same location relative to the gene's transcriptional promoter as it was before the switch took place.

Switch regions are quite long, on the order of 1 to 2 kilobase pairs, but they are composed of many, many copies of very short sequences. The repeated sequences are mostly different for different C_H genes, and they vary in length from 5 to 80 nucleotides. It is not clear whether the recombination events that lead to class switching occur

Figure 23-36
Class switching from IgM to IgG, IgA, or IgE involves DNA recombination at switch signals. In this figure, an assembled *V-D-J* exon is being moved from a position adjacent to the μ constant-region exons to a position next to the $C_{\gamma1}$ exons to create a functional $\gamma1$ heavy-chain gene. For simplicity, the multiple exons of each type of heavy-chain constant region are abbreviated in the figure except for the final picture of $C_{\gamma1}$ and not all the constant regions are indicated. The special sequences involved in switch recombination are indicated as horizontal black boxes. Note that there are no such sequences in front of the C_δ exons, which makes sense, since the switch from μ to δ is accomplished by differential RNA processing rather than by DNA recombination. Note also that the transcriptional enhancer element (black circle) retains its position relative to the gene's promoter after the switch occurs.

Figure 23-37

Diagram of how a helper T cell may be activated to help a B cell produce its antibodies. Antigen is taken up by accessory cells (called antigen-presenting cells) and processed. Fragments of the antigen appear on the surface of the accessory cell. A helper T cell whose antigen-specific T cell receptor can recognize this antigen binds to the accessory cell and becomes activated. The activated T cell then interacts with a B cell whose antibody can recognize the same antigen that stimulated the helper T cell. The B cell also has processed antigen on its surface. The interaction of T and B cells, along with antigen stimulation, triggers the B cell to divide and differentiate to an antibody-producing plasma cell. [After *Immunology,* 2d edition by L. E. Hood, I. L. Weissman, W. B. Wood, and J. H. Wilson. 1984 by The Benjamin/Cummings Publishing Co.]

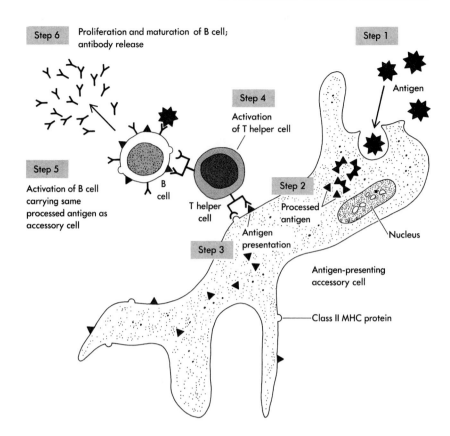

within a single DNA molecule or between different chromosomes. Perhaps both types of events are allowed.

Recently, cells that synthesize both IgM and IgG, or both IgM and IgE, or both IgM and IgA have been observed. Because allelic exclusion forbids the expression of heavy-chain genes from two chromosomes within one cell, these cells are presumably using two heavy-chain C-region "genes" located on the same chromosome. The DNA rearrangements we just described would preclude this situation, and now it is thought that in some cases, instead of DNA rearrangement, extremely long transcripts may be made from a single heavy-chain locus and spliced to allow simultaneous synthesis of two heavy-chain classes (other than IgM with IgD).

Complex Interactions Involving B Cells, T Cells, and Macrophages Help B Cells Produce Their Antibodies[66, 67, 68]

Antigen stimulation of B cells to divide and secrete antibodies is usually a far more complex process than a simple collision between freely circulating antigen and a membrane-bound immunoglobulin receptor. An animal's response to an invading antigen mobilizes a whole network of cells. Much remains to be learned about this process, but a rough sequence of events is clear. First, as described earlier in this chapter, accessory cells, such as macrophages, trap the incoming antigen and *process* it for *presentation* to a subset of T cells called **helper T cells** (**T$_H$ cells**) (Figure 23-37). (As we shall see, helper T cells are distinct from the T cells described earlier, which directly attack and destroy antigens or antigen-bearing cells.) **Antigen processing** seems to involve partially digesting the antigen and returning fragments of it to the cell surface. **Antigen presentation** probably involves positioning these fragments at the cell surface until they are recognized by a helper cell carrying an appropriate T cell receptor. Interaction of the T_H cell with the antigen-charged accessory cell activates the helper. Then the activated helper stimulates B cells to divide and become plasma cells. This interaction is antigen-specific. Apparently, B cells can also process antigens and present them on their surface; so a T helper cell whose receptor first recognized a particular antigen on a macrophage will recognize the same antigen on a B cell.

How the helper T cell stimulates the B cell is not yet clear, but it probably results, at least in part, from the release of growth factors. As will be discussed in Chapter 25, growth factors are polypeptides that bind to specific receptors on the cell surface and by so doing somehow trigger cell division. Different cell types require different growth factors. The factors released by helper T cells that promote the differentiation of B cells to plasma cells are not yet clearly defined. One may be the growth factor called *interleukin 2* (Il-2), which is also a growth factor for T cells. Others, called helper factors, are produced in small amounts and are likely to remain elusive until their genes are cloned. Growth factors are only part of the trigger that sends the B cell on its way. Of course, binding to its cognate antigen is also a trigger. The two stimuli may be related, however, since an important consequence of antigen binding may be an increase in the expression of growth factor receptors on the cell surface.

One clue as to how antigen binding may trigger B and T cells to divide comes from studying the nonspecific stimulation of B and T cell division by mitogens such as lectins and lipopolysaccharides (Table 23-4). These substances bind to the surface of lymphocytes and trigger several rounds of cell division. When they bind, they can induce a phenomenon known as **capping**, in which cell surface receptors become crosslinked and move laterally in the membrane to coalesce as a cap (Figure 23-38). One hypothesis is that capping can send a signal to the cell nucleus that triggers events that lead to cell division. Since many antigens are multimeric and hence able to engage more than one antibody molecule, they, too, can induce capping, and this might be one way they participate in triggering B cell division and maturation.

Why shouldn't interaction with circulating antigen be a sufficient trigger for B cells to differentiate and secrete antibodies? Why do they need helper T cells, too? Speculation is that the need for several cells to interact provides a way to regulate the process of antibody synthesis and also B cell division. If antigen alone were the trigger, then activated B cell clones might divide and differentiate out of control or might not divide enough to cope with the antigen load. As we shall see in the next section, another group of T cells, called suppressor T cells, can block the proliferation and terminal differentiation of B cells. According to another model, the need for T cells to stimulate B cells to produce antibodies may be involved in achieving tolerance.

Although helper T cells are required for all IgG, IgE, and IgA immune responses, some IgM responses are T_H cell–independent. These responses are usually directed against antigens composed of identical repeating determinants; but the biological significance of these T_H-independent responses is obscure.

Table 23-4 Mitogens Specific for B Cells or T Cells

B Cells	T Cells
Lipopolysaccharide (LPS)*, from bacterial cell walls	Concanavalin A (Con A), a plant lectin Phytohemagglutinin (PHA), a plant lectin

*Mitogenic for mouse but not human B cells.

Figure 23-38

Capping. (a) Diagram of how a multimeric antigen can act to cluster mobile immunoglobulin in the cell membrane of a small lymphocyte. (b) Autoradiographs of a B cell binding to its radioactively labeled cognate antigen: (1) Uniform binding of the antigen following incubation at 0°C for 30 minutes.
(2) Aggregation (capping) of antigen at one cell pole after incubation at 37°C for 15 minutes. Such capping results from the cross-aggregation of the surface immunoglobulin receptors by the polymeric antigen. Note that only one of the four cells in this picture has receptors for this particular antigen. [Reproduced with permission from E. Diener and V. H. Paetkau, *Proc. Nat. Acad. Sci.* 69 (1972):2364.]

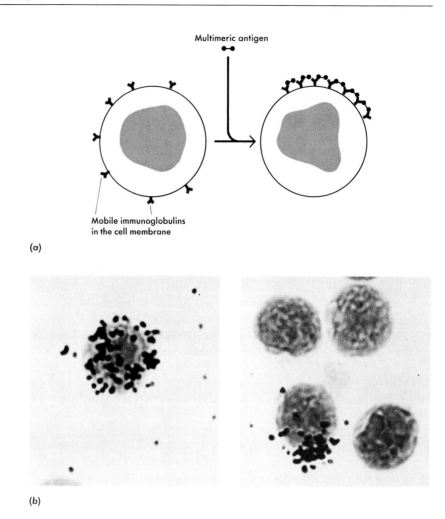

(a)

(b)

THE T CELL RESPONSE

There Are Different Types of T Cells Called Cytotoxic, Helper, and Suppressor Cells[69, 70, 71]

Two very different types of cellular immune responses are mediated by T lymphocytes. One is to destroy cells that display foreign antigens on their surfaces, and it is performed by **cytotoxic T lymphocytes** (**CTLs**). Most often, the targets are the animal's own cells when, for example, they become infected with a virus and so express viral-coded proteins on their surfaces. Cytotoxic T cells can also destroy cells in skin grafts or organ transplants between individuals. Cytotoxic T cells are the cellular immune system's equivalent of antibodies. They recognize antigens via their membrane-bound T cell receptors, which have striking similarities to antibody molecules. Each cytotoxic T cell carries a receptor with a different antigen binding specificity. Interaction with an antigen that specifically binds to the receptor triggers the cell's maturation to a killer and also causes the cell to proliferate and generate memory cells that will respond to future encounters with the same antigen. The primary difference between humoral immunity and cytotoxic T cell responses is that antibodies are secreted, circulate freely, and cope primarily with freely circulating antigens. In contrast, T cell antigen receptors remain

Table 23-5 Some Usefel Cell Surface Antigens on Mature T Cells of Mice and Human

Mouse	Thy-1	L3T4	Lyt-2,3
Helper T cells (T$_H$)	+	+	−
Cytotoxic T cells (CTL)	+	−	+
Suppressor T cells (T$_S$)	+	−	+
Human		**T4**	**T8**
Helper T cells		+	−
Cytotoxic T cells		−	+
Suppressor T cells		−	+

L3T4 and T4 are equivalent, as are T8 and Lyt-2,3.

+ = possesses the antigen. − = lacks the antigen.

membrane-anchored and seem to direct their attention primarily to antigens on cell surfaces.

The second function of cellular immunity is to regulate the responses of B cells and cytotoxic T cells. As we have already seen, specialized T lymphocytes called helper T cells interact with B cells to trigger their maturation to antibody-producing plasma cells. Helper T cells are also needed to help cytotoxic T cells mature into effector cells. A second type of regulatory T cell, the **suppressor T cell (T$_S$)**, has the opposite effect: It blocks B and T cell responses. All these responses are antigen-specific. For example, each helper T cell recognizes a unique antigenic determinant and only stimulates a B cell that recognizes the same determinant. (How helper T cells may interact with B cells to stimulate antibody synthesis was discussed earlier and diagrammed in Figure 23-37). Helper T cells have antigen receptors similar to those of cytotoxic T cells, and they are probably encoded by the same genes. The receptors for suppressor T cells have not yet been clearly defined.

The various types of T lymphocytes (and also B cells) are small, round cells that are difficult to distinguish in the microscope. They can be identified, however, by the presence of specific cell surface proteins. These proteins were discovered serologically, and in most cases their functions are still unknown. They are enormously useful, however, both for identification of lymphocyte subsets and for separating different cell types for experimentation. All mouse T cells carry the Thy-1 antigen on their surface. Helper T cells of humans have the T4 antigen, originally defined by its ability to react with a particular monoclonal antibody prepared against human T cells. The gene encoding T4 has recently been molecularly cloned. Table 23-5 lists antigens that are currently used to identify different types of mouse and human T lymphocytes. The best marker for B cells, of course, is surface immunoglobulin.

Pluripotent stem cells in the bone marrow give rise to all the different types of blood cells: the B and T lymphocytes, as well as the red blood cells and other cells of the myeloid lineage. In humans and mice, immature T cell precursors leave the bone marrow and migrate to the thymus. There they undergo massive proliferation and, after a sequence of events largely shrouded in mystery, emerge as helper, cytotoxic, and suppressor T cells.

Figure 23-39
The experiment that showed that
T cells recognize antigens only in combination with normal cell surface proteins. An inbred mouse of strain A was
immunized with a virus. Its T cells
were then removed and tested for their
ability to kill target cells in vitro. The
target cells were fibroblasts infected
with the same virus used in the immunization, but they were derived either
from a mouse of strain A or from a different inbred strain with a different
MHC gene, designated strain B in the
figure. Despite the presence of the
same viral antigens on both target cells,
the T cells from mouse strain A could
only kill target cells derived from a
mouse of strain A.

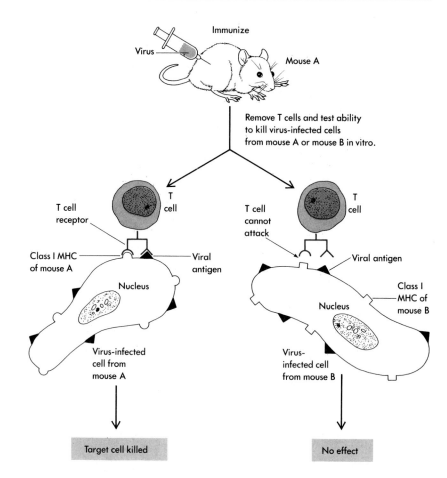

T Cells Recognize Foreign Antigens in Combination with MHC Gene Products[72–76]

Cytotoxic T cells respond to antigens present on cell surfaces, so it
may be important to prevent them from interacting with freely circulating antigens. About ten years ago, an important experiment unexpectedly revealed the mechanism that probably restricts T cells to
interactions with cell-bound antigens. The key is that T cells can only
"see" antigens when they can also "see" certain normal cell surface
proteins simultaneously. In the experiment that led to this discovery
(Figure 23-39), T cells were removed from a mouse that had first been
immunized with a virus so as to activate and stimulate the proliferation of cytotoxic T cells directed against cells infected with this particular virus. These T cells could kill fibroblasts infected with the virus in
tissue culture. Surprisingly, however, they could not kill fibroblasts
infected with the virus if the fibroblasts came from a genetically different mouse. Thus, cytotoxic T cells do not just recognize foreign antigens; they recognize foreign antigens in combination with determinants present on their host's own cells.

By using mouse genetics, it was possible to show that the normal
determinants that T cells see are cell surface proteins coded for by
major histocompatibility complex (MHC) genes, in particular, the
so-called class I *K* and *D* genes. (In humans the corresponding genes
are called *HLA-A, HLA-B,* and *HLA-C.*) These proteins are present on
the surface of all cell types in the body. Thus, cytotoxic T cells are
probably able to eliminate any cell type infected with a virus. MHC

class I proteins are extremely polymorphic. This means that there are many different *K* and *D* alleles in the mouse population. Polymorphism explains why any two mice are likely to have different MHC proteins and why, in the aforementioned experiment, the cytotoxic T cells could not kill the virally infected cells from a different mouse. T cell receptors concentrate on the polymorphism of class I products, using it to distinguish self from nonself. Retrospectively, we see that **MHC restriction** of T cell function makes a lot of sense. It ensures that cytotoxic T cells will restrict their activities to cell surfaces and not clog up their receptors with freely circulating antigens.

The ability of T cells to recognize foreign antigens only in combination with the host's own MHC gene products is acquired as the T cells mature in the thymus, a learning process called **thymic education.** In many ways, it resembles tolerance, the process by which T cells learn not to destroy the host's own normal cells. Tolerance is also learned, rather than being preprogrammed into the T cells. Both processes involve a selection that results in a pool of T cells whose receptors have just the right recognition properties, namely, they see altered self but not self alone and not antigen alone. T cells failing the selection are presumably destroyed before they can leave the thymus or are held in check after they mature.

The fact that T cells recognize antigen + MHC immediately raises the question of whether a single T cell receptor looks at both molecules or whether there are two receptors, one specific for antigen, the other for self MHC (Figure 23-40). Although we do not yet know the answer, at present there is more evidence in favor of a single receptor. One experiment that argues for a single receptor involves fusing two T cells with different specificities to form a hybrid cell. One of the initial cells can recognize antigen A + MHC 1, while the second cell can recognize antigen B + MHC 2. If there were a distinct T cell receptor for each antigen and another T cell receptor for each MHC type, then we would expect the four types of receptors to mix freely, and the hybrid cell would have new recognition capabilities, namely, antigen A + MHC 2 and antigen B + MHC 1, as well as the original capabilities of its parent cells. This is not the case. The hybrid cell has only two specificities, the same ones as its parent cells. Since T cell receptors and the genes that code for them have recently been isolated, the question of whether one or two receptors are involved in T cell recognition of antigens and MHC should soon be answered. If it is true that a single T cell receptor recognizes antigen and MHC simultaneously, then antigens and MHC proteins must lie close together on the cell surface; in fact, they probably have to touch each other. It is hard to see how a vast array of different antigens could all interact with the same MHC proteins, but somehow this would have to be the case, and recently, evidence of such interaction has been obtained. Perhaps only weak, fleeting interactions are required between the antigen and MHC protein in order for them to be seen by the T cell receptors.

Helper T cells are also MHC-restricted, but unlike cytotoxic T cells, they recognize products of class II MHC genes. These genes specify cell surface proteins that are present only on certain cells of the immune system, including antigen-presenting cells. Before helper T cells can regulate the responses of other cells, they must first be activated by their cognate antigens. Accessory cells such as macrophages present the processed antigens to helper cells (see Figure 23-37), and this process requires the presence of class II antigens on the presenting cell. In most cases, helper T cells must also recognize class

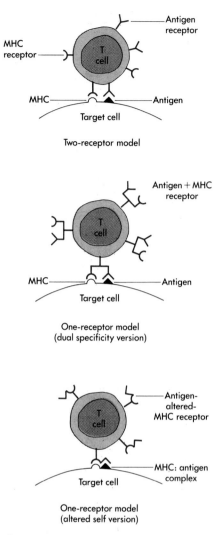

Figure 23-40
Are there separate receptors on T cells for antigen and MHC recognition, or does a single receptor look at both molecules? Two models for the single-receptor hypothesis are shown. In one (dual specificity version), the single receptor is represented as having its two specificities distinctly separate; in the other, the single receptor recognizes antigen because the antigen binds to MHC and so makes the MHC protein look abnormal.

II antigens on the B cells they stimulate. Forcing cells to look at particular cell surface markers seems to be how the immune system controls cellular interactions. MHC gene products are like sign posts that tell cells when they have reached the correct destination.

The T Cell Receptor Is Composed of Two Polypeptide Chains, α and β[77-84]

In striking contrast to antibody molecules, which were among the first proteins whose structure was determined, the T cell receptor proved to be difficult even to find biochemically. This is because there was no T cell equivalent of the myeloma tumor with its abundant supply of single-specificity antibody molecules. For a long time, researchers supposed that T cell receptors were encoded, at least in part, by the same genes that code for antibodies. Since the two types of proteins perform the same function of recognizing innumerable foreign substances, it seemed reasonable to assume that the same genes were involved. The favored hypothesis was that T cell receptors are specified by the same variable-region genes as antibodies, but that they probably have their own constant-region genes, since they require completely different effector functions than antibodies. Like many good ideas, this one turned out to be completely wrong. The genes for T cell receptors are entirely distinct from those that code for antibodies. However, the two families of genes are clearly related evolutionarily.

T cell receptor proteins were finally identified several years ago, when it became possible to grow many types of cloned T cells in tissue culture. When certain cloned T cells are injected into rabbits or mice, they can elicit antibodies that react specifically with the immunizing T cell but not with any other T cell of a similar type. This extreme specificity for a single clone of T cells argues that the molecule(s) recognized by the antisera are the T cell receptor, in particular, its specific antigen binding portion. Furthermore, these antisera can frequently block T cell function, such as helper activity. Some of these antisera can precipitate a protein from the T cells. These proteins, now firmly identified as T cell receptors, are glycoproteins with molecular weights of 80 to 90 kdal. They are dimers that consist of one α polypeptide chain and one β polypeptide chain, each about 40 to 50 kdal, linked by a disulfide bond as well as noncovalent bonds.

The α and β Chains of the T Cell Receptor and a Related Polypeptide Called γ Resemble Light and Heavy Chains of Antibodies

At the same time that some immunologists were trying to purify T cell receptor proteins, others were trying to clone the genes that encode them. This was difficult, since no nucleic acid probes were available to hybridize with the desired genes and since virtually nothing was known about the T cell receptor protein at the time the studies began. The breakthrough came in 1984, when cDNA clones of first β-chain and then α-chain mRNAs were obtained. Figure 23-41 shows how this was done for a mouse β chain using a method called **subtraction hybridization.** The idea behind the experiment was to isolate cDNA clones of mRNAs that are expressed in T cells but not in B cells. These

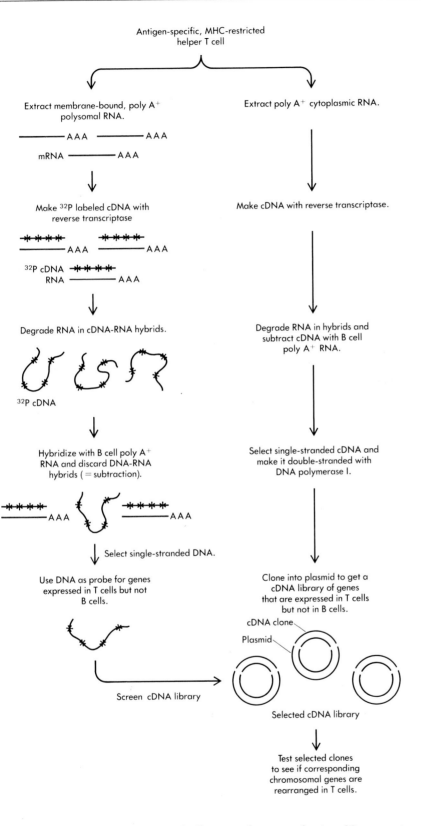

Figure 23-41
The experiment that first led to the isolation of a cDNA clone of the β chain of the mouse T cell receptor. The strategy used makes three assumptions about genes that encode the polypeptides that make up the T cell receptor: Since they encode membrane-anchored polypeptides, their mRNAs will be present on membrane-bound polysomes; they are expressed in T cells but not in B cells; and they undergo DNA rearrangements in T cells relative to embryonic cells or other cell types in the body.

two cell types express a very similar set of genes, sharing 98 percent or more of their mRNAs. Thus, if cDNA prepared from T cell mRNA is hybridized to B cell mRNA and all the cDNA strands that can hybridize are discarded (subtracted), the remaining cDNA should be greatly enriched for copies of T cell–specific mRNAs, including those for the T cell receptor polypeptides.

Figure 23-42

Proposed primary structure of the β chain of the T cell receptor and conformation of the α-β dimer believed to make up the receptor. The figure is based on data derived from nucleotide sequencing of DNA clones of the α- and β-chain genes. The leader peptide is removed from the mature chain. D and J indicate regions of the variable region of the β chain that are encoded by the D and J segments of DNA. So far, α-chain variable regions have been found to contain J but not D segments. The transmembrane and cytoplasmic domains of the chains are indicated. [After M. K. Davis, Y. Chien, N. R. J. Gascoigne, and S. M. Hedrick, *Immunol. Revs.* 8 (1984):235–258.]

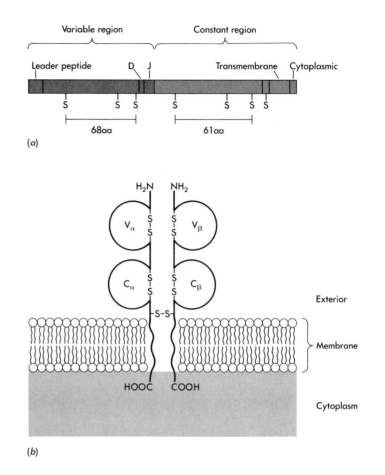

In addition to being expressed exclusively in T cells, genes coding for T cell receptors might be expected to be rearranged in T cell DNA much the way that antibody genes are rearranged in B cells. Some of the cDNA clones obtained in the preceding experiment turned out to correspond to genes with this property. Furthermore, when these clones were sequenced, they were found to encode proteins that had structural similarity and even amino acid sequence homology to antibody light and heavy chains. They could not be antibody genes, however, because they were not expressed in B cells and because their nucleotide sequences were clearly distinct from those of antibody genes. The definitive proof that they corresponded to T cell receptor genes came when amino acid sequences of the recently purified T cell receptor chains were finally obtained. These were very similar to those predicted from nucleotide sequences of the T cell–specific cDNA clones.

As soon as several different cDNA clones of α- and β-chain mRNAs from different T cells were sequenced, it was apparent that both chains have variable regions in their amino-terminal halves and constant regions in their carboxyl halves. These are followed on the carboxyl end by transmembrane domains that anchor the chains to the cell surface and then a short cytoplasmic tail (Figure 23-42). As with antibody molecules, the variable and constant regions are each about 110 amino acids long, as predicted by nucleotide sequences, and each of these homology units has cysteine residues that can form intrachain disulfide bonds at positions similar to those in antibody variable and constant homology units. While the T cell receptor chains are more similar to antibody light chains in size, they are also similar to the membrane-bound form of heavy chains in that they, too, have membrane anchors. Moreover, both the α and β chains of the T cell receptor are glycosylated. From all this evidence, there is virtually no doubt that T cell receptors and antibodies are evolutionarily related proteins.

A completely unexpected finding to emerge from subtraction hybridization experiments was the isolation of another T cell–specific cDNA clone that also encodes an antibody-like protein and whose corresponding genes are also rearranged in T cells. It does not seem to correspond to the α or β chains of the T cell receptor proteins that have been purified, since it lacks canonical Asn-X-Thr/Ser sequences that signal addition of the N-linked sugar residues that are known to be present on these chains. The function of this third T cell receptor-like gene, named the **γ gene,** is a mystery at the moment. An intriguing clue to its function is that γ-gene mRNA is present at high levels in immature fetal thymocytes, while in contrast, α mRNA is low. In populations of mature T cells, the reverse is true. The β-gene mRNA is present in comparable amounts in both cell populations. These results suggest that the γ gene may be important during T cell differentiation. Conceivably, it plays a role in thymic education.

T Cell Receptor Genes Look Like Antibody Genes and Are Assembled by *V-D-J* Joining[85–90]

As soon as cDNA clones were available, it was easy to isolate the corresponding chromosomal genes for the α, β, and γ polypeptides, both in their germ line configurations and in their rearranged, functional forms in T cells. Information about the structure of these multigene families has been accumulating rapidly. The most striking thing about them is how similar their organization is to that of antibody gene families: There are multiple V, D, and J segments that must be assembled by DNA recombination to make functional V-region exons, and these lie upstream from just one or a few constant-region genes. The α, β, and γ multigene families lie on different chromosomes (Table 23-6).

The β-chain gene family resembles heavy-chain genes in having V, D, and J segments, but it is also reminiscent of the λ light-chain locus in having J segments in front of each of its constant-region genes (Figure 23-43). It is easy to see how extensive V-region diversity can

Table 23-6 Chromosome Locations of the T Cell Receptor Genes

Polypeptide Chain	Chromosome	
	Human	*Mouse*
α	14	14
β	7	6
γ	7	13

Figure 23-43
Current picture of the mouse T cell receptor β-chain locus. The D, J, and C segments are drawn approximately to scale, but there is little evidence available concerning the distance or even the orientation of most of the V segments relative to the two D, J, C clusters, although recently one V segment was found on the 3' side of C_{β_2} and in the opposite transcriptioned orientation (not shown). This particular exon is joined to a D segment by an inversion at the DNA level. To draw the top part of the figure approximately to scale, the D and J elements are represented as black lines, as are the second exons of the two constant region genes. A close-up (inset box) shows the orientation of the heptamer-nonamers that flank V, D, and J segments and the nucleotide distance between them (± 1 nucleotide). The structure of a functional β-chain gene obtained after D-J and V-D-J joining is indicated.

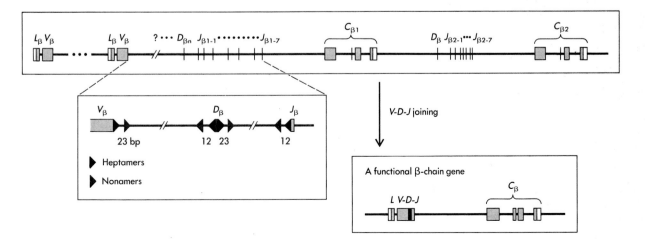

be generated in the β chains. Although they have fewer V segments than heavy-chain genes (there may be as few as 20 V_β segments, compared to 200 to 300 V_H segments), β genes appear to have multiple D_β segments, and they have more J_β segments than do heavy-chain genes. Assuming random association of these segments, great variability is already possible. Furthermore, as with antibody genes, there is variability in the positions of the V-D and D-J joints, which can lead to amino acid sequence variability at these positions. Another mechanism for generating diversity, not seen with antibody genes, is that the D_β segments can be read in all three reading frames. In addition, as with heavy-chain genes, there is evidence of N nucleotide addition at the V-D and D-J joints in rearranged β genes. Whether fully rearranged β-chain genes undergo further mutation (somatic hypermutation) in their V-coding portions is not yet known.

Very similar recombination systems are probably used to join the V, D, and J segments of T cell receptor genes and antibody genes. Similar sequences flank the segments that undergo joining: namely, heptamer-nonamers separated by 12 or 23 base pairs, as we have already described for antibody genes (see Figure 23-27). However, if the same joining rules that have been defined for antibody gene segments are used, then V_β segments would be able to skip over D_β segments and join directly to J_β segments (see Figure 23-43). Whether this is used to increase diversity remains to be seen.

There are two β-chain constant-region genes that encode almost identical constant-region polypeptides. Each is composed of four exons that correspond closely to the structural domains of the constant region: One exon encodes the amino-terminal 125 amino acids that are outside the cell; the second, a mini-exon, encodes the 6-amino-acid hinge region containing the cysteine residue probably involved in disulfide bonding to the α chain; the third specifies 36 hydrophobic amino acids of the transmembrane domain; and the fourth encodes the 5 cytoplasmic amino acids as well as the 3' untranslated region of the mRNA. The β chain of the T cell receptor is expressed in both helper and cytotoxic T cells; however, it does not seem to be used in suppressor T cells.

The α-chain genes include just one constant region "gene," which is composed of four exons. There are multiple V_α segments and an unusually large number of J_α segments (at least 18). So far, there is no evidence that D_α segments are present. Most striking is the enormous distance between J_α segments and the constant-region exons: Some fully assembled α-chain genes would generate primary RNA transcripts of 60 kilobases or more. The γ-chain genes seem to have a smaller number of V segments than α genes and have several J-C clusters. Whether they have D_γ segments is still uncertain. Simplified diagrams of the mouse α and γ genes are shown in Figure 23-44. The β-chain genes and the three antibody gene loci are included for comparison.

Antigen Binding Primes T Cells to Respond to the Growth Factor Interleukin 2[91, 92, 93]

When T cells leave the thymus, they are ready to respond to antigens; but as with B cells, this response usually cannot occur by a simple collision between, for example, a cytotoxic T cell and a virus-infected

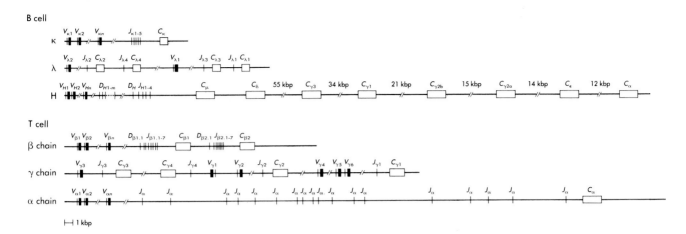

fibroblast. Rather, antigen-stimulated **blast transformation** of a T cell must take place at the surface of an accessory cell such as a macrophage. These accessory cells, which are located throughout the body but particularly concentrated in secondary lymphoid organs such as lymph nodes, present antigens in some special way that primes the T cell for further maturation. As we discussed for B cells, this presentation may resemble the action of lymphocyte mitogens. In any case, seeing antigen in this form induces the T cell to express cell surface receptors for the T cell growth factor interleukin 2. Antigen also stimulates T cells to release Il-2. Helpers are particularly potent producers, and this in part may explain how they stimulate cytotoxic T cells to divide.

Tissue Rejection Antigens and Immune Response Genes Map to the MHC Locus[94–97]

The major histocompatibility complex (MHC) is a cluster of genes that occupies some 2000 to 4000 kilobase pairs of DNA. Four types of genes, discovered independently, map to the complex. The first, discovered more than 45 years ago, encodes the *transplantation antigens* that cause rapid rejection of skin and organ grafts between individuals and thus gives the *MHC* locus its name (*histo* means "tissue"). As we have seen, these are the class I antigens that cytotoxic T cells use to distinguish self from nonself. The second set of genes, originally called *immune response genes* (Ir), determines whether animals can mount a strong or a weak response to particular antigens. Immune response genes encode the class II MHC proteins that helper T cells look at when they see antigens on accessory cells and on B lymphocytes (and, very rarely, class I gene products). Presumably, some antigens cannot interact well with certain MHC proteins. As a result, these antigens are unable to stimulate helper T cells, and the animal is unable to mount a good immune response. The third set of genes, called *Tla* and *Qa* genes, encodes proteins that were discovered as cell surface antigens on lymphocytes. They are classified as class I *MHC* genes because their products have structural similarity to the class I transplantation antigens, but the function of the *Tla* and *Qa* genes is not yet known. The fourth type of gene to map to the *MHC* locus encodes components of the *complement system.* Maps of the *MHC* locus in mouse and human are shown in Figure 23-45. In mice, the

Figure 23-44
Simplified diagram of the genes encoding antibodies and T cell receptors. Boxes indicate V-region-coding segments or C "genes." (The latter have been simplified so that a single box represents the several exons that encode most C regions.) Vertical lines indicate D- or J-gene segments. It is not yet known whether the α and γ genes have D segments. Distances are approximately to scale unless they are too large or are unknown. In these cases, they are indicated by breaks in the lines. [After Winoto et al., *Nature* 316 (1985):832–836.]

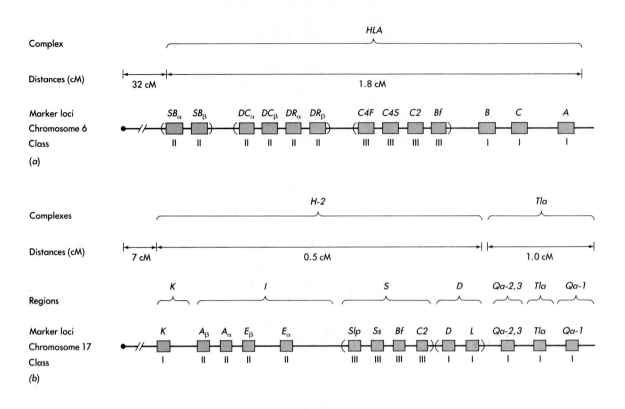

Figure 23-45

(a) Genetic map of the human *HLA* complex located on chromosome 6. Distances are in centimorgans. Black circle at left end of chromosome represents the centromere. Marker loci were defined genetically, and the order of those within parentheses is not known. Note that the boxes do not represent single genes. Class I transplantation antigens are designated A, B, and C. Class II antigens are designated SB, DC, and DR. Class III genes encode the complement components designated C2, C4, and *Bf*. (b) Genetic map of the mouse *H-2* complex located on chromosome 17. Distances in centimorgans. Organization of the locus is strikingly similar to that of the human *HLA*, except that class I genes, *K*, *D*, and *L*, are separated. There are also *Tla* and *Qa* genes, whose possible equivalent is now being discovered in humans. Class I genes encode transplantation antigens. Class II genes encoded by the *I* region include immune response genes. Class III genes of the *S* region, *Ss*, *Slp*, *C2*, and *Bf*, encode some serum proteins of the complement system, important in immune responses. [After *Immunology*, by L. E. Wood, I. L. Weissman, W. B. Wood, and J. H. Wilson. 1984. The Benjamin/Cummings Publishing Co.]

MHC is referred to as *H-2*, while in humans, it is called **HLA (human leukocyte antigen complex)**.

Except for the complement components, *MHC* genes all specify cell surface proteins. Furthermore, many of these proteins serve as signals by which the animal distinguishes self from nonself, and they are intimately involved in responding to nonself. One of the most striking features of this recognition system is that it is incredibly polymorphic. The *K* gene of the mouse, for example, specifies a class I transplantation antigen, and more than 55 alleles of *K* have been identified. Thus, even though each inbred mouse has a single type of *K*-gene product, any two mice are likely to have different *K* genes. This extreme polymorphism is also seen at the mouse class I *D* and *L* genes, the human class I *HLA-A*, *B*, and *C* genes, and certain class II genes. Since cells express several of these genes simultaneously (e.g., *K*, *D*, and *L* in the mouse), and since animals express the *MHC* genes from both chromosomes (one from the mother and one from the father), it is apparent why the cells from each individual are likely to be different.

It seems unlikely that *MHC* genes are polymorphic just to ensure graft rejection between individuals. However, we can imagine one very good reason why it is advantageous for *MHC* genes to be so polymorphic. Antigens are seen in the context of *MHC* gene products, and different antigens would be expected to interact effectively with some but not all MHC proteins. By having extensive polymorphism in MHC genes, there is a better chance that each individual will carry at least one protein that can interact with any particular antigen and so result in a good immune response to this substance. Even more important, polymorphism greatly increases the probability that at least some individuals in the population will survive should a new antigen, such as a killer virus, suddenly appear on the scene.

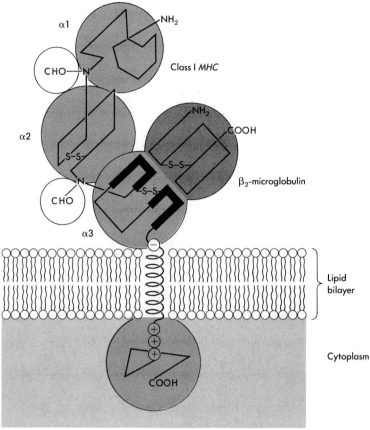

Figure 23-46
Model of a class I molecule in association with β_2 microglobulin. Three external domains and the cytoplasmic domain of the class I polypeptide are indicated by large circles. Small circles represent carbohydrate side chains. Charged amino acids, important in locking the transmembrane domain into the lipid bilayer, are indicated. Intrachain disulfide bridges in the third external domain and in β_2 microglobulin are similar in position to those in antibody polypeptides, and there is amino acid sequence homology to antibody molecules in these two domains. [From Strominger et al., *Scand. J. Immunol.* 11 (1980):573–592.]

Genes and Proteins of the MHC[98–103]

An enormous surprise came when the structure of HLA and H-2 class I proteins were determined. They turned out to have striking structural similarities to antibody molecules! Class I gene products are integral membrane glycoproteins with molecular weights of 40,000–45,000. They are associated noncovalently with a 96-amino-acid-long peptide, **β_2 microglobulin.** β_2 microglobulin has amino acid sequence homology to heavy-chain constant-region domains, although it is encoded by a gene that is entirely separate from antibody genes. The transplantation antigens have five distinct regions: Three are external to the cell, one spans the membrane, and one is the cytoplasmic carboxyl terminus of the protein. Each of the three external domains is 90 amino acids long, about the length of an antibody homology unit, and two of these have the antibody-like disulfide bridges spanning about 65 amino acids (Figure 23-46). β_2 microglobulin associates with the domain closest to the cell surface, and this domain, like β_2, has amino acid sequence homology to a heavy-chain constant-region homology unit. The structure of this paired unit probably resembles a comparable unit in an antibody molecule.

The genes that encode class I proteins have multiple exons, usually eight. As with antibody genes and T cell receptor genes, the coding capacity of these exons corresponds closely to distinct protein domains (Figure 23-47). Class II *MHC* gene products have structural similarities to class I molecules and to antibodies, too. They are heterodimers composed of α and β chains. Each chain has two domains

Figure 23-47
The exons of a class I *MHC* gene correspond closely to structural and functional domains of the encoded protein. White portions of 5′ and 3′ exons encode the 5′ and 3′ untranslated portions of exons, respectively.

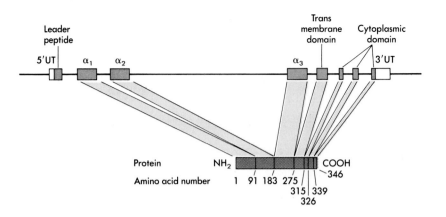

outside the cell that are about 90 amino acids long, and three of these four domains have centrally placed disulfide bridges.

Recently, over 800 kilobase pairs of DNA from the mouse *MHC* locus have been molecularly cloned on cosmids. A surprising result from analyzing this DNA is that there are many more class I *MHC* genes in the *Tla-Qa* region than there are in the *H-2* region (Figure 23-48). It seems likely that these genes, too, will turn out to encode proteins important in cell-cell interactions, perhaps interactions that occur during development.

Evolution of the Immunoglobulin Supergene Family[103]

It has been known for more than two decades that light and heavy chains of antibody molecules are built up by the repetition of structurally related homology units. As we learned, the basic unit is about 110 amino acids long, has a disulfide bond that bridges about 65 amino acids, and can adopt a three-dimensional configuration known as the immunoglobulin fold. Furthermore, pairs of homology units can fit together to make domains with distinct functions, for example, antigen binding or effector functions. In antibody genes, single exons encode approximately one V or one C homology unit, and it is virtually impossible to look at the light- and heavy-chain gene clusters without imagining their evolution by repeated duplications of V and C exons.

Not surprisingly, T cell receptors for antigen also are built up from homology units of the immunoglobulin type. Far more surprising, however, the MHC class I and II molecules also employ this structure, and so does the class I bound polypeptide β_2 microglobulin (Figure 23-49). Thy-1, the surface antigen used for T cell identification in mice, is yet another member of the family, being composed of a single membrane-anchored immunoglobulin-like homology unit. In addition, three other surface proteins—the poly-Ig receptor that me-

Figure 23-48
Location of individual genes that have been cloned within the genetically defined region of the mouse *H-2/Tla* complex. Genes are indicated as color or gray boxes. Color boxes represent functional genes, while gray boxes represent genes that may be nonfunctional. Genes that are linked by a line have been physically linked by cloning. The order of gene clusters in separate regions is not known. Note that there are many more genes in the *Tla/Qa* region than in the *K* and *D* regions. [Adapted from A. Winoto, M. Steinmetz, and L. Hood, *Proc. Natl. Acad. Sci.* 80 (1983):3425.]

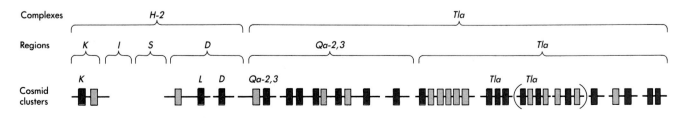

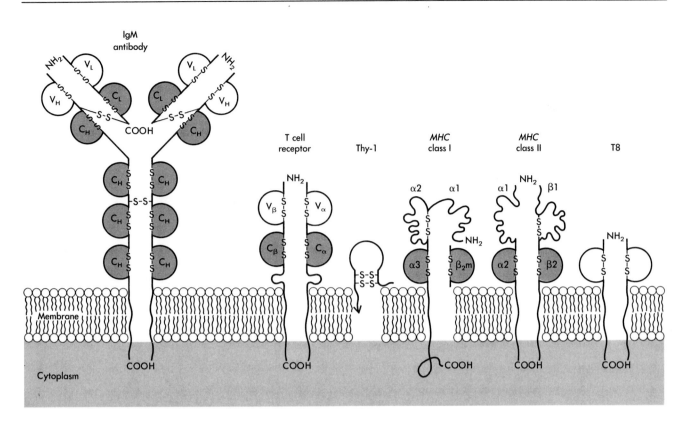

Figure 23-49

A highly schematic representation of the related domains and membrane orientations of six proteins encoded by members of the immunoglobulin supergene family. Each circle represents a domain of about 110 amino acids, usually with an intrachain disulfide bridge (S–S), as indicated. Domains with homology to constant-region domains of antibody polypeptides are shaded. Note that the antibody molecule shown is an IgM type. IgG, a more common immunoglobulin molecule in serum, would have only three C_H domains, the second domain having perhaps diverged to become the hinge region.

diates the transport of polymeric immunoglobulin molecules across endothelial cells and L3T4 and T8, cell surface proteins on certain T cells—are also constructed from this basic unit. These striking structural similarities imply that some primitive gene gave rise to the genes that encode this whole array of proteins.

A multigene family is a group of homologous genes with similar functions. The name **supergene family** has been given to a set of multigene families and single-copy genes that are related by sequence but not necessarily related in function. The proteins described in the preceding paragraphs belong to such a family. Figure 23-50 shows a scheme that has been proposed to explain the evolution of the immunoglobulin supergene family. The scheme is based on similarities in sequence and on the exon-intron structure of the present-day members of the family. The primordial gene is imagined to have encoded a cell surface protein, perhaps with a leader peptide and a transmembrane domain as well as an immunoglobulin homology unit. An early event may have been the partial duplication of this gene to yield separate genes for V and C homology units, since these have little amino acid sequence homology remaining today. This early duplication would allow separate regions to evolve that are specialized for membrane anchoring or for recognition.

Even though members of the immunoglobulin supergene family have evolved to have different functions, many encode proteins that interact with one another. For example, poly Ig receptors bind to antibodies, and T cell receptors may bind to MHC proteins. Did the fact that immunoglobulin homology units have a three-dimensional structure that gives them the ability to bind to one another contribute to the evolution of this very complex multicomponent system? More structural analysis of the members of this family will be needed to answer this intriguing question. The immunoglobulin supergene

Figure 23-50
Hypothetical scheme to explain the evolution of the immunoglobulin supergene family. The earliest member of the family may have encoded a cell surface protein. An early step in the evolution was the duplication of this gene and the divergence of the two resulting genes to encode V and C homology units. The protein structure of the present-day products of the genes are indicated at the bottom of the figure. The T8 molecule of humans has a mouse equivalent called Lyt 2. [From L. Hood, M. Kronenberg, and T. Hunkapiller, *Cell* 40 (1985):225–229.]

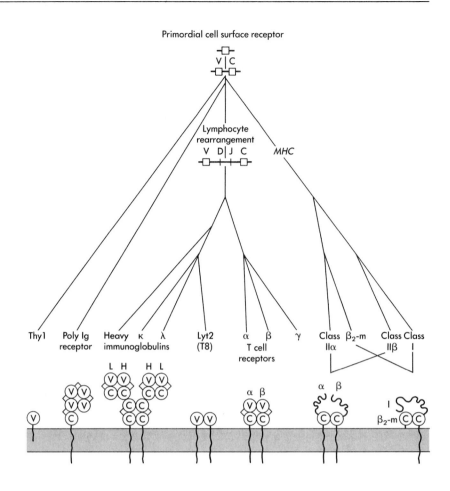

family occupies a significant fraction of the vertebrate genome, perhaps as much as 0.1 to 0.5 percent. If other complex biological systems evolved in a similar way, then we may ultimately be able to trace the evolution of vertebrate protein-coding genes to just a thousand or so ancestral genes.

Summary

Multicellular organisms have many defenses against invasions by microorganisms. Many of these defenses (e.g., an impermeable outer skin or the antiviral agent interferon) are relatively nonspecific. But in addition, vertebrates have evolved an immune system that provides specific defenses against almost any microorganism or other foreign substance that enters the body. Foreign substances that elicit an immune response are called antigens. The immune system has two types of defenses against antigens (microorganisms): Circulating proteins, called antibodies or immunoglobulins (Ig), bind to antigens and inactivate them or cause them to be destroyed; and specialized cells with surface receptors capable of recognizing specific antigens attack and destroy the antigens. The first type of response is referred to as humoral immunity; the second is called cellular immunity.

Immune responses are carried out by lymphocytes. These are small, round cells that originate in the bone mar-

row of adult animals and, after they mature, circulate within the lymph and circulatory systems. There are two major classes of lymphocytes: B lymphocytes synthesize antibodies, while T lymphocytes carry out cellular immune responses using membrane-anchored, antigen-specific T cell receptors. The ability of antibody molecules and T cell receptors to interact specifically with antigens is a consequence of their particular amino acid sequences. The immune system can recognize tens of thousands of different antigens. Even before it meets up with an antigen, the body synthesizes millions of antibody molecules and T cell receptors with slightly different amino acid sequences. Among this vast repertoire of molecules are invariably some that can bind to any incoming antigen.

According to the now-accepted theory of clonal selection, each B lymphocyte synthesizes an antibody of just one specificity, and each T cell has just one type of antigen receptor. Before encountering the antigen capable of binding its particular antibody, each B cell carries its antibodies as membrane-bound receptors. An encounter with antigen stimulates the cell to divide and also to differentiate to a mature plasma cell that secretes antibody capable of binding the stimulating antigen. This sequence of events applies to T cells as well, except that their antigen receptors remain bound to the cell surface after antigen stimulation.

Gene cloning finally provided the explanation for the genetic basis of immunological diversity and specificity, but the first insights came from comparing the amino acid sequences of many different antibody molecules. This was possible because cancers of B cells (called myelomas) yield large populations of cells producing a single type of antibody molecule.

Typical antibody molecules are tetramers composed of two identical light chains, each about 220 amino acids long, and two identical heavy chains, each about 440 amino acids long. The molecule is Y or T shaped and has two identical antigen binding sites in the arms of the Y. These sites are formed by the combination of part of the light chain and part of the heavy chain. When light and heavy chains of different antibody molecules are sequenced, it is apparent that there is extensive amino acid sequence variability in their amino-terminal regions (approximately the first 110 amino acids), while the remainder of the molecule belongs to one of a few types. The amino-terminal portion is called the variable, or V, region; the remainder is the constant, or C, region. Appropriately, it is the V regions of the light and heavy chains that together form the antigen binding sites. The most variable portions of the V regions, called the hypervariable regions or complementarity determining regions (CDRs), line the antigen binding pocket.

In the case of light chains, there are two major types of constant regions that define two types of light chains, κ (kappa) and λ (lambda). In the case of heavy chains, there are five major types of constant regions that define five classes of antibodies (IgM has a μ heavy chain defined by the sequence of its constant region, IgD has a δ heavy chain, IgG has a γ chain, IgA has an α chain, and IgE has an ε chain). The classes defined by the heavy-chain constant regions are important because these regions impart distinct biological functions to the different types of molecules.

Amino acid sequencing made it clear that antibody specificity and diversity are achieved by the joining of thousands and thousands of different variable regions to a few constant regions of either the light- or heavy-chain type. This led some immunologists to postulate that there might be separate genes for the variable and constant regions of the proteins. Genetic studies showed that the genes for heavy chains, κ light chains, and λ light chains lie on different chromosomes and that the genes for V regions are near to the genes for C regions. The central question was, Are there separate genes for all the different V regions, or is the variability in V-region sequence generated by some sort of mutational process within B cells? These two contrasting explanations for antibody diversity were called the germ line theory and the somatic theory.

When the genes encoding λ and κ light chains and heavy chains were cloned, it was clear that the somatic theory was correct. No antibody genes are inherited intact: They are assembled in somatic cells from multiple fragments of DNA and the assembly process generates enormous diversity in the sequences encoding variable regions. Consistent in spirit with the germ line theory, there are many inherited gene fragments that can take part in the assembly, and this, too, contributes to diversity. DNA encoding V regions lies near to, and probably upstream of, DNA encoding constant regions. However, the multigene families that encode the light and heavy chains are very large, with the heavy-chain locus alone probably occupying several thousand kilobase pairs of DNA.

Light-chain V-region "genes" are assembled by a single DNA recombination event that joins a V_L fragment encoding about 90 percent of the V region to a J_L fragment encoding about 10 percent. After this joining, the assembled V-region gene (which is really an exon) is still separated (by about 2 kilobase pairs) from a single exon encoding the constant region. The two are joined at the level of RNA by the standard RNA-splicing mechanism. Heavy-chain V-region genes are assembled by two DNA recombination events. The first joins a small D segment to a J_H segment, while the second joins a V_H segment to the rearranged D-J fragment. The resulting V-D-J exon is separated from the closest constant region exons by several kilobase pairs. There is a separate constant-region gene for each type of heavy chain (8 in the mouse), although these genes really are clusters of exons. In general, each exon encodes a structurally or functionally discrete portion of the constant region.

There are two ways in which information encoding a heavy-chain V region can be joined to a constant-region gene. An assembled V-D-J exon can be joined to either of the two closest C-region genes, μ or δ, by RNA splicing of long precursor transcripts. Alternatively, class switching, the joining of a V-D-J exon to other constant-region exons (γ, α, or ε) can occur by a third DNA rearrangement, the switch recombination, which moves the V region exon near to exons encoding these constant regions.

V-J or V-D-J joining involves recombination at specific sites, and specific sequences govern the event. Each DNA segment that can be joined is bordered by a conserved palindromic sequence of 7 bases, the heptamer, followed by either 12 (± 1) or 23 (± 2) nucleotides whose sequence is unimportant, followed by a conserved 9-base nonamer. Two fragments can only be joined if their heptamer-nonamers are oriented in opposite directions and are separated by 12 or 23 base pairs. These rules prevent incorrect fragments from joining. The recombination event does not always occur at the same nucleotide relative to the heptamer sequence. As a result, joining of the same two DNA segments can lead to different amino acid sequences, a phenomenon called junctional diversity that contributes greatly to V-region diversity. In addition, at D-J and V-D-J joints, N nucleotides can be added in a template-free fashion to further increase diversity.

In mice, at the heavy-chain locus, there are 200 to 300 V_H segments, about 10 D segments, and 4 J_H segments. At the κ light-chain locus, there are 100 to 300 V_κ segments and 4 J_κ segments. At the λ locus, there are 3 V_λ fragments and 3 J_λ fragments. As far as we know, any V segment can join to any D segment, for example. This random association, called combinatorial joining, generates great diversity. Since the D segment and the D-J and V-D-J joints or the V-J joint in light chains encode the third hypervariable regions, a part of the chain that makes the antigen binding cavity, it is clear why the above strategies generate so much diversity in antigen recognition.

A last genetic source of diversity in amino acid sequences is the subjection of fully assembled antibody genes to mutational processes that lead to single amino acid changes scattered randomly about the V region. This process is called somatic hypermutation.

So far, we have listed five sources of diversity in amino acid sequence of light and heavy chains: multiple germ line *V*, *D*, and *J* segments, random association of these fragments (called combinatorial joining) variability in the position of joining (called junctional diversity), N nucleotide addition in heavy chains, and somatic hypermutation. Finally, since antigen binding sites are constructed from the V regions of both light and heavy chains, if any light chain can associate with any heavy chain, then the number of antibody molecules with different antigen binding sites would be the number of different light chains multiplied by the number of heavy chains. In the mouse, this theoretical number greatly exceeds a staggering 300 million!

The rearrangement and expression of antibody genes are carefully regulated in conjunction with B cell differentiation. First, a stem cell undergoes *D-J* rearrangements at both of its heavy-chain loci. Then V_H segments are joined to the *D-J* fragment. As soon as a functional gene is made, further rearrangements stop. The first chains that are made use the μ constant region, so the cell is now a μ-positive pre–B cell. Light chains are then assembled by *V-J* joining. As soon as a functional chain that can associate with the cytoplasmic μ heavy chain is made, further rearrangements cease, and the assembled antibodies move to the cell surface. The cell with its surface immunoglobulin is now called a B_μ cell. At this point, the μ chains terminate with membrane-anchoring amino acids. All of the preceding steps occur in the absence of antigen. The next step seems to be to a $B_{\mu+\delta}$ cell with two types of antibody molecules on its surface. These two types have the same light-chain and heavy-chain V regions, but alternative RNA processing allows the heavy-chain V region to be joined to either the μ or δ constant region. It may be these $B_{\mu+\delta}$ cells, rather than B_μ cells, that are now competent to respond to antigens. In any case, in response to antigen, B_μ or $B_{\mu+\delta}$ cells differentiate to plasma cells and secrete μ antibodies. They also differentiate to produce memory cells. In addition, they will undergo class switching to transfer their heavy-chain V regions to γ, α, and ε constant regions so that antibodies of the same specificity but different classes can be made.

The progression of a competent virgin B cell to an antibody-secreting plasma cell requires more than a collision with its cognate antigen. In addition, T lymphocytes called helper T (T_H) cells must be present in most cases. These cells use their T cell receptors to interact with antigens that have been taken up and presented to them by a class of accessory cells called antigen-presenting cells. Thus, three cells interact to stimulate the final differentiation of the B cell.

Cytotoxic T cells are the cellular immune system's equivalent of antibody molecules, and they carry antigen-specific surface T cell receptors. In contrast to antibodies, however, T cell receptors can only see antigens when the T cell can also see normal cell surface proteins simultaneously. This mechanism ensures that T cells will be restricted to dealing with antigens located on cell surfaces and will not be confused by freely circulating antigens. Whether the same receptor looks at both molecules—antigen and normal cell surface protein—is not yet certain. The normal surface proteins that T cells see are coded for by the major histocompatibility complex (MHC). Cytotoxic T cells look at proteins coded for by the class I *MHC* genes. Helper T cells, which also can see antigens only if they look at normal cell surface proteins simultaneously, look at products of class II *MHC* genes.

Recently, T cell receptors for antigen and the genes that code for them have been isolated. The receptors are dimers composed of one α chain and one β chain. They have variable regions in their amino-terminal halves and constant regions in the carboxyl termini. The genes that encode them look very much like antibody genes and are clearly related evolutionarily. They undergo DNA rearrangement to produce functional genes. In addition to genes for the α and β chains, a third set of genes called γ has been discovered, and these genes also undergo rearrangement in T cells. The function of these genes is unknown. Possibly they are involved in T cell differentiation and thymic education.

The *MHC* is a huge multigene locus that should probably be considered as much a part of the immune system as the three antibody gene loci and the T cell receptor gene loci. It occupies several thousand kilobase pairs of DNA. It encodes the cell surface proteins that are essential to T cell function. Remarkably, the class I and II *MHC* gene products have structural similarity to antibodies. This has led to the notion that there is an extended gene family, called a supergene family, that may have evolved from a single primordial gene.

Bibliography

General References

Benacerraf, B., and E. R. Unanue. 1979. *Textbook of Immunology*. Baltimore: Williams & Wilkins.

Eisen, H. 1980. *Immunology*. 2nd ed. New York: Harper & Row.

Hood, L. E., I. L. Weissman, W. B. Wood, and J. H. Wilson. 1984. *Immunology*. 2nd ed. Menlo Park, Calif.: Benjamin/Cummings.

Kimball, J. W. 1983. *Introduction to Immunology*. New York: Macmillan.

Nisonoff, A. 1985. *Introduction to Molecular Immunology*. 2nd ed. Sunderland, Mass.: Sinauer.

Cited References

1. Hood, L. E., I. L. Weissman, W. B. Wood, and J. H. Wilson. 1984. *Immunology*. 2nd ed. Menlo Park, Calif.: Benjamin/Cummings.

2. Nisonoff, A. 1985. *Introduction to Molecular Immunology*. 2nd ed. Sunderland, Mass.: Sinauer.

3. Eisen, H. 1980. *Immunology*. 2nd ed. New York: Harper and Row.

4. Kimball, J. W. 1983. *Introduction to Immunology*. New York: Macmillan.

5. Benacerraf, B., and E. R. Unanue. 1979. *Textbook of Immunology.* Baltimore: Williams & Wilkins.

6. Gowans, J. L., and E. J. Knight. 1964. "The Route of Recirculation of Lymphocytes in the Rat." *Proc. Roy. Soc. London (B)* 159:257–282.

7. Miller, J. F. A. P. 1961. "Immunological Functions of the Thymus." *Lancet.*

8. Landsteiner, K. 1945. *The Specificity of Serological Reactions.* Rev. ed. Cambridge, Mass.: Harvard University Press. Reprinted by Dover, New York, 1962.

9. Westhof, E., D. Altschuh, D. Moras, A. C. Bloomer, A. Mondragon, A. Klug, and M. H. V. Van Regenmortel. 1984. "Correlation Between Segmental Mobility and the Location of Antigenic Determinants in Proteins." *Nature* 311:123–126.

10. Tainer, J. A., E. D. Getzoff, H. Alexander, R. A. Houghten, A. J. Olson, R. A. Lerner, and W. A. Hendrickson. 1984. "The Reactivity of Anti-Peptide Antibodies Is a Function of the Atomic Mobility of Sites in a Protein." *Nature* 312:127–134.

11. Haber, E. 1964. "Recovery of Antigen Specificity After Denaturation and Complete Reduction of Disulfides in a Papain Fragment of Antibody." *Proc. Nat. Acad. Sci.* 52:1099–1106.

12. Burnet, F. M. 1959. *The Clonal Selection Theory of Acquired Immunity.* Cambridge, Eng.: University Press.

13. Ada, G. L., and P. Byrt. 1969. "Specific Inactivation of Antigen-Reactive Cells with I^{125}-Labeled Antigen." *Nature* 222:1291–1292.

14. Owen, R. D. 1945. "Immunogenic Consequence of Vascular Anastomoses Between Bovine Twins." *Science* 102:400–401.

15. Billingham, R. E., L. Brent, and P. B. Medawar. 1953. "Actively Acquired Tolerance of Foreign Cells." *Nature* 172:603–606.

16. Felton, L. D., G. Kauffmann, B. Prescott, and B. Ottinger. 1955. "Studies on the Mechanism of Immunological Paralysis Induced in Mice by Pneumococcal Polysaccharides." *J. Immunol.* 74:17–26.

17. Edelman, G. M. 1970. "The Structure and Function of Antibodies." *Sci. Amer.* 223:34–42.

18. Porter, R. R. 1973. "Structural Studies of Immunoglobulins." *Science* 180:713–716.

19. Nisonoff, A., J. E. Hopper, and S. B. Spring. 1975. *The Antibody Molecule.* New York: Academic Press.

20. Kohler, G., and C. Milstein. 1975. "Continuous Cultures of Fused Cells Secreting Antibody of Predefined Specificity." *Nature* 256:495–497.

21. Milstein, C. 1980. "Monoclonal Antibodies." *Sci. Amer.* 243:66–74.

22. Potter, M. 1972. "Immunoglobulin-Producing Tumors and Myeloma Proteins of Mice." *Physiol. Rev.* 52:631–719.

23. Hilschmann, N., and L. C. Craig. 1965. "Amino Acid Sequence Studies with Bence-Jones Proteins." *Proc. Nat. Acad. Sci.* 53:1403–1409.

24. Wu, T. T., and E. A. Kabat. 1970. "An Analysis of the Sequences of the Variable Regions of Bence Jones Proteins and Myeloma Light Chains and Their Implications for Antibody Complementarity." *J. Exp. Med.* 132:211–250.

25. Edelman, G. M. 1969. "The Covalent Structure of an Entire γG-Immunoglobulin Molecule." *Proc. Nat. Acad. Sci.* 63:78–85.

26. Amzel, L. M., and R. J. Poljak. 1979. "Three Dimensional Structure of Immunoglobulins." *Ann. Rev. Biochem.* 48:961–967.

27. Silverton, E. W., M. A. Navia, and D. R. Davies. 1977. "Three-Dimensional Structure of an Intact Human Immunoglobulin." *Proc. Nat. Acad. Sci.* 74:5140–5144.

28. Edmonson, A. B., K. R. Ely, E. E. Abola, M. Schiffer, and N. Panagiotopoulos. 1975. "Rotational Allomerism and Divergent Evolution of Domains in Immunoglobulin Light Chains." *Biochemistry* 14:3953–3961.

29. Colman, P. M., J. Diesenhofer, R. Huber, and W. Palm. 1976. "Structure of the Human Antibody Molecule 'Kol', an Electron Density Map at 5 Å Resolution." *J. Mol. Biol.* 100:257–278.

30. Hood, L. E., and D. W. Talmadge. 1970. "Mechanism of Antibody Diversity: Germ Line Basis for Variability." *Science* 168:325–334.

31. Lennox, E. S., and M. Cohn. 1967. "Immunoglobulins." *Ann. Rev. Biochem.* 36:365–406.

32. Dreyer, W. J., and J. C. Bennett. 1965. "The Molecular Basis of Antibody Formation: A Paradox." *Proc. Nat. Acad. Sci.* 54:864–869.

33. Hozumi, N., and S. Tonegawa. 1976. "Evidence for Somatic Rearrangement of Immunoglobulin Genes Coding for Variable and Constant Regions." *Proc. Nat. Acad. Sci.* 73:3628–3632.

34. Fudenberg, H. H., A. C. Wang, J. R. L. Pink, and A. S. Levin. 1971. *Ann. N.Y. Acad. Sci.* 190:501–506.

35. Brack, C., M. Hirama, R. Lenhard-Schuller, and S. Tonegawa. 1978. "A Complete Immunoglobulin Gene Is Created by Somatic Recombination." *Cell* 15:1–14.

36. Seidman, J. G., E. E. Max, and P. Leder. 1979. "A k-Immunoglobulin Gene Is Formed by Site-Specific Recombination Without Further Somatic Mutation." *Nature* 280:370–375.

37. Eisen, H. N., and E. B. Reilly. 1985. "Lambda Chains and Genes in Inbred Mice." *Ann. Rev. Immunol.* 3:337–365.

38. Early, P., H. Huang, M. Davis, K. Calame, and L. Hood. 1980. "An Immunoglobulin Heavy-Chain Variable Region Gene Is Generated from Three Segments of DNA: V_H, D and J_H." *Cell* 19:981–992.

39. Sakano, H., R. Maki, Y. Kurosawa, W. Roeder, and S. Tonegawa. 1980. "Two Types of Somatic Recombination Are Necessary for the Generation of Complete Immunoglobulin Heavy-Chain Genes." *Nature* 286:676–683.

40. Sakano, H., Y. Kurosawa, M. Weigert, and S. Tonegawa. 1981. "Identification and Nucleotide Sequence of a Diversity DNA Segment (D) of Immunoglobulin Heavy-Chain Genes." *Nature* 290:562–565.

41. Sakano, H., K. Huppi, G. Heinrich, and S. Tonegawa. 1979. "Sequences at the Somatic Recombination Sites of Immunoglobulin Light-Chain Genes." *Nature* 280:288–294.

42. Max, E. E., J. G. Seidman, and P. Leder. 1979. "Sequences of Five Potential Recombination Sites Encoded Close to an Immunoglobulin k Constant Region Gene." *Proc. Nat. Acad. Sci.* 76:3450–3454.

43. Kurosawa, Y., and S. Tonegawa. 1982. "Organization, Structure, and Assembly of Immunoglobulin Heavy Chain Diversity DNA Segments." *J. Exp. Med.* 155:201–218.

44. Alt, F., and D. Baltimore. 1982. "Joining of Immunoglobulin Heavy Chain Gene Segments: Implications from a Chromosome with Evidence of Three D-J_H Fusions." *Proc. Nat. Acad. Sci.* 79:4118–4122.

45. Cohn, M., B. Blomberg, W. Geckeler, W. Raschke, R. Riblet, and M. Weigert. 1974. "First Order Considerations in Analyzing the Generator of Diversity." In *The Immune System Genes, Receptors, Signals,* ed. E. Sercarz, A. R. Williamson, and C. F. Fox. New York: Academic Press, pp. 89–117.

46. Bernard, O., N. Hozumi, and S. Tonegawa. 1978. "Sequences of Mouse Immunoglobulin Light Chain Genes Before and After Somatic Changes." *Cell* 15:1133–1144.

47. Kim, S., M. Davis, E. Sinn, P. Patten, and L. Hood. 1981. "Antibody Diversity: Somatic Hypermutation of Rearranged V_H Genes." *Cell* 27:573–581.

48. Siekevitz, M., S. Y. Huang, and M. L. Gefter. 1983. "The Genetic Basis of Antibody Production: A Single Heavy Chain Variable Region Gene Encodes All Molecules Bearing the Dominant Anti-Arsonate Idiotype in the Strain A Mouse." *Eur. J. Immunol.* 13:123–132.

49. Tonegawa, S. 1983. "Somatic Generation of Antibody Diversity." *Nature* 302:575–581.

50. Kindt, T. J., and J. D. Capra. 1984. *The Antibody Enigma.* N.Y.: Plenum Press.

51. Gillies, S. D., S. L. Morrison, V. T. Oi, and S. Tonegawa. 1983. "A Tissue-Specific Transcription Enhancer Element Is Located in the Major Intron of a Rearranged Immunoglobulin Heavy Chain Gene." *Cell* 33:717–728.

52. Banerji, J., L. Olson, and W. Schaffner. 1983. "A Lymphocyte-Specific Cellular Enhancer Is Located Downstream of the Joining Region in Immunoglobulin Heavy Chain Genes." *Cell* 33:729–740.

53. Falkner, F. G., and H. G. Zachau. 1984. "Correct Transcription of an Immunoglobulin-k Gene Requires an Upstream Fragment Containing Conserved Sequence Elements." *Nature* 310:71–74.

54. Alt, F. W., G. D. Yancopoulos, T. K. Blackwell, C. Wood, E. Thomas, M. Boss, R. Coffman, N. Rosenberg, S. Tonegawa, and D. Baltimore. 1984. "Ordered Rearrangement of Immunoglobulin Heavy Chain Variable Region Segments." *EMBO J.* 3:1209–1219.

55. Alt, F., N. Rosenberg, S. Lewis, E. Thomas, and D. Baltimore. 1981. "Organization and Reorganization of Immunoglobulin Genes in A-MULV-Transformed Cells: Rearrangement of Heavy But Not Light Chain Genes." *Cell* 27:381–390.

56. Reth, M. G., and F. W. Alt. 1984. "Novel Immunoglobulin Heavy Chains Are Produced from DJ$_H$ Gene Segment Rearrangements in Lymphoid Cells." *Nature* 312:418–423.

57. Yancopoulos, G. D., and F. W. Alt. 1985. "Developmentally Controlled and Tissue-Specific Expression of Unrearranged V$_H$ Gene Segments." *Cell* 40:271–281.

58. Ritchie, K. A., R. L. Brinster, and U. Storb. 1984. "Allelic Exclusion and Control of Endogenous Immunoglobulin Gene Rearrangement in κ Transgenic Mice." *Nature* 312:517–520.

59. Early, P., J. Rogers, M. Davis, K. Calame, M. Bond, R. Wall, and L. Hood. 1980. "Two mRNAs Can Be Produced from a Single Immunoglobulin mu Gene by Alternative RNA Processing Pathways." *Cell* 20:313–319.

60. Rogers, J., P. Early, C. Carter, K. Calame, M. Bond, and R. Wall. 1980. "Two mRNAs with Different 3' Ends Encode Membrane-Bound and Secreted Forms of Immunoglobulin mu Chain." *Cell* 20:303–312.

61. Maki, R., W. Roeder, A. Traunecker, C. Sidman, M. Wabl, W. Raschke, and S. Tonegawa. 1981. "The Role of DNA Rearrangement and Alternative RNA Processing in the Expression of Immunoglobulin Delta Genes." *Cell* 24:353–365.

62. Blattner, F. R., and P. W. Tucker. 1984. "The Molecular Biology of Immunoglobulin D." *Nature* 307:417–422.

63. Davis, M. M., K. Calame, P. W. Early, D. L. Livant, R. Joho, I. L. Weissman, and L. Hood. 1980. "An Immunoglobulin Heavy-Chain Gene Is Formed by at Least Two Recombinational Events." *Nature* 283:733–739.

64. Maki, R., A. Traunecker, H. Sakano, W. Roeder, and S. Tonegawa. 1980. "Exon Shuffling Generates an Immunoglobulin Heavy Chain Gene." *Proc. Nat. Acad. Sci.* 77:2138–2145.

65. Kataoka, T., T. Miyata, and T. Honjo. 1981. "Repetitive Sequences in Class-Switch Recombination Regions of Immunoglobulin Heavy Chain Genes." *Cell* 23:357–368.

66. Mitchison, N. A. 1971. "The Carrier Effect in the Secondary Response to Hapten-Protein Conjugates. II. Cellular Cooperation." *Eur. J. Immunol.* 1:18.

67. Sprent, J. 1978. "Role of H-2 Gene Products in the Function of T Helper Cells from Normal and Chimeric Mice in Vivo." *Immunol. Rev.* 42:108–137.

68. Unanue, E. R. 1984. "Antigen-Presenting Function of the Macrophage." *Ann. Rev. Immunol.* 2:395–428.

69. Cantor, H., and E. A. Boyse. 1975. "Functional Subclasses of T Lymphocytes Bearing Different Ly Antigens. I. The Generation of Functionally Distinct T-cell Subclasses Is a Differentiation Process Independent of Antigen." *J. Exp. Med.* 141:1377–1389.

70. Gershon, R. K., and K. Kondo. 1972. "Infectious Immunological Tolerance." *Immunology* 21:903–914.

71. Reinherz, E., and S. F. Schlossman. 1980. "Differentiation and Function of Human T Lymphocytes." *Cell* 19:821–827.

72. Zinkernagel, R. M., and P. C. Doherty. 1974. "Restriction of in Vitro T Cell-Mediated Cytotoxicity in Lymphocytic Choriomeningitis Within a Syngeneic or Semiallogeneic System." *Nature* 248:701–702.

73. Zinkernagel, R. M. 1976. "H-2 Compatibility Requirement for Virus Specific, T-Cell-Mediated Cytolysis. The H-2K Structure Involved Is Coded by a Single Cistron Defined by H-2K^b Mutant Mice." *J. Exp. Med.* 143:437–443.

74. Bevan, M. J., and P. Fink. 1978. "The Influence of Thymus H-2 Antigens on the Specificity of Maturing Killer and Helper T Cells." *Immunol. Rev.* 42:3–19.

75. Kappler, J. W., B. Skidmore, J. White, and P. Marrack. 1981. "Antigen-Inducible, H-2-Restricted, Interleukin-2-Producing T Cell Hybridomas. Lack of Independent Antigen and H-2

76. Babbitt, B. P., P. M. Allen, G. Matsueda, E. Haber, and E. R. Unanue. 1985. "Binding of Immunogenic Peptides to Ia Histocompatibility Molecules." *Nature* 317:359–361.

77. Allison, J. P., B. W. McIntyre, and D. Bloch. 1982. "Tumor-Specific Antigen of Murine T-Lymphoma Defined with Monoclonal Antibody." *J. Immunol.* 129:2293–2300.

78. Meuer, S. C., O. Acuto, T. Hercend, S. F. Schlossman, and E. L. Reinherz. 1984. "The Human T-Cell Receptor." *Ann. Rev. Immunol.* 2:23–50.

79. Haskins, K., J. Kappler, and P. Marrack. 1984. "The Major Histocompatibility Complex-Restricted Antigen Receptor on T Cells." *Ann. Rev. Immunol.* 2:51–66.

80. Yanagi, Y., Y. Yoshikai, K. Leggett, S. P. Clark, I. Aleksander, and T. W. Mak. 1984. "A Human T Cell-Specific cDNA Clone Encodes a Protein Having Extensive Homology to Immunoglobulin Chains." *Nature* 308:145–149.

81. Hedrick, S. M., E. A. Nielsen, J. Kavaler, D. I. Cohen, and M. M. Davis. 1984. "Sequence Relationships Between Putative T-Cell Receptor Polypeptides and Immunoglobulins." *Nature* 308:153–158.

82. Chien, Y-h., D. M. Becker, T. Lindsten, M. Okamura, D. I. Cohen, and M. M. Davis. 1984. "Third Type of Murine T Cell Receptor Gene." *Nature* 312:31–35.

83. Saito, H., D. M. Kranz, Y. Takagaki, A. C. Hayday, H. N. Eisen, and S. Tonegawa. 1984. "A Third Rearranged and Expressed Gene in a Clone of Cytotoxic T Lymphocytes." *Nature* 312:36–40.

84. Saito, H., D. M. Kranz, Y. Takagaki, A. C. Hayday, H. N. Eisen, and S. Tonegawa. 1984. "Complete Primary Structure of a Heterodimeric T-Cell Receptor Deduced from cDNA Sequences." *Nature* 309:757–762.

85. Davis, M. M. 1985. "Molecular Genetics of the T Cell-Receptor Beta Chain." *Ann. Rev. Immunol.* 3:537–560.

86. Malissen, M., K. Minard, S. Mjolsness, M. Kronenberg, J. Goverman, T. Hunkapiller, M. B. Prystowsky, Y. Yoshikai, F. Fitch, T. W. Mak, and L. Hood. 1984. "Mouse T Cell Antigen Receptor: Structure and Organization of Constant and Joining Gene Segments Encoding the Beta Polypeptide." *Cell* 37:1101–1110.

87. Hayday, A. C., D. J. Diamond, G. Tanigawa, J. S. Heilig, V. Folsom, H. Saito, and S. Tonegawa. 1985. "Unusual Organization and Diversity of T-Cell Receptor α-Chain Genes." *Nature* 316:828–832.

88. Winoto, A., S. Mjolsness, and L. Hood. 1985. "Genomic Organization of the Genes Encoding Mouse T-Cell Receptor α-Chain." *Nature* 316:832–836.

89. Yoshikai, Y., S. P. Clark, S. Taylor, U. Sohn, B. I. Wilson, M. D. Minton, and T. W. Mak. 1985. "Organization and Sequences of the Variable, Joining and Constant Region Genes of the Human T-Cell Receptor α-Chain." *Nature* 316:837–840.

90. Raulet, D. H., R. D. Garman, H. Saito, and S. Tonegawa. 1985. "Developmental Regulation of T Cell Receptor Gene Expression." *Nature* 314:103–107.

91. Morgan, D. A., F. W. Ruscetti, and R. Gallo. 1976. "Selective in Vitro Growth of T-Lymphocytes from Normal Human Bone Marrows." *Science* 193:1007–1009.

92. Robb, R. J., A. Munck, and K. A. Smith. 1981. "T-Cell Growth Factor Receptors: Quantitation, Specificity and Biological Relevance." *J. Exp. Med.* 154:1455–1474.

93. Smith, K. A. 1984. "Interleukin-2." *Ann. Rev. Immunol.* 2:319–333.

94. Gorer, P. A. 1936. "The Detection of Antigenic Differences in Mouse Erythrocytes by the Employment of Immune Sera." *Br. J. Exp. Pathol.* 17:42–50.

95. McDevitt, H. O., and B. Benacerraf. 1969. "Genetic Control of Specific Immune Responses." *Adv. Immunol.* 11:31–74.

96. Klein, J. 1975. "Biology of the Mouse Histocompatibility-2 Complex." New York: Springer.

97. Klein, J., A. Juretic, C. N. Baxevanis, and Z. A. Nagy. 1981. "The Traditional and New Version of the Mouse H-2 Complex." *Nature* 291:455–460.

98. Orr, H. T., J. A. Lopez de Castro, P. Parham, H. L. Ploegh, and J. L. Strominger. 1979. "Comparison of Amino Acid Se-

Recognition." *J. Exp. Med.* 153:1198–1214.

quences of Two Human Histocompatibility Antigens, HLA-A2 and HLA-B7: Location of Putative Alloantigenic Sites." *Proc. Nat. Acad. Sci.* 76:4395–4399.

99. Nathenson, S. G., H. Uehara, B. M. Ewenstein, T. J. Kindt, and J. E. Coligan. 1981. "Primary Structural Analysis of the Transplantation Antigens of the Murine H-2 Major Histocompatibility Complex." *Ann. Rev. Biochem.* 50:1025–1052.

100. Sood, A. K., D. Pereira, and S. M. Weissman. 1981. "Isolation and Partial Nucleotide Sequence of a cDNA Clone for Human Histocompatibility Antigen HLA-B by Use of an Oligodeoxynucleotide Primer." *Proc. Nat. Acad. Sci.* 78:616–620.

101. Ploegh, H. L., H. T. Orr, and J. L. Strominger. 1980. "Molecular Cloning of a Human Histocompatibility Antigen cDNA Fragment." *Proc. Nat. Acad. Sci.* 77:6081–6085.

102. Hood, L, M. Steinmetz, and B. Malissen. 1983. "Genes of the Major Histocompatibility Complex of the Mouse." *Ann. Rev. Immunol.* 1:529–568.

103. Hood, L., M. Kronenberg, and T. Hunkapillar. 1985. "T Cell Antigen Receptors and the Immunoglobulin Supergene Family." *Cell* 40:225–229.

CHAPTER 24

The Extraordinary Diversity of Eucaryotic Viruses

Viruses are well known agents of disease in humans and other animals, as well as in plants. They can cause relatively benign infections, such as the common cold and warts, or they can induce severe conditions, such as cancer, polio, and AIDS. Quite apart from their importance as pathogens, however, animal viruses have served a critical role in biological research as probes to the inner workings of eucaryotic cells. Long before gene-cloning techniques were available, viruses provided an easy source of pure genes whose functioning obeys the rules of the cell in which they replicate. Since viral replication often occurs at the expense of normal cellular processes, viral gene expression is frequently more efficient, and thus easier to study, than cellular gene expression. Many major discoveries about the molecular biology of higher cells came first from studying their viruses: RNA splicing, 5' caps and 3' poly A tracts on mRNAs, the pathway for synthesis of cell surface proteins, the role of enhancer elements in transcription, and reverse transcription of RNA into DNA are just a few examples. Even today, the remarkably numerous strategies that viruses use to parasitize host cells continue to provide important insights into the biochemical functions of eucaryotic cells. But quite apart from the commendable goals of conquering disease and probing the cell, the study of animal viruses has become an end in itself, illuminating the extraordinary diversity and adaptability of even the simplest organisms.

AN INTRODUCTION TO ANIMAL VIRUSES

Viruses Come in Many Sizes and a Few Shapes[1–5]

A virus is an intracellular parasite that can exist in two fundamentally different forms, one inside the cell and one outside the cell. The extracellular form is called a **virion,** or virus particle. Virions consist of the viral genome, either RNA or DNA, associated proteins, and sometimes other chemical constituents, such as lipids. The proteins (and lipids) protect the genes from the ravages of the extracellular environment, ensure their entry into the host cell, and often carry out some initial steps of the replication cycle. During infection, virion structure is largely or completely destroyed, so that the genome is free to be transcribed and replicated.

At first sight, virions of animal viruses give the impression of great variety (Figure 24-1). Their diameter ranges from about 10 to 200 nm, which corresponds to a volume difference of almost 10,000-fold. They can also have exotic projections at their surface. This first impression is misleading, however, because all but the most complex viruses are constructed according to one of two basic plans. This simplicity is necessitated by the small size of most viruses. To construct a sphere the size of a small virus (like polio) from average-size proteins requires about 5.5 million daltons of protein. However, a genome that can fit within such a sphere is only big enough to code for 300 kdal of protein, and not all of this is available for building the virion. Clearly, the portion of a virus particle that is made of virus-coded proteins, the portion known as the **capsid,** must be made from just one or a few different molecules repeated over and over. Given this restriction, virion capsids are limited to one of two basic organizations: a hollow helix of undefined length with the genome in the center, resembling a rod, or a quasi-spherical *icosahedral* structure (Chapter 17).

Further contributing to the varied appearance of animal virus virions is the presence or absence of an outer membranous coat. Virions that contain only a genome and capsid are considered *naked.* In contrast, some viruses have a lipid bilayer, the **envelope,** surrounding their capsid. In such viruses, the capsid plus genome is called the **nucleocapsid.** The envelope is derived from the cell membrane in a two-step process called **budding** (Figure 24-2). First, virus-coded glycoproteins are inserted into the membrane. Then virion capsids may interact with the cytoplasmic ends of the glycoproteins, causing the membrane to wrap around the capsid and eventually to pinch off from the cell. The cell membrane can seal itself behind the budding virus. Some enveloped viruses can be released continuously by infected cells for many generations without harmful effects.

Unlike phages, animal viruses never possess the elaborate tails needed for injecting genes through the bacterial cell wall. Rather, capsid proteins of naked virions or viral glycoproteins protruding from the surface of enveloped virions interact with receptors on animal cell membranes, causing the virions to be engulfed and subsequently uncoated inside the cell.

Virus Genomes Can Be Single- or Double-Stranded, Linear or Circular, RNA or DNA[6]

In contrast to cells, whose genomes are always double-stranded DNA, viral genomes can have an array of configurations and can be RNA or DNA. Both RNA and DNA genomes can be single- or double-stranded. All known double-stranded RNA genomes are *segmented;* that is, they consist of several RNA molecules, each carrying a different gene, rather than a continuous, covalent polynucleotide. Some single-stranded RNA genomes are also segmented. Double-stranded DNA viral genomes can be either linear or circular. The circular DNA can be covalently closed on both or only one of its strands. One linear double-stranded DNA virus genome, that of the poxvirus vaccinia, has its two strands covalently joined at their ends. As a result, poxvirus DNA forms a single-stranded circle upon denaturation. There are no known segmented DNA viruses, and except for the plant viroids discussed at the end of this chapter, there are no known eucaryotic viruses with circular RNA genomes.

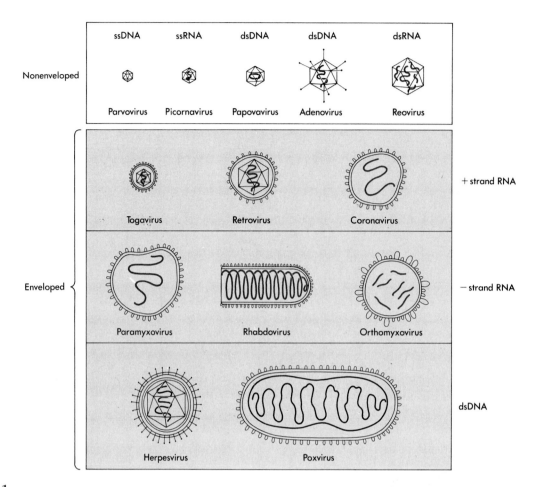

Figure 24-1

Electron micrographs and sketches of representative members of the important families of animal viruses. Sketches show relative sizes of different viruses; ss = single-strand, ds = double-strand. [Sketches after R. E. F. Matthews, *Intervirology* 12 (1979):3–5. Electron micrographs courtesy of Frederick A. Murphy.]

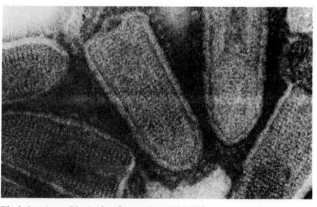

Rhabdorvirus (Vesicular Stomatitis) 216,000 ×

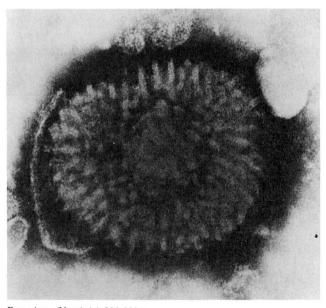

Pox virus (Vaccinia) 200,000 ×

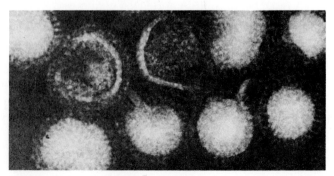

Orthomyxovirus (Influenza A) 135,000 ×

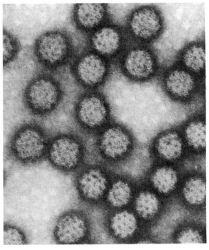

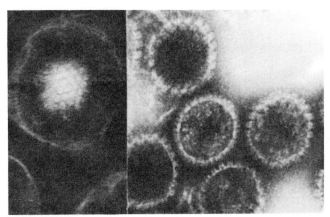

Reovirus 135,000 ×

Herpes (Herpes Simplex Type 1) enveloped (left) and naked capsids (right) 189,000 ×

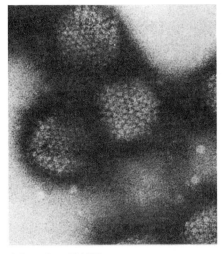

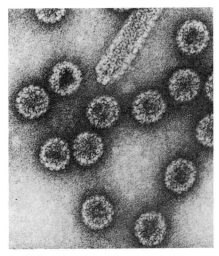

Adenovirus 234,000 ×

Togavirus (Semlikiforest) 120,000 ×

Papovavirus (Papillomavirus; human wart virus) 135,000 ×

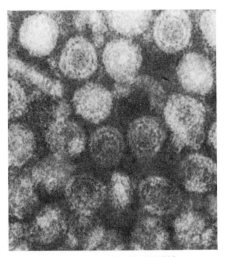

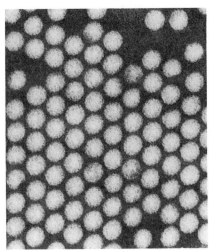

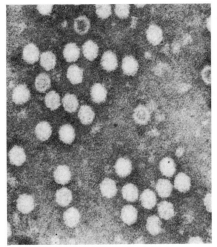

Hepadnavirus (Hepatitis B) 225,000 ×

Picornavirus (Polio virus) 180,000 ×

Parvovirus 171,000 ×

Figure 24-2
Release of virions by budding. Specific association between the cytoplasmic ends of the viral spike glycoproteins and the capsid protein may cause the membrane to enclose around the nucleocapsid.

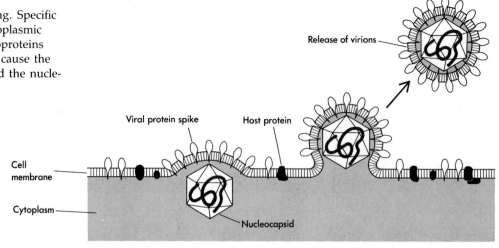

RNA virus genomes vary in size within a narrow range, about 7 to 20 kilobases. DNA virus genomes, on the other hand, can be as small as 3 kilobases or as large as 200 kilobase pairs. There may be a good reason for the upper limit on RNA genome size. Cells have mechanisms to repair DNA but not RNA. Unless the virus were to evolve sophisticated proofreading systems to correct errors made during replication, it would be almost impossible to copy a very large RNA molecule without introducing deleterious mutations. In fact, replication of RNA virus genomes is accompanied by very high mutation rates (10^{-3} to 10^{-4} is common). At this rate, a genome of 100 kilobases would suffer 10 to 100 mutations every time it was copied, and almost no progeny would be viable.

Cells in Culture Revolutionized the Study of Animal Viruses[7, 8, 9]

A number of animal viruses, such as rabies and smallpox, were well known in the last century, long before the discovery of bacterial viruses. A good understanding of animal viruses, however, lagged well behind that of bacteriophages. Essential to the development of modern animal virology was the discovery in the early 1950s that animal viruses can grow on tissue culture cells in vitro. Until then, virtually all work with viruses had to be performed in living animals. Besides being slow and expensive, the use of whole animals made it essentially impossible to carry out quantitative molecular biological studies.

Virologists use three types of cell culture systems. **Primary cultures** consist of cells freshly removed from an animal and placed in a dish with cell culture medium. While closest to the conditions of the living host, these cultures are usually complex mixtures of cell types, many of which are incapable of further division. After growing for a few generations in vitro, primary cultures give rise to a second type of culture, **cell strains,** which usually consist mostly of **fibroblasts,** relatively nondifferentiated precursors to connective tissue. They will grow for about 50 generations in the laboratory, but then they usually die. The third type of culture, **cell lines,** grow indefinitely in vitro. Many frequently used cell lines, such as the human HeLa cell line or mouse L cell line, were derived from naturally occurring tumors, while others, such as the mouse 3T3 cell line, grew out of cell strains, probably as the result of rare mutations. Although the cells of cell lines are not really normal cells, they are by far the easiest to work with of the three types of cultures and they provide the most reproducible results; thus, they are strongly preferred by virologists. In fact, viruses that grow well on established cell lines were the first to be studied extensively at the molecular level.

As with bacteriophages, animal virus replication in cell culture is often detected because the virus kills, or **lyses,** the host cell. A single virus particle can give rise to a **plaque,** just as a phage can. Some animal viruses that do not lyse the cells they grow on can be detected in cell culture by biochemical or immunological methods or by observation of altered cellular morphology.

A Large Number of Animal Viruses Have Been Isolated[10]

By now, there are many thousands of animal virus isolates. At first they seemed to embrace a diverse and confusing array of characteristics, at least in terms of the nature and severity of the diseases they cause, the host animal species they infect, and their mode of transmission. But with the advent of modern biochemical approaches, it became apparent that the thousands of isolates can be classified into a reasonably small number of virus *families* based on fundamental characteristics of the virion: namely, particle morphology, genome structure, and the presence of common antigenic determinants on virion proteins. The important families of animal viruses are presented in Table 24-1.

Although there is little or no relationship between virus families, individual isolates within a given family are strikingly similar to one another. All adenoviruses, for example, have naked icosahedral capsids of 252 subunits with prominent spikes at the vertices and a double-stranded DNA genome about 36,000 base pairs long. This precise description applies to adenoviruses isolated from species as diverse as frogs, turkeys, and humans, and from countries all over the world, although the precise nucleotide sequence of the genome varies considerably. These observations tell us that the evolution of a successful virus type is probably a rare event. Once achieved, however, rapid replication and high mutation rates allow the virus to adapt itself readily to grow in the maximum number of hosts.

Mechanisms of Genome Expression Provide a Molecular Biologist's Classification of Animal Viruses[11–17]

Although the thousands of known animal viruses can be grouped by evolutionary relationships, the resulting number of families, with their diversity of replication and expression strategies, can still overwhelm the beginning virologist. Around 1970, it became clear that although the nature of the genome and its mode of replication vary widely among families, all viruses have a common need for cellular machinery for the synthesis of virus-coded proteins. Thus, viral mRNA is the starting point for a common pathway that all viruses

Table 24-1 Important Families of Animal Viruses

Family	Examples	Genome Size (kb)	Virion
1. dsDNA Viruses			
Papovavirus	Polyoma, SV40, wart viruses	5–8	Naked, icosahedral
Adenovirus	Many human and animal adenoviruses	35–40	Naked, icosahedral
Herpesvirus	Herpes simplex I, II, varicella-zoster (chickenpox), Epstein-Barr virus	120–200	Enveloped, icosahedral
Poxvirus	Variola (smallpox), vaccinia	120–300	Enveloped, complex
Iridovirus	Mostly insect viruses	150–300	Enveloped, icosahedral
2. ssDNA Viruses			
Parvovirus	Adeno-associated virus, minute virus of mice, canine, feline, and human parvovirus	4–5	Naked, icosahedral
3. (+) RNA Viruses			
Picornavirus	Polio, common cold viruses, foot and mouth disease, many enteric viruses	7	Naked, icosahedral
Togavirus	Sindbis virus, Semliki forest virus, yellow fever virus, rubella virus, encephalitis viruses	12	Enveloped, icosahedral
Coronavirus	Human common-cold–like diseases, mouse hepatitis virus	16–21	Enveloped, helical
4. (−) RNA Viruses			
Rhabdovirus	Rabies, vesicular stomatitis virus	12–15	Enveloped, helical
Paramyxovirus	Newcastle disease, measles, mumps, respiratory syncytial virus, Sendai virus	15	Enveloped, helical
Orthomyxovirus	Influenza	14 (8 segments)	Enveloped, helical
Bunyavirus	Bunyamwera, Uukuniemi, LaCrosse, Rift Valley fever viruses	15–20 (3 segments)	Enveloped, helical

Table 24-1   *(Continued)*

Family	Examples	Genome Size (kb)	Virion
Arenavirus	Lassa virus, lymphocytic choriomeningitis virus	12–15 (2 segments)	Enveloped, helical
5. dsRNA Viruses			
Reovirus	Reovirus, human and animal diarrhea viruses	18–30 (10 segments)	Naked, icosahedral
6. RNA-DNA Viruses			
Retrovirus	Rous sarcoma, avian, feline, and murine leukemia, mouse mammary tumor virus, HTLV	7–10 (diploid)	Enveloped, icosahedral
7. DNA-RNA Viruses			
Hepadnavirus	Hepatitis B virus of humans, birds, and rodents	3	Enveloped, icosahedral

ultimately use. Recognition of the central role of mRNA leads to a molecular biological classification of animal viruses based on the relationship between the viral genome and mRNA and also on the genome's mode of replication. The system is shown in Figure 24-3.

It is easy to understand the relationship of double-stranded DNA virus genomes to the synthesis of mRNA, since this process is identical to that used by the host. In fact, most DNA viruses replicate in the nucleus, where they can have access to the cell's RNA polymerase, as well as capping, splicing, and poly A–adding enzymes, to process their transcripts. Single-stranded DNA genomes also have a fairly conventional relationship to mRNA, since they quickly become double-stranded DNA after infection and can follow the same fate as cellular genes. But what about viruses with RNA genomes? Like the RNA phages discussed in Chapter 17, some animal viruses have RNA genomes that correspond to mRNA and that can function as messages even in vitro. These viral genomes are called *positive* or *plus (+) strand* genomes. Many other RNA viruses have *negative* or *minus (−) strand* genomes, meaning they are complementary to the sense or mRNA strand. Since animal cells lack enzymes to copy RNA, and since the negative strands cannot be translated, − strand RNA is functionally dead. Viruses with − strand genomes must encode a **transcriptase** that can make mRNA from a − strand RNA template. Furthermore, the enzyme must be packaged in the virion in association with the viral genome. Then after infection, the genome-associated transcriptase synthesizes viral mRNA, allowing the replication cycle to begin. At the end of the cycle, newly synthesized molecules of transcriptase are again packaged along with the genome, making the next cycle of infection possible.

Viruses with double-stranded RNA genomes must also carry a transcriptase within their particles, since, like − strand RNA, double-

Figure 24-3

Classification of viruses by replication mechanism and relationship of genome to mRNA. Arrows inside the boxes show the information flow during replication. Arrows to mRNA originate at template for mRNA synthesis.

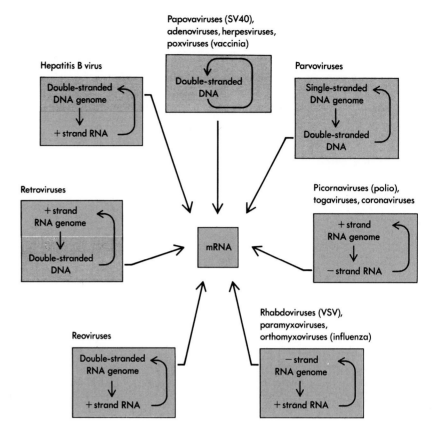

stranded RNA is not a normal template for replication or transcription in animal cells and cannot be translated. That virus-coded enzymes are packaged in virions was one of the most important discoveries in molecular animal virology. Curiously, the first enzyme so discovered was found in a poxvirus, the only known class of DNA viruses that replicates in the cytoplasm instead of in the nucleus, necessitating the production of many enzymes that other DNA viruses borrow from their host.

Using the principles discussed so far, animal viruses can be assigned to seven different groups (see Figure 24-3):

1. *Double-strand DNA viruses* (except the poxviruses) use cellular RNA polymerase for mRNA synthesis. All these viruses encode both capsid proteins and nonvirion proteins required for replication of their genomes.

2. *Single-strand DNA viruses* are either − stranded or made of both + and − strands, each one packaged in a separate virion. Unlike double-strand DNA viruses, single-strand DNA viruses encode capsid proteins only. They rely on cellular systems for both their transcription and their replication.

3. *+ strand RNA virus* genomes, which serve as mRNA soon after infection, have 3′ poly A tails and usually some sort of 5′-end modification like cellular mRNAs. They always specify enzymes that can replicate RNA via a − strand intermediate, since animal cells lack such an enzymatic activity.

4. *− strand RNA virus* genomes serve as a template for the synthesis of mRNA by virion transcriptase soon after infection. Not surprisingly, − strand RNA virus genomes lack the terminal modifica-

tions that characterize eucaryotic mRNAs.

5. *Double-strand RNA viruses* also carry virus-coded enzymes within the virion that can synthesize mRNA from this type of template.

6. *Retroviruses* have single, + strand RNA genomes that can serve as mRNAs. The genome's first act after infection, however, is not to serve as mRNA, but rather to serve as a template for the synthesis of double-stranded DNA. This DNA integrates into host chromosomal DNA, from which it serves as a template for transcription of viral mRNA and progeny genomes. **Reverse transcriptase,** a virus-coded enzyme carried inside the virion, makes viral DNA from the genomic RNA template.

7. *Hepatitis B viruses,* also known as *Hepadnaviruses*, have genomes with a full-length − DNA strand and a short + strand. Virions contain a DNA polymerase that completes the + strand soon after infection, resulting in a double-stranded DNA molecule that can be transcribed by the cell to yield viral mRNAs. As we shall see later, however, the replication of these viruses is quite different from most double-stranded DNA viruses and resembles the replication of retroviruses.

It is often quite easy to categorize a new virus isolate using the preceding information. For example, the presence of a transcriptase, easily detected in virion preparations, indicates a − strand or double-stranded RNA virus. Positive strand viruses or double-stranded DNA viruses have genomes that are infectious in the absence of virion proteins. Thus, just one experiment can sometimes reveal much of the life cycle of a virus.

Animal Viruses Solve Common Problems of Replication and Gene Expression in Diverse Ways[18–21]

In animal cells, in contrast to bacterial cells, most mRNAs are monocistronic. In other words, ribosomes only initiate translation at an AUG codon near the 5' end of the RNA. Consider, for example, a + strand RNA virus where all the genes are encoded on the same strand of RNA. How can such an animal virus synthesize the numerous proteins needed for a successful infection? RNA viruses have several novel solutions to this problem:

- Protein cleavage: Many RNA viruses direct the synthesis of multiple proteins by first synthesizing giant polyproteins, which are then clipped by a protease to release individual proteins. In the most extreme case, that of poliovirus, the entire genome is translated into a single polypeptide, which is cleaved to ten final proteins.

- Segmented genomes: Some RNA viral genomes consist of multiple pieces of RNA, each of which encodes one or a few distinct proteins.

- Nested mRNAs: Some viruses express internal genes by generating a nested set of mRNAs with identical 3' ends but different 5' portions. Each of these yields a different protein, which is translated from the reading frame nearest the 5' end of the mRNA.

- Sequential synthesis: In some cases plus RNA is synthesized as one long unit, and separate mRNAs for each protein are generated by cleavage or termination and reinitiation.

Of course, DNA viruses also face the problem of monocistronic mRNAs. While they can solve this problem by having transcriptional promoters in front of every gene, sometimes they use complex RNA-splicing patterns to join the identical 5' end to multiple transcripts.

DNA viruses face a special problem with replication, which arises from the fact that DNA polymerases cannot initiate synthesis of new DNA strands, but can only extend synthesis from primers. This causes an acute problem at the ends of linear genomes, since removal of the primer, which would normally be a short RNA chain, would always leave a 3'-end single-stranded tail (Chapter 10). Solutions to this problem involve circular genomes, terminal self-complementary structures that can self-prime the synthesis of new DNA chains, and terminal proteins that, when covalently attached to a nucleoside, serve as primers and remain covalently attached to the 5' end of each DNA strand.

Cell Surface Glycoproteins Are Receptors for Animal Viruses[22-24]

The first event in a successful infection is interaction of a virion with a specific *receptor* molecule on the cell surface. This interaction is mediated by virus-coded proteins on the surface of the virion and is highly specific. The presence or absence of appropriate receptors often determines whether a cell is *permissive* for virus infection, so the host range of a virus, including the type of tissue it infects, can be determined by the composition of receptors on the cell surface. Receptors for animal viruses are normal cell proteins and are not present for the convenience of the virus. For example, the ability of AIDS virus to infect a certain type of lymphocyte (T helper cell) most likely reflects its use of a cell surface protein characteristic of this cell type as its receptor.

A cell usually has 10^4 to 10^5 receptors for a particular virus. Most receptors are glycoproteins. Several virus groups (e.g., influenza viruses) recognize the terminal residue of the polysaccharide chain, usually sialic acid, so treatment of a cell with sialidase (**neuraminidase**) destroys its ability to be infected with such viruses (Figure 24-4). The specificity of virion-receptor interactions is demonstrated by the example of influenza virus. Influenza does not interact with receptors that have an $\alpha2{\rightarrow}3$ instead of an $\alpha2{\rightarrow}6$ linkage between the sialic acid and the adjacent sugar (see Figure 24-4). Influenza virus mutants that have acquired the ability to use receptors with the $\alpha2{\rightarrow}3$ linkage can readily be obtained, however, and these have a single amino acid change in their receptor binding protein. Thus, virus-receptor interactions display a stereospecificity similar to that of enzyme-substrate interactions.

Receptors for certain viruses are present on red blood cells. This permits an easy virus assay, since interaction of a virus with more than one cell at a time causes them to visibly **agglutinate,** or clump. In such a case, the viral protein that binds to the receptor is called **hemagglutinin.**

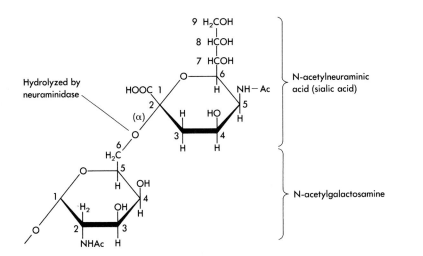

Viruses Enter Cells by Fusion or Endocytosis and Often Uncoat in Acidic Endosomes[25, 26]

Two well-documented mechanisms by which viruses get through the cell membrane and into the cytoplasm are fusion and endocytosis. The fusion mechanism is illustrated by paramyxoviruses, enveloped viruses that include mumps and Sendai viruses. These viruses can catalyze direct fusion between viral and cellular membranes. This ability resides in a viral envelope glycoprotein, called the F or fusion protein. F protein is generated by cleavage of a precursor polypeptide. The cleavage exposes a hydrophobic region in F protein that is necessary for interaction with the cell membrane. Once a virus is bound to its receptor, the fusion activity makes holes simultaneously in the viral envelope and the cell membrane. The two membranes then seal together, and as a result, the capsid is introduced into the cell. Cells that lack a protease capable of cleaving the F protein precursor release virions that are neither infectious nor capable of inducing cell fusion. Treatment of these virions with trypsin in vitro restores both their infectivity and their cell-fusing ability, so fusion activity appears to be essential for infectivity. The simultaneous interaction of a virion that has fusion activity with two or more cells can cause the cells to fuse into a large, multinucleated cell. Such cells are often seen in virus-infected cultures.

The alternative mechanism for virus penetration, endocytosis, uses host cell machinery that is normally used for the uptake of macromolecules. Infection by Semliki Forest virus, a togavirus, has provided a model system for this type of penetration. Semliki Forest virus is an enveloped virus with surface glycoproteins. After attaching to its specific cell surface receptor, the virus, along with its receptor, becomes engulfed by a **coated pit** into a vesicle (Figure 24-5). Coated pits are areas of the cell membrane, lined on the cytoplasmic side with the protein **clathrin,** where endocytosis of receptors and their ligands occurs. Even after the coated vesicle with its adsorbed virion is inside the cytoplasm, the virus itself is still functionally outside the cytoplasm, surrounded by the membranous walls of the pit. Once inside the cell, the pit with its attached virus loses its clathrin coat and fuses with large cellular vesicles called endosomes. The low pH provided

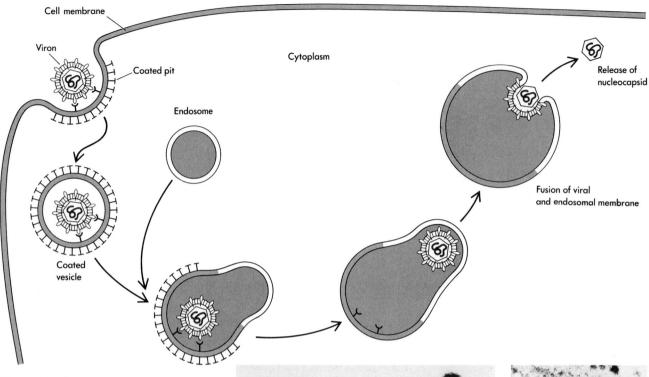

Figure 24-5

Uptake of viruses by endocytosis and fusion. Following receptor-mediated uptake of virus via coated pits, the acidic pH due to fusion of the vesicle with an endosome stimulates fusion of the viral envelope with the vesicle membrane resulting in release of the nucleocapsid into the cytoplasm. [After "How an Animal Virus Gets Into and Out of Its Host Cell," by K. Simons, H. Garoff, and A. Helenius. *Sci. Amer.* 246 (1982). Used with permission.]

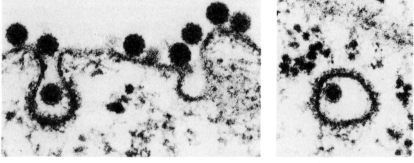

by the contents of these endosomes activates membrane-fusing activity latent in the viral glycoprotein spikes, causing the viral and endosomal membranes to fuse together and release the contents of the virion into the cytoplasm. This fusion activity can be induced in vitro by incubating Semliki Forest virus with cells and lowering the pH. Then direct fusion of the virus envelope with the plasma membrane occurs, in much the same way that paramyxoviruses normally fuse to cells at physiological pH. Infection of cells by viruses that enter by endocytosis can be blocked by chemicals such as chloroquine or by ammonium ions, which can form weak bases that raise the pH of endosomes and so prevent activation of viral fusion proteins.

RNA VIRUSES

Poliovirus: A Model + Strand RNA Virus[27–30]

Poliovirus is the best-studied *picornavirus.* It has a single + strand RNA genome, about 7400 nucleotides long, whose complete nucleotide sequence is known. The naked icosahedral virion is composed of

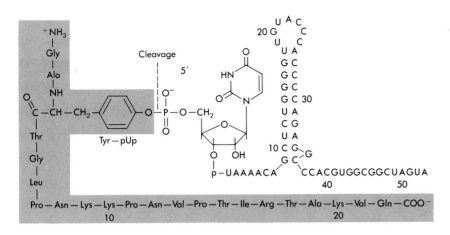

Figure 24-6
The 5′ end of the poliovirus genome. The 22-amino-acid VPg protein is shown linked via a phosphodiester bond between the tyrosine residue and the 5′ uridine of the poliovirus genome. A cellular enzyme removes VPg from polio mRNA by hydrolysis at the position shown. [After N. Kitamura et al., *Nature* 291 (1981):547.]

four virus-coded proteins—VP1, 2, 3, and 4—each present in 60 copies per virion. VP1 is probably responsible for attachment to cellular receptors that are found only on primate epithelial cells, fibroblasts, and nerve cells, accounting for the narrow host range of poliovirus. The entire life cycle of poliovirus occurs in the cytoplasm. This can be shown dramatically by the ability of the virus to replicate in cells whose nucleus has been removed by treatment with the drug cytochalasin B. The poliovirus life cycle is one of the most rapid known for an animal virus. Within 6 to 8 hours, an infected cell releases about 10^5 progeny virions.

Because it is a + strand genome, polio RNA that enters the cell first serves as mRNA for protein synthesis. This is essential, since replication of the viral RNA requires virus-coded enzymes, and none are present in polio virions. As we might expect, RNA isolated from polio virions can serve as mRNA in vitro. Like cellular mRNAs, the polio genome has a poly (A) sequence at its 3′ end. Unlike cellular mRNA and unlike most other viral mRNAs, it is not capped at the 5′ end. Rather, it has a tiny, 22-amino-acid-long protein, called VPg, covalently attached to its 5′ end via a phosphodiester bond between the 5′ uridine nucleotide and a tyrosine residue (Figure 24-6).

The Polio Genome Encodes Only One Polyprotein[31, 32, 33]

When newly made poliovirus proteins inside an infected cell were first examined by gel electrophoresis, it was immediately apparent that the sum of their molecular weights exceeded by two or three times the coding capacity of the genome. This paradox was explained by the finding that most of the proteins represent proteolytic cleavage products of one another. When serving as mRNA, polio RNA is translated into a single polyprotein from which all the viral proteins are derived by cleavage (Figure 24-7). Thus, there is only one ribosome entry point on the entire viral genome. These conclusions are supported by the nucleotide sequence of polio, which reveals a 6620-nucleotide open reading frame occupying the vast majority of the genome (see Figure 24-7). The poliovirus genome was the first large mRNA active in eucaryotic cells that was found to be monocistronic. At the time, the discovery represented a surprising departure from phage and bacterial systems, where polycistronic mRNAs are com-

Figure 24-7
The poliovirus genome and its translation products. The cleavage of the full-length precursor, NCVP, to P1, P2 and P3 is nascent, occurring before translation is complete. All of the cleavage events are catalyzed by polio-encoded proteins.

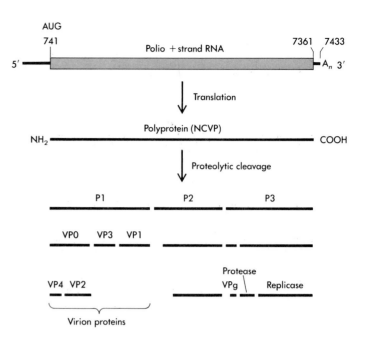

mon; but as we have seen, it turned out to be typical of eucaryotic mRNAs. The protein-coding region in the polio genome is preceded to the 5' side by an untranslated region 740 nucleotides long. While eucaryotic ribosomes frequently initiate translation at the first AUG codon they come to on an mRNA, eight AUG codons at the 5' end of polio RNA are bypassed in favor of the ninth at nucleotide 741. To the 3' side of the protein-coding region is a short untranslated region and finally the poly A tail.

The 240 kdal poliovirus polyprotein precursor does not normally exist in infected cells, because it is proteolytically cleaved at two sites before its synthesis is complete. Only if these **nascent cleavages** are blocked can the complete protein, called NCVP, be observed. The two initial cleavages divide the molecule into three regions. Unexpectedly, the amino-terminal region, which is thus the first part of the protein to be made, contains the virion proteins, while the carboxyl-terminal region is further cleaved to yield proteins essential for replication, including the RNA-dependent RNA polymerase, called **replicase.** In addition to replicase, this region also yields VPg, as well as a protease responsible for most of the polyprotein cleavages, specifically those that occur between glutamine-glycine (Gln-Gly) pairs. This protease is active even while it is still in precursor form, so several cleavages are autocatalytic. Not all cleavages of poliovirus polyproteins are at Gln-Gly pairs. Adjacent tyrosine-glycine (Tyr-Gly) and asparagine-serine (Asn-Ser) pairs are also cleaved, and now evidence suggests that the Tyr-Gly cleavages are made by a second viral protease encoded by the central portion of the genome. Not all Gln-Gly or Tyr-Gly pairs found in the poliovirus polyprotein are cleaved proteolytically. The enzymes involved in cleavage probably recognize specific features of protein conformation as well as particular pairs of amino acids.

The fact that all the poliovirus gene products are obtained by processing a single precursor implies that all the products will be synthesized in equimolar amounts. At most, one replicase or VPg molecule is required to make a new genome, whereas 60 molecules of each capsid protein are required to encapsidate it. This means that each infected cell synthesizes over 5 million unnecessary molecules of replicase in order to generate enough capsid protein for the 10^5 progeny virions. As we shall see, other more complex + strand RNA viruses have devised more efficient strategies of gene expression.

Poliovirus Replication May Be Initiated by a Protein Covalently Attached to the 5' Ends of RNA[34–37]

After serving as mRNA to direct the synthesis of proteins needed in its replication, the incoming polio genome must disengage from the ribosomes in order to serve as a template for RNA replication. Replication begins with the synthesis of − strands, which are generated by copying the + strand from an initiation site near its 3' end. Poliovirus replication has two unusual features. First, since the 3' end of the poliovirus genome consists of poly A, the 5' end of the − strand has the complementary sequence, poly U. Unlike virtually all other eucaryotic mRNAs, the poly A of polio RNA is not added after transcription; instead, it is replicated along with the rest of the genome. Second, unlike most enzymes responsible for RNA synthesis, but like all known DNA polymerases, polio replicase can synthesize RNA by adding nucleotides to an existing primer bound to a template, but it cannot initiate synthesis on a template with a ribonucleoside triphosphate. Thus, replicase is inactive when presented only with poliovirus RNA as template: It must have a primer as well. The primer for polio replicase does not seem to be a short oligonucleotide, the common primer for cellular DNA polymerases. Rather, it is thought to be U residues that have been covalently attached to the tiny protein VPg (see Figures 24-8 and 24-6). This would explain why VPg is covalently attached to the 5' ends of all newly synthesized polio + and − strands.

As infection proceeds, there is an accumulation of up to about 600 **replicative intermediates** per cell. These structures have a full-length − strand template with five or so + strands being made on them simultaneously (see Figure 24-8). The newly made + strands have one of three fates: They can serve as mRNA for further protein synthesis; they can be templates for further − strand synthesis; or they can become genomes in new virions. Probably the amounts of replicase and capsid protein that have built up inside the cell determine which of the pathways will be chosen. Thus, early in infection, + strands predominantly become mRNA after losing their terminal VPg by proteolytic cleavage. As replicase becomes available, + strands are increasingly used to direct − strand synthesis. And finally, as newly synthesized capsids appear, the + strands are incorporated into virions at the expense of the other uses.

Poliovirus Infection Stops the Synthesis of Cell Proteins[38, 39]

Very soon after poliovirus infection, cellular protein synthesis slows down dramatically, decreasing to 10 to 20 percent of normal levels 2 hours after infection. Inhibition of cellular protein synthesis is a common feature of many animal virus infections, but the mechanism has been studied most extensively with poliovirus.

Figure 24-8
Poliovirus RNA replication. Newly synthesized positive strands may serve as mRNA or as templates for further rounds of replication, or they may be packaged into virions. Note that − strands are used only as templates for synthesis of more + strands.

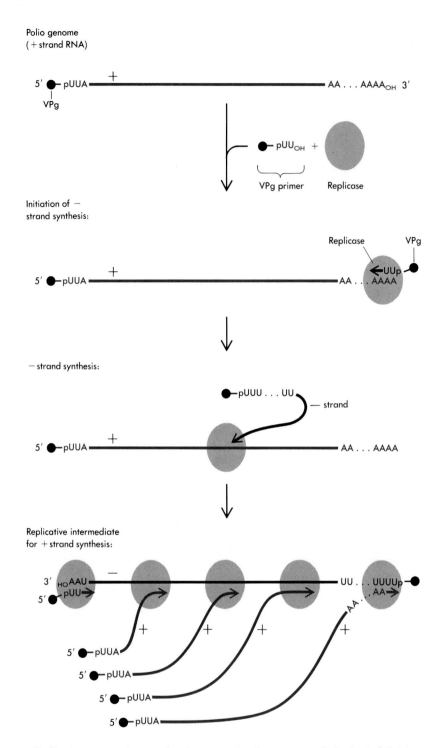

Poliovirus protein synthesis is required to accomplish the inhibition of host translation. Cellular mRNAs are not degraded or modified in polio-infected cells, but their translation is blocked at the initiation step. Polio-infected cell extracts proteolytically cleave and thereby inactivate a translation initiation factor, called eIF-4B, whose normal function is to recognize the capped, methylated 5′ termini of eucaryotic mRNAs and participate in binding mRNA to ribosomes. Since eucaryotic mRNAs are capped but polio mRNAs are not, it is apparent why inactivation of eIF-4B selectively inhibits cellular protein synthesis. Among known animal viruses, only the picornaviruses have capless mRNAs, so other viruses must use different mechanisms to selectively inhibit cellular but not viral protein synthesis.

Poliovirus Assembly Involves Proteolytic Cleavage of Capsid Protein Precursors[40, 41]

One of the earliest nascent cleavages of the giant polyprotein NCVP releases an amino-terminal cleavage product, P1, that gives rise to the four capsid proteins. This occurs in a series of steps that involve proteolytic cleavage and association of the cleavage products with one another (Figure 24-9). Initially, five copies of the P1 precursor associate into pentamers. Then the P1 molecules are cleaved into VP0, VP1, and VP3, after which groups of 12 pentamers aggregate into empty capsidlike structures called **procapsids.** Insertion of a genome into the procapsid leads to an immature structure called a **provirion,** which becomes a mature virion by proteolytic cleavage of VP0 to VP2 and VP4. The early cleavages are all at Gln-Gly bonds, but the final cleavage of VP0 is at an Asn-Ser bond and may be made by VP0 itself. This mode of assembly in which aggregation of precursor molecules is followed by cleavage into distinct proteins is found in other virus groups as well. It provides a simple mechanism for generating sequential changes in conformation as they are needed. For example, the final cleavage of VP0 might change the virion structure from one that is sufficiently open to allow the virus genome to get in, to one that is sufficiently tight to prevent the genome from falling out and to protect it against even extremely low pH.

Other + Strand RNA Viruses Have More Elaborate Strategies of Gene Expression[42–45]

Togaviruses have + strand RNA genomes of about 12,000 bases and icosahedral capsids within an envelope. In contrast to picornaviruses like polio, togavirus genomes are dicistronic (Figure 24-10). The cistron nearest the 5′ end encodes the viral replicase, and the cistron nearest the 3′ end encodes a precursor to the virion proteins. Since the internal initiation site is not seen by ribosomes, the genome itself can serve as mRNA only for the replicase protein. Capsid proteins do not appear until later in infection, and they are translated from a subgenomic mRNA consisting of the 3′ end of the genome. This mRNA is made by initiating replication at an internal site on the − strand. Since this mRNA lacks the entire replicase cistron, the initiation codon for translating capsid protein is now exposed and available for translation. This simple scheme allows the virus to regulate its gene expression both *temporally* (coat protein is not made until it is needed) and *quantitatively* (by the relative amounts of the two mRNAs and their relative efficiency as messengers).

A still more extreme case of nested mRNAs is shown by coronaviruses (Figure 24-11). The 16-kilobase genome of these viruses is expressed via no less than seven distinct mRNAs, each containing the 3′ end of the genome and each serving as messenger for a different gene product. Interestingly the different mRNAs have a common 72-base long, 5′ leader sequence. It is doubtful that this sequence is joined to the body of each mRNA by conventional splicing of a precursor RNA. Instead, newly synthesized leader segments may be used to prime internal initiations of transcription. Thus coronavirus mRNAs may result from discontinuous RNA synthesis.

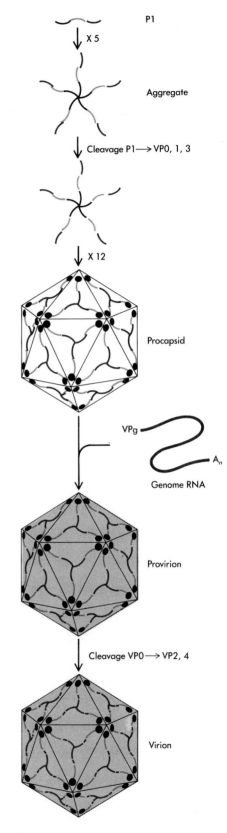

Figure 24-9
Model for polio virion assembly.

Figure 24-10
Expression of the togavirus genome. Early after infection, the + strand RNA genome is translated to yield two nonstructural proteins, including replicase (top portion of figure). Later in infection, a subgenomic mRNA is synthesized by use of an internal initiation site on the − RNA strand. This exposes a previously inaccessible AUG codon, permitting translation of a precursor to all virion proteins (the capsid protein and three membrane glycoproteins).

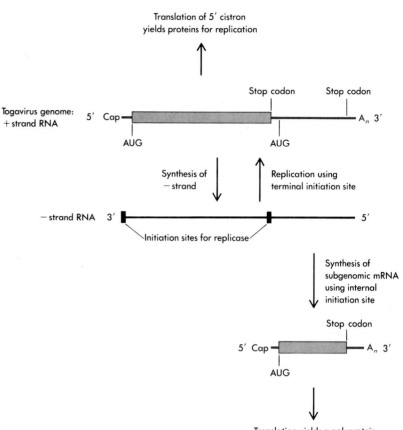

Figure 24-11
Nested mRNAs of coronaviruses. The seven distinct cistrons (not drawn to scale) are each translated from AUG codons near the 5′ end of a distinct subgenomic mRNA, generated by internal initiation from a common − strand template.

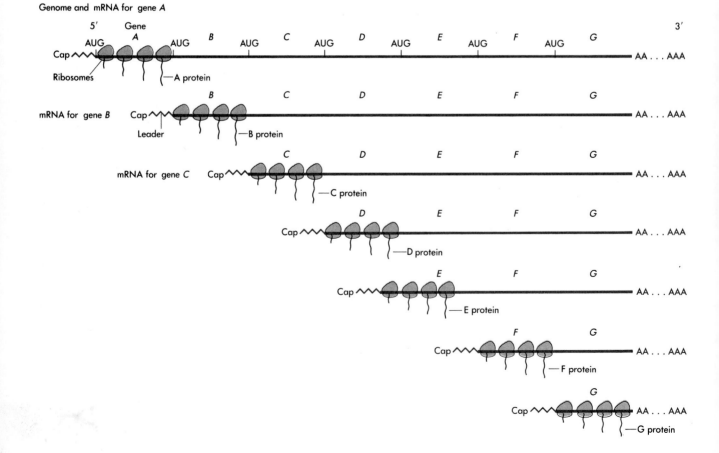

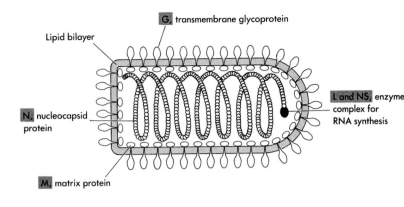

VSV: A Model − Strand RNA Virus[46]

Rhabdoviruses are small viruses with − strand RNA genomes and enveloped virions with a distinctive bullet shape (Figure 24-12). The most well known and deadly member of the family is rabies virus, but the best-studied is vesicular stomatitis virus (VSV), which is a mild animal pathogen that grows well in cell culture. VSV codes for only five proteins, and all of them are found in the virion. The *nucleocapsid (N) protein* encases the genome in a flexible helical capsid found neatly coiled within the virion. The *matrix (M) protein* lies between the capsid and envelope and makes specific contacts with both the N protein and the internal end of the virion *glycoprotein (G)*, which is also the protein responsible for interaction with cellular receptors. Since the genome is − strand RNA, it lacks the cap and poly A characteristic of + strand genomes. It cannot be translated, of course, so enzymes necessary for mRNA synthesis are present in the virus particle. The last two virion proteins, *L (large)* and *NS (nonstructural)*, fill this role. The L-NS complex is called transcriptase, since its role is to synthesize + strand transcripts as soon as the viral genome gets into a cell.

VSV Transcriptase Is a Multifunctional System That Catalyzes Both mRNA Synthesis and Replication[47–51]

With VSV virions, as with most − strand viruses, synthesis of mRNA can be readily duplicated in vitro by using a mild detergent to remove the envelope and then adding an appropriate mix of nucleoside triphosphates, buffers, and salts. Remarkably, the ensuing reaction not only makes + strands, but makes them in exactly the way they are found in the infected cell: as five distinct mRNAs, each encoding one of the virion proteins and each capped, polyadenylated, and in the same proportions relative to one another as found in the infected cell. Thus, the relatively simple L-NS enzyme complex can catalyze six distinct reactions: template directed RNA synthesis, polyadenylation, and the four enzymatic steps involved in capping reactions (Chapter 21).

The template for the L-NS complex is highly specific. Free RNA molecules are inactive, and only VSV RNA complexed with N protein works. Synthesis of the VSV mRNAs proceeds in an ordered fashion

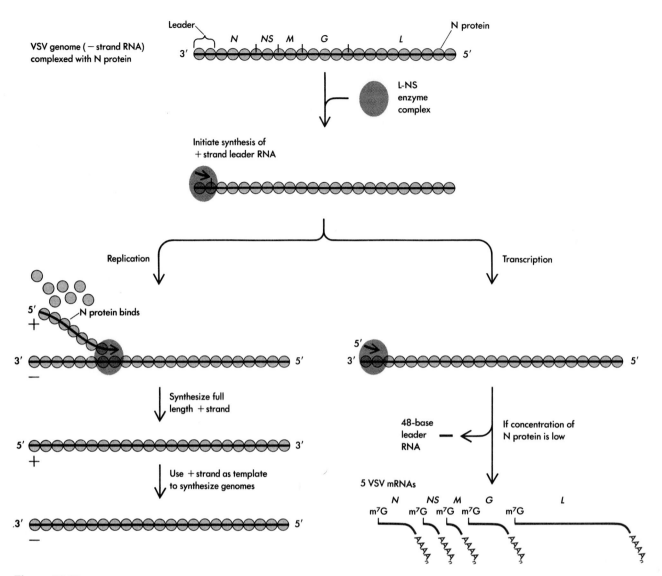

Figure 24-13

Transcription and replication of the VSV genome. The choice between the two processes seems to be determined by the availability of N protein.

from the 3' to the 5' end of the genomic template (Figure 24-13). One transcriptase complex binds to the 3' end of the genome and first synthesizes a 48-base leader fragment of RNA. This fragment is released from the template, and the complex then moves on to synthesize the five mRNAs sequentially, polyadenylating each at its 3' end before initiating synthesis of the next molecule. Between each transcript, the transcriptase has about a 30 percent chance of falling off the template. Since the enzyme can only get on the template at its 3' end, each successive mRNA molecule is about 30 percent less abundant than its predecessor. Examination of the order of the genes along the VSV genome reveals that there will be the most mRNA for N protein and the least for L and NS. This makes sense, since N protein coats the whole genome, but only a few L and NS molecules are needed to copy it.

A novel mechanism is used to add poly A to the five VSV mRNAs. The − strand genome contains a specific sequence at the region corresponding to the junction between each mRNA. This sequence includes a stretch of seven U residues corresponding to the position of

Poly A addition to VSV mRNA

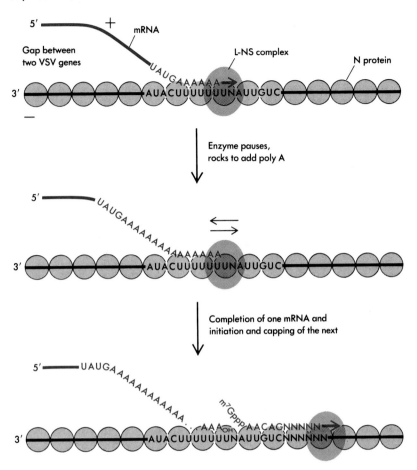

Figure 24-14
Polyadenylation of VSV mRNAs. The sequence shown is found at the border of all VSV genes. It is still uncertain whether the transcriptase terminates following poly A addition and reinitiates synthesis, or whether it synthesizes a continuous precursor, which includes the poly A, that is rapidly cleaved. The result would be the same. Note that two bases (NA) are not copied into mRNA. (N represents any nucleotide.)

the mRNA poly A. The sequence apparently allows the poly A to be added by *slippage* (Figure 24-14), involving a back-and-forth motion of transcriptase over the region with subsequent generation of a long sequence of nothing but A residues.

As VSV replication proceeds, there must be a shift from the synthesis of mRNA to the synthesis of full-length + strands that can serve as templates for the synthesis of progeny genomes. This shift is apparently regulated by the concentration of N protein. The 48-base leader RNA molecule synthesized as the first + strand product contains the specific binding initiation site for N. If sufficient N protein is available, it will bind initially to this site and then rapidly to itself, building a + strand nucleocapsid immediately behind the advancing transcriptase. Coating nascent + strands with N protein prevents the polyadenylation, termination, and reinitiation reactions from occurring and so causes the synthesis of a complete + strand molecule.

The + strand–N protein complexes are not incorporated into virions. Rather, they serve as templates from which L-NS enzyme synthesizes progeny − strand genomes. Since N binding can initiate only on leader RNA, once this fragment is made and released from the template, N can no longer bind, and the transcriptase is committed to synthesizing mRNAs instead of full-length + strands. This regulatory scheme permits the replication to remain nicely in balance: Once N has reached an adequate concentration, genome replication occurs; but should N protein be used up, then the transcriptase will automatically shift back to making more mRNA and hence more N protein.

VSV Nucleocapsids Acquire Envelopes by Budding[52, 53]

VSV G protein reaches the cell membrane by the same route as most cell surface glycoproteins. It is synthesized on membrane-bound polyribosomes, and a short hydrophobic signal sequence causes membrane attachment and then directs the growing protein into the lumen of the rough endoplasmic reticulum. The protein is glycosylated at two sites soon after it passes through the membrane of the endoplasmic reticulum. Twenty hydrophobic amino acids near the carboxyl terminus of the protein serve as a membrane anchor, leaving a short carboxyl-terminal tail protruding on the cytoplasmic face of the rough endoplasmic reticulum. Once G protein is transported through the membrane, the signal peptide is proteolytically removed and the protein is transported to the plasma membrane via the Golgi apparatus, where the oligosaccharide chains are matured.

Patches of cell membrane that will become envelopes are modified by the insertion of G protein, whose C-terminal cytoplasmic end associates with M protein. Most cellular membrane proteins are excluded from the region. Association between nucleocapsids and M proteins is followed by budding of mature virions. The final size of the virion is determined entirely by the size of the nucleocapsid, which itself is a function of genome length. Mutants of VSV with much shorter genomes yield virions of the same diameter but proportionally shorter length than wild-type virions.

Influenza Is a Segmented − Strand RNA Virus[54, 55]

Like VSV, influenza virus is a − strand enveloped virus; it therefore contains a virion transcriptase associated with its genome. In other respects, however, influenza is quite different from VSV and other − strand viruses. Most notably, the genome is divided into eight segments of distinct sizes, each encoding a separate gene product. (There are actually ten gene products, as we will soon see.) Each segment is encased within its own helical nucleocapsid and is associated with its own transcriptase enzyme. The other major difference between influenza and other − strand (and also + strand) RNA viruses is its dependence on the host nucleus for replication. Influenza will not replicate in cells from which the nucleus has been removed by experimental manipulation or which have been treated with drugs (like actinomycin D) that prevent DNA-dependent RNA synthesis. As will become apparent later, this dependence reflects the bizarre method that influenza transcriptase uses to initiate transcription.

Influenza virions are roughly spherical with two distinct types of surface glycoproteins that can be seen as spikes of different appearance in electron micrographs (Figure 24-15). One type of spike, composed of the hemagglutinin (HA) protein, recognizes cell surface receptors and also contains the fusion potential that allows the virus to penetrate the cell membrane after endocytosis. The other, composed

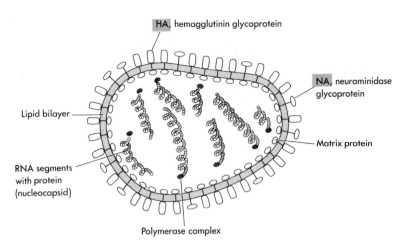

Figure 24-15
The influenza virion.

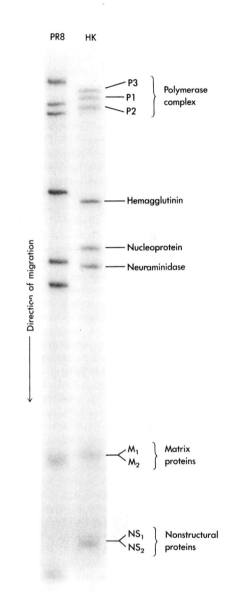

of the neuraminidase (NA) protein, removes terminal sialic acid residues from cell surface proteins. The virions also contain a matrix (M) and nucleocapsid (N) protein like VSV. Four additional proteins together probably constitute the viral transcriptase. Each virus-coded protein has been associated with a specific RNA segment, and there is a rough correspondence in size, so that the larger segments encode the larger proteins (Figure 24-16).

Influenza Initiates Transcription with Stolen Caps[56, 57, 58]

For a long time, the dependence of influenza virus replication on the nucleus was a puzzle. Also unexplained was the fact that the virion transcriptase was essentially inactive in vitro, in sharp contrast to that of VSV, for example. These observations made sense when it was discovered that the influenza transcriptase, like the polio replicase, requires a primer, and the primer is found in the cell's nucleus. The primer for influenza transcriptase can be supplied in vitro by adding full-length cellular mRNAs (such as globin mRNA) to a mixture of disrupted virions and nucleoside triphosphates. The product of this reaction is a faithful copy of each − strand segment, except that the initial few nucleotides have been replaced with a short fragment derived from the 5′ end of the cellular (globin) mRNA, including the capping group. The RNA was initiated with a 5′ capping group stolen from another mRNA!

In infected cells, influenza mRNAs have heterogeneous 5′ ends, with variable numbers of nucleotides preceding the start of the influenza sequence. Rather than having machinery for synthesizing capping groups, the influenza transcriptase has evolved a mechanism for removing them from cellular mRNAs. In the nuclei of infected cells, the transcriptase complex cleaves newly made mRNAs, usually just after an A or a G residue a dozen or so nucleotides from the cap. These fragments then immediately serve as primers by base-pairing between the A or G and the terminal U of each genome segment and thus allow subsequent polymerization using the genome RNA as template. This mechanism explains the dependence of influenza virus replication on cellular RNA synthesis. Only newly made mRNAs can serve as cap donors, so inhibition of cellular mRNA synthesis is followed rapidly by a decline in influenza mRNA synthesis.

Figure 24-16
Electrophoretic separation on a polyacrylamide gel of the eight genomic RNA segments purified from two strains of influenza virus (A/PR/8/34 (PR8) and A/Hong Kong/8/68 (HK)). Corresponding genome segments have slightly different mobilities in the two viruses. Proteins specified by each segment are indicated. Note the double coding of the two smallest RNAs. [Photo courtesy of Dr. Peter Palese.]

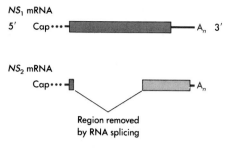

NS₁ mRNA

NS₂ mRNA

Region removed
by RNA splicing

Figure 24-17
Structure and coding of NS1 and NS2 mRNAs. Dotted lines at the 5′ ends represent caps plus a short additional sequence stolen from the 5′ end of cell mRNAs. The translated portions are shown by boxes, with different shading indicating different reading frames.

Since influenza mRNA is made in the nucleus, it is available to cellular enzymes that process normal cellular mRNAs. The virus takes advantage of this unique situation in several ways. Most important, two of its mRNAs are spliced to generate additional messages. The virus does not, however, parasitize its host's poly A–adding equipment. Rather, poly A addition is accomplished by a slippage mechanism similar to that of VSV.

Some Influenza Virus Genome Segments Encode More Than One Protein[59]

RNA splicing explains the observation that although the influenza genome consists of only eight pieces of RNA, more than eight viral proteins can be detected in infected cells, and none is a cleavage product of another. The two smallest genome segments each give rise to two partially overlapping mRNAs and two proteins. In both cases, one of the mRNAs is generated from the other by RNA splicing; and in both cases, splicing occurs so that the carboxyl-terminal portion of the protein encoded by one mRNA is read in a different reading frame than that encoded by the parent mRNA.

Influenza genome segment 8 codes for two nonstructural proteins called NS_1 and NS_2. Excluding the cannibalized cellular 5′ cap fragment, NS_1 mRNA is a colinear, although incomplete, copy of genome segment 8 (Figure 24-17). About 55 virus-specific nucleotides at the 5′ end of NS_2 mRNA are shared with NS_1 mRNA. Then, relative to NS_1 mRNA, there is a 473-nucleotide deletion generated by splicing using host cell machinery. The NS_1 and NS_2 mRNAs are then colinear, although translationally out of phase in their 3′ regions. As a result of these mRNA structures, NS_1 and NS_2 proteins share nine amino-terminal amino acids but differ completely in the remainder of their amino acid sequences (see Figure 24-17). The two mRNAs derived from genome segment 7 have similar relationships and give rise to two proteins that also share nine amino-terminal amino acids but then diverge in sequence.

High-Frequency Recombination Results from Reassortment of Genome Segments[60]

A consequence of having segmented genomes is that in a cell infected by two different influenza viruses, novel viruses can arise by reassortment of segments. As shown in Figure 24-16, equivalent genome segments present in different strains of influenza often have slightly different sizes and encode proteins with slightly different molecular weights or immunological properties. When a cell is infected with two such genetically distinct influenza viruses, a high frequency of the progeny are recombinant for genes located on separate RNA segments. This high-frequency recombination results from the intracellular reassortment of genome segments and provides a rapid means for the evolution of novel strains of influenza.

The segmentation of viral genomes raises the question of how a virus particle selects one copy of each segment for packaging. Clearly, a virion that packaged, say, two copies of segment 1 instead of one copy each of segments 1 and 2, would not be capable of initiating a complete infection cycle. There must be some mechanism for ensur-

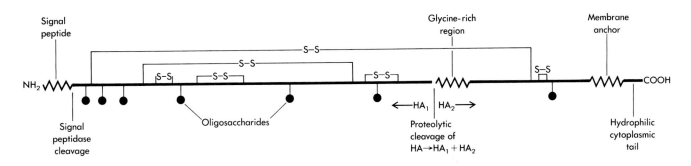

Figure 24-18

Primary structure of the influenza hemagglutinin precursor. The initial product (HA_0) is cleaved by a cellular enzyme to HA_1 and HA_2, which remain linked by a disulfide (S–S) bond. Trimers of this protein make up the HA spikes on the surface of the virion. [After Don C. Wiley, *Virology*. Ed. B. N. Fields et al. New York: Raven Press, 1985.]

ing that each virion gets a complete set of segments. Further, the random assortment of segments following infection with two different viruses shows that the segments do not remain physically associated during replication. The problem is a general one for all viruses with segmented genomes, and at present, virtually nothing is known about how a complete set of genome segments is selected for incorporation into virions.

The Structure of Influenza Hemagglutinin Glycoprotein Is Helping to Reveal the Molecular Basis of Flu Epidemics[61, 62, 63]

The influenza hemagglutinin (HA) glycoprotein is similar to the receptor-recognizing proteins of many viruses. It has two functions essential for infectivity. First, it recognizes and binds sialic acid–containing host cell receptors. Second, after virus uptake into endosomes, it mediates the fusion of virus and vesicle membranes to allow penetration of the influenza nucleocapsids into the cytoplasm. Because of its surface location and role in infection, hemagglutinin is a primary target for the host's immune system. HA elicits antibodies that, by binding to specific regions on the protein, neutralize virus infectivity. Viral mutants that alter the amino acid sequence of these regions arise frequently, and they can infect previously immune individuals. This variability in amino acid sequence of the hemagglutinin (as well as similar changes in the other viral glycoprotein, the neuraminidase) is probably the major cause of recurring flu epidemics.

HA spikes are trimers of identical molecules. Each monomer is synthesized as a typical membrane glycoprotein. After insertion into the membrane, the protein is cleaved to produce two peptides, HA_1 and HA_2, that are held together by disulfide bonds and anchored into the cell membrane via a C-terminal hydrophobic region (Figure 24-18). The amino terminus of HA_2 contains nonpolar amino acids that are not exposed at neutral pH. Acidic pH alters the conformation of HA and must allow the nonpolar sequence of HA_2 to interact with viral and endosomal membranes to promote their fusion.

X-ray crystallography shows that the HA trimer is an elongated cylinder to which each (HA_1 + HA_2) monomer contributes a long fibrous stem and a globular head (Figure 24-19). The receptor binding site is visible as a cleft in the head. Sites for antibody binding are also on the heads, and amino acid substitutions leading to new antigenic types are found in four different regions. Apparently, changes in all four regions are required to produce a new epidemic strain of flu. About every ten years, worldwide outbreaks (called **pandemics**) of influenza occur. These result from flu strains with so many more

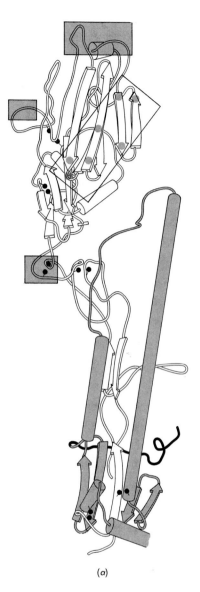

(a)

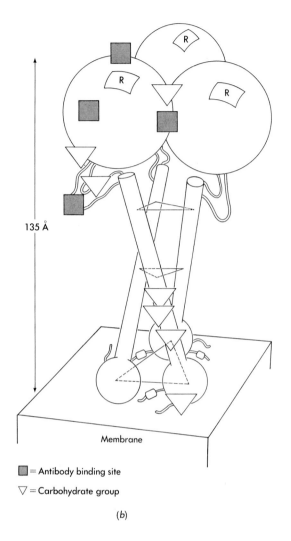

■ = Antibody binding site

▽ = Carbohydrate group

(b)

Figure 24-19
Structure of the influenza hemagglutinin spike based on data from X-ray crystallography. (a) A schematic representation of the peptide backbone of one of the three identical (HA_1 + HA_2) dimers. HA_2 is shaded. Cylinders show regions of α-helical structure. The colored portion indicates the hydrophobic NH_2-terminal region of HA_2. Regions that undergo mutation to create new epidemic strains of flu are boxed. They lie on the head of the molecule, far from the membrane anchor. (b) A highly schematic drawing of the HA trimer. Each of the three units represents a (HA_1 + HA_2) dimer. R indicates receptor binding regions. Other boxes, shown only for one unit of the trimer, as in (a). [After I. A. Wilson, J. J. Skehel, and D. C. Wiley, *Nature* 289 (1981): 366–373.]

amino acid substitutions in their HA that almost no one is immune to infection. At least one such strain arose by reassortment when the HA gene of a duck influenza virus was acquired by a human flu strain.

While we can now begin to understand how changes in influenza HA protein allow the virus to circumvent our immune system, it is not yet clear why a single infection with many other viruses (e.g., mumps or measles) provides life-long immunity. It seems that the receptor binding proteins of these viruses should be able to mutate as rapidly as influenza HA. Conceivably, one explanation is that the most exposed, and hence most antigenic, sites on influenza glycoproteins are relatively unimportant to the function of the proteins, whereas in other viruses, the antigenic parts are close to critical functional domains, thus preventing them from varying rapidly. Also, its segmented genome allowing rapid reassortment is another difference important to the emergence of dramatically new strains of influenza and hence for recurring flu infections.

Reovirus Has a Double-Stranded, Segmented RNA Genome[64, 65, 66]

Reoviruses are characterized by double-stranded RNA genomes in ten segments. The virions are icosahedral and naked, and they contain an unusual two-layered capsid. Since double-stranded RNA cannot serve as mRNA or as a template for mRNA synthesis in uninfected cells, there is an associated virus-coded transcriptase in the reovirus virion. As with influenza virus, each reovirus genome segment encodes a different protein.

Reovirus virions enter cells by endocytosis and are carried to sites near the nucleus. During their passage through endocytic vesicles, about half their capsid proteins are removed by lysosomal proteases. This is the total extent of virion uncoating. The loss and cleavage of the outer capsid proteins activates the transcriptase in the virion and creates pores that permit the entry of nucleotides and the release of newly made mRNAs. This process can readily be duplicated in the test tube by treating reovirus virions with a protease and then adding ribonucleoside triphosphates. A remarkably efficient and prolonged synthesis of mRNAs ensues, with the virions continuing to make mRNAs for many days, extruding the new molecules through pores, at the vertices of the capsid (Figure 24-20). The reaction is asymmetric and conservative: Only + strands are made, and both incoming double strands remain intact and stay within the infecting, partially uncoated, subviral particle. The mRNAs are also unusual in that they are capped but do not have poly A. The virion transcriptase is intimately associated with the capsid, and so far, it has not been possible to purify the enzyme in a soluble form and still retain transcriptase activity.

The + strand RNAs that are spewed out of the reovirus virions not only are highly efficient mRNAs, but also serve as intermediates in replication. Later on in infection, newly made capsid proteins assemble with + strand RNAs into capsidlike structures called **precores.** The + strands are converted to double-stranded form by the virion polymerase within the precores. Precores serve to synthesize new mRNAs, which continue the replication process until large areas of replicating virus, called *factories*, are built up inside the cell. Very late in infection, the precores are converted into new virions by the addition of outer shell proteins.

As with influenza virus, random reassortment of reovirus genome segments is seen when two genetically distinct viruses infect the same cell. Again, the mechanism that directs the encapsidation of a correct set of ten genome segments is completely unknown.

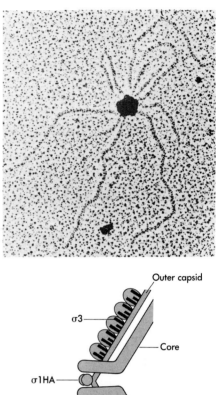

Figure 24-20
Electron micrograph of a stained and shadowed preparation of a reovirus core caught in the process of RNA synthesis. Twelve chains of RNA are thought to be extruded from the vertices of the partially uncoated virion. RNA synthesis took place for 1 minute at 37°C. [From N. M. Bartlett, S. C. Gillies, S. Bullivant, and A. R. Bellamy. 1974. *J. Virology* 14:315–326.] Diagram shows one vertex of a reovirus virion with possible locations of some viral proteins. [From B. N. Fields and M. I. Greene. 1982. *Nature* 300:19.]

DNA VIRUSES

The Challenge of Being a DNA Versus an RNA Virus[67]

There are advantages in an animal virus having a DNA genome instead of an RNA genome. The virus can use its host's transcriptional apparatus and so evolve complex patterns of gene expression. It can also use the DNA repair systems of the host and thereby evolve larger genomes than the RNA viruses. It would seem that another advantage might be the ability to use host enzymes involved in DNA replication. While host enzymes are used in many cases, the relationship of viral and host DNA synthesis is complex and actually presents special problems for the DNA viruses.

DNA synthesis in animal cells takes place in a restricted period of the cell cycle, the S phase. Furthermore, cells in many tissues are in a nondividing state most of the time and thus have concentrations of deoxynucleotides and DNA metabolic enzymes (polymerases, ligases, etc.) too low to support many rounds of DNA virus replication. RNA viruses do not face a comparable problem, since almost all cells, even those that are not dividing, have a constant need for RNA synthesis and thus always have high concentrations of the appropriate precursors (ribonucleoside triphosphates). To overcome the shortage of enzymes and precursors needed for DNA synthesis, DNA viruses must either replicate exclusively in dividing cells or devise means to induce conditions that will permit DNA synthesis. The simplest of DNA viruses, the parvoviruses, are completely incapable of replicating except in cells that are already actively dividing or in which replication of another DNA virus (such as adenovirus) is already occurring. Other more complex DNA viruses that can infect nondividing cells (e.g., papovaviruses and adenoviruses) stimulate cells to initiate DNA synthesis and thus to provide for adequate supplies of precursors and enzymes needed for viral replication. In addition, these viruses encode some proteins used in their replication. The most complex DNA viruses, herpesviruses and poxviruses, also encode a number of enzymes involved in nucleotide metabolism, which allows them to directly stimulate the production of deoxynucleotides by the infected cell. Interestingly, the most complex DNA viruses, which have the most control over their replication needs, also have the fastest life cycles.

SV40: A Small Double-Strand DNA Virus with a Circular Genome[68, 69, 70]

SV40 (simian virus number 40) belongs to the *papovavirus* family, which also includes the wart *(papilloma)* viruses. Another member of the family is the mouse virus polyoma, which is very similar to SV40. SV40 was isolated as a contaminant in poliovirus vaccines. The vaccines were being prepared using monkey cells instead of human cells because the latter were considered a potential source of dangerous cancer-causing viruses. It was a surprise when SV40 turned out to cause tumors in newborn rodents, but it was of great interest to molecular cancer biologists, who made SV40 (and polyoma, which causes tumors in hamsters) among their favorite objects of study (Chapter 26).

SV40 virions are naked icosahedral structures composed of three virus-coded proteins: VP1; the major protein, and lesser amounts of VP2 and VP3. The DNA of SV40 is a closed circular structure about 5200 base pairs in length, and its complete sequence is known. In virions and also inside the cell, the SV40 genome resembles cellular DNA in that it is associated with cellular histones and packaged into nucleosomes (see Figure 9-27 and Figure 21-19). When the SV40 genome is deproteinized, it adopts a negative superhelical configuration

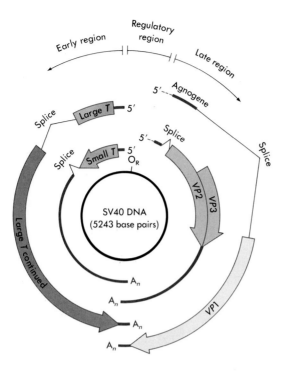

Early region

Regulatory region

Late region

Splice

Large T   5'

5' — Agnogene

Splice

Splice

5'

Small T   5'

O_R

Splice

SV40 DNA
(5243 base pairs)

Large T continued

VP2

VP3

Splice

A_n

A_n

A_n

A_n

VP1

Figure 24-21
SV40 genome and transcription map. The translated portion of each mRNA is shown by large arrows, with different shades of color for each reading frame. The role of the protein encoded by the agnogene at the 5' end of late mRNA is uncertain. O_R is the origin of DNA replication.

(form I DNA). A single nick in either DNA strand can relax the supercoil and result in a circular (form II) molecule, while a double-strand break produces a linear molecule (form III DNA). Supercoils can readily be separated from other viral or cellular DNAs because their condensed shape gives them unusual physical properties (Chapter 9).

Different SV40 Genes Are Expressed at Different Times During Lytic Infection[71, 72, 73]

As with most DNA viruses, expression and replication of the SV40 genome takes place in the nucleus and makes extensive use of cellular polymerases. The genome can be divided into three regions (Figure 24-21). The **early region** and the **late region,** each almost half the genome in length, encode protein. The third region, which lies between them, is a complex regulatory region containing promoters and regulators of expression as well as the site at which DNA synthesis begins, the origin of replication. Soon after SV40 DNA (still associated with histones) moves to the nucleus, cellular RNA polymerase recognizes the efficient early promoter region and transcribes the early region. The late region, which encodes capsid proteins, is silent at this time and is not expressed until after viral DNA replication begins.

The early region provides functions necessary for initiation of viral DNA replication, although it does not encode a DNA polymerase. The region consists of a single transcription unit, but encodes two proteins called **T proteins,** or *T antigens* (T stands for "tumor" and reflects their initial discovery in rodent tumor cells induced by SV40 infection; Chapter 26). As with many other DNA viruses, this multiple use of a single transcript is accomplished by differential RNA splicing of the primary transcript using cellular machinery. The two resulting early mRNAs have identical 5' and 3' ends, but differ in the size of an intron that is removed (see Figure 24-21). Since the larger RNA with the smaller spliced-out region retains a translational terminator, it encodes a relatively small protein, called *small t* (approximately 20 kdal). The other spliced mRNA has a larger region removed, including the terminator, and thus is translated into a much larger protein of about 90 kdal, called *large T.* The two proteins thus share about 80 amino-terminal amino acids, but each has a unique carboxyl-terminal region. The role of the small t protein in the virus life cycle is not well understood. The large T protein is found mostly in the nucleus and is a complex, much-studied protein with multiple functions: It regulates its own synthesis, stimulates cellular DNA synthesis, is involved in initiating viral DNA synthesis, and is needed for late gene expression.

The late region of the SV40 genome, which encodes the capsid proteins, is expressed only after DNA synthesis begins, about 12 to 14 hours after infection. It is transcribed in the opposite direction, hence from the opposite strand, as the early region. Like the T proteins, the late proteins (VP1, VP2, and VP3) are translated from differentially spliced, overlapping RNAs. VP2 and VP3 are translated in the same reading frame, starting at different points, whereas VP1, the major virion protein, is translated from sequences that overlap those encoding VP2 and VP3, but in a different reading frame. A fourth late gene is present in the leader of the late mRNAs. It encodes a protein 61 amino acids long called the *agno* protein, whose function is unknown.

Regulatory Sequences for SV40 Gene Expression[74-82]

A stretch of DNA lying between the 5' ends of the SV40 early and late mRNAs contains sequences that regulate both early and late gene transcription as well as DNA replication. In part because it is transcribed so efficiently, the SV40 early region has been intensively studied as a model for eucaryotic gene expression. Several sets of sequences upstream of the sites of initiation of early SV40 mRNA synthesis are important for early gene transcription (see Figure 21-30). Approximately 20 to 30 bases upstream from the cap sites is an AT-rich region, the Goldberg-Hogness or TATA box, with the sequence TATTTAT in SV40. This sequence is not essential for transcription to occur, but is important for positioning RNA polymerase II. If it is removed, the same number of transcripts is made, but their 5' ends are heterogeneous. Just 5' of the TATA sequence is a nearly perfect triple tandem repeat of 21 base pairs containing six copies of the sequence 5'-GGGCGG-3'. This repeated sequence is essential for efficient transcription and binds a specific cellular transcription factor, Sp1 (see Chapter 21). Between about 100 and 250 bases 5' of the cap site is a perfect tandem repeat of a 72-base-pair sequence that contains a transcriptional enhancer element (see Chapter 21). As with other enhancers, the number of bases between the SV40 enhancer and the early region whose transcription it influences can be varied to a considerable extent without deleterious effect on the level of transcription, and its orientation relative to the early genes can be reversed. Enhancers from other viral or cellular genes can be inserted in place of the SV40 enhancer and the resulting virus still functions.

Even though the transcription of SV40 genes is performed by cellular enzymes and factors, expression of both the early and late regions is regulated by the large T protein. There are three specific binding regions for T in the control segment. Interaction of T with these sites has three effects. First, it shuts down synthesis of early mRNA (and therefore itself); thus, T protein synthesis is autoregulated, just as the synthesis of the phage λ repressor is. Second, it directs the initiation of viral DNA synthesis. Finally, it turns on the synthesis of late mRNAs, either directly or indirectly as a result of stimulating DNA synthesis.

The late SV40 promoter is unusual in that it contains no identifiable TATA sequence. The late SV40 mRNAs are extremely heterogeneous at their 5' ends, presumably reflecting the alternative use of a large number of different initiation sites, consistent with the idea that the role of the TATA sequence is to align the initiation site of transcription, not to provide an initial binding site for RNA polymerase (see Figure 21-30). How sequences in the control region directly regulate late gene expression is not yet well understood.

Large T Protein Is Involved in Initiating Bidirectional DNA Replication from a Unique Origin on the SV40 Genome[83–87]

SV40 replicates its DNA bidirectionally with forks growing in opposite directions from a unique origin of replication. Replicating molecules purified from infected cells contain two circles of equal length composed of replicating DNA, as well as a superhelical loop of still unreplicated DNA (see Figure 10-6). The replication mechanism is similar in many ways to that of bacterial and plasmid genomes. The two replication forks advance from a common origin, and one strand is synthesized as a continuous piece in a 5'→3' direction, whereas the other is synthesized as a series of small RNA-primed Okazaki fragments. Replication is complete when the two moving forks reach each other opposite the origin. The precise sequence at the termination point is unimportant for completion. The initiation signal, on the other hand, has been defined by mutants as a precise sequence about 30 nucleotides long lying between the early and late genes (see Figure 21-30).

SV40 relies almost entirely on cellular enzymes to replicate its DNA, except that large T protein is essential for initiation of replication. However, its role in this process is not yet understood. It specifically and sequentially binds to three tandem sites: One is contained entirely within the initiation sequence for replication, while the other two overlap additional portions of this sequence (see Figure 21-30). Now that a system of in vitro replication that is dependent on T antigen is available, the mechanism of T antigen function should soon be worked out (see Chapter 10).

Adenoviruses Have Linear, Double-Stranded DNA Genomes[88, 89]

Adenovirus genomes are linear, double-stranded DNA molecules about 36 kilobase pairs long. Each strand has a short sequence (about 100 base pairs long) at one end that is complementary to the sequence at the other end, an inverted terminal repeat structure. As a result, when the DNA is denatured, the single strands self-anneal to form panhandle structures. The DNA found in virions has a 55 kdal

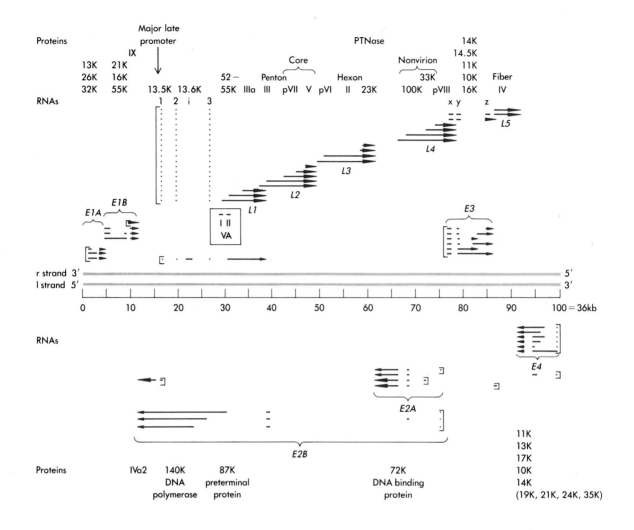

Figure 24-22
The adenovirus genome and its RNA transcripts. (Data are for human adenovirus.) The 40 or so adenovirus transcripts are divided into many families (both early and late) sharing the same promoter, indicated by the bracket at the 5′ end of each family. The proteins encoded by each are listed above and below, named by function or molecular weight. The roman numerals indicate virion proteins. Large arrow heads are present on late transcripts, small arrowheads on early transcripts. [After J. Tooze, ed., *Molecular Biology of Tumor Viruses: DNA Tumor Viruses*, 2nd ed. (Cold Spring Harbor, N.Y.: Cold Spring Harbor Laboratory), with permission.]

protein, the *terminal protein*, covalently attached to the 5′ end of each strand.

Like SV40, adenoviruses replicate their DNA in the nucleus and assemble their progeny virions there; and like SV40, they display both temporal control of gene expression and the use of multiple overlapping spliced mRNAs to pack a large amount of genetic information into their genomes.

Adenoviruses Use Alternative Splicing Patterns, Overlapping Genes, and Both DNA Strands to Maximize Use of Their Genomes[90–99]

Early in infection, during the 6 to 8 hours before viral DNA replication begins, adenovirus mRNAs are made that represent about 25 to 30 percent of the genome. These are derived from scattered regions of the genome called E1 to E4 (E for early) (Figure 24-22). All these regions give rise to multiple transcripts. In each case, the transcripts are derived from common precursor RNAs by differential splicing, and they retain the 5′ and 3′ ends of the precursor mRNAs but have different internal portions. Sometimes, this results in overlapping mRNAs that encode proteins that are a subset of one another, since they use the same reading frame. Sometimes, it results in the use of different overlapping reading frames in such a way that the resulting proteins share amino acids at their amino termini but diverge in their carboxyl portions. Early transcripts are derived from both strands of the DNA. Some, but not all, of the numerous polypeptides predicted to arise from the many early mRNAs have been identified in infected cells or by in vitro translation.

The expression of early genes itself is regulated, with some genes being expressed before others. Transcription of the E1A region is followed by transcription of regions EIB, E3, E4, and, last of all, region E2. Most important in this sequential regulation is the EIA region at the left of the genome. EIA encodes a protein that is required for expression of almost all other adenovirus genes. In the absence of this protein, no RNAs from EIB, E2, E3, or E4 are made, and infection aborts. The EIA gene product stimulates the rate of initiation of transcription, but the mechanism is unknown. Stimulation does not appear to involve an interaction with specific enhancers or promoters and is not even specific for adenovirus genes, since the level of transcription of some other viral and even cellular genes can be affected by EIA.

The E2 region encodes three proteins directly involved in DNA replication: a DNA binding protein, an 80 kdal protein that is cleaved to yield the terminal protein on the ends of the genome, and the 140 kdal adenovirus DNA polymerase. Thus, unlike the simpler papovaviruses, which must rely heavily on cellular enzymes, adenovirus-coded proteins fill many of the major roles in DNA synthesis.

Once viral DNA synthesis begins, late gene transcription becomes dominant in the cell. The late RNAs principally encode the 12 or so virion capsid proteins. Most of the late mRNA originates from a single promoter, the major late promoter (see Figure 24-22). This extremely active promoter directs the synthesis of a primary transcript that can extend almost to the extreme right end of the adenovirus genome, a length of over 28 kilobases. This long primary transcript is processed to yield at least 20 different mRNAs. All share the identical 5' terminus, as well as a 5'-end sequence about 200 nucleotides long known as the tripartite leader. The leader is assembled by RNA splicing that joins three small untranslated regions. The messages differ at their 3' ends owing to differential cleavage and polyadenylation at one of five possible sites that lie downstream of typical poly A addition sequences (AAUAAA) (Figure 24-23). Each of these five poly A–tailed transcripts then serves as the progenitor of a set of spliced mRNAs with common 5' and 3' termini. The different mRNAs in each set are made by joining the leader to one of several different sites on each transcript. This mechanism permits a wide variety of transcripts to be regulated together at the level of transcription and to receive the identical 5' sequence. The sequence encoded by the tripartite leader is a very efficient one for translation, and thus the mRNAs that contain it will be translated at a high level.

The early and late mRNAs described here are all transcribed by RNA polymerase II, and all have 5' capping groups and 3' poly A tails. However, not all adenovirus transcripts are made by RNA polymerase II or serve as mRNAs. Small adenovirus transcripts about 150 nucleotides long called virus-associated, or VA, RNAs are produced by RNA polymerase III, and like other genes transcribed by this enzyme, have promoter sequences within the genes themselves (see Figure 24-22). VA RNAs are made in very large amounts late in infection and seem to be necessary for translation of late mRNAs, possibly by blocking the effects of interferon (discussed later in this chapter).

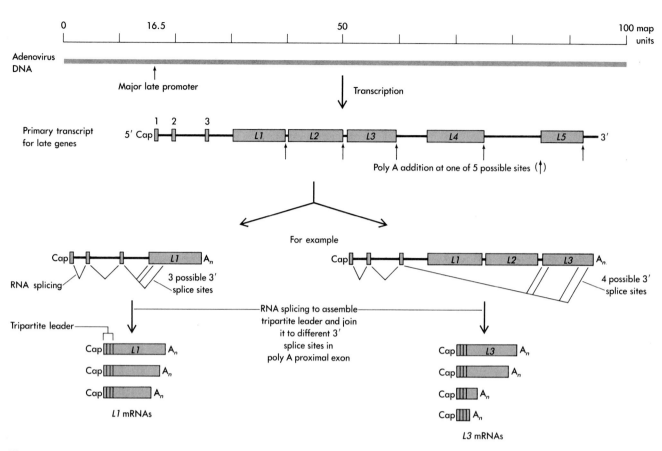

Figure 24-23
A model for adenovirus late mRNA synthesis. Variety is generated both by choice of polyadenylation site and by choice of 3' splicing site to which the tripartite leader is joined. Note that only the late mRNAs derived from the L1 and L3 bodies are shown.

The entire adenovirus genome has been sequenced, and analysis of the sequence reveals numerous additional open reading frames that could encode peptides of up to 10 kdal for which neither mRNAs nor the corresponding proteins have yet been found. Given the dense packing of genetic information in adenoviruses and the fact that RNA splicing makes it difficult to identify all the possible genes from DNA nucleotide sequences alone, it will probably be a long time before all the mRNAs and proteins specified by any adenovirus are known.

An Adenovirus Protein Covalently Binds dCMP to Initiate DNA Replication at the Inverted Terminal Repeats of the Adenovirus Genome[100-104]

Adenovirus DNA replication is unusual in that only one strand is synthesized at a time and all synthesis is continuous. Synthesis begins at either end of the linear molecule (or at both ends of the same molecule) and proceeds 5'→3' until the daughter strand is completed and one parental strand displaced (or until both parental strands are copied) (see Figure 10-31). There is no discontinuous replication involving Okazaki fragments. Rapid progress is being made in identifying proteins involved in adenovirus DNA replication. The first breakthrough came with the discovery of the virus-coded terminal protein of 55 kdal covalently attached to each 5' end of adenovirus DNA and the recognition of the role of this protein as a primer for DNA synthesis. A second breakthrough in the study of adenovirus replication was the finding that if extracts of adenovirus-infected cells were given deoxynucleotides, ATP, and Mg^{2+}, they could initiate and complete semiconservative replication in a test tube, using exogenously added adenovirus DNA as a template. This has permitted the biochemical identification of essential components and a more detailed analysis of the events involved in replication (see Chapter 10 and Figure 10-31).

Parvoviruses Are Small, Single-Strand DNA Viruses That Replicate Using Alternative Hairpin Structures[105, 106]

Parvoviruses are very small animal viruses with single-stranded DNA genomes of about 4500 to 5000 bases. Their virions are composed of three capsid proteins that have overlapping amino acid sequences and that are derived by translation of differentially spliced mRNAs. Unlike other DNA viruses, which either encode enzymes necessary for DNA replication or stimulate the cell to provide them, parvoviruses are dependent on other means. One group of parvoviruses can grow only on cells that are also infected with a more complex virus, such as adenovirus. The remaining parvoviruses are autonomous in that they do not depend on infection by another virus; however, they can only replicate in rapidly dividing cells. In nature, such viruses infect only certain cell types, such as blood-forming cells, which proliferate rapidly. They also infect tumors, since these also contain rapidly dividing cells.

Parvovirus genomes have unusual terminal nucleotide sequences that can fold into special T-shaped terminal structures. These structures provide yet another solution to the problem of replicating the ends of DNA molecules. As shown in Figure 24-24, the 3' end of the genome can serve as a primer. Simple addition of deoxynucleotides generates a copy of the genome, and refolding at the ends creates new primers as necessary. With the introduction of specific single-strand breaks at key points, new genomes can be synthesized.

The Large DNA Viruses Have More Complex Genomes and Replication Cycles[107–115]

The two families of very large DNA viruses, the herpesviruses and the poxviruses, have genomes 80 to 200 kilobase pairs long that encode 50 to 200 different proteins, including many enzymes involved in nucleotide metabolism. Except for size and complexity, the two groups of viruses are quite different. Herpesviruses have enveloped virions with icosahedral capsids and replicate in the nucleus, while poxviruses have complex, seemingly asymmetric capsids and are the only DNA animal viruses that replicate in the cytoplasm. A remarkable technical achievement was the recent determination of the complete nucleotide sequence of the genome of the herpesvirus known as Epstein-Barr virus. Its genome is 172,282 base pairs long!

Herpesviruses include some well-known infectious agents such as herpes simplex types 1 and 2; varicella-zoster virus, which causes chickenpox and shingles; and the Epstein-Barr virus, which causes infectious mononucleosis. Although the virions of all these viruses are indistinguishable from one another, the genomes vary considerably in size and organization (Figure 24-25). All herpesviruses contain

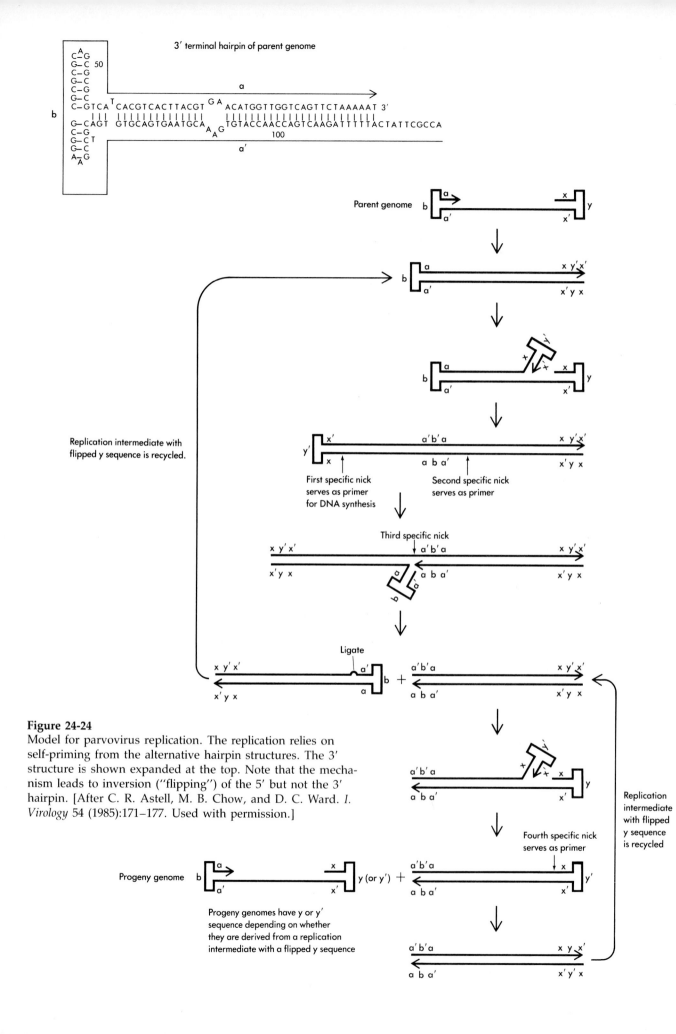

Figure 24-24
Model for parvovirus replication. The replication relies on self-priming from the alternative hairpin structures. The 3′ structure is shown expanded at the top. Note that the mechanism leads to inversion ("flipping") of the 5′ but not the 3′ hairpin. [After C. R. Astell, M. B. Chow, and D. C. Ward. *I. Virology* 54 (1985):171–177. Used with permission.]

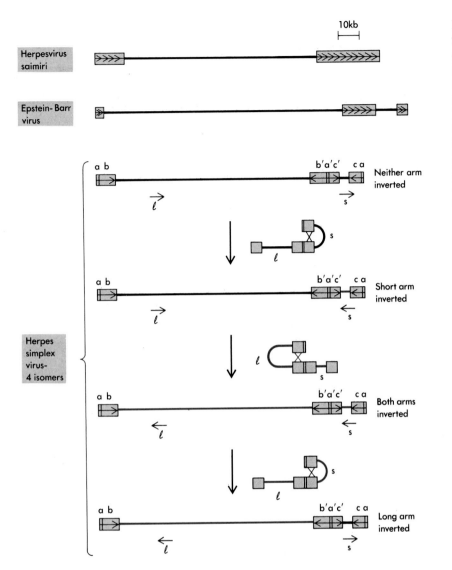

Figure 24-25
Some herpesvirus genomes. The boxes enclose repeated sequences, shown as arrows. Note the extensive short tandem repeats at the ends of the herpes Saimiri and Epstein-Barr virus genomes. The herpes simplex genome can be thought of as consisting of two arms, long (l) and short (s), each flanked by a long inverted repeat. Recombination between the inverted repeat sequences leads to inversion of one arm or the other.

repeated sequences of various kinds. In Epstein-Barr virus, the ends of the genome have a large number of short (500-base-pair) repetitions of the identical sequence as well as an internal sequence consisting of half a dozen repeats of a 3-kilobase-pair sequence. The total number of these repeated sequences does not vary greatly from molecule to molecule, but the distribution between the ends differs from one genome to another. The organization of herpes simplex genomes is quite different. These genomes can be viewed as having two covalently joined parts (arms), each flanked by long inverted repeats (see Figure 24-25). Recombination between the inverted repeat sequences leads to reorganization, or *isomerization*, of the genome. Since each arm can be in one of two orientations, there are four possible isomers, and all four can be found soon after infection with only one isomer.

During infection, the expression of the 50 or so herpes simplex genes is tightly regulated. Three classes of genes are known, called α, β, and γ (or immediate early, early, and late). Expression of the α genes is required to induce β gene expression; expression of β genes both induces γ genes and shuts off α genes; and expression of γ genes turns off β genes. Thus, there are three distinct waves of herpesvirus gene expression during replication. Interestingly, the process seems to be circular in that at least one virion protein (a γ gene product) is required to induce α gene expression soon after infection. This product enters with the virion and so helps to start the cycle.

Poxviruses include the deadly smallpox, but the best-studied member is vaccinia, or cowpox. Poxviruses are unique among DNA viruses in that they replicate in the cytoplasm of infected cells. They thus face the same kind of problem early in infection as − strand or double-stranded RNA viruses: The cytoplasm does not contain enzymes for nucleic acid synthesis, yet viral mRNAs must be made to initiate the infection cycle. The solution is analogous to that used by RNA viruses. Unlike all other DNA animal viruses we know of, poxviruses encode their own RNA polymerase, and this enzyme is found within virions. Soon after infection, vaccinia virions are partially uncoated to cores, and the RNA polymerase transcribes the 100 or so early genes, corresponding to about half the genome. Included among the early gene products are enzymes of nucleotide metabolism, such as thymidine kinase; enzymes for DNA replication, like DNA polymerase; and enzymes involved in synthesis and processing of transcripts, such as RNA polymerase, and capping and poly A−adding systems. Also included are enzymes that complete the uncoating process, releasing free DNA for replication. The function of most of the early genes, however, remains unknown.

The poxvirus genome and its replication are unusual. DNA of vaccinia virus is about 186 kilobase pairs in length. It has an inverted terminal repeat of about 10 kilobase pairs. Its most unusual feature is that the ends are covalently joined by phosphodiester bonds. Thus, if the genome is denatured, it forms a continuous, single-stranded circle. How this genome structure is replicated is still speculative with definitive proof awaiting further experiments (see Figure 10-23).

RETROVIRUSES AND HEPATITIS B VIRUS

Retroviruses: RNA Genomes That Replicate Through DNA Intermediates[116–120]

Retroviruses have a uniquely intimate association with the host cell. Although the viral genome is + strand RNA, the intermediate in replication is a double-stranded DNA copy of the genome, called the **provirus.** The provirus resembles a cellular gene and must even integrate into host chromosomes in order to serve as a template for transcription of new viral genomes. New genomes are therefore synthesized and processed in the nucleus by unmodified cellular machinery. Although viral genome RNA looks like mRNA, it does not serve as such early after infection. Rather, it is used as a template for DNA synthesis. Since an enzyme capable of carrying out this reaction is not usually present in the cell, it must enter with the virion. This virus-coded enzyme is called *reverse transcriptase,* and its presence in virions is diagnostic for retroviruses. Progeny genomes transcribed from integrated proviral DNA move to the cell membrane to assemble into virions. These bud from the cell membrane without killing the cell. A schematic view of the retrovirus life cycle is shown in Figure 24-26.

Retroviruses were first isolated early in this century as agents that could cause cancer in chickens. One of the first viruses found, Rous sarcoma virus (RSV), was isolated by Peyton Rous in 1911 and is still intensively studied as a model cancer-causing agent. Many naturally

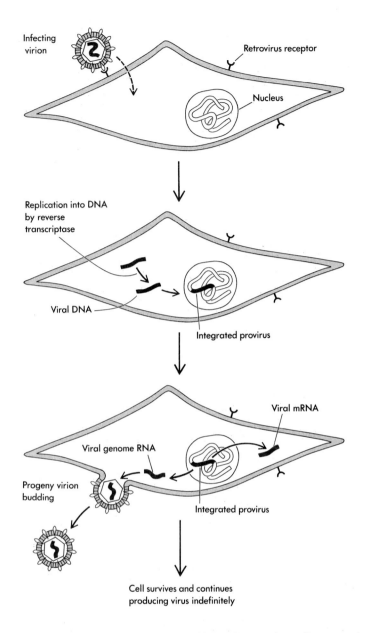

Figure 24-26
An overview of retrovirus replication. Only one DNA provirus is made per infecting virion; this is replicated by the normal cell division process, following integration into the cell genome. An infected cell is resistant to superinfection by another virus once it starts synthesizing viral glycoprotein, which binds to the receptors and so blocks infection. Usually the cell is not killed.

occurring cancers of animals, as well as some other diseases, have been traced to retroviruses. Most recently, several human isolates have been found: one (HTLV-I) causes a relatively rare, but invariably fatal, cancer of T lymphocytes, while another (HIV, also known as HTLV-III or LAV) has been identified as the cause of acquired immunodeficiency syndrome (AIDS). As we shall see in Chapter 26, the study of retroviruses has led to major insights into the molecular basis of cancer and growth control.

Retrovirus virions have icosahedral nucleocapsids surrounded by an envelope. The viral genome is unique in being diploid: Each virion contains two identical copies of the seven- to ten-kilobase genome. The reason for a diploid genome is not clear. Perhaps it permits the repair of damage and high levels of recombination during replication.

The genetic organization of retrovirus genomes is known in great detail, and the complete nucleotide sequence of well over a dozen genomes of different retrovirus isolates have been determined (Figure 24-27). Typical retroviruses have three protein-coding genes: *gag* (so named because it encodes a group-specific antigen) encodes a precur-

Figure 24-27
Some retrovirus genomes. The boxes indicate open reading frames. Boxes on different lines are in different frames. Note that the open reading frames of *tat* are separated by an intron in the AIDS virus. Most naturally occurring retroviruses resemble murine leukemia virus in structure, encoding only *gag, pol,* and *env* genes. mRNAs for *env, src, tat,* and probably *sor* and *orf* are generated by RNA splicing.

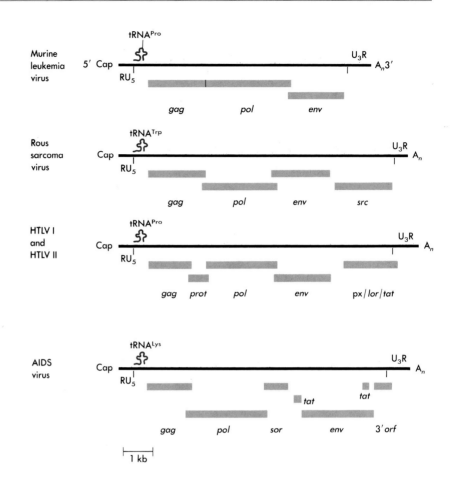

sor polyprotein that is cleaved to yield the capsid proteins; *pol* is cleaved to yield reverse transcriptase and an enzyme involved in proviral integration; and *env* encodes the precursor to the envelope glycoprotein (analogous to the G protein of VSV or the influenza HA protein). Recently a fourth type of gene (called *pX, lor,* or *tat*) has been found at the 3' end of HTLV -I and -II genomes. This gene seems to encode a transcriptional activator (similar at least superficially to the E1A protein of adenovirus), which greatly increases expression of the provirus. The AIDS virus contains still other genes not present in most retroviruses. In addition, a few retrovirus isolates have an additional gene, called an *onc* gene, that gives them the ability to rapidly induce certain types of cancer (Chapter 26).

Retrovirus genomes contain long untranslated regions at both their 5' and 3' ends. These sequences include signals needed for replication and also for transcription. Particularly noticeable is a short, directly repeated sequence (R) at each end of the genome, as well as a tRNA molecule that is held in place by base-pairing to a sequence near the 5' end (see Figure 24-27). The genomic sequence complementary to a 3' portion of the tRNA is referred to as the primer binding (PB) site. The tRNA is stolen from the host cell during replication and serves as a primer for DNA synthesis soon after infection. This is essential because, like all other DNA polymerases, reverse transcriptase cannot initiate DNA chains, but can only extend chains from a 3'-OH primer.

Reverse Transcriptase Generates
Long Terminal Repeats in Proviral DNA[121-128]

Reverse transcriptase that enters the cell in a retrovirus virion has the problem of copying a single-stranded RNA molecule into a double-stranded DNA molecule suitable for integration into host cell DNA and capable of being transcribed by RNA polymerase II. The solution is remarkably complex. The integrated provirus is not a precise copy of the RNA genome. Rather, it is longer at each end owing to the addition of sequences, called U_5 and U_3 (for unique to the 5' or 3' end), derived from the other end of the genome (see Figures 24-27 and 24-28). This results in a perfect direct repeat, called the **long terminal repeat** (LTR), at each end of the provirus. The LTR, with the structure U_3-R-U_5, contains signals for both initiation of transcription by RNA polymerase II and for 3' cleavage and polyadenylation of transcripts.

Viral DNA synthesis takes place in the cytoplasm shortly after the virion enters the cell. Reverse transcriptase begins by adding deoxynucleotides to the tRNA primer. After adding just 100 to 200 nucleotides, corresponding to the R region and to U_5, which lies between R and the tRNA primer, the enzyme comes to the 5' end of the genome. To continue synthesis of this − strand DNA, the RNA that has just been copied is removed by a specific nuclease, RNase H, which is part of the reverse transcriptase molecule. (RNase H degrades the RNA in a DNA-RNA hybrid.) Removal of the RNA permits the newly made DNA copy of R to base-pair with the R sequence at the 3' end of the genomic RNA. Thus, a new template-primer pair is formed, which allows reverse transcriptase to proceed and copy the remainder of the genome (see Figure 24-28). The newly made single-stranded DNA, with its covalently attached primer tRNA, must now be converted to double-stranded DNA. To do this, a specific nick is made in the RNA immediately to the left of U_3. The 3'-OH end of the RNA generated in this way now serves as primer for + strand DNA synthesis. This time, the − strand DNA just made is copied, as is part of the tRNA primer. A second jump then occurs, similar to the first, but this time the + strand DNA copy of the primer can base-pair with the − strand DNA copy of the primer binding site, again following removal of genome RNA by RNase H. Completion of both DNA strands produces the linear double-stranded structure U_3-R-$U_5 \cdots U_3$-R-U_5, where U_3-R-U_5 is the long terminal repeat.

The complicated process of proviral DNA synthesis is carried out entirely by reverse transcriptase. The enzyme needs at least four enzymatic activities for the task: RNA-directed DNA synthesis, degradation of RNA in RNA-DNA hybrids (the RNase H activity), DNA-directed DNA synthesis, and specific cleavage of RNA at the 5' end of U_3. The product of synthesis is linear DNA. This linear DNA is transported to the nucleus, where the ends are joined, possibly by a cellular enzyme, to make covalently closed circles. These serve as precursors for integration. The mechanism of integration of viral DNA is not fully understood. The circle must be opened at a specific site, since the ends of the integrated provirus are always the same. By contrast, the site of integration into the cellular DNA is apparently not sequence specific. At the joints between viral and cellular DNA, two bases are always removed from each end of the provirus, and in addition, four to six bases of cell DNA at the integration site are duplicated. For example, integration of a provirus into the sequence

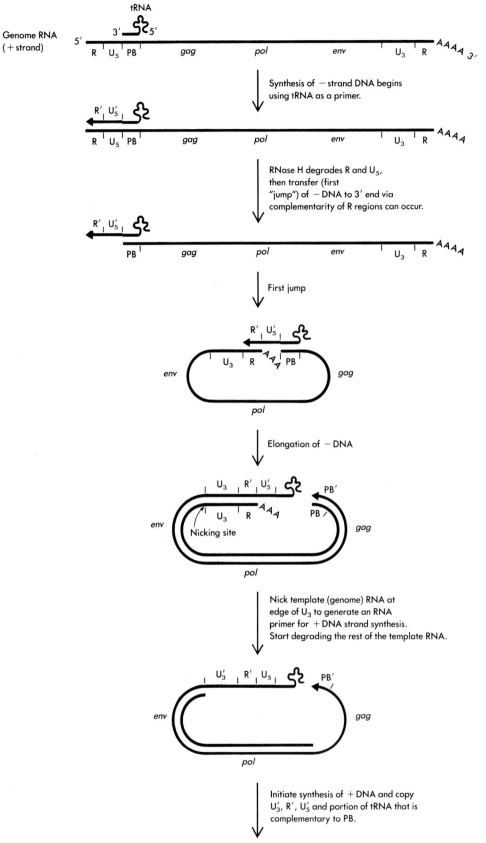

Figure 24-28

Synthesis of retrovirus DNA. All the steps are carried out by reverse transcriptase.

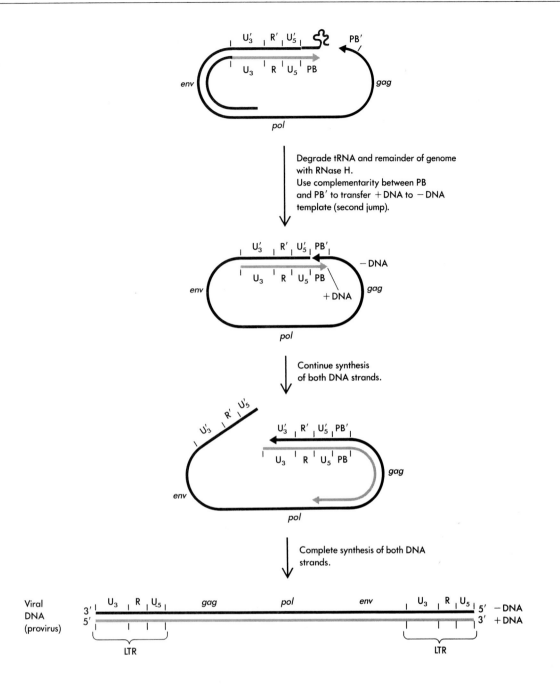

Viral DNA (provirus) — LTR

... TAGTCG ... can give the structure TAGTC-U_3-R-U_5-AGTCG ...
as a result of duplicating the sequence AGTC. This process is poorly
understood, but it might occur as shown in Figure 24-29. At least one
viral enzyme, an endonuclease sometimes called the viral *integrase*, is
essential for the integration process.

Once the provirus is integrated into the cell DNA, it is stable and is
replicated along with the host's DNA. Unlike bacteriophage λ, for
example, proviruses are never excised from the site of integration,
although they can be lost as a result of deletions. As we already
noted, retrovirus infections usually do not harm the cell, and infected
cells continue to divide, with the integrated provirus serving as a
template to direct viral RNA synthesis. A notable exception to this
outcome is seen with the AIDS retrovirus. For some reason, the specific population of helper T lymphocytes that can be infected by this

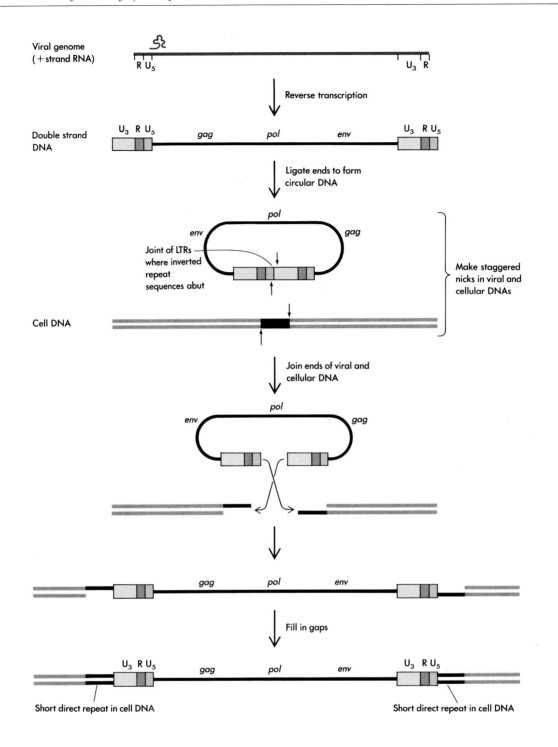

Figure 24-29
A model for integration of viral DNA. These events are inferred from the structures observed. The process is probably irreversible; the virus-coded endonuclease specifically recognizes the sequence formed by joining the inverted repeats, and the last two base pairs from each end are lost during integration, thus this sequence cannot be reassembled.

virus frequently are killed by the infection. It is due to the depletion of helper T cells that infected individuals are unable to mount vigorous immune responses.

LTRs Direct Transcription of the Provirus[129–133]

The LTRs not only provide ends for integration, but also supply transcriptional signals. These can be very efficient. With some retroviruses, as much as 10 percent of the mass of the mRNA in an infected cell can be viral RNA! The U_3 portion of the LTR contains sequences

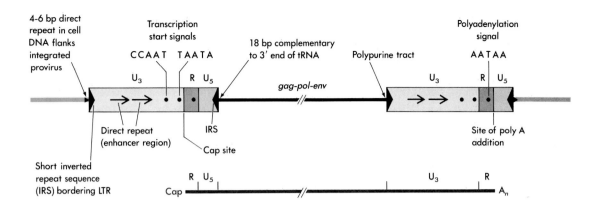

that control initiation of transcription: Typical TATA and CCAAT sequences are about 25 and 80 bases upstream, respectively, of the cap site at the U_3-R junction (Figure 24-30). Still farther upstream is a transcriptional enhancer, often present as a directly repeated sequence 70 to 100 bases long. Retrovirus enhancers are analogous to the enhancer of SV40, and some can even substitute for it.

As with most DNA viruses, the RNA transcripts synthesized by RNA polymerase II from the integrated retrovirus provirus are processed entirely by cell machinery. They are cleaved and polyadenylated at the 3' end of R and have several different fates. A fraction of the full-length RNAs is transported from the nucleus to the cytoplasm and assembled in budding virions to become new viral genomes. Another fraction of full-length RNAs is transported to the cytoplasm to serve as mRNA for the *gag* and *pol* genes (Figure 24-31). Finally, a portion of the RNA is spliced to yield mRNAs for *env* (and for the *pX/lor* gene if present).

Figure 24-30

Features of LTRs. The example shown is from murine leukemia virus. Virtually all retroviruses (and many "transposable elements" of flies and yeast) share these features. Note that the relationship between the cap site and the poly A site (determined by the relevant signals) sets the position of the terminally redundant (R) sequence in the transcript.

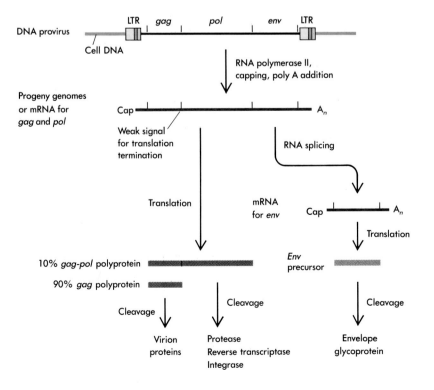

Figure 24-31

Synthesis of retrovirus RNAs and proteins. Only a fraction of the primary transcripts are spliced to yield mRNA for *env*. The remainder serve either as new genomes or as mRNA for *gag*. A small fraction serve as mRNA for the *gag-pol* product by translational suppression of a termination codon at the end of *gag*.

Proviruses Can Become Established in the Germ Line as Normal Genes[134–140]

According to the replication cycle we have described, retroviruses can be propagated not only as infectious agents, but also as cellular genes. If, for example, a virus infection occurs in a germ cell (a cell destined to become a sperm or egg cell), then the resulting provirus can be passed on to the progeny and inherited as if it were a normal cellular gene. Such genetic elements, called **endogenous proviruses,** already exist and in fact are readily detectable in vertebrate DNA, including that of humans. In some animals, such as the mouse, there are literally thousands of endogenous proviruses. In fact, as much as 0.5 percent of mouse DNA is estimated to consist of endogenous retrovirus proviruses.

Endogenous proviruses are identical in structure to proviruses of exogenously infecting retroviruses, although usually they are not transcribed. Some can be induced to yield infectious virus following treatment of the cells with various chemicals (bromodeoxyuridine, or BUdR, and iododeoxyuridine, or IUdR, are particularly effective). The failure to be transcribed is due in part to extensive methylation of endogenous proviruses at CG sequences, and inducing agents frequently reverse this methylation.

As with exogenous retroviruses, endogenous proviruses are found integrated into an apparently random variety of locations in chromosomal DNA. Since different individuals often have the provirus in different locations, endogenous proviruses must have been introduced into the germ line relatively recently. We might expect that insertions of proviruses would sometimes inactivate genes and so give rise to a detectable phenotype. At least one such mutation is known, at the *d* locus in mice. It results in a distinctive coat color known as dilute brown, which is characteristic of a number of strains of inbred mice. Revertants of the *d* mutation are easily detected (by their coat color). They occur when the provirus is deleted, and their frequency indicates that the rate of this event is about 3×10^{-6} per generation.

Now it is possible to insert new proviruses deliberately into the germ line of laboratory mice by infecting early mouse embryos in vitro and reimplanting them into foster mothers, where they mature normally. The resulting animals have proviruses at a variety of locations in many different cell types, sometimes including the germ cells. Further breeding of such mice can lead to strains of mice in which a provirus derived from the original infection is inherited as a Mendelian gene. Once present, such elements behave much like endogenous proviruses. Like endogenous proviruses, newly introduced proviruses sometimes inactivate genes when they integrate, thereby causing mutations.

It should be apparent from the preceding discussion that endogenous proviruses have many characteristics also associated with transposable elements: They form a related group of DNA sequences with a characteristic structure, including terminal repeats, and are found at different locations in the DNA of different individuals. Had they first been discovered as DNA sequences rather than as virus-related sequences, they would have been labeled transposons. Recently, it has been recognized that many elements originally identified as eucaryotic transposable elements have striking similarity to proviruses, including the major hallmarks of LTR structure and function. These

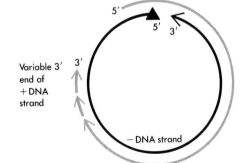

Figure 24-32

The DNA genome of hepatitis B virus (3200 base pairs). The genome is packaged into virions with a DNA polymerase that can further elongate the 3' end of the incomplete + strand. The triangle is a protein molecule covalently bound to the 5' end of the − strand. [After J. Summers and W. S. Mason, *Cell* 29 (1982):403–415.]

elements include the copia family of *Drosophila* and the Ty elements of yeast (Chapter 18) and have been referred to as retrotransposons. In both cases, there is direct evidence that these elements are duplicated and inserted by mechanisms virtually identical to those of retroviruses, except, probably, for the absence of an extracellular virion form.

Hepatitis B Virus Is a DNA Virus That Replicates Like a Retrovirus[17, 141, 142]

Two distinct groups of viruses cause most serious infectious liver disease in humans and some other mammals and birds. Hepatitis A virus, also called infectious hepatitis virus, is an RNA virus belonging to the picornavirus group. Hepatitis B (once known as serum hepatitis) virus is an enveloped virus with a small DNA genome and a unique mode of replication that resembles a retrovirus replication cycle turned inside out. Hepatitis B–like viruses have been isolated from humans, woodchucks, ground squirrels, and ducks, and together these isolates form the Hepadnavirus (for *hepatitis DNA virus*) family.

The hepatitis B genome is circular, but both of its DNA strands are linear (Figure 24-32). A gap in the complete − (or L for long) strand is bridged by an incomplete complementary + (or S for short) strand. The − strand has a protein covalently bound at its 5′ end. The + strands found in virions are in various degrees of completion, resulting in a single-strand gap extending from a fixed point at the 5′ end of the + strand to a variable point at its 3′ end. A DNA polymerase present in the nucleocapsid core can elongate the 3′ ends of incomplete strands.

Cells infected with hepatitis B virus contain large amounts of − DNA strands that are not attached to + DNA, and many are shorter than full genome length. A search for the template for their synthesis resulted in the unexpected finding that − DNA is synthesized in immature corelike particles from an RNA template that is degraded as synthesis proceeds. Thus, hepatitis B viruses replicate by reverse transcription, just like the retroviruses. The RNA template for − DNA strand synthesis presumably is full length and is designated the **pregenome.** Details of this replication cycle remain to be worked out, but a possible scheme for hepatitis B replication is shown in Figure 24-33. Unlike retroviruses, initiation of − strands seems to involve the protein covalently attached to their 5′ ends, a situation similar to adenovirus DNA replication. The + DNA strands would be synthesized using the free single-stranded − DNA as template and probably a + strand RNA fragment as primer.

INTERFERON

Viral Infections Induce Interferons[143–147]

Interferon was discovered accidentally about 30 years ago, when it was observed that culture medium from cells that had been exposed to partially inactivated viruses could protect other cells from infection with live viruses. The active component in the medium was called

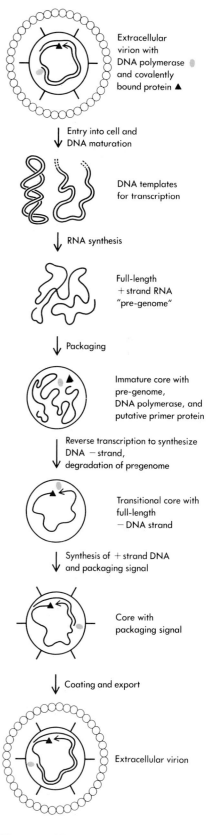

Extracellular virion with DNA polymerase and covalently bound protein ▲

↓ Entry into cell and DNA maturation

DNA templates for transcription

↓ RNA synthesis

Full-length + strand RNA "pre-genome"

↓ Packaging

Immature core with pre-genome, DNA polymerase, and putative primer protein

↓ Reverse transcription to synthesize DNA − strand, degradation of pregenome

Transitional core with full-length − DNA strand

↓ Synthesis of + strand DNA and packaging signal

Core with packaging signal

↓ Coating and export

Extracellular virion

Figure 24-33
A model for hepatitis B virus DNA replication. Details of process not yet clarified. [After J. Summers and W. S. Mason, *Cell* 29 (1982):403–415.]

interferon and was subsequently identified as a family of glycoproteins. Early experiments revealed that interferon synthesis is frequently induced in cells by virus infection. Although the infected cells usually die, they release interferons, which act on adjacent cells and protect them against viral infections. Interferon is an important defense mechanism in vertebrates. It is induced within hours of virus infection, a far more rapid response than the highly specific immune responses involving antibodies and activated cytotoxic T cells (Chapter 23). Without interferon, we would probably not survive many virus infections.

Infection of cells with most DNA and RNA viruses induces the synthesis and release of similar interferons. This similarity of response to a very diverse set of viral agents raises the question of what common aspect of viral infection signals the cell to produce interferon. Although the answer is not fully understood, we know that at least in some cases, the key component is double-stranded RNA. Almost all virus infections involve the synthesis of at least some RNA of both + and − strands. Although double strands are rarely directly involved in virus replication, they can form by chance if complementary single strands are present. Supporting the hypothesis that double-stranded RNA is the key inducer of interferon is the fact that synthetic double-stranded RNAs are very efficient inducers of interferon.

Interferons are extraordinarily active proteins. They are usually detected by their ability to reduce the number of virus plaques obtained following infection with some standard virus (often VSV). Concentrations of interferon as low as 3×10^{-14} M are detectable by this assay, and fewer than 50 molecules are apparently sufficient to protect a single cell.

There are three known types of interferons, and they are produced by different cell types under different conditions. In response to viral infection, lymphocytes synthesize primarily, but not exclusively, α interferon, sometimes called leukocyte interferon, whereas infection of fibroblasts usually induces β interferon. The two species of interferon are related proteins that share approximately 20 percent of their amino acid sequences. By contrast, γ interferon, sometimes called immune interferon, is scarcely related to the other two in amino acid sequence and is not induced by virus infection. Rather, it is synthesized by lymphocytes in response to *mitogens* (agents that stimulate cells to divide; Chapter 23).

Because they are released in such small amounts and because of their heterogeneity, for a long time interferons were difficult to study. Fortunately, the problem was overcome, first by advances in large-scale cell culture and purification techniques and, most importantly, by recombinant DNA techniques, which made it possible to synthesize very large amounts of active interferon in *E. coli*. These advances have raised hopes that interferon may one day provide effective treatment for viral infections.

Interferon Induces the Antiviral State[148, 149]

Interferon does not directly protect cells from virus infection. Rather, it alters the metabolism of the cell in a number of ways that make it less suitable as a host for viral replication—a complex condition called the **antiviral state.** Interferon that is chemically coupled to large beads

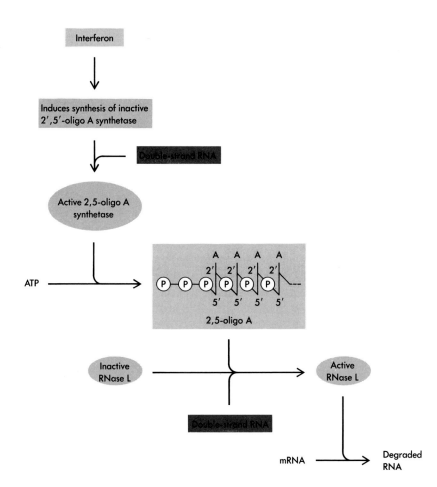

Figure 24-34
The 2,5-oligo A pathway for interferon action. Note that the final effect depends both on the induction of 2,5-oligo A synthetase by interferon and on the presence of double-stranded RNA.

so it cannot get into cells can still induce the antiviral state. Apparently, interferons act by binding to specific receptors on the cell surface. The activated receptors then transfer a signal to the inside of the cell, a situation analogous to the way hormones or growth factors trigger cellular responses. How this occurs in the case of interferon to bring about the antiviral state is unknown.

Viral infection in interferon-treated cells is blocked at various stages of infection following attachment and penetration. The level of viral genome and mRNA synthesis is reduced, and translation is also depressed. Whether replication and virus assembly are also directly affected by the antiviral state is not clear. Two molecular mechanisms have been deduced that help to explain the antiviral state. Both mechanisms involve the induction of a cellular enzyme, and in both cases, the enzyme is activated by double-stranded RNA.

Interferon induces a protein kinase, which, in the presence of double-stranded RNA, specifically phosphorylates the α subunit of initiation factor eIF-2. This causes eIF-2B to be sequestered in eIF-2-eIF-2B complexes (see Chapter 21) and hence to become inactive in protein synthesis.

The second enzyme induced by interferon is called 2′, 5′-oligo A synthetase. It is part of a complex pathway that ultimately leads to the degradation of single-stranded RNA, and hence of mRNA (Figure 24-34). The 2′, 5′-oligo A synthetase polymerizes ATP residues into short oligomers consisting of 2 to 15 adenosine monophosphate residues linked by phosphodiester bonds from the 2′ position of one ribose to the 5′ position of the next (in contrast to the common 3′–5′

linkage). Like the protein kinase induced by interferon, 2′, 5′-oligo A synthetase is only active in the presence of double-stranded RNA. The 2′, 5′-oligo A does not affect the cell directly. Rather, it binds to and activates a specific ribonuclease, called RNase L (L for latent, since the enzyme is always present in the cell but normally inactive). RNase L then cleaves single-stranded RNA. It seems to be equally active on both cellular and viral mRNAs. This lack of specificity is probably not a disadvantage, however, since the host will benefit if a virus-infected cell dies rather than releasing even a small burst of progeny virus particles.

The Interferon Multigene Family[150–154]

The genes encoding α and β interferons are very similar in organization. Each encodes a glycoprotein 166 to 169 amino acids long after removal of signal peptides. In humans, neither α nor β interferon genes contain introns; the absence of introns is a feature shared with only a few other genes. Regions of 200 or so bases lying immediately 5′ of the coding regions of the α and β genes are also related and contain sequences that regulate gene expression. Remarkably, when a fragment as short as 46 bases from this region is placed in front of an unrelated gene (such as globin) and then introduced back into a cell, it renders the gene inducible by virus infection, just like the interferon gene from which the fragment was derived.

Humans and most other mammals contain multiple α interferon genes that encode closely related but distinct proteins with slightly different biological properties. In humans, there are 15 functional α genes belonging to two major types, α_I and α_{II}, as well as 10 or more pseudogenes. (Pseudogenes are closely related to functional genes, but have suffered mutations so that they cannot be expressed.) All of these genes, as well as the single β interferon gene, are located on chromosome 9. The nucleotide sequences of many of these genes are known, and from their relationships, we can infer the evolutionary relationships between interferon genes. The α and β genes probably diverged from a common ancestor 250 to 500 million years ago, about the time of the divergence of mammals and birds. About 120 million years ago, around the time of the origin of mammals, the α_I and α_{II} subtypes separated. Most recently, since the divergence of mammalian species, the α_I gene family has been generated, presumably by tandem duplication. Similar sorts of diversity have arisen independently among interferon genes of other species as well.

The diverse biological properties of the different interferons include the specificity of interaction with different types of cells and the potency of induction of the antiviral state. Presumably, this diversity is important for ensuring an effective response to viral infections.

Some Viruses Fight Interferon by Synthesizing Small RNAs[155]

Whatever mechanisms host organisms devise to combat viral infections, viruses seem to come up with ways to circumvent these defenses. So it is with certain viruses and interferon. Recent evidence indicates that both adenoviruses and Epstein-Barr virus produce small

RNAs, such as the adenovirus VA RNAs discussed earlier (see Figure 24-22), whose role may be to reverse the effects of interferon. The situation has been studied most with adenovirus VA RNAs.

VA RNAs are present in exceedingly high amounts late in infection with adenoviruses (probably 10^8 VA_I and over 10^6 VA_{II} molecules per cell). Adenovirus mutants from which the VA_I genes have been deleted have a defect in the initiation of translation of their late mRNAs. In vitro, this initiation defect can be overcome by the addition of the translation factor eIF-2B to infected cell extracts. These observations suggest that the recycling of initiation factor eIF-2 is somehow affected by the presence of the VA_I RNAs.

The following scenario is consistent with all the known facts and could explain why adenovirus infections are relatively refractile to interferon. Adenovirus infection induces interferon, which binds to its receptors and induces the antiviral state, activating the double-stranded RNA dependent kinase. Although the kinase requires double-stranded RNA to function, only *low* levels of such RNA and only RNAs that have perfect double-stranded regions of a certain minimum length are effective. In fact, high levels of RNA or imperfect duplex RNA can inhibit the activation of the enzyme. VA RNAs possess much double-stranded character but are interrupted by many bulges and loops. Furthermore, they are made at very high levels. So the present hypothesis is that VA RNA prevents the activation of the interferon-induced protein kinase that would otherwise inhibit translation by phosphorylating the α subunit of eIF-2.

PLANT VIRUSES

The Viruses of Plants Also Present a Variety of Genomic Organizations[156]

Like animal viruses, plant viruses exist in a variety of forms; they contain either DNA or RNA as their genetic material, have either rod or polyhedral shapes, and can be transmitted either by insect bites or by contact upon wounded regions. Most known plant viruses contain single + stranded RNA as their genetic material. Some like tobacco mosaic virus (TMV) have rod shapes while other have polyhedral shapes: for example, tomato bushy stunt virus is structurally remarkably similar to poliovirus, suggesting a common origin long ago in the evolutionary past. + stranded plant viruses can further be divided into those which possess a single RNA chain and those which have several RNA chains, each necessary for viral infectivity and which are separately encapsulated into separate virions. Cowpea mosaic virus, for example, contains two RNAs, one encoding several proteins including a terminal protein and a protease, with the other chain coding for capsid products. There also exist segmented double-strand RNA plant viruses whose genome structure somewhat resembles that of the reoviruses. The best known of these is wound tumor virus (WTV) which contains 12 different segments and which can replicate in either insect or plant cells. While infection of insect cells by WTV is nonsymptomatic, its replication within its plant vectors gives rise to a variety of symptoms including tumor formation. Many passages of WTV in plant cells often cause it to lose its ability to multiply in insect

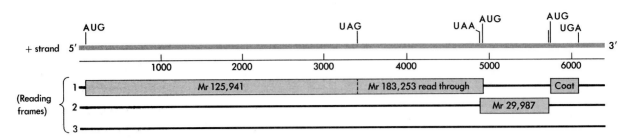

Figure 24-35

TMV coding sequences. The three reading frames of the positive strand are presented, with their respective coding regions, drawn to scale as open boxes with indicated polypeptide molecular weights. Initiation and termination codons of these coding sequences are displayed above the positive strand. (After Goelet et. al., 1982.)

cells, a state that correlates with genetic deletion in two RNA segments which encode outer capsid polypeptides.

Less variety exists in known DNA plant viruses. Only two known classes exist, one of which contains double strand DNA and which is polyhedral in shape. The best understood of this class is cauliflower mosaic virus (CaMV). The second class of DNA plant viruses are the geminiviruses that consist of paired capsids held together like twins (gemini) with each capsid containing a circular single-stranded DNA molecule of some 2500 nucleotides. In some cases, the two paired genomes are genetically identical while in other cases, the two DNA genomes bear almost no sequence resemblance.

TMV Remains the Best-Known Plant Virus[157, 158]

The single + stranded TMV RNA of 6395 nucleotides encodes four polypeptides. Sixty-nine residues from the 5′ capped end is an AUG initiation codon that is followed by an open reading frame (terminated by an UAG stop codon) that specifies a 125,941 MW polypeptide. Occasionally (2–3%) readthrough occurs past this stop codon to produce a larger 183,253 MW polypeptide (Figure 24-35). Both proteins are thought to be part of the enzymatic complex responsible for RNA replication. Slightly overlapping the terminal codons of the readthrough polypeptide is the reading frame coding for the third polypeptide (MW 29,987) whose function remains unknown. The third gene stops two nucleotides before the initiation codon of the fourth protein, the coat protein (17,604 MW). Synthesis of these two latter polypeptides initiates at internal AUG codons using two nested sub-genomic mRNAs with identical 3′ ends. These sub-genomic mRNAs may be made like those of the animal toga viruses, by initiating replication at internal sites on the − strand template. In this way the AUG initiation codons for translating the 29,987 and 17,604 (coat protein) genes are exposed.

Assembly of the coat protein subunits around the newly made + strands to produce progeny TMV particles occurs in a non-obvious way (Figure 24-36). A unique internal site within the gene coding for the 29,987 polypeptide serves as a nucleation center which binds to disk-like aggregates of the coat protein. Subsequent disk addition leads to growth in the 5′ direction while assembly in the 3′ direction involves addition of single coat protein monomers. There are sequence homologies between the nucleation assembly site and the section of the coat protein gene which codes for those amino acids which bind to the RNA. So, conceivably the 29,987 polypeptide also can bind to RNA.

Replication of the DNA Containing Cauliflower Mosaic Virus Involves Reverse Transcription[156]

The chromosome of the double-stranded DNA virus, cauliflower mosaic virus (CaMV), contains some 8000 bp, the sequence of which has been totally worked out. Its two chains (α and β) both contain single-stranded breaks with the 3' ends slightly overlapping the 5' ends to produce triple-stranded regions. Two major transcripts, one of which is slightly more than full length, are made off the α strand. Six major open reading frames exist in the full-length transcript. At first, it seemed likely that CaMV might replicate like SV40, but now it appears that DNA replication starts by reverse transcription of the larger mRNA transcript to yield a DNA-RNA hybrid helix that in turn is converted into double helical DNA identical to that found in the infectious parental CaMV virion. The replication of CaMV thus resembles that of the hepatitis B viruses.

Frontiers of Virology: Viruses Without Genes?[159–164]

By now, we understand the nature of most viruses and their role in disease. There remain, however, a few infectious diseases whose cause has not been resolved. Most prominent of these are a set of related conditions of different species that lead to progressive degeneration of brain tissue and eventually to death. The group includes two human diseases known as *kuru* and *Creutzfeldt-Jakob disease*, as well as a disease of sheep, called *scrapie*. These are clearly infectious diseases: Kuru was transmitted by cannibalism until the custom ceased, and Creutzfeldt-Jakob disease is usually passed on by improperly sterilized surgical instruments or by organ transplants from an affected donor. Scrapie is extremely difficult to work with in the laboratory. It has been adapted to mice and hamsters, decreasing the assay time for disease from years to months, but no cell culture system yet exists. It is readily transferred from animal to animal, however, by injections of infected brain.

The most remarkable feature of scrapie is that no one has been able to obtain biochemical evidence that it contains nucleic acid. The infectious agent in scrapie preparations is highly resistant to treatments such as ultraviolet light or chemical crosslinking reagents, potent inactivators of RNA or DNA genomes. It is, however, sensitive to chemicals such as phenol, which denature protein. Furthermore, the scrapie agent seems to have a molecular weight far below the usual virus when determined chromatographically. The agent that causes scrapie has been highly purified (7000-fold) from crude brain extracts. This material contains no detectable nucleic acid and consists largely of a single protein of about 27 kdal.

These observations have led to the suggestion that scrapie is a novel kind of infectious agent, called a **prion** (for proteinaceous infectious particle). The 27 kdal protein is called PrP (prion-related protein). Although PrP seems to be inseparable from the infectious scrapie agent, it has recently been shown that PrP is probably a normal cell protein. DNA complementary to mRNA for a protein closely

Figure 24-36
A possible mechanism for nucleation of assembly of TMV. The protein and RNA are drawn diagrammatically, as is the predicted configuration of the RNA backbone at each stage.

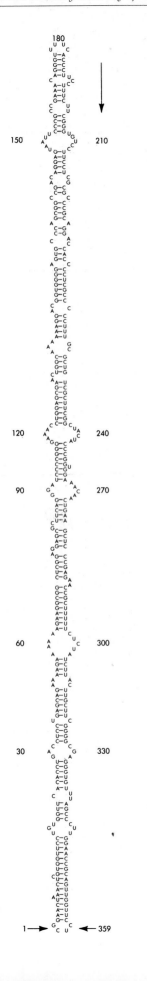

related or identical to PrP has been cloned. Both infected and uninfected brains express the RNA. Given this, what can scrapie be?

If, indeed, scrapie contains no nucleic acid, then conceivably it is a cell protein that stimulates its own synthesis, even to a level high enough to kill the cell. Since PrP is expressed in uninfected cells, however, this possibility seems rather unlikely. Further complicating the hypothesis that scrapie lacks nucleic acid is the apparent existence of distinct strains of the agent with different biological properties that breed true upon repeated passage. This fact suggests some sort of genetic system able to generate, propagate, and express such alterations. Why, then, has the nucleic acid in scrapie not been detected? To both escape detection and be resistant to treatments that inactivate nucleic acid, such a "genome" must be exceptionally small, and probably otherwise unusual in structure as well.

There is a precedent for extremely small viruses with unusual genomes. They are the well-characterized **viroids** of plants. Viroids violate many of the principles established for animal viruses. They consist only of RNA, about 240 to 350 nucleotides long. The RNA is a single-stranded circle that can form extensive base pairs within itself so that it assumes a stiff, nearly double-helical structure (Figure 24-37). The small size means that viroids present only very small targets for inactivation by ultraviolet light, and are thus highly UV resistant. The double-stranded structure makes them resistant to ribonuclease, which usually cleaves only at unpaired bases. The nucleotide sequence displays little or no potential for coding proteins: In potato spindle tuber viroid (PSTV), for example, there are no AUG initiation codons, and frequent termination codons occur in all reading frames.

Clearly, viroid RNAs cannot code for proteins to replicate their genomes. Unlike animal cells, plant cells must therefore contain enzymes capable of replicating RNA. Indeed, such enzymes have been detected. Synthesis of RNA strands both identical and complementary to viroid genomes can readily be observed in infected cells in culture. These strands include full-length genomes as well as multiple-genome-length molecules. Although details are scanty, replication seems to occur, at least in part, by a rolling circle type of mechanism in which synthesis continues around and around the template. The enzymes involved in viroid RNA replication and their role in the life of the plant cell remain to be explored.

The extent to which viroids may be models for diseases like scrapie remains unknown. At the very least, such small and seemingly improbable entities as viroids serve to reemphasize the great diversity of eucaryotic viruses.

Figure 24–37
The structure of a viroid. The nucleotide sequence shown is that of the potato spindle tuber virus (PSTV). Extensive base pairing allows this sequence to form an almost correct double helix. [After H. L. Sänger, in *The Microbe 1984. I. Viruses*, ed. B. W. J. Mahy and J. R. Pattison (New York: Cambridge University Press, 1984), pp. 282–334.]

Plate 7

Immunoglobulin and Viral Hemagglutinin

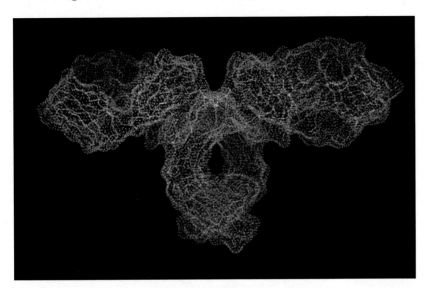

A. Representation of an antibody molecule, human immunoglobulin G, showing its solvent accessible surface (light chain, blue; heavy chain, purple) and α-carbon backbone (light chain, yellow; heavy chain, red).

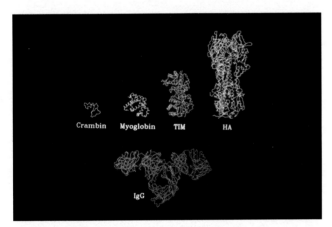

B. Size comparison of immunoglobulin G (IgG) with various other proteins. TIM is triose phosphate isomerase; HA is influenza hemagglutinin.

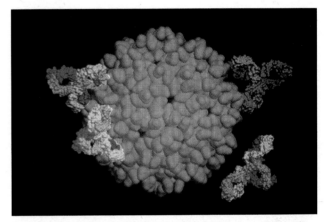

C. Model of antibody molecules interacting with a virus particle. The heavy chains of each immunoglobulin are light and dark blue; the light chains are purple. The virus is tomato bushy stunt virus, whose capsid (yellow) is made up of 180 identical protein subunits.

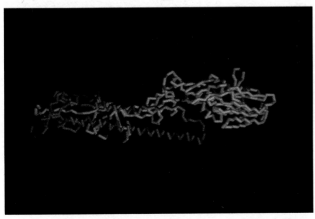

D. Hemagglutinin monomer of the influenza virus. Polypeptide HA$_1$ is blue, and HA$_2$ is red. (See pages 923–924.)

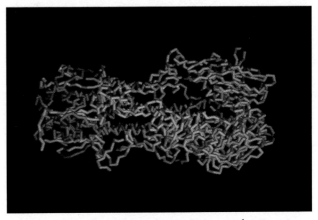

E. Trimer of the influenza hemagglutinin. Images on plate 7 courtesy of A.J. Olson, Scripps Clinic and Research Foundation.

Plate 8

Capsid Structure of Icosahedral Viruses

A. Architecture of the tomato bushy stunt virus. Large spheres represent protein subunits on the surface of the virus. The colors blue, aqua, and red indicate the three different symmetry-packing environments of the subunits. Half of the spheres are removed to reveal the amino terminal arms (yellow) of the red subunits linked together to form an internal scaffold.

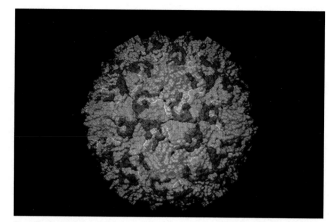

B. Surface of the poliovirus. The solvent accessible surface of poliovirus consists of four capsid proteins, 60 copies each. Proteins VP1 (blue), VP2 (yellow), and VP3 (red) can be seen from the outside of the particle. The maximum radius of the virus particle is 160 Å. (See page 915.)

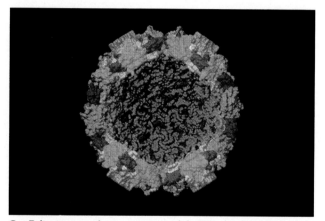

C. Poliovirus capsid cut away to reveal the VP4 capsid protein (green) in the interior. The other capsid proteins are colored as in B. The minimum inside radius is 100 Å.

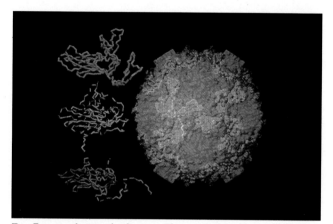

D. Protein subunits of poliovirus. The α-carbon backbones of the three surface protein subunits (capsid proteins) are shown. VP1 is blue; VP2 is yellow; VP3 is violet.

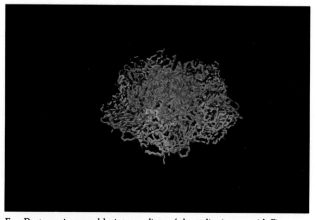

E. Pentameric assembly intermediate of the poliovirus capsid. Five assembled protomers are shown, with 4 subunits in each (color coded as in D). This image shows a view from the outside of the virus.

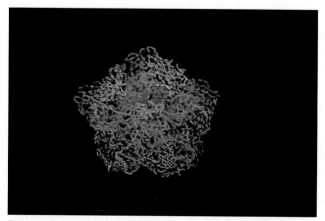

F. As in E, but viewed from the inside of the virus. Images on plate 8 courtesy of A.J. Olson, Scripps Clinic and Research Foundation.

Summary

Eucaryotic viruses have been studied intensively by molecular biologists both because of their ability to cause disease and because they serve as probes to the study of eucaryotic gene expression. The development of cell cultures on which viruses could grow made the molecular biology of many eucaryotic viruses feasible.

The extracellular form of a eucaryotic virus is called a virion. It consists of a protein coat, the capsid, surrounding the nucleic acid genome. In some viruses, the capsid is surrounded by an envelope that is derived from cellular membranes by budding. Capsids can have one of two basic shapes: rodlike or roughly spherical (icosahedral). Rod-shaped capsids can be folded up within an outer envelope to yield virions of varied appearance, and surface spikes can add further diversity.

The virus life cycle consists of attachment to specific receptors on the cell surface, entry of the virion into the cell by fusion or endocytosis, uncoating, transcription and replication of the viral genome, and virion assembly.

Thousands of animal viruses have been isolated. They can be classified into families based on their particle morphology and similarity of genome structure. Viral genomes can be RNA or DNA, single-stranded or double-stranded, linear or, in the case of some double-stranded DNA genomes, circular. Single-stranded RNA virus genomes can correspond to the mRNA or sense strand, in which case they are called positive or plus (+) strand RNA, or they can correspond to the complement of mRNA, in which case they are called negative or minus (−) strand RNA. So far, the known single-stranded DNA viruses are either − strand, or they consist of complementary strands, but each strand is packaged in a separate virion.

A useful system for classifying animal viruses is based on the chemical nature of the genome, the way in which it replicates, and the relationship of the genome to the synthesis of viral mRNA. This classification scheme emphasizes molecular biological principles that underlie animal virus life cycles. According to the system, there are seven major types of animal viruses: (1) Double-strand DNA viruses use cellular enzymes to transcribe their mRNAs; (2) single-strand DNA viruses convert their DNA to double-stranded DNA and then use cellular enzymes to synthesize their mRNAs; (3) + strand RNA viruses can act as mRNA and so direct the synthesis of enzymes that can replicate RNA and also transcribe RNA to yield viral mRNAs; (4) − strand RNA viruses must carry an enzyme inside their virions that can synthesize mRNA from the − strand genome; (5) double-strand RNA viruses must carry an enzyme in the particle that can synthesize mRNA from a double-strand RNA template; (6) retroviruses have a + strand RNA genome, but carry a reverse transcriptase in the virion that converts the RNA to double-stranded DNA; the DNA integrates into a host chromosome and is transcribed by cellular RNA polymerase to make viral mRNA and progeny genomes; (7) hepatitis B virus has a double-stranded DNA genome, but the DNA replicates via an RNA intermediate, thus resembling a retrovirus.

Most RNA viruses (except influenza) replicate in the cytoplasm. They can grow to high titers in a short period of time on many cell types, since they encode their own enzymes for replication and since all cells, even those that are not dividing, are always synthesizing RNA and thus have high concentrations of the needed triphosphates.

Most DNA viruses (except poxviruses) replicate in the nucleus where they can use the cell's transcriptional machinery, and most use cellular enzymes to replicate their DNA as well. However, they face a problem with DNA replication: Most cells in an animal are not dividing at any one time and thus lack the high concentrations of enzymes and precursors needed for extensive DNA replication. To overcome this problem, most DNA viruses have devised ways to turn on the replicative capacity of the host cell.

A problem common to RNA and DNA viruses is how to express numerous genes when most eucaryotic mRNAs are monocistronic.

Poliovirus (a picornavirus) has a + strand RNA genome about 7.4 kilobases long in a naked icosahedral capsid. The genome serves as mRNA (usually after removal of the tiny 5′ terminal VPg protein). The genome encodes a single protein, the polyprotein, from which about a dozen proteins are generated by sequential proteolytic cleavages. Critical to the early events of the life cycle are the virus-coded proteins involved in synthesizing RNA. These include VPg and replicase. VPg is thought to become covalently attached to U residues, which then can prime synthesis of polio RNA from either a + or − strand template. This is important, since replicase, like DNA polymerases, can extend but not initiate polynucleotide chains on a template. In contrast to the poly A on cellular mRNAs, the polio poly A is copied during RNA replication. Replicative intermediates consist of a single − strand with several + strands in various states of completion dangling off them. Virion assembly occurs when aggregates of a precursor undergo successive proteolytic cleavages to yield four final virion polypeptides: VP1, 2, 3, and 4.

Vesicular stomatitis virus (VSV) (a rhabdovirus) is a − strand RNA virus. The genome is encased in viral coded N protein, and the rodlike capsid is wound up inside an envelope to form a bullet-shaped particle. Four other virus-coded proteins are present in the virion: G, an envelope glycoprotein that binds the virus to specific receptors on cells; M, a matrix protein that lies between the envelope and the capsid; and L and NS, which together form an enzyme complex that synthesizes mRNAs and replicates RNA. VSV virions that are treated with a mild detergent (to allow triphosphates to get in) can synthesize mRNAs in vitro. Five mRNAs are produced, each with a 5′ cap and 3′ poly A tail. Thus, the L-NS complex can carry out the multiple enzymatic reactions involved in capping, RNA synthesis, and poly A addition. Poly A addition to the viral RNA is performed by a method entirely different from cellular poly A addition. Between each VSV gene is a conserved nucleotide sequence that includes a string of seven U residues. When the L-NS complex comes to this sequence, it pauses and stutters to add a long string of A residues. Then it either falls off the template (30 percent of the time) or moves on to synthesize the next mRNA.

In vivo, the choice of whether to synthesize VSV mRNAs or to synthesize a full-length + strand that can serve as a template for the synthesis of more progeny genomes is determined by the concentration of N protein that has built up inside the cell. N protein initiates binding to + strands at a site near the 5′ end on a short leader RNA. In the absence of N, the leader RNA is freed, and mRNAs are synthesized. In the presence of N protein, binding to the leader occurs and continues immediately behind the advancing L-NS enzyme, preventing the synthesis of mRNAs and allowing the synthesis of full-length + strand RNA. VSV virions are assembled by budding.

Influenza virus (an orthomyxovirus) is a segmented, − strand RNA virus. Influenza virions are enveloped and have two surface, virus-coded glycoproteins, a hemagglutinin and a neuraminidase. The genome consists of eight fragments of RNA. In contrast to other RNA viruses, influenza replicates in the nucleus. There it initiates viral RNA transcription by clipping 5′ caps from newly made cellular RNAs and using them to initiate transcription of its eight genome fragments. It also uses cellular RNA-splicing machinery to generate spliced mRNAs from two of its genome segments, so the eight segments of flu give rise to ten mRNAs and ten virus-coded proteins. The three-dimensional structure of the HA (hemagglutinin) protein (and also of the neuraminidase) has been determined. The primary HA polypeptide is cleaved to yield two polypeptides (HA_1 and HA_2) that are held together by a disulfide bond. Three ($HA_1 + HA_2$) monomers associate to form the mature HA protein, which has a cylindrical shape. Grooves on the top of the cylinder recognize receptors on cells, and four sites have been identified on the surface of the protein that are altered to produce antigenically new strains of flu. The ability to change its coat, sometimes by genome reassortment, allows flu to induce recurring epidemics.

Reovirus has a double-stranded, segmented RNA genome in a naked complex double capsid. When reovirus virions enter cells, they are only partly uncoated. They lose proteins from the outer shell, but the inner capsid remains intact, and RNA synthesis commences inside of it. Synthesis is asymmetric and conservative. Only + strands are made at this point, and the incoming double strands remain intact. The + strands are extruded through pores at the 12 vertices of the capsid. + strands serve as mRNA and as intermediates in replication. They assemble with new capsid proteins into precores, where they are converted to double-stranded RNA. The precores can synthesize additional + strand RNA until late in infection, when they acquire their outer shell.

SV40 (a papovavirus) has a small, double-stranded, circular DNA genome. The genome is about 5.2 kilobases long and is divided into three regions: About half is the early region; the other half is the late region; and a small segment between the two is a control region containing the origin of DNA replication and transcriptional signals to regulate early and late gene expression. Transcription of the early and late genes occurs in opposite directions from different DNA strands. The shift from early to late gene expression occurs after DNA replication. Two early gene proteins, small t and large T proteins, are made from overlapping spliced mRNAs. The function of small t is unknown, while large T has multiple functions. It turns on cellular DNA synthesis, is needed for initiation of viral DNA replication, shuts down synthesis of early genes, and is needed to help turn on late gene expression. It binds to three adjacent sites in the regulatory region of the genome. Transcriptional signals that control early gene expression include a TATA sequence, a GC-rich 21-base-pair repeat, and an enhancer element.

Adenoviruses have linear, double-stranded DNA genomes, about 36 kilobase pairs long, in a naked icosahedral virion with spikes. The virus replicates in the nucleus. At least seven regions of the genome are expressed early in infection, and both strands of DNA are used. The transcription map for early and late genes is extraordinarily complex. Multiple mRNAs are generated from single segments of DNA by complex splicing patterns. Almost all late genes are expressed from a single very strong promoter, the major late promoter. A primary transcript, about 28 kilobases long, gives rise to more than 20 late mRNAs by differential poly A addition and RNA splicing. The 5′ end of each strand of adenovirus DNA is covalently attached to a 55 kdal terminal protein. Its precursor, preterminal protein, becomes covalently attached to a C residue, which then serves as a primer for DNA replication. During replication, only one strand at a time is synthesized, and synthesis is continuous. Adenovirus DNA replication can now be achieved in vitro, so the various proteins involved are rapidly being identified and purified.

Parvoviruses have single-stranded, linear DNA genomes. These simple viruses do not encode proteins that can stimulate host DNA synthesis and so can only replicate in dividing cells. They have unusual sequences at their termini that allow the ends to fold and form primers for DNA synthesis.

Poxviruses have huge double-stranded DNA genomes. The poxvirus genome is linear, but its two strands are covalently joined at each end. These are the only known DNA viruses that replicate in the cytoplasm. Their genomes are so large that they can encode all the enzymes needed to make DNA and RNA transcripts without the help of cellular enzymes in the nucleus.

Retroviruses have a single + strand RNA genome, but for unknown reasons, they are diploid: There are two identical copies of the RNA in each virion. Virions are enveloped. After entering the cell, reverse transcriptase within the virion makes a double-stranded DNA copy of the genome. This linear DNA moves to the nucleus, where it is circularized and integrated into a nonspecific site in the cellular DNA. Integration occurs at a specific site on the viral genome. The viral DNA, called the provirus, is longer than the RNA genome because of the presence of long direct repeats (several hundred nucleotides long) at each end of the DNA. These are called the long terminal repeats or LTRs, and they contain the signals for initiation and termination of transcription of viral RNA. RNA polymerase II transcribes the provirus from its integrated position in the host's DNA. Full-length transcripts become progeny genomes or serve as mRNA for the synthesis of two polyproteins. Proteolytic cleavage of the *gag* gene product gives rise to the capsid proteins, while proteolytic cleavage of the *pol* gene product releases reverse transcriptase, a protease involved in the cleavages, and a nuclease believed to serve as an integrase for proviral DNA. A spliced mRNA is translated to yield a surface glycoprotein responsible for attachment of the virus to specific receptors on cells. Progeny virions are budded from the cell membrane without killing the cell, and most virus-infected cells grow indefinitely without any apparent harm from the virus.

All vertebrates inherit multiple copies of retrovirus proviruses in their genomes. Most of these are silent, but some are expressed to yield viral proteins and in some cases complete infectious viruses. Had proviruses been discovered in genomic DNA before they were known to be retroviruses, they would have been called transposable elements, since their structure is similar to a class of transposable elements found in yeast and *Drosophila*.

Hepadnaviruses (hepatitis B virus) have a double-stranded, circular DNA genome, but both strands in the circle are linear. The virus replicates its DNA through an RNA intermediate. Thus, it resembles a retrovirus, except that the DNA rather than the RNA phase of the life cycle is packaged in the virions.

Viral infections often induce the synthesis of a set of cellular glycoproteins called interferons. Double-stranded RNA is also a powerful inducer of interferon, and the double-stranded RNA formed during many types of viral replication is probably the actual inducer of interferon in viral infections. Interferons are released from the infected cell, bind to receptors on this cell and on adjacent cells, and induce a complex series of changes that render cells less infectable by many types of viruses. This set of changes is called the antiviral state. Two aspects of the antiviral state are understood in some detail. First, interferon induces a protein kinase, which, when activated by double-stranded RNA, specifically phosphorylates the α subunit of the initiation factor eIF2, thereby inactivating it. Second, interferon induces the synthesis of an enzyme called 2',5'-oligo A synthetase. The enzyme is activated by double-stranded RNA and then synthesizes 2',5'-oligo A from ATP. 2',5'-oligo A binds to a specific ribonuclease, RNase L, and thus activates it to cleave single-stranded RNA and hence mRNA. The enzyme does not seem to discriminate between cellular and viral mRNA; however, it may not matter that the infected cell dies, as long as it does not release a large burst of progeny viruses. Interferons are like hormones in many ways. They are effective at very low concentrations, and they exert complex effects on cells by binding to specific cell surface receptors. There are three types of interferons, called α, β, and γ. There are multiple genes for α interferons.

There are certain neurological diseases (scrapie in sheep, kuru and Creutzfeldt-Jakob disease in humans) that seem to be caused by viruses, but the viruses have not yet been identified. These diseases may be caused by agents resembling plant viroids, tiny RNA molecules without capsids and apparently without protein-coding genes as well, or they may be caused by viruses that consist of protein but no nucleic acid (prions).

Bibliography

General References

Armstrong, J., E. J. Atencio, B. E. Bergman, H. S. Bilofsky, L. B. Brown, C. Burks, G. N. Cameron, M. J. Cinkosky, U. Elbe, C. E. England, J. W. Fickett, B. T. Foley, W. B. Goad, G. H. Hamm, J. A. Hayter, D. Hazledine, M. Kanehisa, L. Kay, G. G. Lennon, F. I. Lewitter, C. R. Linder, A. Leutzenkirchen, P. McCaldon, M. J. McLeod, D. L. Melone, G. Myers, D. Nelson, J. L. Nial, H. M. Perry, W. P. Rindone, L. D. Sher, M. T. Smith, G. Stoesser, C. D. Swindell, and C.-S. Tung, eds. 1985. *Nucleotide Sequences 1985. IV. Viral and Synthetic Sequences.* Washington, D.C.: IRL Press.

Fields, B. N., D. M. Knipe, R. M. Chanock, J. L. Melnick, B. Roizman, and R. E. Shope, eds. 1985. *Virology.* New York: Raven Press.

Mahy, B. W. J., and J. R. Pattison, eds. 1984. *The Microbe 1984. I. Viruses.* New York: Cambridge University Press.

Tooze, J., ed. 1980. *Molecular Biology of Tumor Viruses: DNA Tumor Viruses.* 2nd ed. Cold Spring Harbor, N.Y.: Cold Spring Harbor Laboratory.

Weiss, R. A., N. Teich, H. E. Varmus, and J. M. Coffin, eds. 1985. *Molecular Biology of Tumor Viruses: RNA Tumor Viruses.* 2nd ed. 2 vols. Cold Spring Harbor Laboratory, N.Y.

Cited References

1. Wiley, D. C. 1985. "Viral Membranes." In *Virology,* ed. B. N. Fields, D. M. Knipe, R. M. Chanock, J. L. Melnick, B. Roizman, and R. E. Shope. New York: Raven Press, Chap. 4, pp. 45–68.

2. Harrison, S. 1985. "Virus Structure." In *Virology,* ed. B. N. Fields, D. M. Knipe, R. M. Chanock, J. L. Melnick, B. Roizman, and R. E. Shope. New York: Raven Press, Chap. 3, pp. 27–44.

3. Hogle, J. M., M. Chow, and D. J. Filman. 1985. "Three-Dimensional Structure of Poliovirus at 2.9 Å Resolution." *Science* 229:1358–1365.

4. Rossmann, M. G., E. Arnold, J. W. Erickson, E. A. Frankenberger, J. P. Griffith, H. J. Hecht, J. E. Johnson, G. Kramer, M. Luo, A. G. Mosser, R. R. Rueckert, B. Sherry, and G. Vriend. 1985. "Structure of a Human Common Cold Virus and Functional Relationship to Other Picornaviruses." *Nature* 317:145–153.

5. Simons, K., H. Garoff, and A. Helenius. 1982. "How an Animal Virus Gets Into and Out of Its Host Cell." *Sci. Amer.* 246:58–66.

6. Reanney, D. 1984. "The Molecular Evolution of Viruses." In *The Microbe 1984. I. Viruses,* ed. B. W. J. Mahy and J. R. Patterson. New York: Cambridge University Press, pp. 175–196.

7. Enders, J. F., T. H. Weller, and F. C. Robbins. 1949. "Cultivation of the Lansing Strain of Poliomyelitis Virus in Cultures of Various Human Embryonic Tissues." *Science* 109:85–87.

8. Todaro, G., and H. Green. 1963. "Quantitative Studies on the Growth of Mouse Embryo Cells in Culture and Their Development into Established Lines." *J. Cell. Biol.* 17:299–313.

9. Dulbecco, R. 1952. "Production of Plaques in Monolayer Tissue Cultures by Single Particles of an Animal Virus." *Proc. Nat. Acad. Sci.* 38:747–752.

10. Matthews, R. E. F. 1982. "Classification and Nomenclature of Viruses." 4th report of the ICTV. *Intervirology* 17:1–199.

11. Baltimore, D. 1971. "Expression of Animal Virus Genomes." *Bacteriol. Revs.* 35:235–241.

12. Kates, J. R., and B. R. McAuslan. 1967. "Poxvirus DNA-Dependent RNA Polymerase." *Proc. Nat. Acad. Sci.* 58:134–141.

13. Mayor, H. D., K. Torikai, J. Melnick, and M. Mandel. 1969. "Plus and Minus Single-Stranded DNA Separately Encapsidated in Adeno-Associated Satellite Virions." *Science* 166:1280.

14. Shatkin, A. J., and J. D. Sipe. 1968. "RNA Polymerase Activity in Purified Reovirus." *Proc. Nat. Acad. Sci.* 61:1462–1469.

15. Temin, H. M., and S. Mizutani. 1970. "RNA-Dependent DNA Polymerase in Virions of Rous Sarcoma Virus." *Nature* 226:1211–1213.

16. Baltimore, D. 1970. "RNA-Dependent DNA Polymerase in Virions of RNA Tumor Viruses." *Nature* 226:1209–1211.

17. Kaplan, P. N., R. L. Greenman, J. L. Gerin, R. H. Purcell, and W. S. Robinson. 1973. "DNA Polymerase Associated with Human Hepatitis B Antigen." *J. Virology* 12:995–1005.

18. Summers, D. F., and J. V. Maizel. 1968. "Evidence for Large Precursor Proteins in Poliovirus Synthesis." *Proc. Nat. Acad. Sci.* 59:966–971.

19. Duesberg, P. H. 1968. "The RNA's of Influenza Virus." *Proc. Nat. Acad. Sci.* 59:930–934.

20. Ball, L. A., and C. N. White. 1976. "Order of Transcription of Genes of Vesicular Stomatitis Virus." *Proc. Nat. Acad. Sci.* 73:442–446.

21. Simmons, D. T., and J. H. Strauss, Jr. 1972. "Replication of Sindbis Virus. I. Relative Size and Genetic Content of 26S and 49S RNA." *J. Mol. Biol.* 71:599–614.

22. Dalgleish, A. G., P. C. L. Beverley, P. R. Clapham, D. H. Crawford, M. F. Greaves, and R. A. Weiss. 1984. "The CD4(T4) Antigen Is an Essential Component of the Receptor for the AIDS Retrovirus." *Nature* 312:763–767.

23. Klatzmann, D., F. Barre-Sinoussi, M. T. Nugeyre, C. Dauguet, E. Vilmer, C. Griscelli, F. Brun-Vezinet, C. Rouzioux, J. C. Gluckman, J.-C. Chermann, and L. Montagnier. 1984. "Selective Tropism of Lymph-Adenopathy Associated Virus (LAV) for Helper-Inducer T Lymphocytes." *Science* 225:59–63.

24. Rogers, G. N., J. C. Paulson, R. S. Daniels, J. J. Skehel, I. A. Wilson, and D. C. Wiley. 1983. "Single Amino Acid Substitutions in Influenza Hemagglutin Change Receptor Binding Specificity." *Nature* 304:76–78.

25. Scheid, A., and P. W. Choppin. 1974. "Identification of Biological Activities of Paramyxovirus Glycoproteins. Activation of Cell Fusion, Hemolysis and Infectivity by Proteolytic Cleavage of Inactive Precursor Protein of Sendai Virus." *Virology* 57:475–490.

26. White, J., and A. Helenius. 1980. "pH Dependent Fusion Between the Semliki Forest Virus Membrane and Liposomes." *Proc. Nat. Acad. Sci.* 77:3273–3277.

27. Kitamura, N., B. Semler, B. G. Rothberg, G. R. Larsen, C. J. Adler, A. J. Dorner, R. A. Emini, R. Hanecak, J. J. Lee, S. van der Werf, C. W. Anderson, and E. Wimmer. 1981. "Primary Structure, Gene Organization and Polypeptide Expression of Poliovirus RNA." *Nature* 291:547–553.

28. Racaniello, V. R., and D. Baltimore. 1981. "Molecular Cloning of Poliovirus cDNA and Determination of the Complete Nucleotide Sequence of the Viral Genome." *Proc. Nat. Acad. Sci.* 78:4887–4891.

29. Crocker, T. T., E. Pfendt, and R. Spendlove. 1964. "Poliovirus Growth in Non-Nucleate Cytoplasm." *Science* 145:401–403.

30. Nomoto, A., Y. F. Lee, and E. Wimmer. 1976. "The 5' End of Poliovirus mRNA Is Not Capped with m^{7}G(5')ppp(5')-Np." *Proc. Nat. Acad. Sci.* 73:375–380.

31. Jacobson, M., and D. Baltimore. 1968. "Polypeptide Cleavages in the Formation of Poliovirus Proteins." *Proc. Nat. Acad. Sci.* 61:77–84.

32. Palmenberg, A. C., M. A. Pallansch, and R. R. Rueckert. 1979. "Protease Required for Processing Picornaviral Coat Protein Resides in the Viral Replicase Gene." *J. Virology* 32:770–778.

33. Palmenberg, A. C., and R. R. Rueckert. 1982. "Evidence for Intramolecular Self-Cleavage of Picornaviral Replicase Precursors." *J. Virology* 41:244–249.

34. Baltimore, D., and M. Girard. 1966. "An Intermediate in the Synthesis of Poliovirus RNA." *Proc. Nat. Acad. Sci.* 56:741–748.

35. Flanegan, J. B., and D. Baltimore. 1977. "Poliovirus-Specific Primer Dependent RNA Polymerase Able To Copy Poly(A)." *Proc. Nat. Acad. Sci.* 94:2677–2680.

36. Lee, F. Y., A. Nomoto, B. M. Detjen, and E. Wimmer. 1977. "A Protein Covalently Linked to Poliovirus Genome RNA." *Proc. Nat. Acad. Sci.* 74:59–63.

37. Yogo, F., and E. Wimmer. 1973. "Poly(A) and Poly(U) in Poliovirus Double-Stranded RNA." *Nature New Biol.* 242:171–174.

38. Penman, S., and D. Summers. 1965. "Effects on Host Cell Metabolism Following Synchronous Infection with Poliovirus." *Virology* 27:614–620.

39. Etchison, D., J. Hanse, E. Ehrenfeld, I. Edery, N. Sonenberg, S. Milburn, and J. W. B. Hershey. 1984. "Demonstration *in Vitro* That Eukaryotic Initiation Factor 3 Is Active but That a Cap-Binding Protein Complex Is Inactive in Poliovirus-Infected HeLa Cells." *J. Virology* 51:832–837.

40. Fernandez-Tomes, C. B., and D. Baltimore. 1973. "Morphogenesis of Poliovirus. II. Demonstration of a New Intermediate, the Provirion." *J. Virology* 12:1122–1130.

41. Rueckert, R. R., A. K. Dunker, and C. M. Stoltzfus. 1969. "The Structure of Maus-Elberfeld Virus: A Model." *Proc. Nat. Acad. Sci.* 62:912–919.

42. Strauss, E. G., C. M. Rice, and J. H. Strauss. 1984. "Complete Nucleotide Sequence of the Genomic RNA of Sindbis Virus." *Virology* 133:92–110.

43. Wirth, D. F., F. Katz, B. Small, and H. F. Lodish. 1977. "How a Single Sindbis Virus mRNA Directed the Synthesis of One Soluble Protein and Two Integral Membrane Proteins." *Cell* 10:253–263.

44. Stern, D. F., and S. I. T. Kennedy. 1980. "Coronavirus Multiplication Strategy. II. Mapping of the Avian Infectious Bronchitis Virus Intracellular RNA Species to the Genome." *J. Virology* 36:440–449.

45. Lai, M. M. C., C. D. Patton, R. S. Baric, and S. A. Stohlman. 1983. "Presence of Leader Sequences in the mRNA of Mouse Hepatitis Virus." *J. Virology* 46:1027–1033.

46. Schubert, M., G. G. Harmison, and E. Meier. 1984. "Primary Structure of the Vesicular Stomatitis Virus Polymerase (L) Gene: Evidence for a High Frequency of Mutations." *J. Virology* 51:505–514.

47. Ball, L. A., and G. W. Wertz. 1981. "VSV RNA Synthesis: How Can You Be Positive?" *Cell* 26:143–144.

48. Iverson, L. E., and J. K. Rose. 1981. "Localized Attenuation and Discontinuous Synthesis During Vesicular Stomatitis Virus Transcription." *Cell* 23:447–484.

49. Emerson, S. U., and Y.-H. Yu. 1975. "Both NS and L Proteins Are Required for *in vitro* RNA Synthesis by Vesicular Stomatitis Virus." *J. Virology* 15:1348–1356.

50. Blumberg, B. M., M. Leppert, and D. Kolakofsky. 1981. "Interaction of VSV Leader RNA and Nucleocapsid Protein May Control VSV Genome Replication." *Cell* 23:837–845.

51. Herman, R. C., M. Schubert, J. D. Keene, J. D. and R. Lazzarini. 1980. "Polycistronic Vesicular Stomatitis Virus RNA Transcripts." *Proc. Nat. Acad. Sci.* 77:4662–4665.

52. Katz, F. N., J. E. Rothman, V. R. Lingappa, G. Blobel, and H. F. Lodish. 1977. "Membrane Assembly *in Vitro*: Synthesis, Glycosylation, And Asymmetric Insertion of a Transmembrane Protein." *Proc. Nat. Acad. Sci.* 74:3278–3282.

53. Huang, A. S., J. W. Greenwalt, and R. R. Wagner. 1966. "Defective T Particles of Vesicular Stomatitis Virus. I. Preparation, Morphology and Some Biologic Properties." *Virology* 30:161–172.

54. Palese, P., and J. L. Schulman. 1976. "Mapping of the Influenza Virus Genome: Identification of the Hemagglutinin and Neuraminidase Genes." *Proc. Nat. Acad. Sci.* 73:2142–2146.

55. Barry, R. D., D. R. Ives, and J. G. Cruickshank. 1962. "Participation of Deoxyribonucleic Acid in the Multiplication of Influenza Virus." *Nature* 194:1139–1140.

56. Bouloy, M., S. J. Plotch, and R. M. Krug. 1978. "Globin mRNA's Are Primers for the Transcription of Influenza Viral RNA *in Vitro*." *Proc. Nat. Acad. Sci.* 75:4886–4890.

57. Braam, J., I. Ulmanen, and R. M. Krug. 1983. "Molecular Model of a Eukaryotic Transcription Complex: Functions and Movements of Influenza P Proteins During Capped RNA Primed Transcription." *Cell* 34:609–618.

58. Robertson, J. S., M. Schubert, and R. A. Lazzarini. 1981. "Polyadenylation Sites for Influenza mRNA." *J. Virology* 38:157–163.

59. Lamb, R. A., and P. W. Choppin. 1979. "Segment 8 of the Influenza Virus Genome Is Unique in Coding for Two Polypeptides." *Proc. Nat. Acad. Sci.* 76:4908–4912.

60. Burnet, F. M., and M. Edney. 1951. "Recombinant Viruses Obtained from Double Infections with the Influenza A Viruses MEL and Neuro-WS." *J. Exp. Biol. Med. Sci.* 39:353–362.

61. Wilson, I. A., J. J. Skehel, and D. C. Wiley. 1981. "Structure of the Hemagglutinin Membrane Glycoprotein of Influenza Virus at 3 Å Resolution." *Nature* 289:366–373.

62. Wiley, D. C., I. A. Wilson, and J. J. Skehel. 1981. "Structural Identification of the Antibody-Binding Sites of Hong Kong Influenza Hemagglutinin and Their Involvement in Antigenic Variation." *Nature* 289:373–378.

63. Skehel, J. J., P. M. Bayley, E. B. Brown, S. R. Martin, M. D. Waterfield, J. M. White, I. A. Wilson, and D. C. Wiley. 1982. "Changes in the Conformation of Influenza Virus Hemagglutinin at the pH Optimum of Virus Mediated Membrane Fusion." *Proc. Nat. Acad. Sci.* 79:968–972.

64. Gomatos, P. J., and I. Tamm. 1963. "Animal and Plant Viruses with Double Helical RNA." *Proc. Nat. Acad. Sci.* 50:878–885.

65. Skehel, J. J., and W. K. Joklik. 1969. "Studies on the *in Vitro* Transcription of Reovirus RNA Catalyzed by Reovirus Cores." *Virology* 39:822–831.

66. Acs, G., H. Klett, M. Schonberg, J. Christman, D. H. Levin, and S. C. Silverstein. 1971. "Mechanism of Reovirus Double-Stranded Ribonucleic Acid Synthesis *in Vitro* and *in Vivo*." *J. Virology* 8:684–689.

67. Roizman, B. 1985. "Multiplication of Viruses: An Overview." In *Virology*, ed. B. N. Fields, D. M. Knipe, R. M. Chanock, J. L. Melnick, B. Roizman, and R. E. Shope. New York: Raven Press, Chap. 5, pp. 69–76.

68. Sweet, B. H., and M. R. Hillerman. 1960. "The Vacuolating Virus SV40." *Proc. Soc. Exp. Biol. Med.* 105:420.

69. Fiers, W., R. Contreras, G. Haegeman, R. Rogiers, A. van de Voorde, H. Van Heuverswyn, J. Van Herreweghe, G. Volkaert, and M. Ysebaert. 1978. "Complete Nucleotide Sequence of SV40 DNA." *Nature* 273:113–120.

70. Reddy, V. B., B. Thimmappaya, R. Dhar, K. N. Subramanian, B. S. Zain, J. Pan, P. K. Ghosh, M. L. Celma, and S. M. Weissman. 1978. "The Genome of Simian Virus 40." *Science* 200:494–502.

71. Crawford, L. V., C. N. Cole, A. E. Smith, E. Paucha, P. Tegtmeyer, K. Rundell, and P. Berg. 1978. "Organization and Expression of Early Genes of SV40." *Proc. Nat. Acad. Sci.* 75:117–121.

72. Sambrook, J., B. Sugden, W. Keller, and P. A. Sharp. 1973. "Transcription of Simian Virus 40. III. Orientation of RNA Synthesis and Mapping of 'Early' and 'Late' Species of Viral RNA Extracted from Lytically Infected Cells." *Proc. Nat. Acad. Sci.* 70:3711–3715.

73. Jay, G., S. Nomura, C. W. Anderson, and G. Khoury. 1981. "Identification of the SV4-Agno Gene Product: A DNA Binding Protein." *Nature* 291:346–349.

74. Benoist, C., and P. Chambon. 1981. "The SV40 Early Promoter Region: Sequence Requirements *in Vivo*." *Nature* 290:304–310.

75. Ghosh, P., P. Lebowitz, F. Frisque, and Y. Gluzman. 1981. "Identification of a Promoter Component Involved in Positioning the 5' Termini of SV40 Early mRNAs." *Proc. Nat. Acad. Sci.* 78:100–104.

76. Fromm, M., and P. Berg. 1982. "Deletion Mapping of DNA Regions Required for SV40 Early Region Promoter Function *in Vivo*." *J. Mol. Appl. Genetics* 1:457–481.

77. Dynan, W. S., and R. Tjian. 1985. "Control of Eukaryotic Messenger RNA Synthesis by Sequence-Specific DNA-Binding Proteins." *Nature* 316:774–778.

78. Banerji, J., S. Rusconi, and W. Schaffner. 1981. "Expression of a Beta-Globin Gene Is Enhanced by Remote SV40 Sequences." *Cell* 27:299–308.

79. Gruss, P., R. Dhar, and G. Khoury. 1981. "Simian Virus 40 Tandem Repeated Sequences as an Element of the Early Promoter." *Proc. Nat. Acad. Sci.* 78:943–947.

80. Tjian, R. 1978. "The Binding Site on SV40 DNA for a T-Antigen Related Protein." *Cell* 13:165–179.

81. Ghosh, P., V. Reddy, J. Swinescoe, P. Lebowitz, and S. Weissman. 1978. "Heterogeneity and 5' Terminal Structure of the Late RNA's of SV40." *J. Mol. Biol.* 126:813–846.

82. Ryder, K., E. Vakalopoulou, R. Mertz, I. Masriangelo, P. Hough, P. Tegtmeyer, and E. Fanning. 1985. "Seventeen Base Pairs of Region I Encode a Novel Tripartite Binding Signal for SV40 T Antigen." *Cell* 42:539–548.

83. Sebring, E. D., T. J. Kelly, Jr., M. M. Thorsen, and N. P. Salzman. 1971. "Structure of Replicating Simian Virus 40 Deoxyribonucleic Acid Molecules." *J. Virology* 8:478–490.

84. Danna, K., and D. Nathans. 1972. "Bidirectional Replication of Simian Virus 40 DNA." *Proc. Nat. Acad. Sci.* 69:3097–3100.

85. Tegtmeyer, P. 1972. "Simian Virus 40 Deoxyribonucleic Acid Synthesis: The Viral Replicon." *J. Virology* 10:591–598.

86. Fareed, G. C., and N. P. Salzman. 1972. "Intermediate in SV40 DNA Chain Growth." *Nature New Biol.* 238:274.

87. Shortle, D. R., R. F. Margolskee, and D. Nathans. 1979. "Mutational Analysis of the SV40 Replicon: Pseudorevertants of Mutants with Defective Replication Origin." *Proc. Nat. Acad. Sci.* 76:6128–6131.

88. Garon, C. F., K. W. Berry, and J. A. Rose. 1972. "A Unique Form of Terminal Redundancy in Adenovirus DNA Molecules." *Proc. Nat. Acad. Sci.* 69:2391–2395.

89. Rekosh, D. M. K., W. C. Russell, A. J. D. Bellet, and A. J. Robinson. 1977. "Identification of a Protein Linked to the Ends of Adenovirus DNA." *Cell* 11:283–295.

90. Sharp, P. A., P. H. Gallimore, and S. J. Flint. 1974. "Mapping of Adenovirus 2 RNA Sequences in Lytically Infected Cells and Transformed Cell Lines." *Cold Spring Harbor Symp. Quant. Biol.* 39:457–474.

91. Berget, S. M., C. Moore, and P. A. Sharp. 1977. "Spliced Segments at the 5' Terminus of Adenovirus 2 Late mRNA." *Proc. Nat. Acad. Sci.* 74:3171–3175.

92. Chow, L. T., R. E. Gelinas, T. R. Broker, and R. J. Roberts. 1977. "An Amazing Sequence Arrangement at the 5' Ends of Adenovirus 2 Messenger RNA." *Cell* 12:1–8.

93. Chow, L. T., T. R. Broker, and J. B. Lewis. 1979. "Complex Splicing Patterns of RNAs from the Early Regions of Adenovirus 2." *J. Mol. Biol.* 134:265–303.

94. Ricciardi, R. P., R. L. Jones, C. L. Cepko, P. A. Sharp, and B. E. Roberts. 1981. "Expression of Early Adenovirus Genes Requires a Virus-Coded Acidic Polypeptide." *Proc. Nat. Acad. Sci.* 78:6121–6125.

95. Stillman, B. W., J. B. Lewis, L. T. Chow, M. B. Mathews, and E. Smart. 1981. "Identification of the Gene and mRNA for the Adenovirus Terminal Protein Precursor." *Cell* 23:497–508.

96. Montell, C., E. F. Fisher, M. H. Caruthers, and A. J. Berk. 1982. "Resolving the Functions of Overlapping Genes by Site-Specific Mutagenesis at a mRNA Splice Site." *Nature* 295:380–384.

97. Ziff, E. B., and R. M. Evans. 1978. "Coincidence of the Promoter and Capped 5' Terminus of RNA from the Adenovirus 2 Late Transcription Unit." *Cell* 15:1463–1476.

98. Thomas, G. P., and M. B. Mathews. 1980. "DNA Replication and the Early to Late Transition in Adenovirus Infection." *Cell* 22:523–533.

99. Price, R., and S. Penman. 1972. "A Distinct RNA Polymerase Activity Synthesizing 5.5S and 4S RNA in Nuclei Isolated from Adenovirus 2 Infected HeLa Cells." *J. Mol. Biol.* 70:435–450.

100. Pearson, G. D., and P. C. Hanawalt. 1971. "Isolation of DNA Replication Complexes from Uninfected and Adenovirus-Infected HeLa Cells." *J. Mol. Biol.* 62:65–80.

101. Sussenbach, J. S., P. C. Van der Vliet, D. J. Ellens, and H. S. Jansz. 1972. "Linear Intermediates in the Replication of Adenovirus DNA." *Nature* 239:47–49.

102. Challberg, M. D., and T. J. Kelly. 1979. "Adenovirus DNA Replication *in Vitro*." *Proc. Nat. Acad. Sci.* 76:655–659.

103. Challberg, M. D., S. V. Desiderio, and T. J. Kelly. 1980. "Adenovirus DNA Replication *in Vitro*: Characterization of a Protein Covalently Bonded to Nascent Chains." *Proc. Nat. Acad. Sci.* 77:5105–5109.

104. Richy, J. H., M. S. Horowitz, and J. Hurwitz. 1981. "Formation of a Covalent Complex Between the 80,000 Dalton Adenovirus Terminal Protein and 5' dCMP *in Vitro*." *Proc. Nat. Acad. Sci.* 78:2678–2682.

105. Astell, C. R., M. Smith, M. B. Chow, and D. C. Ward. 1982. "Structure and Replication of Minute Virus of Mice DNA." *Cold Spring Harbor Symp. Quant. Biol.* 47:751.

106. Tattersall, P., and D. C. Ward. 1976. "Rolling Hairpin Model for Replication of Parvovirus and Linear Chromosomal DNA." *Nature* 263:106–109.

107. Hayward, G. S., R. J. Jacob, S. C. Wadsworth, and B. Roizman. 1975. "Anatomy of Herpes Simplex Virus DNA: Evidence for Four Populations of Molecules That Differ in the Relative Orientations of Their Long and Short Components." *Proc. Nat. Acad. Sci.* 72:4243–4247.

108. Baer, R., A. T. Bankier, M. D. Biggin, P. L. Deininger, P. J. Farrel, J. J. Gibson, G. Hatfull, G. S. Hudson, S. C. Satchwell, C. Seguin, P. S. Tofnell, and B. G. Barell. 1984. "DNA Sequence and Expression of the B95-8 Epstein-Barr Virus Genome." *Nature* 310:207–211.

109. Honess, R. W., and B. Roizman. 1974. "Regulation of Herpesvirus Macromolecular Synthesis. I. Cascade Regulation of the Synthesis of Three Groups of Viral Proteins." *J. Virology* 14:8–19.

110. Batterson, W., and B. Roizman. 1983. "Characterization of the Herpes Simplex Virion-Associated Factor Responsible for the Induction of α Genes." *J. Virology* 46:371–377.

111. Kates, J., and B. R. McAuslan. 1967. "Messenger RNA Synthesis by a 'Coated' Viral Genome." *Proc. Nat. Acad. Sci.* 57:314–320.

112. Oda, K., and W. K. Joklik. 1967. "Hybridization and Sedimentation Studies on 'Early' and 'Late' Vaccinia Messenger RNA." *J. Mol. Biol.* 27:395–419.

113. Belle Isle, H., S. Venkatesan, and B. Moss. 1981. "Cell-Free Translation of Early and Late mRNAs Selected by Hybridization to Cloned DNA Fragments Derived from the Left 14 Million to 72 Million Daltons of the Vaccinia Virus Genome." *Virology* 112:306–317.

114. Geshelin, P., and K. Berns. 1974. "Characterization and Localization of the Naturally Occurring Cross-Links in Vaccinia Virus DNA." *J. Mol. Biol.* 88:785–796.

115. Baroudy, B. M., S. Bankatesan, and B. Moss. 1982. "Incompletely Base-Paired Flip-Flop Terminal Loops Link the Two DNA Strands of the Vaccinia Virus Genome into One Uninterrupted Polynucleotide Chain." *Cell* 28:315–324.

116. Rous, P. 1911. "A Sarcoma of the Fowl Transmissible by an Agent Separable from the Tumor Cells." *J. Exp. Med.* 13:397–411.

117. Poeisz, B. J., F. V. Ruscetti, A. F. Gazdar, P. A. Bunn, J. D. Minna, and R. C. Gallo. 1980. "Detection and Isolation of Type C Retrovirus Particles from Fresh and Cultured Lymphocytes of a Patient with Cutaneous T-Cell Lymphoma." *Proc. Nat. Acad. Sci.* 77:7415–7419.

118. Barré-Sinoussi, F., J. C. Chermann, F. Rey, M. T. Nugeyre, S. Chamaret, J. Gruest, C. Danguet, C. Axler-Blin, F. Vézinet-Brun, C. Rouzioux, W. Rozenbaum, and L. Montagnier. 1983. "Isolation of T-Lymphotropic Retrovirus from a Patient at Risk for Acquired Immune Deficiency Syndrome (AIDS)." *Science* 220:868–870.

119. Popovic, M., M. G. Sarnagadharan, E. Read, and R. C. Gallo. 1984. "Detection, Isolation, and Continuous Production of Cytopathic Retroviruses (HTLV-III) from Patients with AIDS and Pre-AIDS." *Science* 224:497–500.

120. Billeter, M. A., J. T. Parsons, and J. M. Coffin. 1974. "The Nucleotide Sequence Complexity of Avian Tumor Virus RNA." *Proc. Nat. Acad. Sci.* 71:3560–3564.

121. Varmus, H. E. 1982. "Form and Function of Retroviral Proviruses." *Science* 216:812–821.

122. Dahlberg, J. E., R. C. Sawyer, J. M. Taylor, A. J. Faras, W. E. Levinson, H. M. Goodman, and J. M. Bishop. 1974. "Transcription of DNA from the 70S RNA of Rous Sarcoma Virus. I. Identification of a Specific 4S RNA Which Serves as Primer." *J. Virology* 13:1126–1133.

123. Coffin, J. M., and W. A. Haseltine. 1977. "Terminal Redundancy and the Origin of Replication of Rous Sarcoma Virus RNA." *Proc. Nat. Acad. Sci.* 74:1908–1912.

124. Sodroski, J. G., C. A. Rosen, and W. A. Haseltine. 1984. "Trans-Acting Transcriptional Activation of the Long Terminal Repeat of Human T Lymphotropic Viruses in Infected Cells." *Science* 225:381–385.

125. Shank, P. R., S. Hughes, H. J. Kung, J. E. Majors, N. Quintrell, R. V. Guntaka, J. M. Bishop, and H. E. Varmus. 1978. "Mapping Unintegrated Forms of Avian Sarcoma Virus DNA: Both Termini of Linear DNA Bear a 300 Nucleotide Sequence Present Once or Twice in Two Species of Circular DNA." *Cell* 15:1383–1395.

126. Gilboa, E., S. W. Mitra, S. Goff, and D. Baltimore. 1979. "A Detailed Model of Reverse Transcription and Tests of Crucial Aspects." *Cell* 18:93–100.

127. Panganiban, A. T., and H. M. Temin. 1984. "Circles with Two Tandem LTRs Are Precursors to Integrated Retrovirus DNAs." *Cell* 36:673–679.

128. Moelling, K., D. Bolognesi, H. Bauer, W. Büsen, H. W. Plassmann, and P. Hausen. 1971. "Association of Viral Reverse Transcriptase with an Enzyme Degrading the RNA Moiety of RNA-DNA Hybrids." *Nature New Biol.* 234:240–243.

129. Yamamoto, T., B. deCrombrugghe, and I. Pastan. 1980. "Identification of a Functional Promoter in the Long Terminal Repeat of Rous Sarcoma Virus." *Cell* 22:787–798.

130. Laimins, L. A., G. Khoury, C. Gorman, B. Howard, and P. Gruss. 1982. "Host-Specific Activation of Transcription by Tandem Repeats from Simian Virus 40 and Moloney Murine Sarcoma Virus." *Proc. Nat. Acad. Sci.* 79:6453–6457.

131. Gorman, C. M., G. T. Merlino, M. C. Willingham, I. Pastan, and B. H. Howard. 1982. "The Rous Sarcoma Virus Long Terminal Repeat Is a Strong Promoter When Introduced into a Variety of Eukaryotic Cells by DNA-Mediated Transfection." *Proc. Nat. Acad. Sci.* 79:6777–6781.

132. Hayward, W. S. 1977. "Size and Genetic Content of Viral RNAs in Avian Oncovirus Infected Cells." *J. Virology* 24:47–63.

133. Vogt, V. M., and R. Eisenman. 1973. "Identification of a Large Polypeptide Precursor of Avian Oncornavirus Proteins." *Proc. Nat. Acad. Sci.* 70:1734–1738.

134. Vogt, P. K., and R. R. Friis. 1971. "An Avian Leukosis Virus Related to RSV(0): Properties and Evidence for Helper Activity." *Virology* 43:223–234.

135. Lowy, D. R., W. P. Rowe, N. Teich, and J. W. Hartley. 1971. "Murine Leukemia Virus: High-Frequency Activation in Vitro by 5-Iododeoxyuridine and 5-Bromodeoxyuridine." *Science* 174:155.

136. Groudine, M., R. Eisenman, and H. Weintraub. 1981. "Chromatin Structure of Endogenous Retroviral Genes and Activation by an Inhibitor of DNA Methylation." *Nature* 292:311–317.

137. Astrin, S. 1978. "Endogenous Viral Genes of White Leghorn Chickens: A Common Site of Residence as Well as Sites Associated with Specific Phenotypes of Viral Gene Expression." *Proc. Nat. Acad. Sci.* 75:5941–5945.

138. Jenkins, N. A., N. G. Copeland, B. A. Taylor, and B. K. Lee. 1981. "Dilute (d) Coat Colour Mutation of DBA/2J Mice Is Associated with the Site of Integration of an Ecotropic MuLV Genome." *Nature* 293:370–374.

139. Breindl, M., K. Harbers, and R. Jaenisch. 1984. "Retrovirus-Induced Lethal Mutations in Collagen I Gene of Mice Is Associated with an Altered Chromatin Structure." *Cell* 38:9–16.

140. Boeke, J. D., D. J. Garfinkel, C. A. Styles, and G. R. Fink. 1985. "Ty Elements Transpose Through an RNA Intermediate." *Cell* 40:491–500.

141. Summers, J. A., A. O'Connell, and I. Millman. 1975. "Genome of Hepatitis B Virus: Restriction Enzyme Cleavage and Structure of DNA Extracted from Dane Particles." *Proc. Nat. Acad. Sci.* 72:4597–4601.

142. Summers, J., and W. S. Mason. 1982. "Replication of the Genome of a Hepatitis B-Like Virus by Reverse Transcription of an RNA Intermediate." *Cell* 29:403–415.

143. Isaacs, A., and J. Lindenmann. 1957. "Virus Interference. I. The Interferon." *Proc. Roy. Soc. London (B)* 147:258–263.

144. Lengyel, P. 1982. "Biochemistry of Interferons and Their Actions." *Ann. Rev. Biochem.* 51:251–282.

145. Lampson, G. P., A. A. Tytell, A. K. Field, M. M. Nemes, and M. R. Hilleman. 1967. "Inducers of Interferon and Host Resistance. 1. Double-Stranded RNA from Extracts of *Penicillium Funiculosum*." *Proc. Nat. Acad. Sci.* 58:782–789.

146. Taniguchi, T., M. Sakai, Y. Fujii-Kurigama, M. Muramatsu, S. Kobayashi, and T. Sudo. 1979. "Construction and Identification of a Bacterial Plasmid Containing the Human Fibroblast Interferon Gene Sequence." *Proc. Jpn. Acad.* 55:464–469.

147. Nagata, S., T. H. Taira, A. Hall, L. Johnsrud, M. Streuli, J. Escodi, W. Ball, K. Cantell, and C. Weissmann. 1980. "Synthesis in *E. coli* of a Polypeptide with Human Leukocyte Interferon Activity." *Nature* 284:316–320.

148. Kerr, I. M., and R. E. Brown. 1978. "pppA2' p5'A2'p5'A: An Inhibitor of Protein Synthesis Synthesized with an Enzyme Fraction from Interferon Treated Cells." *Proc. Nat. Acad. Sci.* 73:520–523.

149. Lebler, B., G. C. Sen, S. Shaila, D. Cabrer, and P. Lengyel. 1976. "Interferon, Double-Stranded RNA and Protein Phosphorylation." *Proc. Nat. Acad. Sci.* 73:3107–3111.

150. Owerbach, D., W. J. Rutter, T. B. Strauss, P. Gray, D. V. Goeddel, and R. M. Lawn. 1981. "Leucocyte and Fibroblast Interferon Genes Are Located on Human Chromosome 9." *Proc. Nat. Acad. Sci.* 55:1133–1140.

151. Fujita, T., S. Ohno, H. Yasumitsu, and T. Taniguchi. 1985. "Delimitation and Properties of DNA Sequences Required for the Regulated Expression of Human Interferon-β Gene." *Cell* 41:489–496.

152. Ryals, J., P. Dierks, H. Ragg, and C. Weissmann. 1985. "A 46-Nucleotide Promoter Segment from an IFN-α Gene Renders an Unrelated Promoter Inducible by Virus." *Cell* 41:497–507.

153. Goodbourn, S., K. Zinn, and T. Maniatis. 1985. "Human β-Interferon Gene Expression Is Regulated by an Inducible Enhancer Element." *Cell* 41:509–520.

154. Miyata, T., and H. Hayashida. 1982. "Recent Divergence from a Common Ancestor of Human IFN-α Genes." *Nature* 295:165–168.

155. Kitajewski, J., R. J. Schneider, B. Safer, S. M. Munemitsu, C. E. Samuel, B. Thimmappaya, and T. Shenk. 1986. "Adenovirus VAI RNA Antagonizes the Antiviral Action of Interferon by Preventing Activation of the Interferon-Induced elF-2α Kinase." *Cell* 45:195–200.

156. Robertson, H. D., S. H. Howell, M. Zaitlin, and R. L. Malmberg, editors. 1983. "Plant Infectious Agents," *Viruses, Viroids, Virusoids, and Satellites.* Cold Spring Harbor Laboratory Press.

157. Goelet, P., G. P. Lomonossoff, P. J. G. Butler, M. E. Akam, M. J. Gait, and J. Karn. 1982. "Nucleotide Sequence of Tobacco Mosaic Virus RNA." *Proc. Nat. Acad. Sci.* 79:5818–5822.

158. Klug, A. 1979. The Assembly of Tobacco Mosaic Virus: Structure and Specificity." The Harvey Lectures 74:141.

159. Gajdsek, C. 1985. "Unconventional Viruses Causing Subacute Spongiform Encephalopathies." In *Virology*, ed. B. N. Fields, D. M. Knipe, R. M. Chanock, J. L. Melnick, B. Roizman, and R. E. Shope. New York: Raven Press, Chap. 63.

160. Priusner, S. B. 1982. "Novel Proteinacious Infectious Particles Cause Scrapie." *Science* 216:136–144.

161. Oesch, B., D. Westaway, M. Wälchli, M. P. McKinley, S. B. H. Kent, R. Abersold, R. A. Barry, P. Tempst, D. B. Teplow, L. E. Hood, S. B. Prusiner, and C. Weissmann. 1985. "A Cellular Gene Encodes Scrapie, Pr-P 27-30 Protein." *Cell* 40:735–746.

162. Diener, T. 1971. "Potato Spindle Tuber Virus." IV. A Replicating, Low Molecular Weight RNA. *Virology* 45:411–428.

163. Gross, H. J., H. Domdey, C. Lossow, P. Jank, M. Raba, H. Albern, and H. L. Sänger. 1978. "Nucleotide Sequence and Secondary Structure of Potato Spindle Tuber Viroid." *Nature* 273:203–208.

164. Branch, A. D., H. D. Robertson, and E. Dickson. 1981. "Longer-Than-Unit-Length Viroid Minus Strands Are Present in RNA from Infected Plants." *Proc. Nat. Acad. Sci.* 78:6381–6385.

X

CANCER AT THE GENETIC LEVEL

CHAPTER 25

The Control of Cell Proliferation

Biologists have long known that the control of cell division is one of the most basic aspects of multicellular existence. Throughout embryological development, as well as through all of adult life, many differentiated cells have the choice to divide or not, and only if a programmed series of correct decisions is made can the respective organisms continue to function normally. But mere desire to come to grips with the important questions does not necessarily generate clean answers; we would be less than candid if we did not admit that understanding the control of cell proliferation has been one of the more slowly moving fields in biology. Today, however, key facts are at last emerging, and it is likely that the next few years of research will provide the intellectual framework to understand proliferation control in a large variety of seemingly unrelated animal cell types. An early key to progress in higher eucaryotes was the realization that experiments need not be restricted to intact organisms, but that real answers can come from growing animal cells in culture. Now much hope lies in our ability to use molecular cloning techniques to identify and isolate genes that control cell growth.

Always intertwined with any study of cell proliferation is the nature of cancer. This collection of horrific diseases by definition involves cells that divide when they should not, usually to produce the contiguous cellular masses called **tumors.** The different classes of cancer arise by changes in the various forms of differentiated cells, with the resulting cancer cells exhibiting many of the morphological and functional characteristics of their normal precursors. Most often, they arise in cells undergoing frequent division (e.g., epithelial cells of the skin), and so their essence is not the fact that they can divide rapidly, but that they lack normal control systems to shut off unwanted cell division. So a prime scientific goal for most of this century has been to find out the essential biochemical differences between normal and cancer cells.

USE OF TISSUE CULTURE CELLS TO STUDY CELL PROLIFERATION

Establishing Cells in Culture[1, 2]

Until the past decade, there was much mysticism about growing cells in culture. Success apparently demanded very precise experimental conditions; moreover, not everyone seemed to possess the necessary magic touch. For a long time, only masses of excised tissue would consistently grow, and then only when implanted upon very undefined serum clots or embryonic extracts. Such "tissue cultures" occasionally multiplied indefinitely if microbial contamination could be prevented and if regular changes were made in the surrounding nutrient medium. Although there were repeated efforts to grow cultures from dissociated cells, successes were very infrequent until small amounts of the proteolytic enzyme trypsin were used to dissociate tissue masses into their component individual cells. Then the dissociated cells (**primary cells**) usually grew well when seeded onto culture plates at high cell densities. In contrast, initial attempts to induce single cells to grow into clones did not succeed. At best, the individual cells multiplied several times and then died. It is thought that the failure of these early cloning attempts is due to the inherent leakiness of most vertebrate primary cells in culture. Because vital nutrients diffuse out, growth occurs best when cells are seeded at high density, making cross-feeding possible. Hence, the practice arose of placing single primary cells in very small drops of culture fluid. Here, extensive dilution is not possible, and tiny clones result. Another way to increase cloning efficiency is to place dissociated single cells upon a "feeder" layer of cells sterilized by large X-ray doses. None of the irradiated feeder cells can multiply, but they remain metabolically active and release sufficient nutrients to allow most of the seeded primary cells to form visible colonies.

Cell cultures arising from the transfer and subsequent multiplication of primary cells are called **secondary cells.** Such cells generally do not have infinite lives, but divide only a finite number of times before dying out. For example, most secondary chicken cells multiply 20 to 40 times and then invariably die. Likewise, cultures of human skin cells usually divide no more than 50 to 100 times before they die. Interestingly, this number coincides with the number of divisions that such cells make during the life span of their respective vertebrate source. Furthermore, skin cells or fibroblasts derived from older individuals have shorter lifetimes, as measured in cell generations, than do those from younger donors. The temptation thus exists to wonder whether this seemingly programmed cell death is related to the normal vertebrate aging process.

Fortunately, not all animals yield primary cultures all of whose progeny die. Although most mouse and hamster cells also die after dividing some 30 to 50 times, a few progeny pass through the "crisis period" and acquire the ability to multiply indefinitely. Progeny derived from such exceptional cells form the continuous **cell lines** (Table 25-1). Cells that can grow indefinitely in culture are said to be **established** and frequently are referred to as **immortal.** Although no "normal" human cell line has yet been developed, a number of cell lines have been derived from human tumors. Similarly, a few chicken tumors have been established into cell lines. The possession of the cancerous phenotype often allows easier adaptation to the conditions of cell culture. In part, this property may relate to the fact that both cancer cells and cells adapted to permanently grow in culture often have an increased chromosome number, frequently nearly twice the normal diploid number. For example, cells from the human cancer cell line *HeLa* have some 70 to 80 chromosomes, compared to the normal 46 chromosome complement (Figure 25-1). Likewise, mouse 3T3 cell lines, although derived from normal mouse embryos, have about 65 chromosomes rather than the normal diploid number of 40. Apparently, the possession of extra chromosomes (**aneuploidy**) frequently helps cells to grow in culture.

Table 25-1   Some Properties of Several of the More Important Cell Lines

| | Origin | | Characteristics | | |
	Species	Tissue	Morphology	Growth in Suspension	Chromosomal Makeup
3T3	Mouse 2N = 40	Endothelial	Fibroblast	No	Heteroploid (differs from cell to cell) (mode = 65)
L	Mouse 2N = 40	Connective	Fibroblast	Yes	Heteroploid (mode = 65)
CHO	Chinese hamster 2N = 22	Ovary	Epithelial	Yes	Pseudodiploid
BHK 21	Syrian hamster 2N = 44	Kidney	Fibroblast	Yes	Diploid
BSC	Monkey 2N = 42	Kidney	Epithelial	No	Diploid
MPC	Mouse 2N = 40	Bone marrow myeloma cell	Lymphoid	Yes	Heteroploid
RPH	Frog 2N = 26	Egg	Epithelial	No	Haploid (N = 13)
HeLa	Human 2N = 46	Cervical tumor	Epithelial	Yes	Heteroploid (mode = 75)
KB	Human 2N = 46	Nasopharyngeal tumor	Epithelial	Yes	Heteroploid (mode = 75)

The Obscure Origin of Many Cell Lines[1, 3]

Usually primary cultures are prepared from large tissue masses, if not from organs or whole embryos. Initially, they consist of a large variety of differentiated cells (*fibroblasts, macrophages, epithelial* and *endothelial* cells, etc.), and so the cellular origin of a given cell line is frequently obscure. Under the conditions that have been used most often for growth in culture, the cells that multiply best have the spindle shape of connective tissue cells and so are called fibroblasts (Figure 25-2a). In the past decade, however, much effort has gone toward the development of culture conditions that selectively promote the growth of other cell types and inhibit the growth of fibroblasts. For example, an early success with epithelial cells (Figure 25-2b) took advantage of their ability to use D-valine as their sole valine source (epithelial cells possess the enzyme D-amino-acid oxidase, which allows the transformation of D-valine into L-valine). Fibroblasts, however, do not contain this enzyme and so cannot grow when D-valine is the only source of valine in the growth medium.

The shape that cells of a given line assume in culture is not invariant. For example, some strains of HeLa cells growing in rich (high-serum-content) media have the stretched, elongated fibroblast shape, whereas in poorer media, they pack close together and have a rounded, cobblestone-like appearance (Figure 25-3). Thus, while we may refer to one cell line as epithelial cells and another as fibroblasts, we must not always assume that their origin was from epithelial (connective) tissue.

Sometimes, however, the presence of specific cell products makes the origin of certain cell lines unambiguous. In particular, the origin of cells secreting specific hormones is easy to ascribe. Cells from tumors of the testes secrete androgens, while ACTH and growth hormones are made by cell lines derived from pituitary tumors. Like-

Figure 25-1
(a) The 73 chromosomes of a HeLa cell contrasted with (b) the normal 46 complement of a cell from a normal human male. (Courtesy of Dr. T. T. Puck, University of Colorado, Denver.)

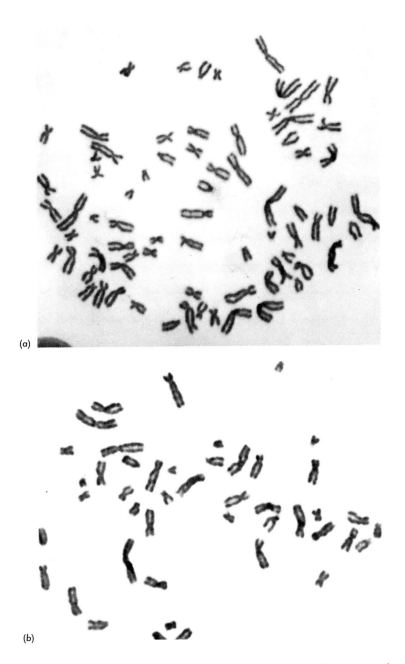

(a)

(b)

wise, cells derived from many liver tumors preferentially secrete the major liver product albumin. Equally clear is the cellular origin of cultured **neuroblastomas** (tumors of nervous tissue), since their cells have excitable membranes and under certain nutritional conditions send out numerous axonal processes (Figure 25-4). Lymphoid cells growing in culture can be distinguished by the presence of antibodies (B cells) or characteristic surface antigens (T cell types or B cells), and the epithelial origin of many cell lines now can be routinely established through their possession of one to several keratins, the fibrous proteins used to construct their intermediate (100 A) filaments.

Not all the properties of differentiated cells are retained in their cultured progeny. Many established cell lines seem less differentiated than their in vivo precursors and often show similarities to the undifferentiated embryonic cells present in early development. Many so-called "normal" cell lines thus closely resemble cancer cells, many types of which also represent partially dedifferentiated or incom-

Figure 25-2
Outgrowth of secondary cells starting from primary human lung tissue. Selective proliferation of (a) fibroblasts in L-valine and (b) epithelial cells in D-valine. [Reproduced with permission from S. Gilbert and B. Migeon, *Cell* 5 (1975):11.]

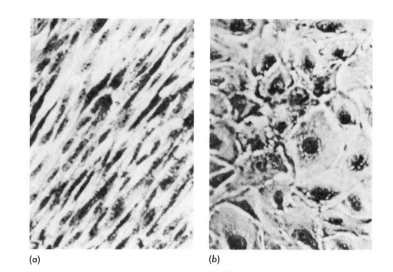

(a)   (b)

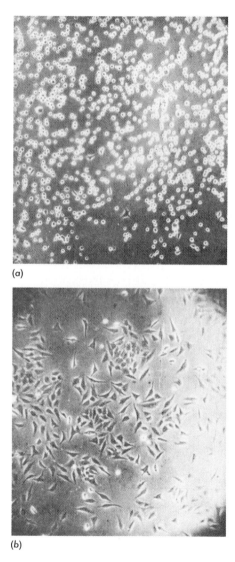

(a)

(b)

Figure 25-3
Dependence of the shape of HeLa cells on growth conditions. (a) In low serum, they have a rounded cobblestone shape, while (b) in enriched serum, they have the spindle shape of fibroblasts. (Reproduced with permission from T. Puck, *The Mammalian Cell as a Microorganism*, Holden-Day, 1972.)

pletely differentiated states. For example, most liver-derived hepatomas fail to make many normal liver enzymes. This becomes more plausible when we realize that continued cell growth in culture may be achieved by a loss of some of the devices that slow down, if not inhibit, cell division in the multicellular organism, and whose loss may result in the cancerous state. The isolation of many cell lines may thus have been dependent on the selection of cell mutants less able to respond to signals halting cell division.

Attachment to Solid Surfaces Versus Growth in Suspension

Most "fibroblasts" and "epithelial" cells grow best, if not exclusively, when attached to solid surfaces, forming thin cell lawns one to several layers thick. Under optimal conditions, primary epithelial cells form layers one cell thick (monolayers), while the lawns formed by fibroblasts are frequently two to three cells thick. Only rarely can cell lines be derived that multiply well in suspension. Such lines usually originate from tumors (transformed cells). In general, cultures containing cells that grow well in suspension produce many more cells than those whose growth is restricted to surfaces. So the cell lines that have been most frequently used to probe the molecular biology of vertebrates are those like HeLa and KB, which have suspended growth patterns. Recently, it has been possible to increase the yield of cells that can only grow when they are attached to surfaces by using microcarrier beads (Figure 25-5). Microcarrier beads provide a large surface area for cell attachment within a small volume of medium, thereby producing high concentrations of dividing cells, similar to a suspension culture.

Most lymphoid cells grow best, or exclusively, in suspension, either as single cells or in tiny clumps (Figure 25-6). Generally, resting lymphocytes can be coaxed to divide only in the presence of certain mitogens and growth factors or following infection with certain viruses. Many resulting lymphoid lines are immortal, and gram amounts of such cells are easily obtainable. Growth of other blood cell

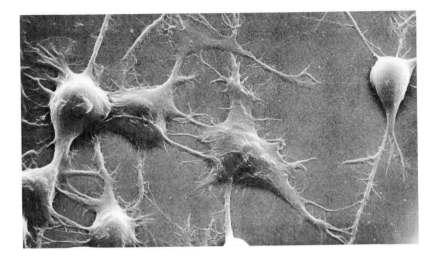

Figure 25-4
Phase-contrast photomicrograph of mouse neuroblastoma cells growing attached to a solid surface. Many elongated processes (neurites) extend from each cell. When such cells grow in suspension, they have the rounded cancerous appearance characteristic of the cells that form tumors; but when they attach to a solid surface, they become flattened and begin to send out large numbers of neurites. (Courtesy of Dr. Gunter Albrecht-Bühler, Cold Spring Harbor Laboratory.)

types in culture has become possible with the discovery of factors that promote the differentiation of erythrocytes, granulocytes, and macrophages.

"Normal" Cell Lines[4, 5]

In the early days of cell culture, the suspicion existed that cells became capable of growing in culture only when they acquired mutations that gave them the cancer cell's capability for unrestricted growth. This belief arose not only because it was much easier to make cell lines out of cancer cells, but also because many so-called "normal" cell lines had the roundish cancerous morphology, had the property of disorganized growth (Figure 25-7), and could cause tumors when injected into animals. Ideas flip-flopped 20 years ago, however, when the growth properties of mouse cell lines were found to depend on the cell density at which secondary cells are grown. If secondary cultures are always grown to high cell density, then cell lines that emerge after crisis form cultures of disorganized cells several layers thick. Such cells, though of normal origin, generally look like cancer cells and form tumors when injected into immunologically identical mice. If, however, the secondary mouse cultures are always kept at the low densities in which cell-cell contact is infrequent, the resulting cell lines have normal fibroblast morphology and form the thin organized monolayers characteristic of "normal" cells (Figure 25-8). Correspondingly, these cells only infrequently form tumors when injected into appropriate mice.

Realization of the powerful selective forces that exist during the "crisis period" now allows isolation of cell lines with desired growth properties. By keeping the cell number low, we remove a selection in favor of cancerlike cells, which can grow well even under crowded conditions. Thus, there are now a number of cell lines, the prototype being mouse 3T3 lines, that have many of the growth properties of the normal cell and so can be used for in vitro studies of transformation to the cancerous state. However, it is clear that most established cell lines, even the 3T3 type, are not completely normal. Besides their enhanced growth potential relative to primary cells, they frequently have highly abnormal chromosome numbers, as mentioned earlier.

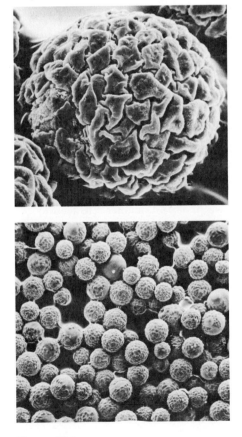

Figure 25-5
Scanning electron micrographs of pig kidney cells growing on microcarrier beads. The cells are used for the production of foot-and-mouth disease vaccine. (Courtesy of B. Meignier and J. Tektoft, Institut Merieux, Lyon, France.)

Figure 25-6
Human lymphoid cells growing in suspension. Such cells grow into tiny clumps that do not stick well to solid surfaces. (Courtesy of Dr. John Hlinka, Sloan-Kettering Institute.)

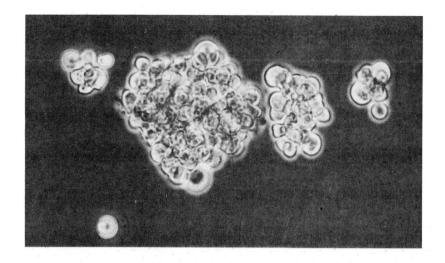

Figure 25-7
The morphologies of normal (cell line 9) versus transformed (cell line 14b) hamster embryo cells. (a, b) The phase-contrast photomicrographs (250×) show the spread-out, flat character of the normal hamster cell growing on a flat surface. The cell in the center is moving toward the right with its fan-shaped leading edge containing many ruffles (arrow). Five hours elapsed between the two photos. (c, d) Corresponding views (300×) of cancerous hamster cells resulting from transformation by adenovirus 2. The cells have a much rounder appearance and exhibit very little motility (arrow points to cells that have not moved in the 19 hours separating the two photos). Daughter cells tend to stay close together, resulting in tiny islands of cells. [Reproduced with permission from R. D. Goldman, C.-M. Chang, and J. Williams, *Cold Spring Harbor Symp. Quant. Biol.* 39 (1975):601.]

Cell Transformation[6, 7, 8]

The sudden change of a cell possessing normal growth properties into one with many of the growth properties of the cancer cell is called **cell transformation.** Spontaneously, this phenomenon occurs only rarely, although the frequency can be significant (10^{-5} to 10^{-6}) in certain established 3T3-like cell lines. As we shall see in Chapter 26, transformation occurs at much higher frequencies after infection with a tumor virus or after exposure to a carcinogenic (cancer-causing) chemical or radiation source. Irrespective of its origin, however, the transformed cell, besides often having the potential to grow into tumors, usually possesses additional characteristics: It has a much rounder shape, reflecting less organized arrangement of cytoplasmic microfilaments and microtubules; it is much less subject to contact inhibition of movement; it grows to higher cell densities; it is often less restricted to growth on solid surfaces; and it appears nutritionally less fastidious, particularly in its serum requirements. Not all transformed cells, however—not even those derived from the same cell line—are morphologically identical. As we shall detail later, the cancerous phenotype probably arises by a number of quite distinct biochemical changes.

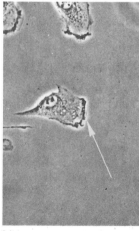
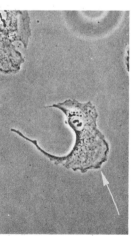
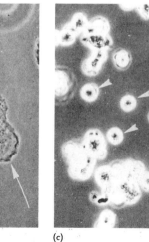
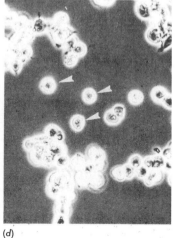

(a)   (b)   (c)   (d)

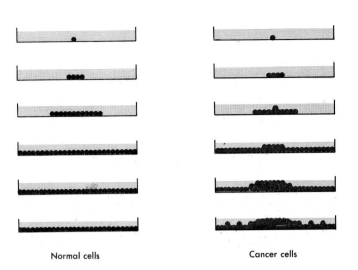

Normal cells   Cancer cells

Figure 25-8
Schematic comparison of the multiplication of a normal cell and of a cancer cell upon a solid surface. The normal cells divide until they form a solid monolayer. Cancer cells, however, can form irregular masses several layers deep.

Working Out Nutritional Requirements[9, 10]

Well-characterized cell lines allow controlled experiments on the nutritional requirements of vertebrate cells. A semisynthetic medium consisting of 13 different amino acids, 8 vitamins, and a mixture of inorganic salts, together with glucose as an energy source and dialyzed horse serum (Table 25-2), permits HeLa and L cells to grow optimally to high densities. Additional requirements (e.g., serine, inositol, and pyruvate) sometimes emerge when cells grown at low density leak too many essential metabolites. Continuously passed cell lines tend to grow better at lower densities than newly derived lines,

Table 25-2 Basal Media for Growth of the HeLa Cell Line in Culture

L-Amino Acid (mM)		Vitamins (mM)		Salts (mM)	
Arginine	0.1	Biotin	10^{-3}	NaCl	100
Cysteine	0.05	Choline	10^{-3}	KCl	5
Glutamine	2.0	Folic acid	10^{-3}	$NaH_2PO_4 \cdot H_2O$	1
Histidine	0.05	Nicotinamide	10^{-3}	$NaHCO_3$	20
Isoleucine	0.2	Pantothenic acid	10^{-3}	$CaCl_2$	1
Leucine	0.2	Pyridoxal	10^{-3}	$MgCl_2$	0.5
Lysine	0.2	Thiamine	10^{-3}		
Methionine	0.05	Riboflavin	10^{-4}		
Phenylalanine	0.1				
Threonine	0.2				
Tryptophan	0.02				
Tyrosine	0.1				
Valine	0.2				

Miscellaneous

Glucose	5 mM
Penicillin	0.005%*
Streptomycin	0.005%*
Dialyzed horse serum	5%

*To prevent microbial contamination.
SOURCE: After H. Eagle, *Science* 122 (1955):501.

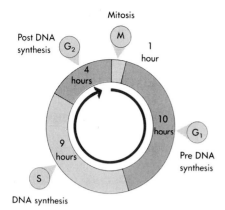

Figure 25-9

Phases in the life cycle of a mouse hepatoma cell growing in tissue culture and dividing once every 24 hours.

suggesting that cloning probably exerts selection pressure for diminished leakiness.

The requirement for serum is almost universal for all vertebrate cell lines, hinting that one or more serum factors may play key roles in regulating in vivo cell multiplication. In vitro, the final cell densities reached by specific cell lines are often a function of the serum concentration. Certain cell lines that form a one-cell-thick monolayer in 1 percent serum grow layers three to four cells thick in 10 percent serum. Also very important is the exact pH at which cells grow. The optimal pH can vary from one cell line to another, though most normal cells grow best at a pH of around 7.6.

Growth of Cells in Completely Defined Media[11, 12]

Although amino acid, vitamin, salt, and energy requirements for growing most vertebrate cells in culture were worked out over 30 years ago, the identification of factors in serum that allow cells to grow has taken far longer. Even today, most cell culture is performed with serum-containing medium, since despite some successes, completely defined media remain difficult to work with. Once again, the existence of cell lines with their great hardiness and often reduced serum requirement plays an important part in the development of completely defined media. Serum can sometimes be replaced by a combination of specific polypeptide growth factors and hormones, carrier proteins, and, for certain cell types, lipids and proteins that facilitate cell attachment. For example, mouse 3T3 cells, which are fibroblastic cell lines, will divide in basal medium (Table 25-2) supplemented with transferrin, platelet-derived growth factor (PDGF), insulin or insulin-like growth factors (IGF) called somatomedins, and epidermal growth factor (EGF). Transferrin, a protein that transports ferric iron (Fe^{3+}) into cells, is required for growth of almost all cultured cells and is present at a high concentration in serum. Growth factors are polypeptides that stimulate the growth of specific cell types. Thus, different factors are often needed for different cells.

Different Requirements for Growth of Different Cell Types[12]

The reason that fibroblastic cells frequently take over primary cultures is that the 10 percent fetal calf or calf serum employed in standard culture conditions contains PDGF and other factors for fibroblast growth, but lacks specific growth factors required by many other cell types and even contains substances that inhibit the growth of some cells. Little is known about the inhibiting substances, but many specific growth factors have been identified now. For example, growth of certain T lymphoid cells requires a T cell growth factor called interleukin-2. Thus, the correct combination of protein supplements, in combination with an appropriate basal medium, will allow the growth of different cell types in vitro. Whether the cells have the same requirements in vivo is not known in most cases. Although defined media will support the growth of specific cell lines or cell types, most researchers still prefer to grow T cells, for example, in a medium containing serum to which interleukin-2 was added.

Despite considerable progress in tissue culture systems, we are probably still a long way from identifying all the stimulatory and inhibitory macromolecules that regulate the growth of all the different vertebrate cell types in vivo.

The Cell Cycle[13–16]

The length of the cell cycle under optimal nutritional conditions varies from one cell line to another. At 37°C, a few lines can divide once every 10 to 12 hours, but the majority require some 18 to 24 hours. Now, most biologists divide cell cycles into the M phase (mitosis), G_1 phase (period prior to DNA synthesis), S phase (period of DNA synthesis), and G_2 phase (period between DNA synthesis and mitosis) (Figure 25-9). Collectively, G_1, S, and G_2 are called **interphase,** the period of the cell cycle distinct from division of the nucleus (*mitosis*) and cytoplasm (**cytokinesis**). M is the shortest stage, usually lasting only an hour or so. All M phase cells, irrespective of their previous shapes, become completely spherical with accompanying depolymerization of the microtubule skeleton. The length of the S phase is remarkably similar in many different cells, as is G_2, while the greatest variation is seen in the length of G_1. Often, G_1 and S are about equal, with an 8- to 10-hour duration; but in some cells, G_1 lasts only 1 to 2 hours. At some point late in G_1, called the **restriction** or **R point,** a cell becomes committed to traverse the remainder of the cell cycle. Thus, variations in cell cycle time are mostly due to variations in the length of G_1 up to the R point.

A most important tool in the analysis of the cell cycle is the synchronization of growth to place virtually all cells in the same phase. One of the best methods of synchronization uses the fact that spherical M phase cells (Figure 25-10) attach much less firmly to glass surfaces than do interphase cells. Thus, by shaking the cultures, we can isolate large numbers of uncontaminated M phase cells and subsequently measure the rates at which different important compounds are made throughout the cell cycle. In HeLa cells, RNA and protein synthesis occur throughout interphase (G_1, S, G_2), but both show abrupt decrease during the mitotic phase, when the chromosomes contract and line up along the spindle fibers. While the cessation of RNA synthesis may be a consequence of shape changes in chromatin concomitant with chromosome contraction, why protein synthesis declines is not clear. The effect is primarily due to translational control, and synthesis resumes promptly at the start of G_1.

Fusion of Cells in Different Phases of the Cell Cycle[17, 18, 19]

Fusion of a cell in interphase (G_1, S, G_2) with an M phase cell quickly results in the appearance of condensed chromosomes in the nucleus of the interphase cell (Figure 25-11). Mitotic cells thus possess a chemical factor that can promote the condensation of interphase chromosomes. When G_1 cells are used in such experiments, their chromosomes are thinner upon condensation than those of their M phase partners, reflecting their single chromatid construction. In contrast, when G_2 cells are fused, their chromosomes, upon condensation, more closely resemble the normal double-chromatid mitotic chromosome. And when S phase cells are fused, the resulting condensed chromosomes are usually highly fragmented, perhaps because they were caught in the process of chromosome duplication and so were highly susceptible to nuclease attack.

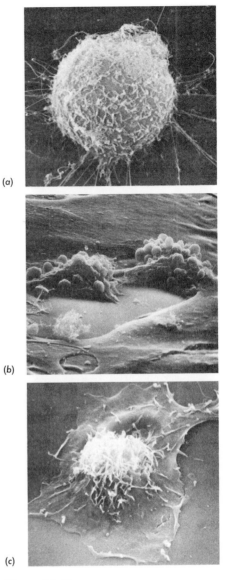

(a)

(b)

(c)

Figure 25-10
Scanning electron micrographs of normal hamster cells (cell line 9). (a) A cell rounded up for mitosis. Its surface is covered with microvilli, a characteristic feature of all M phase cells. (b) Two daughter cells in the process of separation. Note that both are covered with blebs, very characteristic protrusions of the late M, early G_1 period of normal cells. (c) A cell in late G_1. No blebs are visible; instead, a number of microvilli are present, especially in the more central, less spread out region. When the spreading is complete and S phase commences, the outer surface becomes very smooth and only the occasional microvillus can be seen. [Reproduced with permission from R. D. Goldman, C.-M. Chang, and J. Williams, *Cold Spring Harbor Symp. Quant. Biol.* 39 (1975):601.]

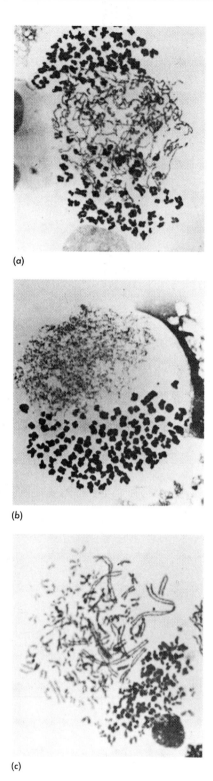

(a)

(b)

(c)

Figure 25-11

Appearance of the chromosomes in hybrid cells formed by the fusion of two HeLa cells phased in different parts of the cell cycle. (a) M and G_1. (b) M and S. (c) M and G_2. (Reproduced with permission from T. Puck, *The Mammalian Cell as a Microorganism,* Holden-Day, 1972.)

Initiation of DNA Synthesis[14, 20]

Once the decision is made to reenter the proliferation cycle, a sequence of preprogrammed operations is initiated long before actual DNA synthesis begins. For example, the levels of the deoxyribonucleoside triphosphate precursors (dATP, dGTP, dCTP, and dTTP) are very low at the beginning of G_1 and remain that way until some 30 minutes prior to initiation of DNA synthesis. Then they rise rapidly to peak amounts during S phase, remaining at relatively high levels until M ends and G_1 commences. How these levels are enzymatically controlled remains to be worked out. Likewise, the enzymology of eucaryotic DNA synthesis itself is still poorly understood. We do know that multiple replication forks, moving bidirectionally, are needed to complete the replication of all the DNA within the 8- to 10-hour S phase (see Figure 21-8). In reproducible succession, distinct areas of chromosomes, called **replication units,** become populated with multiple replication forks.

How does the cell keep track of all the DNA and thus ensure that each chromosome is replicated just once during each S phase? Fusion of cells in the various phases of the cell cycle has provided a preliminary answer to this question. When an S phase nucleus and a G_1 nucleus are brought together, the G_1 nucleus also begins to make DNA, indicating the presence of a positive initiating factor(s). However, fusion between an S phase cell and a G_2 cell does not lead to DNA synthesis in the G_2 nucleus or to its inhibition in the S nucleus. The refractoriness of the G_2 chromosomes might conceivably to be due to the rapid coating of newly synthesized DNA by concomitantly synthesized nondiffusible chromosomal proteins. These proteins would then be the mechanism for keeping track of replicated DNA and preventing it from being duplicated again during the S phase. Cell cycle–related changes would have to occur in one or more of these chromosomal proteins before a new round of DNA synthesis could start. Since G_1 nuclei, fresh out of mitosis, will undergo DNA synthesis when fused with S phase nuclei, the change must occur during M phase.

Mutations in Cells Growing in Culture[21, 22, 23]

One approach to studying the control of cell growth is to isolate mutants conditionally blocked in their growth or possessing altered growth properties. Because cell lines are diploid (heteroploid), we would anticipate that most mutations that are not sex-linked will be only infrequently expressed. Thus, much effort was initially spent on finding subdiploid cell lines, where one or more of the chromosomes of the normal diploid chromosomal complement are missing. Much work has been done with the Chinese hamster, which not only has a normal $2N$ complement of only 22 chromosomes, but also frequently gives rise to cell lines in which one or more of the autosomes are missing. Using these lines, researchers have found mutations that block the biosynthesis of several key metabolites, such as glycine, adenine, and inositol. Moreover, the routine employment of specific mutagens greatly increases the frequency of such mutations.

Vital to such experimentation is the employment of techniques that selectively isolate desired mutants from a large background of normal cells. The best technique utilizes the fact that exposure to visible light rapidly kills cells that have incorporated the base analog 5-BUdR into their DNA (Figure 25-12). Since only actively multiplying cells incorporate 5-BUdR (or thymidine) into their DNA, those mutagenized cells that are unable to grow are not killed by an exposure to visible light.

The BU visible light technique has been particularly successful in the isolation of **temperature-sensitive (ts) mutants,** which are unable to multiply at high (or low) temperature, but which multiply normally at low (or high) temperature. This class of mutants should include virtually all cellular functions, since their creation depends only on induction of amino acid changes that lead to loss of protein function at either the high or low temperatures. Already there are ts mutants defective in DNA replication, RNA synthesis, protein synthesis, and ion transport, as well as others unable to traverse specific stages in the cell cycle. There are, in addition, mutant cells that can express the transformed state only at low temperature (30°C), as well as other mutants that can express the transformed phenotype only at high temperature (37°C) (Figure 25-13).

Stopping in Early G₁[14, 15]

Examination of fibroblasts that have stopped dividing upon formation of a confluent monolayer reveals that all the cells are blocked in early G₁. (This resting point for quiescent cells is sometimes referred to as G₀.) Similarly, when the growth of such cells is prematurely blocked by removal of essential growth factors like isoleucine, glutamine, phosphate, or serum, the starved cells do not stop randomly in the cell cycle, but always proceed through mitosis, to yield cells with the 2N amount of DNA. In each case, growth is stopped very early in G₁, for DNA synthesis is commenced about 8 hours after reintroduction of the limiting growth factors (about the length of the G₁ stage in these cells). When so blocked in early G₁, quiescent cells remain healthy, even if the needed nutrients are withheld for many generation times. This picture parallels the situation of most in vivo animal cells, which can remain viable and metabolically active even after many months in a nonproliferating state. In contrast, if poisons like colchicine or hydroxyurea are added to the cells, they do not block the cells at G₁: Colchicine stops cells in M phase, while hydroxyurea interrupts S phase DNA synthesis. Cells so inhibited do not remain viable for long and usually die within 24 hours after exposure to these poisons.

Many cancer cells have lost the control mechanism that sends nutritionally limited normal cells into G₁ quiescence. So when, for one reason or another, such cancer cells do not divide, a spectrum of cancer cells blocked in all phases of the cell cycle is produced. As such, they are much less healthy than their normal equivalents and die off more rapidly. This concept provides a rationale for the frequent observation that the presence of antimetabolites differentially kills off cancer cells.

Activation of G₁ Quiescent Cells by Mitogenic Stimuli[24]

Exposure to any of a large number of different stimuli, that collectively are called **mitogens,** triggers many quiescent G₁-blocked cells into division. Some mitogens, like certain hormones, or growth

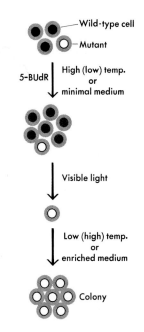

Figure 25-12
Use of the 5-BUdR visible light technique for the isolation of mutant cells unable to multiply as the result of either a nutritional deficiency or the possession of a temperature-sensitive vital enzyme or structural protein. A mutagenized mixed population is exposed to 5-BUdR under conditions where only wild-type cells grow and incorporate 5-BUdR into DNA. Subsequent exposure to visible light kills only the wild-type cells, leaving the mutant cells untouched. Subsequently, an enriched medium is added (or the temperature lowered or raised) to allow the mutant cells to grow into colonies.

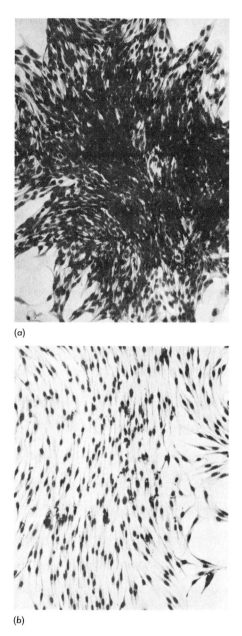

(a)

(b)

Figure 25-13
Photomicrograph of temperature-sensitive mutants of a transformed Chinese hamster lung cell line that prevent the expression of the transformed state at high temperature. (a) The transformed growth pattern of cells growing at 34°C. (b) Normal growth pattern of cells growing at 39°C. [Reproduced with permission from K. Miyashita and T. Kakynaga, *Cell* 5 (1975):131.]

factors present in normal serum, are probably normal physiological effectors. But other strong mitogens, like some bacterial lipopolysaccharides and certain cell agglutinating proteins (*lectins*) isolated from plant seeds, normally may never provoke cell division. Yet their addition invariably sets certain quiescent cells into division. Likewise, proliferation often results when certain cell types are treated with small amounts of the proteolytic enzyme trypsin or with neuraminidase, the enzyme that removes the complex amino sugar sialic acid from surface proteins. Much effort now goes toward understanding how mitogens stimulate cell proliferation. Once again, the use of tissue-culture-grown cells makes the studies possible.

FACTORS THAT REGULATE CELL PROLIFERATION, FUNCTION, AND DIFFERENTIATION

Platelet-Derived Growth Factor (PDGF), Insulin-Like Growth Factor (IGF), and Epidermal Growth Factor (EGF) Together Stimulate the Growth of 3T3 Cells[25–34]

Much of our current understanding of how growth factors work comes from the study of tissue-culture-grown 3T3 fibroblasts and peripheral blood lymphocytes. A key concept to come from these studies is that multiple growth factors are required for optimal cell growth and division. In the case of 3T3 cells, it was found that animal sera prepared from unclotted blood did not support the growth of cells nearly as well as did animal sera prepared from clotted blood. A mitogen found in platelets, the **platelet-derived growth factor (PDGF),** can restore most of the growth-promoting activity to the serum prepared from unclotted blood. This factor is released from platelets during the clotting reaction. However, PDGF by itself is a rather poor mitogen for 3T3 cells. Only in combination with the growth factors found in platelet-poor plasma does it give an optimal mitogenic response.

Treatment of 3T3 cells with PDGF makes the cells *competent* to respond to the growth factors found in platelet-poor plasma. Cells remain competent for several hours, even after PDGF is removed, but they do not progress any closer in time to the onset of DNA synthesis (S phase). Exposure to platelet-poor plasma allows the competent cells to progress through the 10- to 12-hour G_1 phase of the cell cycle and on to DNA synthesis. For 3T3 cells, the progression factors present in platelet-poor plasma can be replaced with the defined growth factors, **epidermal growth factor (EGF)** and **insulin-like growth factor (IGF).**

Most polypeptide growth factors are small proteins with molecular weights that vary from 1 to 40 kdal. PDGF is a glycoprotein dimer of about 30 kdal, and genes that encode it have been cloned (Table 25-3). EGF is a 53-amino-acid-long polypeptide. A cDNA that encodes it has been molecularly cloned and sequenced. Interestingly, this sequence reveals that EGF is encoded as part of a large polypeptide, about 1200 amino acids in length, containing several EGF-like sequences (Figure 25-14). The sequence also reveals that the precursor may be able to

Table 25-3 Properties of Some Polypeptide Growth Factors Whose Corresponding Genes/mRNAS Have Been Cloned (See also Table 25-5)

Name	Structure	Biological Properties
PDGF (platelet-derived growth factor)	Two disulfide-bonded chains; glycoprotein	Mitogen for cultured cells of mesenchymal origin; repair of vascular system in vivo
EGF (epidermal growth factor)	53 amino acids (preproEGF 1200 amino acids)	Mitogen for cells of ectodermal and mesodermal origin; role in early development in vivo?
TGF-α (transforming growth factor)	50 amino acids; (preproTGF-α 160 amino acids); homology with EGF	Mitogen like EGF but can make normal cells grow in agar; binds to EGF receptor; high levels in some cancer cells
TGF-β (transforming growth factor)	Homodimer of MW 25,000; no homology with TGF-α or EGF	Mitogen for some cells, growth inhibitor for others; stimulates growth in agar; synergy with TGF-α
NGF (nerve growth factor)	Two identical β chains of 118 amino acids; low homology to insulin	Development and maintenance of sympathetic and embryonic neurones
IGF-I (insulin-like growth factor-I; somatomedin C)	70 amino acids (preproIGF-I 130 amino acids); 45% homology with insulin	Intermediate in the action of growth hormone; skeletal elongation; mitogen for cells in vitro
IGF-II (insulin-like growth factor-II)	67 amino acids (preproIGF-II 180 amino acids); homology with insulin	In vivo role uncertain
GH (growth hormone; somatotropin)	191 amino acids	Synthesized in anterior pituitary; stimulates liver to produce somatomedins
GHRF (growth hormone releasing factor; somatoliberin)	44 amino acids (precursor about 100 amino acids	Hypothalamic peptide; regulates synthesis of growth hormone in the pituitary
Insulin	Two disulfide-bonded chains; A is 21 amino acids, B is 30 amino acids; A, B cleaved from preproinsulin of 110 amino acids	Stimulates glucose uptake, protein synthesis, and lipid synthesis; can be mistaken for growth factor in vitro

travel to the plasma membrane and be anchored there like a receptor molecule, but the significance of this is unknown. The IGF that promotes 3T3 growth, called IGF-1 or somatomedin C, is a protein of about 7 kdal. It is closely related to insulin and shares some of its biological effects (Figure 25-15 and Table 25-3). Insulin and IGF-1 are about 45 percent homologous at the amino acid level and can bind to each other's receptors on the cell surface. The affinity of the two proteins for each other's receptors is about a hundred- to a thousandfold

Figure 25-14
Diagram of mRNA for epidermal growth factor (EGF) as deduced from the nucleotide sequence of a cDNA clone of the mouse EGF mRNA. The boxed region represents the open reading frame that encodes the 1217-amino-acid-long preproEGF from which the 53-amino-acid-long EGF is proteolytically cleaved. Colored areas to the 5' side of EGF encode EGF-like sequences. [After A. Gray, T. J. Dull, and A. Ullrich, *Nature* 303 (1983):722–725; and J. Scott et al., *Science* 221 (1983):236–240.]

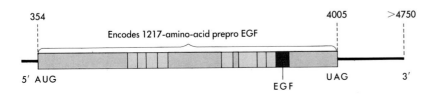

354   Encodes 1217-amino-acid prepro EGF   4005   >4750

5' AUG   EGF   UAG   3'

Figure 25-15
Sequence and schematic representation of the structure of IGF-1 and comparison to insulin. (a) Amino acids in white circles are the same as those in identical positions in the A and B chains of human insulin. (b) Representation of insulin is based on X-ray analysis of insulin crystals; representation of proinsulin and IGF-1 is based on model building. Note that insulin consists of an A and a B chain that are generated by proteolytic cleavage. [After T. L. Blundell, S. Bedarkar, S. Rinderknecht, and R. E. Hubel, *Proc. Nat. Acad. Sci.* 75 (1978)183.]

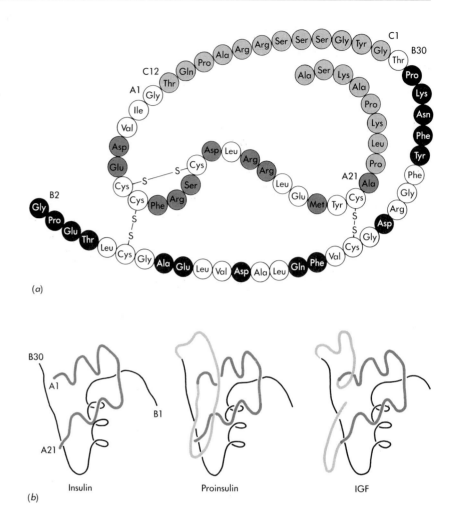

(a)

(b)   Insulin   Proinsulin   IGF

less than for their own, but at high concentrations, insulin mimics IGF-1 and so can easily be mistaken for a growth factor.

In Vivo PDGF May Promote Wound Healing, While IGF Mediates the Action of the Pituitary Growth Hormone[35, 36, 37]

The ability of a polypeptide to stimulate growth of a particular cell type in vitro does not prove that it serves the same function in vivo. The role that many growth factors play in the whole organism is only beginning to be understood.

In vitro, platelet-derived growth factor (PDGF) is the major polypeptide mitogen in serum for cells of mesenchymal origin, such as fibroblasts, smooth muscle cells, and glial cells. In vivo, PDGF does not circulate freely in blood, but is stored in the α granules of blood platelets. During blood clotting and platelet adhesion, the contents of these granules are released, often at sites of injured blood vessels. PDGF probably has an important role in the repair of blood vessels. PDGF can stimulate the migration of arterial smooth muscle cells from the medial to the intimal layer of the artery, where they then proliferate as an early response to injury.

Insulin-like growth factor IGF-1 (somatomedin C) is a polypeptide growth factor that controls animal size and height. Growth hormone from the pituitary had long been known to control animal size, but its mechanism remained elusive. Then, when hypophysectomized animals (those from whom the pituitary gland had been removed) were treated with growth hormone, their serum was found to contain a factor that could stimulate the incorporation of SO_4^{-2} into chondroitin (cartilage) slices, as well as the oxidation of fat and the synthesis of DNA. Since growth hormone itself generates no such activity when added to serum from pituitary-free animals, searches were made for the intermediate. It turned out to be a factor of about 8 kdal and was first named somatomedin because of its ability to regulate the growth of most forms of somatic tissue (Figure 25-16). Somatomedin turned out to be a mixture of polypeptides. The polypeptides have significant homology to the pancreatic hormone insulin and so were renamed insulin-like growth factors, or IGFs. Besides stimulating somatomedin production, growth hormone may contribute to tissue growth by promoting the differentiation of certain types of precursor cells, which then become targets for somatomedin.

The in vivo role of EGF is unclear. It was first identified and purified on the basis of its effects on prenatal and neonatal mouse tissues, including ability to promote precocious eye opening and incisor eruption in mouse embryos. It also inhibits release of hydrochloric acid (HCl) from the intestinal mucosa. EGF is present at particularly high concentration in the submaxillary gland and in human urine.

Polypeptide Growth Factors Bind to Specific Cell Surface Receptors[38-45]

How polypeptide growth factors stimulate cell division has always been a major question in growth factor research. The process begins when the factors bind to specific cell surface receptors located at the plasma membrane. Most factors can exert mitogenic functions at very low concentrations, in the nanogram-per-milliliter range. To act at such low concentrations, growth factors have receptors of very high affinity. For a given growth factor, the number of receptors on the cell surface ranges from about 5000 to 500,000. Not every cell expresses the same receptors on its surface, which accounts for the specificity of growth factors for some cell types but not others. Growth factor receptors are frequently large glycoproteins. Recently, genes encoding a number of growth factor receptors and the related insulin receptor have been molecularly cloned (Figure 25-17).

Growth factor receptors usually have three identifiable domains (see Figure 25-17). There is an extracellular domain, usually highly glycosylated, to which the polypeptide hormone (the ligand) binds; a short hydrophobic domain, which is the transmembrane portion of the receptor; and an intracellular domain, which may be responsible for transmitting signals to the cell that a ligand has bound to the receptor. One model of receptor function holds that a receptor is similar to an allosteric enzyme: When the growth factor (allosteric effector) binds to the extracellular domain of the receptor, it causes a change in the three-dimensional structure such that the enzymatic activity of the receptor is activated in the cytoplasmic domain (Chapter 5).

In many cases, the enzymatic activity stimulated by growth factor binding is a tyrosine kinase activity. Upon ligand binding, the receptors for PDGF, EGF, IGF, and the hematopoietic factor CSF-1 all become phosphorylated on tyrosine residues located in their cytoplasmic domains (Figure 25-18). The kinase activity responsible for this

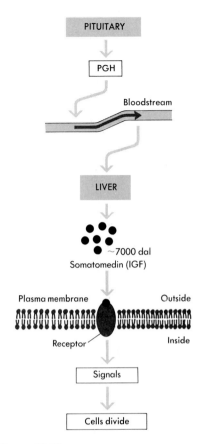

Figure 25-16
Schematic view of how the pituitary growth hormone (PGH) acts through somatomedins (IGFs). PGH also acts directly to increase the number of target cells that respond to somatomedins.

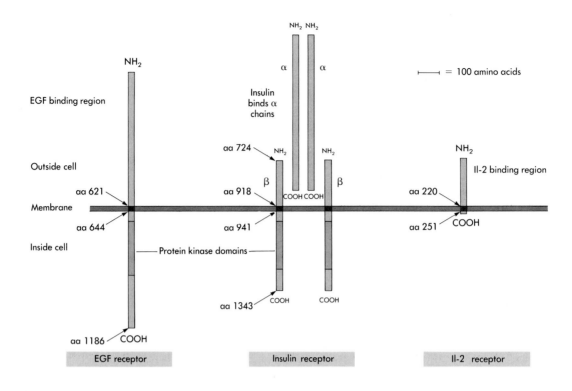

Figure 25-17
Diagrams of the EGF, insulin, and Il-2 cell surface receptors based on nucleotide sequences of cDNA clones of their mRNAs. All three are glycoproteins. The EGF and insulin receptors have protein (tyrosine) kinase activity in their cytoplasmic domains.

phosphorylation is in the receptor itself. Thus, the reaction is autocatalytic. This result is in striking contrast to most cellular protein phosphorylations, which occur on serine or threonine rather than tyrosine residues.

Growth factor receptors that possess tyrosine kinase activity may be capable of phosphorylating other cellular proteins as well as themselves. Although this has not yet been directly demonstrated, the number of cellular proteins phosphorylated on tyrosine dramatically increases after the addition of growth factors whose receptors have tyrosine kinase activity.

After binding, the receptor–growth factor complex is internalized, transported to lysosomes, and degraded (Figure 25-19). By this mechanism, occupied receptors are cleared from the cell surface. With time, new receptors are synthesized and expressed on the cell surface. It is thought that the important signals resulting from growth factor–receptor binding are not the result of internalization of the receptor-ligand complex but rather are transmitted from the cell surface. This is because in some cases, antibody to a receptor can mimic the effect of the receptor's normal ligand. However, there is no definitive evidence that would rule out the possibility that some of the effects of growth factors result from the internalization and degradation of the factor-receptor complexes.

Growth Factor–Receptor Interactions Have Pleiotropic Effects on Cells, Including Changes in Gene Expression[34, 46–49]

The responses of cells to growth factors are numerous. Table 25-4 lists some responses of fibroblasts to the platelet-derived growth factor. Some effects are "immediate," being detectable within seconds or minutes of growth factor addition. These changes include rapid ionic

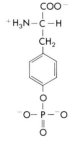

Figure 25-18
Phosphotyrosine.

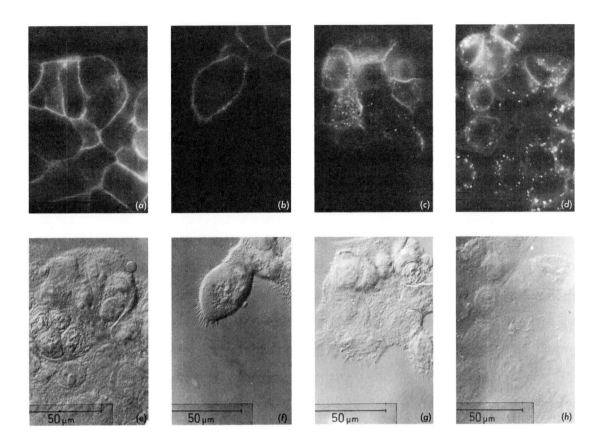

changes, cytoskeletal redistributions, protein phosphorylations, membrane lipid metabolism changes, and alterations in the protein synthetic machinery. Other effects of growth factors are not quite so rapid. These are the so-called early events, which can be measured within the first few hours of growth factor addition. They include induction of specific new gene transcription and expression of new proteins.

Although different growth factors elicit different cellular responses in a given cell, there are many overlapping responses as well. For example, in 3T3 cells, insulin, IGF-1, EGF, and PDGF all stimulate the phosphorylation of the ribosomal protein S6. However, only PDGF leads to a reorganization of the actin filaments.

A single growth factor can elicit different responses from different cell types. Furthermore, in addition to stimulating cell growth, some growth factors can trigger cellular differentiation, enhance cell survival, initiate cell migration, and stimulate the secretion of tissue-specific hormones from differentiated cell types.

It is estimated that the early PDGF induced genes are about 0.1 to 0.3 percent of the total genes transcribed in 3T3 cells. This would correspond to roughly 10 to 100 genes. An important question is, Which of these genes are critical to the mitogenic response? A strong candidate for such a gene has come from the finding that a variety of mitogens stimulate expression of a cellular gene, called c-*myc*, in several different cell types. An increase in c-*myc* mRNA is seen after PDGF treatment of quiescent 3T3 cells, after con A treatment of T lymphocytes, after lipopolysaccharide treatment of B cells, after EGF treatment of an epidermal carcinoma cell line, and as an early response in liver regeneration. And as we shall learn in Chapters 26 and

Figure 25-19
Binding and internalization of EGF monitored with fluorescein-labeled EGF on A-431 cells. Cells were incubated with fluorescein-conjugated EGF for 45 minutes at 6°C, washed to remove unbound EGF, then either fixed (a and b) or incubated at 37°C for 2 minutes (c and d), 10 minutes (e and f), or 20 minutes (g and h) and then fixed. Photos (a), (c), (e), and (g) are fluorescence micrographs taken under epi-illumination. Photos (b), (d), (f), and (h) are photographs of the same cells by Nomarski optics. [Courtesy of H. Haigler, J. F. Ash, S. J. Singer, and S. Cohen, *Proc. Nat. Acad. Sci. 75* (1978):3317.]

Table 25-4   The Mitogenic Response to PDGF

Immediate Events: 1–10 min and/or Transcription-Independent	Early Events: 30–180 min and/or Transcription-Dependent
1. Tyrosine-specific phosphorylations	1. PDGF can be removed from culture medium and cells still divide
2. Inhibition of EGF binding	2. Acquisition of intracellular signal as shown by cell fusion
3. Inhibition of guinea pig insulin binding	3. Induction of rare gene sequences encoding low-abundance mRNAs
4. Stimulation of phospholipase A_2 and prostaglandin release	4. Appearance of rare cytoplasmic proteins
5. Stimulation of polysome formation	5. Stimulation of somatomedin binding
6. Stimulation of phosphatydylinositol turnover	6. Increase in low-density lipoprotein receptor content
7. Reorganization of actin filaments	7. Stimulation of amino acid A transport system

SOURCE: C. D. Stiles, "The Molecular Biology of Platelet-Derived Growth Factor," *Cell* 33 (1983):653–655.

27, c-*myc* expression is altered in many types of cancer cells, strongly implicating c-*myc* as an important gene in cell growth control.

Further experimental confirmation that the c-*myc* gene is at least partly responsible for the mitogenic response of certain growth factors comes from the observation that microinjection of *myc*-encoded protein into quiescent 3T3 cells causes them to enter the cell cycle even in the absence of PDGF. However, just like cells treated with PDGF, the presence of the progression-type growth factors present in platelet-poor plasma is also needed for entry of the microinjected cells into S phase. The *myc* gene product is located in the nucleus, but how it may function to make cells competent to respond to progression factors is not known.

Another gene whose expression increases rapidly in response to PDGF is c-*fos*. Like c-*myc*, c-*fos* encodes a nuclear protein and is also known to be involved in the cancerous transformation of cells (Chapter 26).

What Are the Second Messengers for Growth Factor Receptors?[50–57]

At least three types of mechanisms are now being considered to explain how growth factor–receptor interactions exert such pleiotropic effects. These include the generation of ion fluxes, the activation of protein kinases, and the enzymatic generation of small hydrophilic molecules that can freely diffuse away from the plasma membrane. That these mechanisms are operating as growth factor second signals or messengers is suggested by the observations that experimental perturbation of these metabolic pathways can have dramatic effects on cell growth.

T lymphocytes can be activated independently of antigen or concanavalin A treatment by simultaneous treatment of the cells with calcium ionophores and phorbol esters. Calcium ionophores allow calcium to cross the cell membrane, increasing the intracellular concentration of calcium. The increased calcium then activates the calcium binding regulatory protein calmodulin. Calmodulin, with its bound calcium, can then bind to and regulate the activity of many different cell proteins, including kinases. That such calcium fluxes are physiologically relevant to the growth of T cells is further indicated by

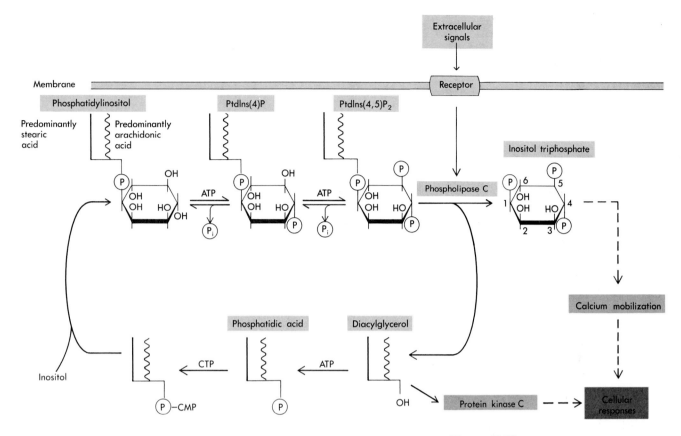

Figure 25-20
Phosphatidylinositol turnover leading to the appearance of diacylglycerol and inositol triphosphate, two second messengers for hormone stimulation. Extracellular signals (e.g., hormones or factors) may transmit signals that alter cell behavior by stimulating the turnover. Inositol triphosphate causes release of Ca^{2+} from intracellular stores and also activates a kinase. Diacylglycerol activates protein kinase C.

recent findings that activation of T cells by antigen opens up calcium channels in the cell membrane.

As already mentioned, calcium ionophores are not sufficient for the mitogenic activation of T lymphocytes. They must be used in conjunction with phorbol esters to achieve full mitogenic effect. Phorbol esters are compounds extracted from croton oil that have long been known to act as tumor promoters (agents that in combination with other so-called initiating carcinogens can cause cancer). Phorbol esters have a wide variety of effects on a great many cell types. In some fibroblasts, phorbol esters have a mitogenic effect very similar to that of PDGF. They even activate the cellular *myc* gene. At least some of their effects on cells are mediated by binding to and activating the calcium-dependent, lipid-dependent **protein kinase C** at the cell membrane. This protein kinase has a specificity for serine and threonine residues on protein substrates. Protein kinase C is found in most mammalian tissues, but its physiological substrates are largely uncharacterized.

The normal endogenous cellular activator of protein kinase C is diacylglycerol (Figure 25-20). Diacylglycerol is produced transiently in cell membranes as a result of the breakdown of phosphatidylinositol. Phosphatidylinositol itself is a minor membrane phospholipid whose breakdown is catalyzed by a membrane-bound phospholipase C activity (Figure 25-20). Phospholipase C is in turn activated by a variety of extracellular stimuli, many of which are growth factors. Thus, PDGF treatment of fibroblasts, antigen or lectin treatment of T cells, and fertilization of eggs all lead to an activation of phospholipase C and a concomitant rapid turnover of membrane phosphatidylinositol.

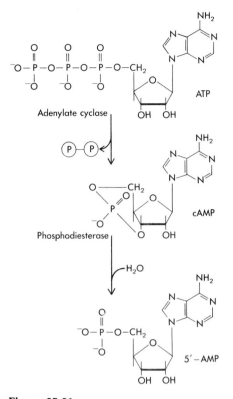

Figure 25-21
Synthesis and degradation of cAMP.

The diacylglycerol released from the breakdown of phosphatidylinositol is then able to transiently activate protein kinase C. By this mechanism, protein kinase C is thought to act as one of the signal transduction pathways for polypeptide growth factors.

The other product formed by the breakdown of phosphatidylinositol is the hydrophilic group inositol triphosphate (Figure 25-20). This compound diffuses away from the plasma membrane and also acts as a second messenger. Inositol triphosphate can provoke the release of intracellular calcium from the endoplasmic reticulum. The result is a transient increase of the calcium concentration within the cell and indirect effects on other ion fluxes as well. Inositol triphosphate also seems to be able to activate cellular protein kinases.

Another important second messenger for polypeptide hormones is *cyclic AMP (cAMP)*. Some polypeptide hormones can exert their effects by either activating or inactivating the inner plasma membrane–bound enzyme *adenylate cyclase* (Figure 25-21). This alters the level of cAMP inside the cell, which then determines the activity of yet another set of kinases known as the *cAMP-dependent protein kinases*. These enzymes transfer phosphate from ATP to the serine or threonine residues of a small group of specific target proteins. These proteins, as a result of being phosphorylated, can be either activated or inactivated and so bring about changes in cellular metabolism, including changes in gene expression. The level of cAMP also regulates the enzyme *phosphoprotein phosphatase*, which dephosphorylates the substrates of the cAMP-dependent protein kinases. Apparently, different cells can express different sets of kinases or different sets of target proteins, so changes in cAMP levels can have very different effects on different cell types.

The effect of hormone receptor complexes on adenylate cyclase is indirect. It is mediated by a third membrane protein called the GTP binding protein, or *G protein*. G protein binds GDP on a cytoplasmic face until it collides with a specific hormone-receptor complex. Then it is activated, exchanges its GDP for GTP, and as a result binds to and alters the activity of adenylate cyclase. The reaction terminates when a GTPase activity intrinsic to the G protein hydrolyzes the GTP back to GDP. The activation of phospholipase C by ligands is also thought to be mediated by a G-type protein.

Despite much effort and precedent from many biological systems, efforts to relate cAMP levels to growth factor–induced responses have not yet yielded clear-cut answers.

A Different Mechanism of Action for Steroid Hormones[58, 59]

The mechanism of action of steroid hormones is different from that of the polypeptide growth factors, even though the types of changes they induce in cells can appear similar. The steroid hormones, such as estrogen, do not have any direct effect on cyclic nucleotide levels, nor are their receptors latent kinases. Their hydrocarbon nature allows them to pass through the plasma membrane to interact with their specific protein receptors, which are located inside the cells. The resulting steroid-receptor complexes then combine with specific chromosomal sites. The activation of specific RNA sythesis that follows is the direct result of the binding of the receptor-steroid complex to chromatin.

Molecular Cloning of the Polypeptide Factors That Stimulate Growth, Maturation, and Function of Blood Cells[60–66]

Most of the red and white cells that circulate in the blood have a limited life span. They are constantly replaced through the division of cells located in the bone marrow. The process of blood cell formation is called **hematopoiesis,** and its scale is enormous. For example, red cells are formed at a rate of more than 100 million cells a minute in an adult human. There are many different types of blood cells belonging to distinct cell lineages (Figure 25-22). Along each lineage are cells at different stages of maturation, with many of the different types existing in close proximity within the bone marrow.

All the different types of blood cells arise from a small population of stem cells formed early in the embryo. The existence of these **pluripotent stem cells** can be demonstrated by heavily irradiating a mouse so that it can no longer form its own blood cells. The animal will die if left untreated; but it can be saved by injection of bone marrow cells from a second mouse. A tiny fraction of the donor cells settle in the bone marrow of the recipient animal and differentiate to produce cells belonging to all the blood cell lineages. If the donor cells are marked by lightly irradiating them to introduce random chromosome breaks into different cells, a specific chromosome abnormality can sometimes be found in cells of all lineages within a recipient animal. This result indicates that all the cell types can arise from a single donor cell.

Pluripotent stem cells injected into an irradiated recipient can migrate to the spleen and give rise to colonies containing cells of the different blood cell lineages. Hence, they are called **colony-forming units–spleen (CFU-S).** As can be seen in Figure 25-22, pluripotent stem cells give rise to **progenitor cells** that will yield a more limited array of cell types. A major breakthrough in the study of hematopoiesis was the finding, beginning about 20 years ago, that many types of progenitor cells will grow and differentiate in vitro. Progenitor cells are named according to the type of progeny they yield. For example, a colony-forming unit–erythroid (CFU-E) is a cell capable of forming a colony containing only cells of the erythroid lineage.

Maintaining the correct ratios of blood cell types in an animal would seem to be an awesome task. Only rarely does the process become deranged and result in illnesses like anemia or leukemia. Some of the mechanisms that control hematopoiesis involve cell-cell interactions within the complex cellular architecture of the bone marrow. In addition, a group of glycoprotein factors have been shown to control cell growth, commitment to differentiation, and the functioning of hematopoietic cells.

Factors that stimulate hematopoietic cells have been identified by removing the supernatant fluids from different cultured cell types and showing that they can promote the growth or function of a particular type of blood cell in vitro. Often, the factors were named first according to the biological effect they elicit: for example, mast cell growth factor, macrophage-granulocyte inducer, macrophage-inhibiting factor, and thymocyte mitogenic factor. In many cases, the cells producing the factors and the target cells the factors affect are not pure populations. It is not always possible to be certain how many

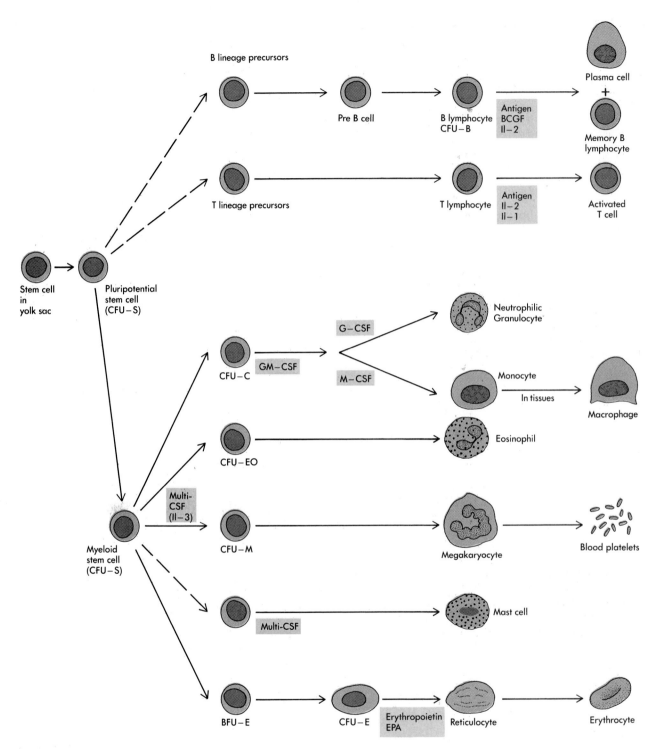

Figure 25-22
Schematic representation of hematopoiesis. Stages thought to be regulated by particular growth factors are indicated. The scheme is still undergoing revisions and additions.

different factors are present in the culture fluids or which cell type produces them. Gradually, researchers have managed to purify hematopoietic growth and differentiation factors and more recently to molecularly clone the genes that encode them. Since some factors are active at low concentrations and since it is difficult to purify the proteins biochemically, the ability to clone the corresponding genes and

Table 25-5 Protein Factors That Stimulate and/or Are Made by Blood Cells

Name	Structure	Biological Properties
Il-2 (interleukin 2; also known as T cell growth factor, TCGF)	133 amino acids; MW 15,000	Proliferation of activated T cells; also stimulates B cells and thymocytes; made by activated T cells
Multi-CSF or Il-3 (multi-lineage colony stimulating activity; also known as BPA, HCGF, MCGF, and PSF)	133 amino acids (mouse) (from 166-amino-acid precursor); glycoprotein	Stimulates growth of stem cells, myeloid progenitor cells of all types, and mast cells
GM-CSF (granulocyte-macrophage colony-stimulating factor)	118 amino acids (mouse); glycoprotein; MW 23,000	Needed continuously for proliferation and differentiation of progenitors of granulocyte-macrophages
G-CSF (granulocyte colony-stimulating activity)	glycoprotein; MW 25,000	Stimulates formation of granulocytes
M-CSF (macrophage colony-stimulating factor; also known as CSF-1)	glycoprotein; MW 70,000	Stimulates production of macrophages from bipotential cells
Erythropoietin (Epo)	166 amino acids; glycoprotein	Growth and differentiation of erythrocyte precursors
Il-1 (interleukin 1); defined in mouse, same as Il-1α of human	271 amino acids; may be processed to active 159-amino-acid form	Diverse effects; stimulates Il-2 release, B cell growth, prostaglandin release; may be endogenous pyrogen
Lymphotoxin	171 amino acids; forms multimers; glycoprotein	Released by activated lymphocytes; antitumor activity
TNF (tumor necrosis factor; cachectin)	157 amino acids; forms multimers; glycoprotein; 30% homology with lymphotoxin	Produced by activated monocytes; antitumor activity; cachectin inhibits enzymes of fat utilization; causes emaciation
γ-interferon	146 amino acids; glycoprotein	A lymphokine with antitumor activity; member of a family of proteins with antiviral activity

then to produce purified proteins by recombinant DNA technology is a major advance in the field. Often, factors that were previously assigned different names turn out to be identical. Combined with systems for in vitro cell growth, it is at last becoming possible to develop clean systems to study the function and mode of action of many of the factors that influence the behavior of blood cells. A list of some of the cleanly identified hematopoietic growth factors is presented in Table 25-5.

Erythropoietin Induction of Red Blood Cell Production[67, 68]

The level of red blood cells (erythrocytes) and hence the level of oxygen in the blood is under the specific control of **erythropoietin**, a glycoprotein with a molecular weight of about 37 kdal. Somehow, cells in the kidney and liver that synthesize erythropoietin can detect a drop in the level of erythrocytes in the blood and respond by releasing the factor. Erythropoietin raises the red cell number by inducing the selective proliferation and differentiation of erythrocyte precursors. An early precursor that will grow in vitro in response to high concentrations of erythropoietin plus a second factor, erythroid-potentiating activity (EPA), is called an **erythrocyte burst-forming unit,** or **BFU-E** (Figure 25-22). The BFU-E undergoes about 12 divisions before it becomes a mature erythrocyte. **Erythrocyte colony-forming units (CFU-Es)** can also grow in vitro and respond to lower concentrations of erythropoietin. These cells are only about 5 cell divisions from a mature erythrocyte. The erythropoietin-stimulated cascade of divisions produces an ever-increasing number of red blood cells from a small number of progenitors. It is estimated that in vivo one BFU-E can give rise to 2^6 CFU-Es and 2^{11} mature erythrocytes!

An early effect of erythropoietin on cells is the selective stimulation of RNA synthesis. It occurs within 1 hour of erythropoietin addition and involves synthesis of tRNA, rRNA, and mRNA. The mRNA molecules that code for globin itself appear later, indicating that hemoglobin formation is the end result of an involved differentiation process that leads not only to hemoglobin formation but also to the mitotic events that increase the number of erythropoietin-sensitive cells.

The gene for erythropoietin has been cloned, and a cDNA clone of its mRNA has been isolated and sequenced. The first step in the cloning was to purify erythropoietin from human urine. Then the protein was digested with trypsin, and the amino acid sequence of tryptic fragments was determined. These sequences made it possible to synthesize corresponding oligonucleotide probes that could hybridize to the erythropoietin gene, and these probes were used to select the gene from a library of human DNA.

Glycoprotein Factors That Control the Production, Differentiation, and Function of Granulocytes and Macrophages[69, 70]

The granulocyte and macrophage components of blood form by a differentiation-mitotic process starting from a common bipotential precursor cell in the bone marrow (Figure 25-22). These events can be studied in vitro, where the progenitor cells will grow into clones in semisolid media if they are given specific protein inducers, called **colony-stimulating factors (CSFs).** Progenitor cell division and survival is completely dependent on the continued presence of one or another of the CSFs.

So far, four CSFs from mice have been purified—G-CSF, M-CSF, GM-CSF, and multi-CSF—and DNA encoding all of these have been cloned (see Table 25-5). All four factors are glycoproteins, and so far they show no amino acid sequence homology to one another. Each factor binds to a specific receptor. A given granulocyte-macrophage progenitor cell can respond to more than one CSF, with different cells requiring different concentrations of the factors for proliferation. Although the factors were discovered because they promote cell proliferation, they also have three other important effects on responsive cells: maintenance of cell survival, promotion of differentiation commitment, and stimulation of differentiated cell functions. How the factors influence differentiation may be concentration-dependent. For example, a high concentration of GM-CSF causes a bipotential cell to enter the granulocytic pathway, while a low concentration favors formation of macrophages. CSFs can stimulate the functions of mature cells, for example, phagocytosis of bacteria by granulocytes and macrophages. How survival, proliferation, differentiation, and function are separately controlled by these factors is not yet understood.

At least some CSFs have additional targets besides granulocyte-macrophage precursors. In particular, multi-CSF, also known as interleukin 3 or Il-3, can stimulate proliferation and differentiation of multipotential stem cells leading to the production of all the major blood cell types.

Many different cell types seem to be capable of making one or another CSF, but the biologically important sources and the mechanisms that ensure correct CSF levels in vivo are not yet known.

T Cell Growth Factor (Interleukin 2) Stimulates Growth of Activated T Lymphoid Cells[71-75]

For a long time, it was thought that mitogen (antigen) alone could stimulate mature T or B cells to divide in the presence of standard media containing serum; but now it is believed that additional specific polypeptide factors are also needed. One example is the **T cell growth factor (TCGF),** called *interleukin 2* (Il-2), which is required for the proliferation of mature T cells. The term *interleukin* denotes factors made by leukocytes (white blood cells) that also act on leukocytes. Interleukin-2 can also be classified as a *lymphokine,* a term used to describe factors made by lymphocytes, one type of white blood cell.

Mature T cells taken from peripheral blood will grow indefinitely in culture if they are stimulated with antigen (mitogen) and supplied with Il-2. Antigen stimulates the appearance of Il-2 receptors, and it also stimulates the synthesis of some Il-2. This explains why antigen alone can appear to be a complete mitogen for T cells (Figure 25-23). How antigen binding to its receptor (the T cell antigen receptor discussed in Chapter 23) triggers expression of Il-2 receptor genes and Il-2 is not known, but, as noted, an early event following antigen binding may be the change in the level of intracellular calcium.

Both Il-2 and its receptor have been molecularly cloned (see Figure 25-17). How Il-2 binding triggers T cell growth is unknown. The receptor has such a small cytoplasmic domain (13 amino acids) that it is doubtful that there is an intracellular enzymatic activity intrinsic in the cytoplasmic portion of the molecule, such as the kinase activity associated with the cytoplasmic portion of the EGF, PDGF, or insulin receptors.

Figure 25-23
Stimulation of T cell growth by antigen and Il-2. Antigen (often a preparation of heterologous spleen cells) induces expression of Il-2 receptors.

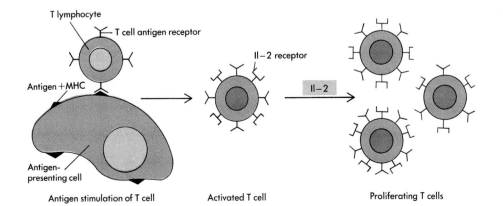

T lymphocyte

T cell antigen receptor

Il–2 receptor

Antigen + MHC

Antigen-presenting cell

Il–2

Antigen stimulation of T cell Activated T cell Proliferating T cells

Transforming Growth Factors Permit Cells to Grow Without Attachment[76, 77]

Normal fibroblasts or 3T3 cells must be attached to a surface to grow in vitro. Some transformed cells produce factors that can reversibly allow fibroblasts to grow in suspension. Since the ability to grow in suspension is often associated with the tumorigenic phenotype, these factors are called **transforming growth factors.** Two types have been separated: TGF-α and TGF-β. TGF-α is structurally related to the epidermal growth factor (EGF) and binds to the same receptor. EGF can substitute for TGF-α in culture. TGF-β is a distinct growth factor that, like PDGF, is found in platelets; yet, it binds to a receptor distinct from PDGF, EGF, or IGFs. TGF-α and TGF-β are not by themselves sufficient to support the growth of fibroblasts in suspension: PDGF is required along with the TGFs.

Do Negative Growth Factors Exist?[78–81]

Although much cell proliferation in vivo may be regulated by controlling the availability of growth factors to target cells, in some cases it seems that it would be useful for an organism to be able to shut off cell proliferation directly. This idea has prompted the search for physiological factors that may prevent cells from dividing. The identification of such factors has been arduous and often controversial because of the difficulties in distinguishing between a negative growth regulator and a compound that simply has a toxic effect on the cell. However, there is now widespread agreement that negative growth regulators do exist.

Astonishingly, TGF-β is a growth inhibitory factor. Although TGF-β can enhance the growth of fibroblasts in culture, it can be inhibitory to other cell types. TGF-β inhibits the growth of some epithelial cell lines in culture. Interestingly, it is also able to inhibit the growth of various cancer cells in culture, raising hopes that it might have use in the treatment of cancer.

Another negative growth factor may be γ-interferon, a member of the interferon family of proteins whose synthesis is induced in response to viral infections (Chapter 24). One effect of interferon is to cause target cells to stop cycling and make them refractory to the actions of mitogenic stimuli. This growth-arrested state is reversible, and the cells return to their normal growth factor responsiveness a short time after removal of interferon.

What Are Lymphotoxin and Tumor Necrosis Factor (Cachectin)?[82–85]

Sometimes, cancer patients that get severe bacterial infections experience tumor regressions, and this led to attempts, almost a hundred years ago, to cure cancer by injections of bacteria or bacterial toxins. A paradox was that these substances did not destroy tumor cells in vitro. The reason became clear when it was found that they probably work indirectly by inducing the synthesis of naturally occurring proteins with antitumor activity. Mice or rabbits injected with an immunostimulatory bacterium (*Mycobacterium bovis* strain bacillus Calmette-Guerin, or **BCG**) and subsequently with endotoxin produce a protein in their serum called **tumor necrosis factor** (**TNF**), which preferen-

to the fact that some cancer cells secrete TGFs that somehow allow cells to bypass the requirement for anchorage.

Conversion of Fibroblasts into Adipose Cells Is Enhanced by Azacytidine[91, 92, 93]

The conversion of fibroblasts into adipose (fat) cells depends on two stimuli: polypeptide factors, called adipogenic factors, and conditions where cell division is not possible (e.g., excessive crowding). While multiplying, 3T3 cells always maintain a fibroblast shape. But upon reaching confluence, they have an increasing probability of converting into very large, round adipose cells (Figure 25-26). Not all arrested fibroblasts become so differentiated, but certain 3T3 strains (preadipose strains) have high probabilities of undergoing the adipose conversion. It is lack of growth, not contact between arrested cells, that is important for fat accumulation. This is shown by suspending single 3T3 cells in methylcellulose gels. Under such conditions, they remain small and cannot make new DNA or divide. If the cells differentiate into adipose cells, they enlarge and synthesize neutral fat (triglyceride) until the center of the cell is filled with a huge fat droplet, which pushes the nucleus into an eccentric position near the cell membrane (Figure 25-27). As long as the adipose cell maintains its huge fat droplet, it is incapable of cell division. But under the right conditions, young adipose cells can lose their triglyceride content and regain their fibroblast appearance. Then they are capable of further cell division.

While cessation of cell division is a requirement for the fibroblast-to-fat cell conversion, it is not a sufficient signal. Conversion also requires a polypeptide with the properties of an adipogenic factor. One such factor is growth hormone.

An intriguing nonphysiological stimulus of the fibroblast-to-fat cell conversion is azacytidine (aza C), which can substitute for cytosine residues in cellular DNA. Treatment of embryonic fibroblasts with aza C leads to the appearance of fat cells, as well as muscle cells and chondrocytes. The mechanism for this effect is thought to be a change in gene expression brought about by a change in methylation patterns of the DNA. DNA that is substituted with aza C cannot be properly methylated, and the extent of methylation is known to be correlated with gene expression (Chapter 21). Which particular genes are important in bringing about the conversion of fibroblasts to fat cells remain to be determined.

Maintenance of Myoblasts in Continuous Cell Culture[94]

It is possible to routinely culture **myoblasts,** the cellular precursors of the multinucleated striated muscle cells. Although myoblasts appear essentially undifferentiated, in culture they usually divide only several times before they begin to aggregate and fuse into postmitotic, multinucleated muscle cells that soon acquire a cross-striated appearance and develop a contractile potentiality (Figure 25-28). By this stage, the differentiation process is irreversible, and no further DNA synthesis is possible. The resulting multinucleate cells, however, retain the capacity to make the mRNA for the many proteins whose continued synthesis is essential for normal muscle functioning.

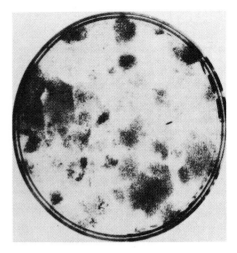

Figure 25-26
Formation of fat cells within a confluent layer of mouse fibroblasts (3T3 L1). Three weeks after the colonies became confluent, the culture was stained with Oil Red O (shows black in this photograph), which selectively stains fat deposits red. The marked variation in size and staining intensity reflects differences in the proportion of cells undergoing adipose conversion in the different colonies. [Reproduced with permission from H. Green and M. Meuth, *Cell* 3 (1974):127.]

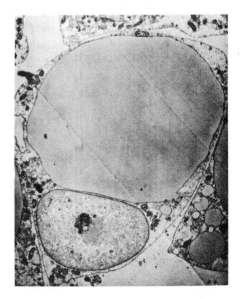

Figure 25-27
A thin section of a mature adipose cell, showing a thin cytoplasmic rim surrounding the large central fat droplet. The nucleus has the characteristic eccentric location of adipose cells. [Electron micrograph by Elaine Lenk of MIT; reproduced with permission from H. Green and M. Meuth, *Cell* 3 (1974):127.]

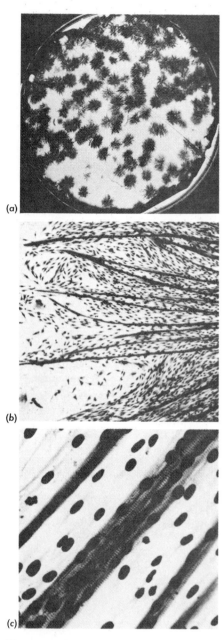

Figure 25-28
Fusion of myoblasts and the development of striated muscle fibers set within colonies of multiplying myoblasts. (a) View of a Petri dish upon which 150 myoblasts had been plated 13 days previously. (b) Higher magnification of a portion of a single colony, showing large numbers of nuclei surrounding each developing fiber. (c) A still higher magnification, showing the striated nature of the individual fibers. [Reproduced with permission from D. Yaffee, *Proc. Nat. Acad. Sci.* 61 (1968):477.]

Such irreversible differentiation, however, is not a necessary fate for all dividing myoblasts. Circumstances have been found that can maintain myogenic cells in continuous cultures. The key elements are conditions that inhibit the cell-fusing process (e.g., low Ca^{2+} or periodic additions of trypsin), together with employment of highly enriched serum of fetal origin. Thus, the transmission of a specific differentiation potential to progeny cells is not necessarily dependent on a visible morphological expression. So all the descendants of a single cloned myoblast can give rise to muscle-forming colonies.

Skin in Culture[95, 96]

It is now possible to grow epidermal and other types of epithelial cells in culture. Keratinocytes derived from any type of stratified squamous epithelium, those of the epidermis, cornea, mouth, tongue, etc., will divide in vitro to form colonies and eventually a continuous epithelium (Figure 25-29). One of the tricks to getting the cells to grow was the addition of irradiated fibroblast feeder cells that presumably supply needed growth factors. Also important was to add agents that increase the intracellular content of cAMP.

In real skin, keratinocytes in the basal layer divide, and one daughter cell moves up and out of the layer to begin terminal differentiation. The cells enlarge as they differentiate, and they alter the pattern of keratins they synthesize. Finally, as they reach the outer layers of the skin, the cells destroy most of their proteins and organelles, including the nucleus, and become flattened **squames** packed with keratin fibrils. Squames are ultimately sloughed off from the outer skin layers. A very similar process occurs in vitro, except that the final stages involving loss of the nucleus do not occur unless special steps are taken. One way to ensure that complete differentiation occurs is to detach the cells and place them in a suspended state. Another method is to reduce the level of vitamin A in the medium. Yet another method is to enzymatically detach the entire sheet of cells from the plastic surface of the tissue culture dish and graft it onto an animal.

The ability to lift up sheets of cultured skin cells and graft them has led to the remarkable use of tissue-culture-grown skin to treat people with severe burns. Because skin cells grow so well in culture, it is possible to take a small piece of an individual's skin and within two or three weeks, grow enough skin in the laboratory to cover the entire body. Since the skin is derived from the patient's own cells, there is no danger of immune rejection. However, the skin is not completely normal, since it lacks hair follicles and sweat glands.

CANCER CELLS

Searching for Chemical Differences Between Normal and Cancer Cells[1]

Almost as soon as biologists began to describe the molecules within normal cells, they looked to see whether those same molecules could be found in cancer cells. Likewise, when a new chemical reaction (or enzyme) was described in a cancer cell, a search was often made to see if the same reaction occurred in the normal cell. Often, these searches initially seemed to lead somewhere, with the tumor cells containing much more (or less) of a particular compound than their normal equivalents. Further analysis, however, usually failed to justify the initial hope that the respective change was at the heart of the cancerous phenotype. A fundamental difficulty in this approach has always been the inability to decide whether a given change is the primary metabolic disturbance that creates the cancer, or whether it is a minor secondary response to the altered metabolism caused by the primary change. Moreover, in many such experiments, it is impossible to select a good control with which to compare cells from a growing tumor: Often, we cannot be sure in what type of normal cell the cancerous transformation occurred, and often the normal cell equivalent is difficult to obtain in sufficient quantity or purity to study. Also, comparisons have frequently been made between cells from actively growing animal tumors and their so-called normal counterparts growing in culture. But many of the so-called "normal" cell lines may have undergone a number of genetic changes during their adaptation to growth in the artificial environment of cell culture.

Therefore, many of the differences we associate with the cancerous transformation were discovered by comparing a "normal" cell line (e.g., 3T3) with the transformed cell line created by exposure of cells from the "normal" cell line to a cancer-causing virus. Even here, it has been necessary to guard against being misled by new sets of spontaneous mutations that occur during the further growth in cultures of the respective normal and transformed cell lines. One of the most reliable systems for probing cancer biochemistry has been cell lines transformed by mutated cancer viruses (Chapter 26). The resulting cells have a cancerous behavior when grown at a low temperature (say, 32°C) but a normal behavior when grown at a higher (e.g., 38°C) temperature.

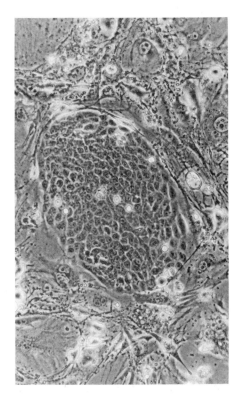

Figure 25-29
Six-day colony consisting of about 300 human epidermal cells. Supporting peripheral 3T3 cells are lethally irradiated. (Courtesy of Y. Barrandon and H. Green.)

Warburg and the Meaning of Increased Glycolysis

The first convincing biochemical difference between normal cells and cancer cells was discovered over 60 years ago by the German biochemist Otto Warburg. He observed that virtually every type of cancer cell that forms a solid tumor (as opposed to cancers of the blood, like leukemia) excretes much larger quantities of lactic acid than does its normal counterpart. The cancer cells do this both when growing as solid tumors in animals and when multiplying as single cells in culture. Since the original observation in the early 1920s, the meaning of lactic acid overproduction (often called the **Warburg effect**) has been studied again and again, but its real significance remains elusive. The excessive lactic acid, however, arises from glucose via the glycolytic pathway (Chapter 2). But this increase in fermentation (anaerobic metabolism) does not result from an insufficiency of any enzymes involved in the various oxidative pathways (e.g., oxidative phosphorylation proceeds normally). Nonetheless, much more glucose is consumed by these tumor cells than they need to grow and multiply. This suggests the loss of a normal control device that regulates the rate at which glucose is taken into a cell, and in fact, many transformed cells can be shown to have an increased rate of glucose transport.

Contact Inhibition of Movement[97]

Strong evidence that the plasma membranes of cancer cells are very different from those of normal cells comes from comparing their behaviors after their ruffled edges touch each other. Such contacts between normal cells, but not between cancer cells, frequently result in cell adhesion, accompanied by a slowing down of random amoeboid movements by which the cells move along surfaces (contact inhibition of movement). Early in this process (10 to 20 seconds after contact), actin-containing cables begin to form in the seemingly disorganized cytoplasmic region beneath the ruffled edges (Figure 25-30). At the same time, the ruffled edges start to lose their morphological identity, soon becoming indistinguishable from less active regions of the plasma membrane. Generation of new intracellular actin-containing cables in response to cellular contacts may thus be an essential aspect of contact inhibition of movement.

Malignancy as a Loss of Normal Cellular Affinities[98]

The "sticky" quality of normal cells that leads to adhesion displays considerable specificity. A given type of cell (e.g., a liver cell) prefers to stick to others of its own kind (e.g., other liver cells) and shows very little, if any, affinity for other types (e.g., kidney cells). This type of specificity has been elegantly demonstrated in experiments in which small amounts of the proteolytic enzyme trypsin are used to break apart organs such as the liver or the kidney into their single-cell components. If these isolated cells are then incubated in the absence of trypsin, they reaggregate to form tissue fragments similar to those in the intact organ, that is, small fragments of liver tissue and small fragments of kidney tissue. When kidney and liver cells are mixed together, small fragments of liver and kidney are again detected. Thus, a kidney cell prefers to stick to a kidney cell, and a liver cell to a liver cell.

If this experiment is repeated with cancer cells, however, the normal cellular affinities no longer hold. For example, the mixing of cells from a skin cancer with normal kidney cells results in aggregates in which the skin cancer cells are interspersed among the kidney cells. Conceivably, the inability of cancer cells to form tight adhesive junctions allows them to lie adjacent to almost any cell type. This may be one reason why malignant cells invade a variety of normal organs.

The nature of the sticky substance(s) that normal cells use to attach to each other is still unclear. As already discussed, however, fibronectin appears to be a glue at least for cell-matrix interactions, and it provides a good substrate for cell attachment and spreading in vitro. Fibronectin is frequently decreased or absent in transformed cells, and this may contribute to the cells' loss of adhesiveness. Electron micrographs of cells that have stuck to glass (or to each other) reveal dense adhesive regions (plaques) 60 to 80 Å thick that are located just inside the plasma membrane (see Figure 25-30). One component of adhesion plaques is a protein called vinculin, which may serve as an anchor for cytoskeletal cables.

Alterations of the Cytoskeleton of Transformed Cells[99–101]

Examination of many cancer cells reveals that they usually display a much less organized arrangement of the various cytoskeletal proteins than is found in their normal equivalents. Normal fibroblasts, for example, contain large numbers of fibers, cables in which the various

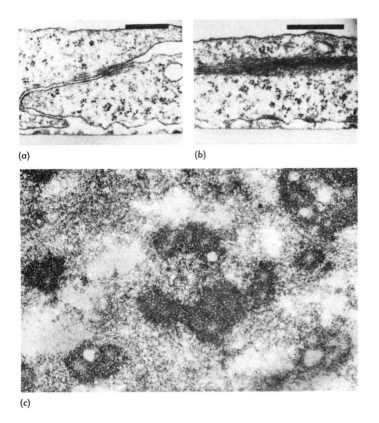

(a)   (b)

(c)

Figure 25-30

(a, b) Electron microscopic views of the "adhesive plaques" (arrows) that form within 10 to 20 seconds of contact between ruffled edges. Organized groups of actin-containing filaments form beneath the plaques within less than a minute. Bar represents 0.5 μm. [Reproduced with permission from J. Haeysman, *CIBA Foundation Symp.* 14 (new series) (1973):190.] (c) An electron micrograph in which the plane of thin section cuts just above the region of hamster embryo cell surface contact. This section reveals the crystalline-like lattices of the adhesive plaques (dark patches) (42,000×). (Courtesy of R. D. Goldman, Carnegie-Mellon University.)

cytoskeletal proteins (e.g., actin, myosin, and tropomyosin) are regularly arranged. In contrast, transformed cells contain many fewer such fibers, and those that are present are generally much thinner (Figure 25-31). Correspondingly, the touching of the cancer cells does not lead to cessation of cell movement, but instead frequently induces

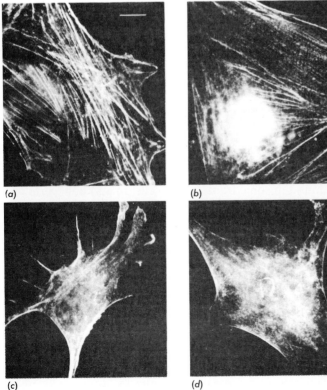

(a)   (b)

(c)   (d)

Figure 25-31

A comparison of the organization of muscle proteins within normal and transformed mouse cells. Immunofluorescent techniques are used to visualize (a) actin within a "normal" 3T3 cell, (b) myosin within a 3T3 cell, (c) actin in a 3T3 cell transformed by SV40 virus (SV3T3), and (d) the myosin of an SV3T3 cell. [Reproduced with permission from R. Pollack, M. Osborn, and K. Weber, *Proc. Nat. Acad. Sci.* 72 (1975):994.]

Figure 25-32
Blebs on the surface of cancer cells.
(a, b) In these cancerous hamster cells that originated through transformation by adenovirus 5, blebs are a characteristic feature throughout the cell cycle except during M phase. As shown in these photos taken 1 minute apart, blebs are very unstable protrusions, generally having lifetimes of less than a minute. In other types of cancer cells, blebs are less frequent. In their place are found large numbers of microvilli and ruffles, which, like blebs, are constantly thrown out and then retracted. [Reproduced with permission from R. D. Goldman, C.-M. Chang, and J. - Williams, *Cold Spring Harbor Symp. Quant. Biol.* 39 (1975):601.] (c) An electron micrograph of a thin section of a portion of a chicken cell transformed by Rous sarcoma virus, showing a large bleb. It is largely filled with free polyribosomes. (Courtesy of Lan Bo Chen and Elaine Lenk, MIT.)

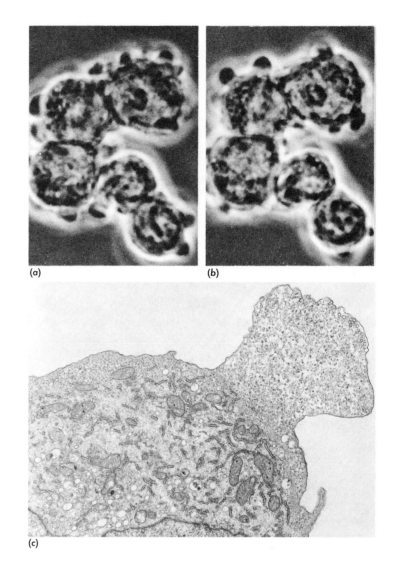

(a)   (b)

(c)

the uncoordinated generation of many more ruffles and blebs than are found on their normal cellular equivalents (Figure 25-32). Why cancer cells are less able to form large numbers of highly organized fibers is not yet clear. The reason, however, may be due to alterations in the expression of some of the cytoskeletal proteins. For example, alterations in the expression of tropomyosin are a feature of some transformed cells. These alterations in tropomyosin expression appear to correlate with the rearrangement of stress fibers and this may in part be responsible for the morphological alterations observed in transformed cells. Furthermore, expression of a mutant actin gene has been found in some transformed cells. In addition, certain cancer cells may have an alteration in one or more components at the plasma membrane (e.g., vinculin), which may hinder the attachment of individual actin filaments to the inner membrane surface.

Molecular Changes at the Cell Surface That Accompany Cell Transformation[87, 102, 103]

Only recently have the chemical techniques been available for us to critically compare the plasma membranes of normal cells and transformed cells. So, despite intense interest in this problem, current knowledge is still very incomplete. Making this problem inherently difficult is the complexity of most plasma membranes, most of which at a minimum must easily contain hundreds of different components. Of necessity, attention has been focused either on the major structural components or on those relatively few membrane-bound enzymes that we can conveniently assay. However, an essential change that defines the cancer cell membrane might occur in a minor structural component, so it is not clear whether we could reasonably expect to find critical cancer changes by membrane biochemistry alone.

Moreover, it is becoming increasingly clear that normal plasma membranes are not invariant in structure. The exact groupings that protrude on their outer surfaces not only change with the different phases of the cell cycle, but are very much a function of the contacts that they make with other cells. Great caution must be shown in comparing normal cells with their transformed equivalents unless we are sure that both types have been harvested in the same point in the cell cycle, at the same density, and under identical conditions of growth.

Now it seems fairly certain that the cancerous transformation does not change the relative amounts of the four main phospholipids that form the basic lipid bilayer. Differences, however, and sometimes seemingly quite specific ones, have been seen in both the nature and kind of the glycolipids and glycoproteins that are inserted into the bilayer. In particular, much study has gone toward the gangliosides and other glycolipids. Not only are the more complex of them specifically diminished in amounts in many cancers, but the enzymes involved in their biosynthesis are also specifically reduced. It is not at all clear, however, that such changes are primary events in the formation of a cancer cell. Not only do such changes often lag behind the assumption of the distinctive cancer cell morphology when viral transformation occurs, but, perhaps even more important, other transformed cell lines have a perfectly normal glycolipid content. So we must postulate either that changes in a variety of quite distinct molecular components each can produce the distinctive cancer cell membrane, or that the alterations in ganglioside content are irrelevant features that are somehow selected for by a prior primary change.

Most recently, use has been made of monoclonal antibodies to probe the difference between normal and transformed cell surfaces. About 10 percent of the monoclonal antibodies raised against cell surfaces are directed against glycolipids. Interestingly, it is some of these antibodies that are found to be specific for the transformed cell surface. The antigens detected are not strictly limited to tumor cells, however, since they are frequently found on embryonic cells as well.

Much attention has gone toward examination of the major proteins that are inserted in the outer plasma membrane, especially the glycoproteins whose exterior protruding surfaces are covered with sugar groups. Here at least one change common to many types of cancer cells has already been found: the reduction or loss of the large external glycoprotein, fibronectin. This protein, composed of disulfide-bonded subunits of about 230 kdal, is found as a fibrillar network above, beneath, and around cells in culture. It disappears upon transformation of many cell types by RNA or DNA tumor viruses. Readdition of purified fibronectin to transformed cells in culture causes them to flatten and elongate, to adhere more tenaciously, and to acquire more organized microfilament bundles (Figure 25-33). However, the cells can still grow in agar and to high cell densities. So whether fibronectin plays a critical role in tumorgenicity, possibly adhesion and metastasis, remains to be seen.

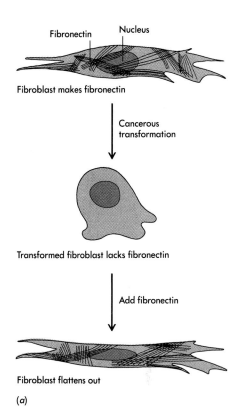

Fibronectin   *Nucleus*

Fibroblast makes fibronectin

Cancerous transformation

Transformed fibroblast lacks fibronectin

Add fibronectin

Fibroblast flattens out

(a)

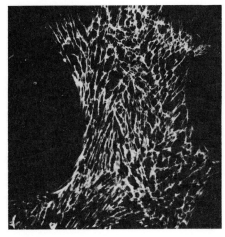

(b)

Figure 25-33

(a) Diagram showing how the synthesis or external addition of fibronectin is associated with the flattened shape of the fibroblast. (b) Immunofluorescent micrograph of fibronectin on the surface of a Ni1 8 hamster fibroblast growing in vitro. Antibodies to fibronectin were conjugated to fluorescein to make this picture. [Courtesy of Richard O. Hynes.]

Decreased Serum Requirements of Cancer Cells and Secretion of Growth Factors[76, 104–107]

Many cancer cells can grow in culture medium supplemented with much less serum than that required by the corresponding normal cells. This difference reflects a lower need for one or more of the growth-controlling factors that serum supplies. The mouse cell line 3T3, for example, multiplies faster and to higher density in growth medium containing 10 percent serum than in a 1 percent serum-supplemented medium. But transformed 3T3 cells grow equally well in 1 percent serum and 10 percent serum.

Many years ago, it was found that some cells that require little or no serum to grow produce their own growth factors. Culture fluids from these cells can stimulate the growth of serum-requiring cells. This finding and the fact that cancer cells require lower concentrations of serum led to the notion that cancer may arise when cells begin to uncontrollably produce growth factors for which they also express the corresponding receptors, the autocrine hypothesis of tumor formation. Now there is evidence that at least some tumor-virus-induced cancers arise by this mechanism (Chapter 26). Other cancers may involve mechanisms that allow cells to bypass the need for specific growth factors.

TGFs were first found in the medium of certain cell lines transformed by tumor viruses. As discussed earlier in this chapter, not only do they stimulate growth, but TGF-βs confer the cancerous property of anchorage-independent growth when added to the culture medium of normal cells (Figure 25-34). The effects of TGF-βs are completely reversible: Twenty-four hours after they are removed from the medium, normal cells revert to a normal phenotype.

What the normal function of TGF-β might be is anyone's guess, but it has been postulated that certain embryological processes might transiently require cells to have cancerlike phenotypes. Whether secretion of TGF-β is a common feature of spontaneously arising human cancers remains to be seen.

Many human small-cell lung cancer cell lines secrete polypeptides that belong to the family of bombesin-like peptides (BLPs), and these cells also possess receptors for BLPs. Bombesin was originally isolated from the skin of the frog *Bombina bombina*. It is a peptide that can stimulate pancreatic enzyme secretion. The related peptides of mammals, which include gastrin-releasing peptide, cause a variety of physiological responses: They can induce gastrin cell hyperplasia, increase pancreatic DNA content in vivo in rats, and stimulate proliferation of mouse 3T3 cells in vitro. Interestingly, monoclonal antibodies that bind to bombesin and block its binding to cellular receptors can inhibit the growth of human small-cell lung cancer cells growing in vitro or growing in vivo following their subcutaneous injection into mice. These results have been interpreted as further support for the autocrine hypothesis of tumor formation.

Differential Gene Expression in Normal Versus Transformed Cells[47, 108, 109]

Powerful new tools of molecular biology are providing brute force approaches to defining biochemical differences between normal cells and cancer cells. Popular and promising is the preparation of cDNA

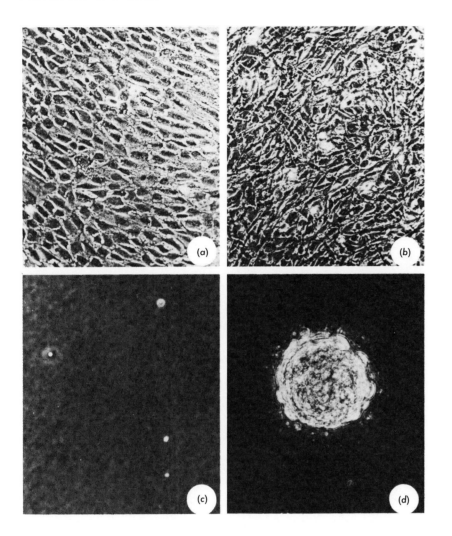

Figure 25-34
Effect of TGFs on the morphology and growth in soft agar of normal rat kidney cells. (a) Cells growing in standard medium versus (b) similar cells with partially purified TGF-α added to the medium for 6 days. (c) Cells fail to grow in agar unless (d) treated with partially purified TGF-α, in which case they grow into colonies of well over 500 cells within 2 weeks. [Courtesy of G. J. Todaro, J. E. DeLarco, H. Marquardt, M. L. Bryant, S. A. Sherwin, and A. H. Sliski, *Hormones and Cell Culture*, ed. G. H. Soto and R. Ross (Cold Spring Harbor, N.Y.: Cold Spring Harbor Laboratory, 1979).]

clones that represent mRNAs present exclusively or more abundantly in transformed cells than in their normal cell counterparts. Approximately 3 to 5 percent of mRNAs, corresponding to roughly a few hundred genes, seem to fall in this category. How long it will be before this method of defining cancer proves fruitful is hard to guess.

Another modern approach to defining cancer cells is the use of high-resolution two-dimensional gel electrophoresis to analyze the proteins of normal cells and cancer cells. This remarkable technique, which separates proteins on the basis of their isoelectric points and on their sizes, can resolve more than a thousand proteins in mammalian cells. When appropriate pairs of normal and transformed cells are compared, specific differences often can be seen (Figure 25-35); but determining the importance of such differences to the cancerous phenotype is usually difficult.

The Daunting Incompleteness
of Eucaryotic Cell Biochemistry

The fact that only a small percent of the biochemical reactions of a cell are known might make us wonder if we will soon be able to understand biological processes as complex as growth control and cancer.

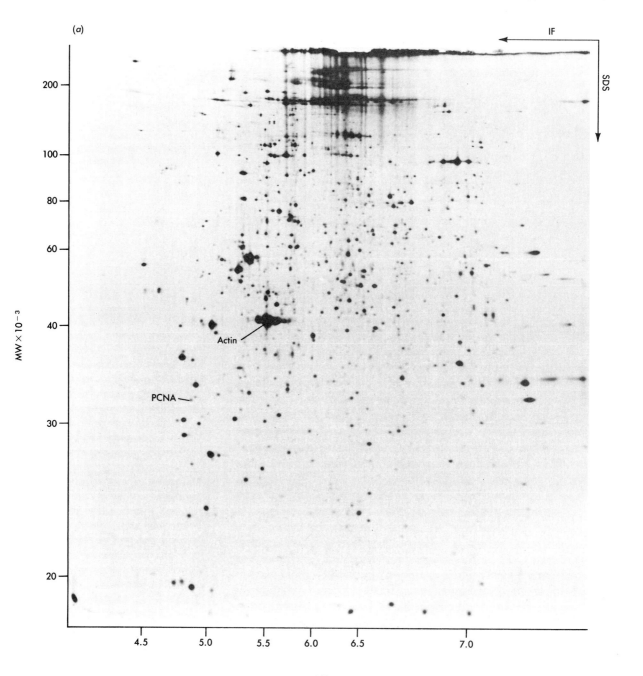

Figure 25-35
Proteins in (a) normal and (b) SV40-transformed cells of the rat fibroblast cell line REF52 analyzed by two-dimensional gel electrophoresis. Cells were grown in the presence of 35S-methionine for 2 hours prior to the analysis. Direction of migration in the first and second dimensions of the gels is indicated. In (b) some of the proteins whose amounts increase in SV40-transformed cells are indicated by upward pointing arrows, while those whose amount decreases are indicated by downward pointing arrows. About 12% of the proteins are altered two-fold or more by viral transformation, but no entirely new proteins can be detected and no proteins disappear completely. Viral T antigens are present in too low an amount to be seen on gel. PCNA, proliferating cell nuclear antigen: the rate of synthesis of this nuclear protein is tightly coupled to the rate of proliferation in normal cells but not in transformed cells. [Courtesy of James I. Garrels and B. Robert Franza, Jr., Cold Spring Harbor Laboratory.]

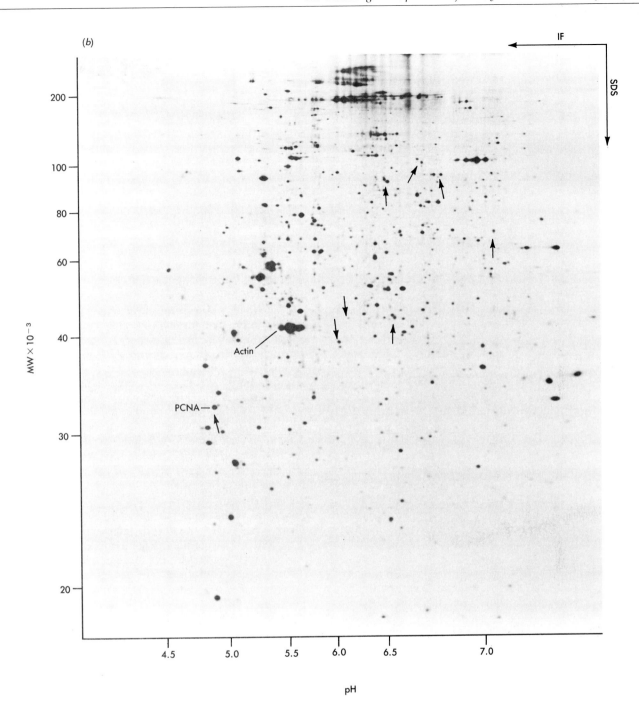

As we have seen, however, reproducible biochemical differences between normal cells and cancer cells have been found; and as we shall learn in Chapters 26 and 27, genes that can cause cancer and so induce at least some of these changes have been isolated. Most incredibly, some cancer-causing genes have been identified as genes that were already known to be involved in growth control, namely, genes encoding growth factors and their receptors. Thus, the incompleteness of eucaryotic biochemistry—while daunting—is not insurmountable when the right experiments are performed. However, we must admit that the many black boxes of cellular biochemistry still pose a substantial challenge to those who wish to fully understand growth control and cancer.

Summary

To learn how the cells of multicellular animals control their proliferation rates has long been a key objective of experimental biology. With such understanding, we will be in a much stronger position to probe both the embryological development and the correct adult functioning of higher animals. And we may at last be able to come to grips with the essence of the various forms of cancer, hereditary changes in cells that lead to uncontrolled cell proliferation.

Now almost all productive approaches to the control mechanisms governing cell proliferation utilize the in vitro growth of cells in culture. The cells excised from an animal to start a culture are called primary cells, while their descendants present several division cycles later are known as secondary cells. Most secondary cell cultures fail to multiply more than 20 to 50 times, and it is only the rare cell that acquires the ability to multiply indefinitely to form a cell line. Why only the exceptional cell has unlimited growth potential is unclear. Up to now, cell lines have been obtained only from selected animal species (e.g., mouse, rat, hamster, human, and rarely the chicken). And in general, it proves to be much easier to obtain cell lines from tumor cells than from their normal equivalents.

The greater potential of tumor cells to yield cell lines led for many years to the belief that perhaps all cells that grow as cell lines do so because they have acquired by mutation(s) many of the essential properties of the cancer cell. However, we now realize that only those fibroblasts (epithelial cells) continuously passaged at high cell density later give rise to cell lines that grow to the high cell density characteristic of cancer cells. If cells are selected at low density, the respective cell lines grow to much lower densities and generally stop dividing after they have formed organized cell layers. Such cell lines, the prototype of which is mouse 3T3 lines, are not completely normal, however, since they have enhanced growth potential relative to normal cells and they often have highly abnormal numbers of chromosomes.

Possession of well-defined cell lines, like the human cancer cell line HeLa and the mouse L cell line, made it possible to work out the detailed nutritional requirements of cells growing in culture. In general, growth not only requires a well-defined collection of amino acids and vitamins, but also depends on protein factors present in blood serum. The nature of the "serum factors" long remained a mystery; but now it is apparent that most serum factors are polypeptide hormones that act by binding specific receptors on the surface of their target cells. Each cell type requires two to five specific factors that may have to function in a specific sequence. Some, like transferrin, which transports ferric iron into cells, are almost universally required; in contrast, PDGF affects cells of mesodermal origin, and interleukin 2 appears to be highly specific for certain lymphoid cells. Polypeptide growth factors stimulate cells by binding to specific cell surface receptors. The receptors for some growth factors (PDGF and EGF) and hormones (insulin) have tyrosine kinase activity that is activated when their corresponding ligands bind. This may result in phosphorylation and thus activation of cellular proteins, which in some unknown way triggers altered gene transcription and ultimately cell division.

Some polypeptide hormones, by binding to their receptors, modify the activity of the membrane-bound enzyme adenylate cyclase, using the GTP binding protein (G protein) as an intermediate, so that the level of cAMP in the cell is altered. This, in turn, affects the activity of cAMP-dependent protein kinases that phosphorylate specific target proteins, thereby regulating their activity. How cAMP levels are related to the cell division cycle is still unclear. Other possible second messengers to carry signals from growth factor receptors to the cell include calcium and diacylglycerol.

In the absence of growth-promoting factors (mitogens), most cells enter a resting state in the G_1 phase of the life cycle and so contain a diploid amount of DNA. Once a cell has begun to replicate its DNA, it continues through the G_2 phase and obligatorily passes into mitosis. Once into G_1, a new mitogenic stimulus must be applied before the cell will start another cell cycle. Otherwise, the cell enters a quiescent state, which often leads into a differentiation process that may create a highly specialized cell (e.g., nerve) unable to initiate a new cycle of growth and cell division. Differentiation per se, however, does not lead to an inability to divide. For example, the commonly used fibroblasts, epithelial cells, and lymphoblasts all represent clearly defined states of differentiation. So it is only when the physiological role of a given cell makes further division unwanted (as with nerve cells and red blood cells) that the capacity to divide is lost.

Transformation is the process by which normal cells acquire many of the morphological as well as growth properties of cancer cells. In general, transformed cells have a more spherical shape than their normal counterparts and show many similarities to cells just emerging from mitosis. For example, many tumor cells show excessive blebbing throughout the cell cycle, a characteristic of normal cells only in early G_1. Likewise, many of the membrane proteins of cancer cells appear more mobile than those found in normal interphase cells, conceivably reflecting a much less organized arrangement of their cytoskeletons. Transformed cells also have lost many of the normal adhesive properties, leading to an inability to be sited next to appropriate cellular neighbors. In addition, many tumor cells have lost the normal control of glucose utilization. They consume much more glucose than can be metabolized efficiently, thereby leading to massive lactic acid secretion.

Many tumor cells have lower serum requirements than their normal cell equivalents, and some tumor cells secrete factors that alter the growth, and in some cases the morphology, of normal cells. To what extent the turning on of these factors may mediate the transformed phenotype in many cancers is unknown.

Molecular biologists are now applying powerful new technologies to analyze and clone genes whose expression is altered in transformed cells relative to normal cells, or in cells following mitogen stimulation. One very interesting finding so far is that PDGF added to quiescent 3T3 cells leads to increased levels of mRNA for genes called c-*myc* and c-*fos*, which, as we shall learn in the next chapter, are aberrantly expressed in certain cancers.

Bibliography

General References

Alberts, B., D. Bray, J. Lewis, M. Raff, K. Roberts, and J. D. Watson. 1983. *Molecular Biology of the Cell.* New York: Garland.

Baserga, R. 1985. *The Biology of Cell Reproduction.* Cambridge, Mass., and London: Harvard University Press.

Feramisco, J., B. Ozanne, and C. Stiles. 1985. *Growth Factors and Transformation.* Cold Spring Harbor, N.Y.: Cold Spring Harbor Laboratory.

Littlefield, J. W. 1976. *Variation, Senescence, and Neoplasia in Cultured Somatic Cells.* A Commonwealth Fund Book. Cambridge, Mass., and London: Harvard University Press.

Metcalf, D. 1984. *The Hematopoietic Colony Stimulating Factors.* Amsterdam, London, and New York: American Elsevier.

Mitchison, J. M. 1971. *The Biology of the Cell Cycle.* New York: Cambridge University Press.

Paul, J. 1975. *Cell and Tissue Culture.* 5th ed. New York: Churchill Livingstone.

Pollack, R., ed. 1975. *Readings in Mammalian Cell Culture.* 2nd ed. Cold Spring Harbor, N.Y.: Cold Spring Harbor Laboratory.

Cited References

1. Pollack, R., ed. 1975. *Readings in Mammalian Cell Culture.* 2nd ed. Cold Spring Harbor, N.Y.: Cold Spring Harbor Laboratory.
2. Hayflick, L. 1965. "The Limited in Vitro Lifetime of Human Diploid Cell Strains." *Exp. Cell Res.* 37:614–636.
3. Buonassisi, V., G. Sato, and A. I. Cohen. 1962. "Hormone-Producing Cultures of Adrenal and Pituitary Tumor Origin." *Proc. Nat. Acad. Sci.* 48:1184–1190.
4. Todaro, G. J., and H. Green. 1963. "Quantitative Studies of the Growth of Mouse Embryo Cells in Culture and Their Development into Established Lines." *J. Cell Biol.* 17:299–313.
5. Aaronson, S. A., and G. J. Todaro. 1968. "Basis for Acquisition of Malignant Potential by Mouse Cells Cultivated in Vitro." *Science* 162:1024–1026.
6. Temin, H. M., and H. Rubin. 1958. "Characteristics of an Assay for Rous Sarcoma Virus and Rous Sarcoma Cells in Tissue Culture." *Virology* 6:669–688.
7. Todaro, G. J., and H. Green. 1964. "An Assay for Cellular Transformation by SV40." *Virology* 23:117–119.
8. Macpherson, I., and L. Montaignier. 1964. "Agar Suspension Culture for the Selective Assay of Cells Transformed by Polyoma Virus." *Virology* 23:291–294.
9. Eagle, H. 1955. "Nutrition Needs of Mammalian Cells in Tissue Culture." *Science* 122:43–46.
10. Ham, R. G. 1965. "Clonal Growth of Mammalian Cells in a Chemically Defined, Synthetic Medium." *Proc. Nat. Acad. Sci.* 53:288–293.
11. Hayashi, I., and G. Sato. 1976. "Replacement of Serum by Hormones Permits Growth of Cells in Defined Medium." *Nature* 259:132–134.
12. Sato, G., A. Pardee, and D. Sirbasku, eds. 1982. "Growth of Cells in Hormonally Defined Media." *Cold Spring Harbor Conference on Cell Proliferation* 9. Cold Spring Harbor Laboratory, Cold Spring Harbor, New York.
13. Alberts, B., D. Bray, J. Lewis, M. Raff, K. Roberts, and J. D. Watson. 1983. *Molecular Biology of the Cell.* New York: Garland.
14. Baserga, R. 1985. *The Biology of Cell Reproduction.* Cambridge, Mass., and London: Harvard University Press.
15. Mitchison, J. M. 1971. *The Biology of the Cell Cycle.* New York: Cambridge University Press.
16. Pardee, A. B., R. Dubrow, J. L. Hamlin, and R. F. Kletzien. 1978. "Animal Cell Cycle." *Ann. Rev. Biochem.* 47:715–750.
17. Harris, H., and J. F. Watkins. 1965. "Hybrid Cells Derived from Mouse and Man: Artificial Heterokaryons of Mammalian Cells from Different Species." *Nature* 205:640–646.
18. Rao, P. N., and R. T. Johnson. 1970. "Mammalian Cell Fusion: Studies on the Regulation of DNA Synthesis and Mitosis." *Nature* 225:159–164.
19. Rao, P. N., and R. T. Johnson. 1974. "Induction of Chromosome Condensation in Interphase Cells." *Adv. Cell & Mol. Biol.* 3:136.
20. Huberman, J. A., and A. D. Riggs. 1968. "On the Mechanism of DNA Replication in Mammalian Chromosomes." *J. Mol. Biol.* 32:327–341.
21. Kao, F. T., and T. T. Puck. 1968. "Genetics of Somatic Mammalian Cells. VII. Induction and Isolation of Nutritional Mutants in Chinese Hamster Cells." *Proc. Nat. Acad. Sci.* 60:1275–1281.
22. Patterson, D., F. T. Kao, and T. T. Puck. 1974. "Genetics of Somatic Mammalian Cells: Biochemical Genetics of Chinese Hamster Cell Mutants with Deviant Purine Metabolism." *Proc. Nat. Acad. Sci.* 71:2057–2061.
23. Basilico, C. 1978. "Selective Production of Cell Cycle Specific ts Mutants." *J. Cell Physiol.* 95:367–376.
24. Nowell, P. C. 1960. "Phytohemagglutinin: An Initiator of Mitosis in Cultures of Normal Human Leukocytes." *Cancer Res.* 20:462–466.
25. Cohen, S. 1962. "Isolation of a Mouse Submaxillary Gland Protein Accelerating Incisor Eruption and Eyelid Opening in the New-Born Animal." *J. Biol. Chem.* 237:555–562.
26. Balk, S., J. F. Whitfield, T. Youdale, and A. C. Braun. 1973. "Roles of Calcium, Serum, Plasma, and Folic Acid in the Control of Proliferation of Normal and Rous Sarcoma Virus-Infected Chicken Fibroblasts." *Proc. Nat. Acad. Sci.* 70:675–679.
27. Ross, R., J. A. Glomset, B. Kariya, and L. Harker. 1974. "A Platelet-Dependent Serum Factor That Stimulates the Proliferation of Arterial Smooth Muscle Cells in Vitro." *Proc. Nat. Acad. Sci.* 71:1207–1210.
28. Kohler, N., and A. Lipton. 1974. "Platelets as a Source of Fibroblast Growth-Promoting Activity." *Exp. Cell Res.* 87:297–301.
29. Pledger, W. J., C. D. Stiles, H. N. Antoniades, and C. D. Scher. 1977. "Induction of DNA Synthesis in BALB/c 3T3 Cells by Serum Components: Reevaluation of the Commitment Process." *Proc. Nat. Acad. Sci.* 74:4481–4485.
30. Pledger, W. J., C. D. Stiles, H. N. Antoniades, and C. D. Scher. 1978. "An Ordered Sequence of Events Is Required Before BALB/c-3T3 Cells Become Committed to DNA Synthesis." *Proc. Nat. Acad. Sci.* 75:2839–2843.
31. Antoniades, H. N., C. D. Scher, and C. D. Stiles. 1979. "Purification of Human Platelet-Derived Growth Factor." *Proc. Nat. Acad. Sci.* 76:1809–1813.
32. Gray, A., T. J. Dull, and A. Ullrich. 1983. "Nucleotide Sequence of Epidermal Growth Factor cDNA Predicts a 128,000-Molecular Weight Protein Precursor." *Nature* 303:722–725.
33. Scott, J., M. Urdea, M. Quiroga, R. Sanchez-Pescador, N. Fong, M. Selby, W. Rutter, and G. I. Bell. 1983. "Structure of a Mouse Submaxillary Messenger RNA Encoding Epidermal Growth Factor and Seven Related Proteins." *Science* 221:236–240.
34. Cochran, B. 1985. "The Molecular Action of Platelet-Derived Growth Factor." *Adv. Cancer Res.* 45:183–216.
35. Evans, H. M., and J. A. Long. 1921. "The Effect of the Anterior Lobe Administered Intraperitoneally Upon Growth, Maturity and Oestrous Cycles of the Rat." *Anat. Rec.* 21:62.
36. Clemmons, D. R., and J. J. Van Wyk. 1981. "Somatomedin: Physiological Control and Effects on Cell Proliferation." In *Handbook of Experimental Pharmacology,* vol. 57, ed. R. Baserga. New York: Springer-Verlag, pp. 161–208.
37. Zapf, J., E. R. Froesch, and R. E. Humbel. 1981. "The Insulin-Like Growth Factors (IGF) of Human Serum: Chemical and Biological Characterization and Aspects of Their Possible Physiological Role." *Curr. Top. Cell Regul.* 19:257–307.
38. Hunter, T., and B. M. Sefton. 1980. "Transforming Gene Product of Rous Sarcoma Virus Phosphorylates Tyrosine." *Proc. Nat. Acad. Sci.* 77:1311–1315.
39. Ushiro, H., and S. Cohen. 1980. "Identification of Phosphotyrosine as a Product of Epidermal Growth Factor-Acti-

vated Protein Kinase in A-431 Cell Membranes." *J. Biol. Chem.* 255:8363–8365.

40. Kasuga, M., Y. Zick, D. L. Blithe, M. Crettaz, and C. R. Kahn. 1982. "Insulin Stimulates Tyrosine Phosphorylation of the Insulin Receptor in a Cell-Free System." *Nature* 298:667–669.

41. Ek, B., B. Westermark, A. Wasteson, and C. H. Heldin. 1982. "Stimulation of Tyrosine-Specific Phosphorylation by Platelet-Derived Growth Factor." *Nature* 295:419–420.

42. Nishimura, J., J. S. Huang, and T. F. Deuel. 1982. "Platelet-Derived Growth Factor Stimulates Tyrosine-Specific Protein Kinase Activity in Swiss Mouse 3T3 Cell Membranes." *Proc. Nat. Acad. Sci.* 79:4303–4307.

43. Jacobs, S., F. C. Kull, J. H. S. Earp, M. E. Svoboda, J. J. Van Wyk, and P. Cuatrecasas. 1983. "Somatomedin-C Stimulates the Phosphorylation of the Beta-Subunit of Its Own Receptor." *J. Biol. Chem.* 258:9581–9584.

44. Stoscheck, C. M., and G. Carpenter. 1984. "Down Regulation of Epidermal Growth Factor Receptors: Direct Demonstration of Receptor Degradation in Human Fibroblasts." *J. Cell Biol.* 98:1048–1053.

45. Sherr, C. J., C. W. Rettenmeier, R. Sacca, M. F. Roussel, A. T. Look, and E. R. Stanley. 1985. "The c-*fms* Proto-Oncogene Product Is Related to the Receptor for the Mononuclear Phagocyte Growth Factor CSF-1." *Cell* 41:665–676.

46. Armelin, H. A., M. C. S. Armelin, K. Kelly, T. Stewart, P. Leder, B. H. Cochran, and C. D. Stiles. 1984. "Functional Role for c-*myc* in Mitogenic Response to Platelet-Derived Growth Factor." *Nature* 310:655–660.

47. Cochran, B. H., A. C. Reffel, and C. D. Stiles. 1983. "Molecular Cloning of Gene Sequences Regulated by Platelet-Derived Growth Factor." *Cell* 33:939–947.

48. Kelly, K., B. H. Cochran, C. D. Stiles, and P. Leder. 1983. "Specific Regulation of the c-*myc* Gene by Lymphocyte Mitogens and Platelet-Derived Growth Factor." *Cell* 35:603–610.

49. Chambard, J. C., A. Franchi, A. LeCam, and J. Pouyssegur. 1983. "Growth Factor-Stimulated Protein Phosphorylation in G_0-G_1-Arrested Fibroblasts." *J. Biol. Chem.* 258:1706–1713.

50. Castagna, M., Y. Takai, K. Kaibuchi, K. Sano, U. Kikkawa, and Y. Nishizuka. 1982. "Direct Activation of Calcium-Activated, Phospholipid Dependent Protein Kinase by Tumor-Promoting Phorbol Esters." *J. Biol. Chem.* 257:7847–7851.

51. Cooper, J. A., D. F. Bowen-Pope, E. Raines, R. Ross, and T. Hunter. 1982. "Similar Effects of Platelet-Derived Growth Factor and Epidermal Growth Factor on the Phosphorylation of Tyrosine in Cellular Proteins." *Cell* 31:263–273.

52. Schuldiner, S., and E. Rozengurt. 1982. "Na^+/H^+ Antiport in Swiss 3T3 Cells: Mitogenic Stimulation Leads to Cytoplasmic Alkalinization." *Proc. Nat. Acad. Sci.* 79:7778–7782.

53. Rozengurt, E., P. Stroobant, M. D. Waterfield, T. F. Deuel, and M. Keehan. 1983. "Platelet-Derived Growth Factor Elicits Cyclic AMP Accumulation in Swiss 3T3 Cells: Role of Prostaglandin Production." *Cell* 34:265–272.

54. Berridge, M. J., J. P. Heslop, R. F. Irvine, and K. D. Brown. 1984. "Inositol Triphosphate Formation and Calcium Mobilization in Swiss 3T3 Cells in Response to Platelet-Derived Growth Factor." *Biochem. J.* 222:195–201.

55. Gilman, A. G. 1984. "G Proteins and Dual Control of Adenylate Cyclase." *Cell* 36:577–579.

56. Nishizuka, Y. 1984. "The Role of Protein Kinase C in Cell Surface Signal Transduction and Tumor Promotion." *Nature* 308:693–698.

57. Truneh, A., F. Albert, P. Goldstein, and A.-M. Schmitt-Verhulst. 1985. "Early Steps of Lymphocyte Activation Bypassed by Synergy Between Calcium Ionophores and Phorbol Ester." *Nature* 313:318–320.

58. Yamamoto, K. R., and B. M. Alberts. 1976. "Steroid Receptors: Elements for Modulation of Eukaryotic Transcription." *Ann. Rev. Biochem.* 45:721–746.

59. Payvar, F., O. Wrange, J. Carlstedt-Duke, S. Okret, J.-A. Gustafsson, and K. R. Yamamoto. 1981. "Purified Glucocorticoid Receptors Bind Selectively in Vitro to a Cloned DNA Fragment Whose Transcription Is Regulated by Glucocorticoids in Vivo." *Proc. Nat. Acad. Sci.* 78:6628–6632.

60. Metcalf, D. 1984. *The Hematopoietic Colony Stimulating Factors.* Amsterdam, London, and New York: Elsevier.

61. Till, J. E., and E. A. McCulloch. 1961. "A Direct Measurement of the Radiation Sensitivity of Normal Mouse Bone Marrow Cells." *Radiat. Res.* 14:213–222.

62. Bradley, T. R., and D. Metcalf. 1966. "The Growth of Mouse Bone Marrow Cells in Vitro." *Aust. J. Exp. Biol. Med. Sci.* 44:287–300.

63. Wu, A. M., J. E. Till, L. Siminovitch, and E. A. McCulloch. 1967. "A Cytological Study of the Capacity for Differentiation of Normal Hemopoietic Colony-Forming Cells." *J. Cell Physiol.* 69:177–184.

64. Abramson, S., R. G. Miller, and R. A. Phillips. 1977. "The Identification in Adult Bone Marrow of Pluripotent and Restricted Stem Cells of the Myeloid and Lymphoid Systems." *J. Exp. Med.* 145:1567–1579.

65. Johnson, G. R., and D. Metcalf. 1977. "Pure and Mixed Erythroid Colony Formation in Vitro Stimulated by Spleen Conditioned Medium with No Detectable Erythropoietin." *Proc. Nat. Acad. Sci.* 74:3879–3882.

66. Keller, G. M., and R. A. Phillips. 1982. "Detection in Vitro of a Unique Multipotent Hematopoietic Progenitor." *J. Cell Physiol. Suppl.* 1:31–36.

67. Goldwasser, E. 1975. "Erythropoietin and the Differentiation of Red Blood Cells." *Fed. Proc.* 34:2285–2292.

68. Heath, D. S., A. A. Axelrad, D. L. McLeod, and M. M. Shreeve. 1976. "Separation of the Erythropoietin-Responsive Progenitors BFU-E and CFU-E in Mouse Bone Marrow by Unit Gravity Sedimentation." *Blood* 47:777–792.

69. Metcalf, D. 1985. "The Granulocyte-Macrophage Colony-Stimulating Factors." *Science* 229:16–22.

70. Ihle, J. N., J. Keller, L. Henderson, F. Klein, and E. Palaszynski. 1982. "Procedures for the Purification of Interleukin-3 to Homogeneity." *J. Immunol.* 129:2431–2436.

71. Morgan, D. A., F. W. Ruscetti, and R. C. Gallo. 1976. "Selective in Vitro Growth of T-Lymphocytes from Normal Human Bone Marrows." *Science* 193:1007–1008.

72. Taniguchi, T., H. Matsui, T. Fujita, C. Takaoka, N. Kashima, R. Yoshimoto, and J. Hamuro. 1983. "Structure and Expression of a Cloned cDNA for Human Interleukin-2." *Nature* 302:305–309.

73. Cantrell, D. A., and K. A. Smith. 1984. "The Interleukin-2 T-Cell System: A New Cell Growth Model." *Science* 224:1312–1316.

74. Leonard, W. J., J. M. Depper, G. R. Crabtree, S. Rudikoff, J. Pumphrey, R. J. Robb, P. Kronke, P. B. Svetlik, N. J. Peffer, T. A. Waldman, and W. C. Greene. 1984. "Molecular Cloning and Expression of cDNAs for the Human Interleukin-2 Receptor." *Nature* 311:626–631.

75. Nikaido, T., A. Shimizu, N. Ishida, H. Sabe, K. Teshigawara, M. Maeda, T. Uchiyama, J. Yodoi, and T. Honjo. 1984. "Molecular Cloning of cDNA Encoding Human Interleukin-2 Receptor." *Nature* 311:631–635.

76. DeLarco, J. E., and G. J. Todaro. 1978. "Growth Factors from Murine Sarcoma Virus Transformed Cells." *Proc. Nat. Acad. Sci.* 75:4001–4005.

77. Roberts, A. B., C. Frolik, M. A. Anzano, and M. B. Sporn. 1983. "Transforming Growth Factors from Neoplastic and Non-Neoplastic Tissues." *Fed. Proc.* 42:2621–2626.

78. Holley, R. W., R. Armour, and J. H. Baldwin. 1978. "Density-Dependent Regulation of Growth of BSC-1 Cells in Cell Culture: Growth Inhibitors Formed by the Cells." *Proc. Nat. Acad. Sci.* 75:339–341.

79. Tucker, R. F., G. D. Shipley, H. L. Moses, and R. W. Holley. 1984. "Growth Inhibitor from BSC-1 Cells Closely Related to Platelet Type β Transforming Growth Factor." *Science* 226:705–707.

80. Iversen, O. H. 1981. "The Chalones." In *Tissue Growth Factors*, ed. R. Baserga. New York: Springer-Verlag, pp. 491–450.

81. Sager, R. 1985. "Genetic Suppression of Tumor Formation." *Adv. Cancer Res.* 44:43–68.

82. Granger, G. A., and W. P. Kolb. 1968. "Lymphocyte in Vitro Cytotoxicity: Mechanisms of Immune and Nonimmune Small Lymphocyte Mediated Target L Cell Destruction." *J. Immunol.* 101:111–120.

83. Carswell. E. A., L. J. Old, R. L. Kassel, S. Green, N. Fiore, and B. Williamson. 1975. "An Endotoxin-Induced Serum Fac-

tor That Causes Necrosis of Tumors." *Proc. Nat. Acad. Sci.* 72:3666–3670.

84. Pennica, D., G. E. Nedwin, J. S. Hayflick, P. H. Seeburg, R. Derynck, M. A. Palladine, W. J. Kohr, B. B. Aggarwal, and D. V. Goeddel. 1984. "Human Tumor Necrosis Factor: Precursor Structure, Expression and Homology to Lymphotoxin." *Nature* 312:724–729.

85. Beutler, B., D. Greenwald, J. D. Hulmes, M. Chang, Y.-C. E. Pan, J. Mathison, R. Ulevitch, and A. Cerami. 1985. "Identity of Tumor Necrosis Factor and the Macrophage-Secreted Factor Cachectin." *Nature* 316:552–554.

86. Levi-Montalcini, R. 1976. "The Nerve Growth Factor: Its Role in Growth, Differentiation and Function of the Sympathetic Adrenergic Neuron." *Prog. Brain Res.* 45:235–258.

87. Hynes, R. O. 1982. "Fibronectin and Its Relation to Cellular Structure and Behavior." In *Cell Biology of Extracellular Matrix,* ed. E. D. Hay. New York: Plenum.

88. Hynes, R. O., and K. M. Yamada. 1982. "Fibronectins: Multifunctional Modular Glycoproteins." *J. Cell Biol.* 95:369–377.

89. Folkman, J., and A. Moscona. 1978. "Role of Cell Shape in Growth Control." *Nature* 273:345–349.

90. Ben-Ze'ev, A., S. Farmer, and S. Penman. 1980. "Protein Synthesis Requires Cell-Surface Contact While Nuclear Events Respond to Cell Shape in Anchorage-Dependent Fibroblasts." *Cell* 21:365–372.

91. Green, H., and M. Meuth. 1974. "An Established Pre-Adipose Cell Line and Its Differentiation in Culture." *Cell* 3:127–133.

92. Taylor, S. M., and P. A. Jones. 1979. "Multiple New Phenotypes Induced in 10T1/2 and 3T3 Cells Treated with 5-Azacytidine." *Cell* 17:771–779.

93. Nixon, B. T., and H. Green. 1984. "Growth Hormone Promotes the Differentiation of Myoblasts and Preadipocytes Generated by Azacytidine Treatment of 10T1/2 Cells." *Proc. Nat. Acad. Sci.* 81:3429–3432.

94. Yaffe, D. 1968. "Retention of Differentiation Potentialities During Prolonged Cultivation of Myogenic Cells." *Proc. Nat. Acad. Sci.* 61:477–483.

95. Rheinwald, J. G., and H. Green. 1975. "Serial Cultivation of Strains of Human Epidermal Keratinocytes: The Formation of Keratinizing Colonies from Single Cells." *Cell* 6:331–344.

96. Gallico, G. G. III, N. E. O'Connor, C. C. Compton, O. Kehinde, and H. Green. 1984. "Permanent Coverage of Large Burn Wounds with Autologous Cultured Human Epithelium." *N. Eng. J. Med.* 311:448–451.

97. Abercrombie, M., and J. E. M. Heaysman. 1954. "Observations on the Social Behavior of Cells in Tissue Culture." *Exp. Cell Res.* 6:293–306.

98. Hynes, R. O., ed. 1979. *Surfaces of Normal and Malignant Cells.* New York: Wiley.

99. Pollack, R. E., M. Osborn, and K. Weber. 1975. "Patterns of Organization of Actin and Myosin in Normal and Transformed Cultured Cells." *Proc. Nat. Acad. Sci.* 72:994–998.

100. Matsumura, F., J. J. C. Lin, S. Yamashiro-Matsumura, G. P. Thomas, and W. C. Topp. 1983. Differential expression of tropomyosin forms in the microfilaments isolated from normal and transformed rat cultured cells. *J. Biol. Chem.* 258:13954–13964.

101. Leavitt, J., G. Bushar, T. Kakunaga, H. Hamada, T. Hirakawa, D. Goldman, and C. Merril. 1982. Variations in mutant α-actin expression accompanying incremental increases in human fibroblast tumorigenicity. *Cell* 28:259–268.

102. Hynes, R. O. 1973. "Alteration of Cell-Surface Proteins by Viral Transformation and by Proteolysis." *Proc. Nat. Acad. Sci.* 70:3170–3174.

103. Hakomori, S., 1986. Glycosphingolipid. *Sci. Am.* 254:44–53.

104. Temin, H. M. 1967. "Control by Factors in Serum of Multiplication of Uninfected Cells and Cells Infected and Converted by Avian Sarcoma Virus." *Wistar Inst. Symp.* 7:103.

105. DeLarco, J. E., and G. J. Todaro. 1978. "Growth factors from Murine sarcoma virus transformed cells." *Proc. Nat. Acad. Sci.* 75:4001–4005.

106. Sporn, M. B., and A. B. Roberts. 1985. "Autocrine Growth Factors and Cancer." *Nature* 313:745–747.

107. Cuttitta, F., D. N. Carney, J. Mulshine, T. W. Moody, J. Fedorko, A. Fischler, and J. D. Minna. 1985. "Bombesin-Like Peptides Can Function as Autocrine Growth Factors in Human Small-Cell Lung Cancer." *Nature* 316:823–826.

108. Linzer, D. I. H., and D. Nathans. 1983. "Growth-Related Changes in Specific mRNAs of Cultured Mouse Cells." *Proc. Nat. Acad. Sci.* 80:4271–4275.

109. O'Farrell, P. H. 1975. "High Resolution, Two-Dimensional Electrophoresis of Proteins." *J. Biol. Chem.* 250:4007–4021.

CHAPTER 26

The Genetic Basis of Cancer

For the first half of this century, much of cancer research was focused on identifying agents that can induce cancers in experimental animals. The first agents that were found were viruses. Eventually, certain chemicals and also radiation were shown to induce tumors in animals and transform cells in tissue culture. With a deeper understanding of the mechanism of action of viruses and of chemicals and radiation as mutagens, it became apparent that the majority of agents that can induce cancer act at the level of DNA. Today, it is widely accepted that cancer is a disease of malfunctioning cellular genes or unwanted viral gene expression. Until very recently, when cloning technologies allowed the isolation of cellular genes from spontaneously arising tumors, the only way to get at cancer-causing genes was to study those present in the genomes of certain DNA and RNA tumor viruses. The study of these genes and the mechanisms by which they transform cells provided a key to understanding the genetic basis of cancer.

AGENTS THAT CAUSE CANCER IN ANIMALS

Cancer as a Hereditary Change[1-4]

When a cancer cell divides, the two progeny cells are usually morphologically identical to the parental cell. The factors that give cancer cells their essential quality of unrestrained growth are thus regularly passed on from parent to progeny cells. These changes persist not only in tumors growing in intact animals but also in tumor cells growing in tissue culture. Hundreds of generations of growth can occur in tissue culture without appreciable reversion to a normal state. The permanence of such changes is shown not only by perpetuation of a typical morphology, but also by the ability of progeny cells to cause new tumors when injected into a tumor-free animal of genetic composition similar to the one from which the original tissue culture was obtained.

The heritability of the changes allowing unrestrained growth suggests that genetic changes within chromosomal DNA may underlie the cancerous phenotype. An early line of support for the notion of cancer as a genetic change came from experiments in which cancer cells were fused with normal cells. The 4N tetraploid cells that resulted from nuclear fusion often, though not always, had a normal phenotype and did not form tumors when injected into genetically identical hosts. Most 4N cells, however, generally are not stable and frequently divide to produce cells with lower chromosome numbers. Many of the subtetraploid cells that arose from noncancerous hybrids of normal and cancer cells were cancerous, indicating that the loss of one or more cellular chromosomes leads to the reexpression of the cancerous phenotype. Besides indicating that the cancerous phenotype is a stable property encoded in the genes of cancer cells, this result also implies that the phenotype is a recessive genetic trait, perhaps resulting from a loss of normal genetic material. Now, however, we know that some genetic changes leading to cancer are dominant while others are recessive. In some cases dominant genes have been identified that can make cells cancerous, and normal chromosomes that can suppress the cancerous phenotype have also been identified.

Somatic Mutations Are Probable Causes of Cancer[5]

The essential changes that make cells cancerous are now thought to be primarily somatic mutations, that is, mutations occurring in cells not destined to become sex cells. One line of evidence for this hypothesis comes from the fact that many cancers have one or more distinctive, abnormal chromosomes, usually resulting from the breakage and rejoining of two other chromosomes. In these cases, every cell in the tumor possesses the aberrant chromosome, but usually none of the individual's normal cells do. Presumably, the abnormal chromosome arises in a somatic cell and contributes to the cell's development into cancer. Since somatic mutations, including chromosome abnormalities, do occur, it would be surprising if some of them did not upset the normal control devices regulating cell division and thus cause cancer.

There is now substantial evidence that several somatic mutations are necessary to cause cancer. This idea arose in part from the fact that the incidence of cancer greatly increases with age. This phenomenon could be explained if a particular cell had to accumulate several mutations, each occurring randomly in time, before becoming a full-fledged cancer cell (Chapter 27).

Cancer Can Be Induced by Radiation[6]

Indirect evidence that somatic mutations are important causal factors in the induction of cancer also came from the realization that a very large number of the agents that increase the frequency of cancer are strong mutagens. Collectively, all cancer-causing agents are called **carcinogens**. Among the most potent carcinogens are the ultraviolet and ionizing forms of radiation, agents long known to be highly mutagenic. For example, exposure of skin to ultraviolet light causes skin cancer, while X-rays applied to the thyroid induce thyroid cancer. None of these various forms of radiation, however, can be used to cause all the cells in an exposed population to become cancerous. This is because they also cause many other deleterious changes, many of which lead to cell death. A radiation dose large enough to make virtu-

Figure 26-1
Benzpyrene (benzo[a]pyrene), benzanthracene, and aflatoxin B_1 (found in moldy peanuts).

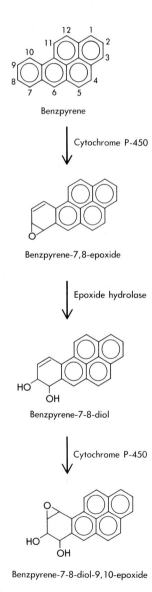

Benzpyrene

↓ Cytochrome P-450

Benzpyrene-7,8-epoxide

↓ Epoxide hydrolase

Benzpyrene-7-8-diol

↓ Cytochrome P-450

Benzpyrene-7-8-diol-9,10-epoxide

Figure 26-2
Conversion of benzpyrene by enzymes of the liver into a carcinogenic (mutagenic) epoxide derivative.

ally all cells cancerous would also kill practically every cell. Not only can radiation cause simple mutations, but the ionizing form also breaks chromosomes, frequently resulting in large deletions of genetic material. Thus, radiation causes a variety of genetic changes, many of which may contribute to the cancerous state.

The deleterious effect of the ultraviolet component of sunlight on skin is exaggerated in certain individuals with a rare inherited disorder called *xeroderma pigmentosum*. These people are homozygous for a recessive mutation that inactivates a gene involved in repairing thymine dimers, the major damage caused by ultraviolet radiation, and so they develop multiple skin cancers. Presumably sufferers of xeroderma pigmentosum are unable to repair somatic mutations, some of which cause a cell to become cancerous. The increased incidence of skin cancers in people with xeroderma pigmentosum provides further support for the mutational origin of cancer.

Chemical Carcinogens Can Be Converted into Strong Mutagens in Vivo[7,8,9]

An extremely diverse set of seemingly unrelated organic and inorganic compounds have long been known to be potential carcinogens, but a firm connection between many chemical carcinogens and mutagens did not emerge until the 1960s. While certain compounds (e.g., the nitrogen mustards) that mimic ionizing radiation in their physiological consequences had long been known to be both mutagens and carcinogens, many of the most powerful carcinogens long appeared devoid of any mutagenic potential. Particularly striking was the absence of any apparent mutagenicity within a variety of highly carcinogenic fused-ring hydrocarbons, such as benzpyrene, benzanthracene, and aflatoxin (Figure 26-1). These have no special affinity for DNA, and it was therefore postulated that their carcinogenicity somehow resulted from their binding to and inactivation of key proteins. More recent evidence, however, shows that when these carcinogens enter certain cells, they become metabolized into derivatives that are not only highly mutagenic but also much more powerfully carcinogenic than their precursors. For example, benzpyrene itself causes no genetic changes, but in cells is converted into the very potent mutagen benzpyrene-7,8-diol-9,10-epoxide (Figure 26-2).

Much of the mutagenicity of such activated carcinogens arises from their ability to chemically modify DNA. These modifications lead to base substitutions during DNA replication and also to breaks in the polynucleotide backbones, which may in turn generate more extensive genetic rearrangements. In general, compounds that are transformed into strong carcinogens likewise become strong mutagens, and vice versa. There thus seems to be little doubt that much carcinogenesis is the result of changes in DNA.

Long unexplained were strikingly different responses to the same carcinogen often shown by different individuals within a species. Recent evidence indicates that such differences are due in part to genetically determined levels of the enzymes that metabolize these carcinogens. Among the most important of these carcinogen-metabolizing enzymes are the cytochromes P-450, a family of hydroxylases. While these enzymes act on most hydrocarbons and other foreign compounds to produce harmless metabolites, occasionally their metabolites are chemically reactive and, as mentioned above, subsequently react with DNA. Now there is intense interest in the regulation of expression of the cytochromes P-450. Understanding this regulation may help to explain the wide variation of metabolizing activity within the human population. Study of this variation is commonly referred to as pharmacogenetics.

The discovery that carcinogens are often activated to mutagens by enzymes present in the liver led to the development of several important assays for the potential carcinogenicity of various chemicals. The first and still most widely used of these is the Ames test (Figure 26-3; see also Chapter 12). In this test, the ability of a compound to be activated to a mutagen is estimated by incubating an indicator strain of bacteria with the compound in the presence of an extract of liver cells. The most frequently used bacteria are a strain of *Salmonella* with a mutation at the *his* locus. Bacteria that can grow and form colonies on histidine-free medium arise from revertants at this locus induced by the carcinogen, which in turn was activated by the enzymes present in the liver extract. Since almost all known carcinogens are revealed as mutagens in this test, it is now used extensively for preliminary screening of compounds that will come in contact with humans, replacing much more expensive and time-consuming animal testing.

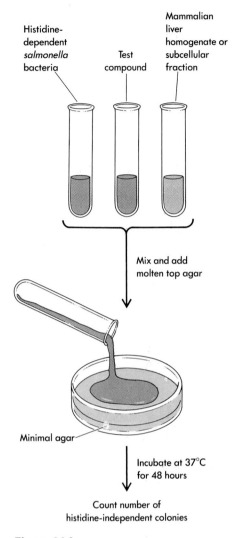

Figure 26-3
Schematic representation of the bacterial mutagenesis assay used in the Ames test.

Tumor Promoters Act Synergistically with Carcinogens[10]

While most of the chemicals described above are probably carcinogenic because they induce changes in DNA (either directly or after chemical modification), other chemicals contribute to cancer induction by a different mechanism. One group of such agents was first identified in experiments involving carcinogenesis of mouse skin. If mouse skin is painted with a carcinogen like benzpyrene, then few tumors will develop, even with multiple treatments. If, however, a single exposure to benzpyrene is followed by multiple applications of one of a variety of agents, such as the phorbol ester *12-O-tetradecanoyl phorbol-13-acetate (TPA)* (Figure 26-4), which itself is not a carcinogen, tumors will eventually develop. In this situation, the benzpyrene is

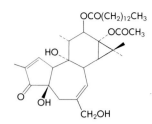

Figure 26-4
The phorbol ester tumor promoter 12-O-tetradecanoyl phorbol-13-acetate (TPA).

referred to as an *initiator* and the TPA as a *promoter*. Initiators are carcinogens that act by inducing mutations in the cell's DNA. Promoters, on the other hand, are not mutagens and alone they never induce tumors. Rather they mimic the effects of growth factors in that they transiently stimulate both cell division and altered gene expression (Chapter 25). In the case of phorbol esters this is thought to occur by activation of the cellular enzyme protein kinase C, which acts as a receptor for phorbol esters. Protein kinase C, which can phosphorylate appropriate substrate proteins, is normally activated by phospholipids, Ca^{2+}, and diacylglycerol (Chapter 25). However, phorbol esters can substitute for diacylglycerol (with which they have some structural homology), apparently by increasing the affinity of protein kinase C for Ca^{2+}. Since the role of protein kinase C in cell growth is still poorly understood, how its activation may cause an initiated cell to become a malignant one is not known. Perhaps it allows the initial mutation to become fixed in a cell and its descendants and provides a setting appropriate for the occurrence of additional mutations.

Viruses as a Cause of Cancer[11,12]

Athough the first known tumor virus, Rous sarcoma virus, was discovered in 1911, over 70 years ago, examples of additional viruses that could induce tumors in animals at first appeared so infrequently that it was generally believed that viruses could not be a major cause of cancer. But as soon as newborn animals began to be used as test systems during the early 1950s, it became much easier to show that a virus has oncogenic potential. Also initially preventing many people from accepting a viral causation of cancer was the fact that electron micrographs do not reveal the presence of viruses in most tumor cells. Now, however, we realize that under many circumstances, the genetic material of a virus may be inserted into a host chromosome in a "proviral" form without any production of virus particles. Thus, the absence of detectable viruses does not provide evidence either for or against a viral origin for a given cancer cell. We also realize now that not all viruses spread horizontally as highly contagious agents and also that viruses that are widespread in the population may contribute to cancer in only a small percent of infected individuals, so that epidemiological evidence alone cannot easily rule out viral involvement in many cancers. Furthermore, the latent period between infection and appearance of the cancer is often far greater with virus-induced cancer than with many other infectious diseases, so that the connection may be very obscure.

Viruses and Chemicals Can Transform Cells in Culture[13,14,15]

It is highly desirable to use cell culture systems to study the transformation of normal cells into malignant cells. Fortunately, some tumor viruses and chemical carcinogens can induce the appearance of cells with altered phenotypic characteristics consistent with properties of cancer cells (Chapter 25). Of particular importance from an experi-

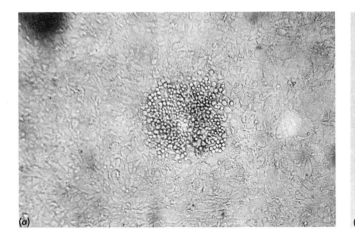

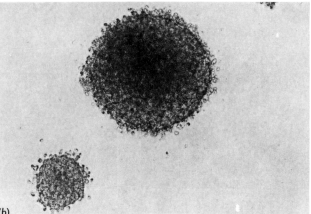

mental standpoint are the alterations in morphology and the loss of various kinds of growth control, since these allow transformed cells to be recognized and separated from their normal neighbors. Most carcinogenic agents (in contrast to some very efficient tumor viruses) can transform only a very small fraction of the cells exposed to them (sometimes one in a million, or less), so analysis of the changes they induce would be impossible if the transformed cells could not be readily isolated.

Transformed cells are usually isolated in one of two ways: either as **foci** of cells arising from a single transformed cell that overgrows its neighbors, or as a colony of cells grown in suspension culture (Figure 26-5). These two methods rely on two different specific alterations: loss of **contact inhibition** of growth in the case of foci, and loss of **anchorage independence** of growth in the case of suspension cells. A third property that can be used to isolate transformed cells in culture is their immortality. Primary cultures of animal cells usually have a limited life in culture and die after a maximum of about 50 generations. Many tumor cells and transformed cells, by contrast, will grow indefinitely in culture. A fourth property of transformed cells that permits their selective growth relative to normal cells is their ability to grow in low concentrations of serum. These four properties of altered growth do not always occur together during transformation, and some are due to the action of distinct genes.

Figure 26-5
The two most frequently used assays for transformation of animal cells by tumor viruses. (a) Focus assay. Photograph of a group (focus) of chicken cells transformed by RSV infection. The spherical transformed cells are easily distinguished from the background of normal cells. (Courtesy of H. Rubin, Department of Molecular Biology, University of California, Berkeley.) (b) Colony formation in agar or methocel. A colony of rat embryo cells (REF52 line) transformed by the *ras* and adenovirus *E1A* genes. Individual cells can be seen at the borders of the colonies. (Photograph by Kazuo Maruyama.)

Use of Newborn Animals and Nude Mice to Demonstrate the Oncogenic Potential of Cancer Cells and Viruses[16,17]

A breakthrough in studying the tumorgenicity of viruses and virus-transformed cells was the use of very young, even newborn, animals. The reason is that virus-coded proteins are usually expressed on the surface of virus-transformed cells, making the cells targets for destruction by the immune system. In many species, the immunological system does not become very effective until just after birth, so in newborn animals, the oncogenicity of the virus can be detected in the absence of immunological interference.

Certain tricks can be used to temporarily suppress the immune system and thus study the oncogenicity of antigenic tumor cells in older animals as well. Large doses of ionizing radiation or immunosuppressive drugs like the cortical steroids can be applied before cells of uncertain oncogenic potential are to be tested. In recent years, increasing use has been made of *nude mice,* a mutant strain of bald mice that congenitally contain very little thymus tissue and so are incapable of mounting the strong T cell responses involved in rejecting tumor cells. In these mice, the tumorgenicity of cells (including human cells) that have different histocompatibility antigens on their surface can be tested. Interestingly, even when tumor cells are injected into genetically identical animals or into nude mice, it is usually necessary to inject over a million tumor cells to provoke a tumor. Whether this is due to residual immunological responses or to other factors that involve the survival of tumor cells in a new environment is unclear.

Antigens that appear on the cell surface and that mediate tumor rejection are called **tumor antigens** or, if they are unique to tumor cells and absent from normal cells, **tumor-specific transplantation antigens** (**TSTAs**). Tumors induced by the same virus often possess the same TSTA, while different viruses induce distinct TSTAs. It is possible to determine whether two different tumors have the same TSTA by determining whether killed cells of one tumor can elicit immune protection against live cells of a second tumor. First, the killed cells of a tumor are used to immunize an animal; then, several weeks later, live tumor cells of a second tumor are injected into the animal. If the prior immunization leads to decreased growth (transplantation rejection) of the live cells, then the two tumors are said to share TSTA. If there is no effect on the growth of the live tumor cells, then the two tumors are considered to possess distinct TSTAs.

An interesting finding is that tumors induced in rodents by chemicals or radiation possess strong tumor antigens and that each antigen is unique. The molecular basis of these carcinogen-induced TSTAs and thus the basis for their diversity is unknown.

A question of considerable importance is whether spontaneously arising tumors, particularly in humans, possess tumor-specific surface antigens and thus whether the immune system plays an important role in protecting us from cancer. At present, the best evidence for **immune surveillance** against cancer comes from studying the incidence of cancer in individuals whose immune system is defective, either as a result of a genetic defect, as a result of deliberate drug-induced suppression to prevent rejection of transplanted organs, or as a result of diseases such as AIDS that destroy immune function. The finding is that the most common forms of cancer are not more frequent in these individuals but that they have significantly higher incidences of cancers that are not widely seen in the immunologically competent population. These include Kaposi's sarcoma and B cell lymphoma. Now it is suspected that these tumors may have a viral origin and so possess tumor antigens that normally cause their rejection. In contrast, more common cancers may not possess strong TSTAs, and this may be why we are more susceptible to them. In any case, it is clear that the evolution of vertebrates with their relatively long lives must have been dependent on the development of an immunological response against certain forms of cancer.

Viral Transformation Is Mediated by Oncogenes[18]

Although studies of carcinogen- and radiation-induced tumors have yielded much information on the chemical mechanisms involved in the alteration of DNA, until recently they revealed nothing about the specific genes responsible for transforming normal cells into cancer cells. Far more insight into these genes came first from the study of viruses that cause cancer in experimental animals and transform cells in culture. This is because the viruses usually carry the cancer-inducing genes, or *oncogenes*, as part of their genome.

So far, viruses that can cause cancer and also transform cells in culture include several groups with DNA genomes—papovaviruses (SV40, polyoma, and papilloma viruses), adenoviruses, and herpesviruses—as well as the RNA-containing retroviruses. In general, the oncogenes of DNA viruses are virus-coded gene products that are also required for the normal replication of the virus. In contrast, most of the retrovirus oncogenes that have been studied are not required for replication. Rather, they are cellular genes that were transduced by chance onto the viral genome. Normal cellular genes with the potential to become oncogenes are called **proto-oncogenes** or **cellular oncogenes (c-oncs)**.

Besides carrying oncogenes, tumor viruses that induce stable transformation usually must have the capacity to form stable associations with the infected cell, often with the cell's genome. In addition, they must not impair the ability of the cell to divide. As discussed in Chapter 24, these characteristics are typical of retrovirus infections. With most oncogenic DNA viruses, however, integration of viral DNA into the cell genome is not part of the normal replication cycle, and replication of the virus usually leads to cell death. In this case, stable transformation occurs only when the cell is nonpermissive for virus DNA replication or when the viral genome is damaged so that replication cannot occur. Also, in this case, integration of viral DNA occurs only rarely and by mechanisms that are not well understood.

DNA TUMOR VIRUSES

The Small DNA Viruses SV40 and Polyoma Induce Tumors in Newborn Animals[11,19,20]

Polyoma and SV40 were discovered by accident about 30 years ago. Polyoma turned up as a contaminant in an extract from a leukemic mouse. Rather than inducing leukemias when injected into newborn mice, the extract induced parotid (salivary gland) adenocarcinomas. The active agent was identified as a virus and named polyoma ("many tumors") because of its ability to induce tumors occasionally in the thymus, ovaries, mammary glands, adrenals, and liver, as well as inducing the more common parotid tumors. Polyoma is common in wild mice, where it apparently behaves as a harmless passenger in adults. It causes tumors only when injected into newborn mice and also newborn hamsters, rats, and rabbits.

SV40 was a contaminant in the rhesus monkey cell cultures used to prepare the Salk and Sabin polio vaccines in the late 1950s. Although it has no visible effects on rhesus monkey cells in culture, it kills cultured cells derived from African green monkeys. Ironically, monkey cells were used to prepare polio vaccines because human cells were considered more likely to harbor potentially dangerous viruses! Although SV40 was soon found to induce tumors when injected into newborn hamsters, no one has ever found evidence that it induced cancer in the early recipients of polio vaccine.

Lytic Versus Transforming Response[21,22]

To be transformed by a virus, a cell must survive the infection in good health; however, papovaviruses are frequently lytic. Cells in which a specific virus multiplies are called *permissive* cells, whereas cells in which viral growth does not occur are known as *nonpermissive* cells. Permissive cells are usually derived from an animal that is the natural host for a virus. In contrast, nonpermissive cell lines usually originate from animals incapable of multiplying a given virus. Thus, depending on the specific virus, a cell can be either permissive or nonpermissive. For example, the very well known 3T3 mouse cell lines are permissive for polyoma, a mouse virus, but nonpermissive for SV40, a monkey virus. When SV40 infects a 3T3 cell, it never multiplies but occasionally induces cell transformation. When polyoma infects the nonpermissive cell line BHK (baby hamster kidney), it can transform but not multiply, and so BHK cells are frequently used in polyoma transformation experiments.

Both the lytic and transforming responses require that only a single virus particle infect a cell. This can be deduced from the fact that when SV40 particles are added to cells, the number of cells transformed (or lysed) is directly proportional to the amount of input virus. Provided enough particles are added, virtually all the exposed cells can be transformed, indicating that all the cells growing in the culture are capable of being transformed. The probability of a single SV40 particle causing transformation, however, is very, very low. Some tens of thousands of infecting particles ordinarily must be present to give the average cell a high probability of conversion to the cancer state.

Only the Early Region of the SV40 or Polyoma Genome Is Needed for Cell Transformation[23–28]

Lytic growth of SV40 or polyoma on permissive cells can be divided into an early stage, which lasts some 8 to 10 hours and culminates in the onset of viral DNA replication, and a late stage, which commences with viral DNA replication, involves synthesis of the viral capsid proteins that package the viral DNA, and terminates with the destruction of the cell. In the early stage of the cycle, only about half the viral genome, the so-called early region, is transcribed. SV40 has two early gene products, large and small T proteins, while polyoma has three, the large, middle, and small T proteins.

The SV40 and polyoma early gene products are involved in viral DNA replication, both directly by acting on viral DNA and indirectly by inducing cellular genes needed for DNA replication. Soon after viral infection of either permissive or nonpermissive cells, there is a dramatic increase in the amount of many of the host enzymes involved in DNA synthesis (Figure 26-6). This increase occurs even when SV40 infects G_1-blocked cells whose DNA synthesis has stopped. Growth in such inhibited cells can occur despite the fact that before infection, these cells lack many of the enzymes (such as CDP reductase, dTMP synthetase, and DNA polymerase) used to make various DNA nucleotide precursors and to link them together. Thus, successful SV40 infection must somehow unlock the cellular control device that normally shuts off the synthesis of the DNA enzymes when their presence is no longer wanted. The ability to induce these

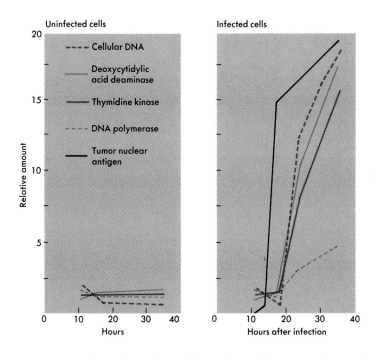

Figure 26-6
Induction of enzymes involved with DNA synthesis following infection of permissive monkey cells by SV40 virus. Before these cell-coded enzymes appear, synthesis of the virus-specific large T protein has begun. (After Dulbecco, *Sci. Amer.*, April 1967.)

enzymes means the virus can replicate in quiescent cells as well as cells that are already dividing. In the case of SV40, it is the large T antigen that is responsible for the induction of host enzymes, and purified T protein microinjected into cells can transiently produce the effect. Induction of the DNA enzymes initiates not only SV40 DNA synthesis, but also synthesis of host DNA.

Only the early region of the SV40 (or polyoma) viral genome is needed to transform cells. Numerous temperature-sensitive mutants have been isolated that prevent viral growth at high temperature. Among these mutants, those located in late genes do not block cellular transformation, while many of those in early-region genes do. Even more convincing are experiments in which fragments of the SV40 (or polyoma) genome prepared by restriction endonuclease digestion of viral DNA are introduced into cells. Fragments that encompass the early region of the genome are sufficient to induce transformation.

Cells that are nonpermissive for growth of SV40 or polyoma can only express viral early genes. Furthermore, viral DNA replication can not occur in these cells. It is the block to a successful lytic cycle and thus the failure to be lysed that allows nonpermissive cells to become transformed into cancer cells. Very rarely, SV40 (or polyoma) succeeds in transforming permissive cells. In such cases, the virus involved turns out to carry mutations that block its ability to undergo DNA replication. Without these mutations, the cell would be killed and transformation would not occur.

SV40 or Polyoma DNA Is Integrated into Chromosomal DNA in the Transformed Cell[29]

One can imagine a number of different mechanisms by which viruses might be carcinogenic. For example, the transient expression of viral gene products in a cell might initiate a cascade of events, including altered cellular gene expression, that end in transformation. Alternatively, viruses whose genomes integrate into cellular DNA might

Figure 26-7
Schematic illustration of how SV40 may integrate into a host chromosome to produce a transformed cell. Integration within the late region has the important consequence that intact late SV40 mRNA cannot be synthesized. In contrast, intact early mRNA molecules can be made, and so the cell is transformed.

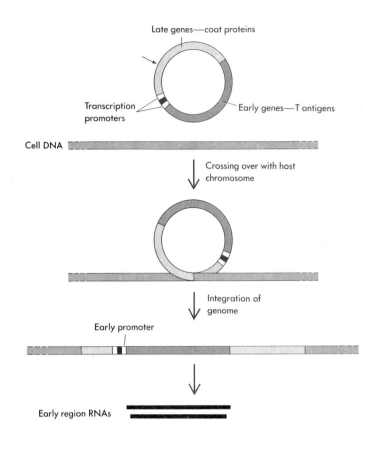

cause cancerous mutations in cellular genes in the process of integration. Yet another possibility is that viral genes become permanently associated with the cell either as independently replicating plasmids or as genes integrated into the host chromosomes and that their gene products are needed continuously to maintain the cancerous phenotype of the cell. SV40 and polyoma transform cells by the last mechanism.

DNA-RNA hybridization studies first revealed that cells transformed by SV40 or polyoma always contain viral DNA integrated into chromosomal DNA. The integrated SV40 DNA is present within the nucleus covalently bound to chromosomal DNA. The number and organization of integrated viral genomes vary from cell line to cell line, with different transformed cells containing from 1 to 20 SV40 genomes. The site of SV40 (or polyoma) DNA integration within the cell's DNA is random and does not involve sequence homology with the virus. Furthermore, when whole viral genomes integrate, the integration point on the circular viral DNA is random. However, only integration events that interrupt viral late genes can produce transformed cells, since the early genes must be intact and functioning to achieve transformation (Figure 26-7).

Stable transformation is a fairly rare event requiring large numbers of infecting virus particles. Far more common, however, is abortive transformation. In this case, viral infection of nonpermissive cells is followed by early gene expression, and the cell acquires a transformed appearance, but only transiently. Then the cell reverts to a normal phenotype and viral DNA is lost. Thus, integration of viral DNA is not necessary for transformation per se, but it is necessary for stable transformation.

Tandem copies of SV40 DNA

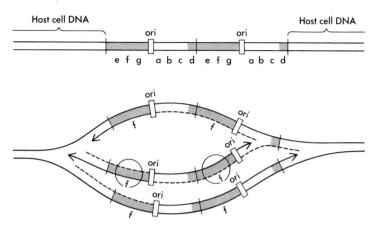

Figure 26-8
Tandem copies of SV40 DNA that could release complete viral genomes by recombination, and replication of the integrated SV40 DNA by the onionskin model following fusion of a transformed nonpermissive cell with a permissive cell. Recombination at identical sequences within tandem copies of the SV40 genome, for example at the two sequences within the circles, would release a complete viral genome that could replicate normally. (After P. Bullock and M. Botchan, *Gene Amplification.* Edited by R. T. Schimke, Cold Spring Harbor Laboratory. 1982. 215–224.)

Infectious Particles Are Liberated Following Fusion of Some Transformed Nonpermissive Cells with Nontransformed Permissive Cells[30,31,32]

Although SV40 (or polyoma)-transformed cells may contain a complete viral genome, replication of free viral DNA does not occur. That the viral genome is competent for replication is shown by rescue of infectious virus particles following fusion with normal permissive cells. For example, when mouse 3T3 cells transformed by complete SV40 viral genomes are fused with monkey cells, many of the fused cells yield SV40 progeny within 24 hours. Thus, something present in the permissive cell allows SV40 to carry out a complete multiplication cycle. This result is somewhat surprising, since most integrations of the SV40 genome interrupt late genes essential to the viral life cycle (see Figure 26-7). Furthermore, to yield infectious virus, a complete SV40 genome must be able to excise from the host chromosome and recircularize. Probably this can occur when more than one SV40 genome is present in tandem (Figure 26-8), a situation that could result from integration of a circular dimer. Recombination within such an integrated structure could release a complete circular viral genome, capable of replicating in the environment provided by the permissive cell. In addition, fusion with a permissive cell is thought to result in the "onionskin replication" of SV40 genomes, in which initiation occurs at the usual SV40 origin but on viral DNA still trapped within the host chromosome (Figure 26-8). The multiple copies of SV40 DNA that would result from these extra initiations would provide additional substrates for recombination events that could release complete viral genomes.

Two or Three Polyoma-Coded Early Proteins (One or Two SV40 Proteins) Maintain the Transformed State[22,33–37]

Not only is polyoma (or SV40) DNA invariably present in virus-transformed cells, but viral mRNA and early proteins can always be detected as well. These observations originally suggested that the continuing presence of viral gene products may be necessary to maintain transformation. Proof came when it was shown that when SV40 (or polyoma) early gene products are inactivated, the cell loses its trans-

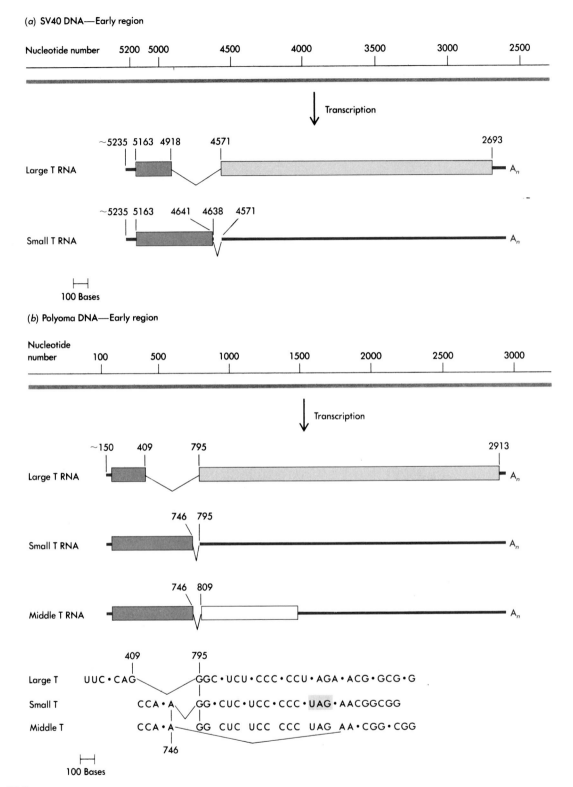

Figure 26-9

Early regions and early-region transcripts of SV40 and polyoma. (a) SV40 early region. Nucleotide numbers shown on the DNA are those assigned from the nucleotide sequence of the genome. The transcripts for large and small T proteins are shown just prior to splicing. Boxes represent coding portions of RNAs; lines represent untranslated portions. The splice in large T RNA joins two segments in different reading frames; the splice in small T RNA occurs after the stop codon that terminates translation of small T. [Based on data from V. B. Reddy et al., *Science* 200 (1978):494.] (b) Polyoma early region. All three reading frames are used to encode small, middle, and large T proteins. The nucleotide sequences at the splice points of the three transcripts are shown below the figure. The last few amino acids of small T are encoded by sequences lying just beyond the splice acceptor. [After Rassoulzadegan et al., *Nature* 300 (1982):713.]

formed phenotype. Inactivation of the viral early gene products can be accomplished if the virus used to transform the cell carries a temperature-sensitive mutation. The first successful experiments involved temperature-sensitive (ts) mutants in SV40 large T protein. Cells transformed at low temperature by some ts mutants of SV40 large T lose their transformed phenotype when placed in high temperatures and become transformed again when the temperature is lowered. This result, also accomplished for a number of other tumor virus transforming genes, is most revealing. Given the innumerable biochemical changes that distinguish normal cells from cancer cells, we might have thought that once begun, the tangle of events that result in a cancer cell could not be reversed. Instead, a few critical proteins may be sufficient to maintain these many changes.

More than 20 years have already gone into identifying the early gene products of SV40 and polyoma and determining which are needed for transformation and why. Despite this effort, we still do not have all the answers. Much of this work took place before the discovery of RNA splicing and overlapping genes. Now, sequencing of cDNAs from the early polyoma and SV40 mRNAs, in vitro translation of early mRNAs, and direct analysis of the proteins specified by the early mRNAs have led to the maps of the viral early regions shown in Figure 26-9. The three polyoma virus early proteins—small, middle, and large T—share amino-terminal amino acids, but differ in their carboxyl termini because of different splicing patterns employing different overlapping reading frames. The two SV40 early proteins—large and small T—also share amino-terminal amino acids and differ at their carboxyl termini. In this case, however, small T protein-coding sequences are contiguous, with a splice in the mRNA occurring after the stop codon that terminates small T protein.

The extensive overlapping of the polyoma and SV40 early genes has made it difficult to isolate viral mutants that affect just one gene product. Consider, for example, the much-studied polyoma mutants called **hr-t mutants.** They alter or abolish both small and middle T proteins, and although these mutants can still replicate in certain cell types, they are unable to transform cells. This result indicates that either small T, middle T, or both are involved in transformation. Also complicating the analysis is the finding that different cell types respond differently to the viruses. For example, deletion mutants that affect the SV40 small T protein can transform rat cells to produce dense-growing, morphologically altered cells, but the cells are unable to grow in the semisolid medium methocel. However, when hamster cells are used, these same mutants transform just as well as wild-type virus. While analysis of viral mutants has implicated all three polyoma early proteins and the two SV40 T proteins in the transformation of at least some cell types, a deeper understanding of the role of these proteins in transformation has come only recently from studying the transforming ability of the individual proteins separately.

Polyoma Large T Protein Immortalizes Cells, and Polyoma Middle T Protein Transforms Immortalized Cells[38-41]

The solution to separating the functions of the SV40 (and polyoma) transforming proteins from one another was accomplished by the technique of cDNA cloning, that is, making a DNA copy of an mRNA using reverse transcriptase and then cloning this DNA in a plasmid vector. Since the different T proteins are translated from differently spliced RNAs, any cDNA derived from an early-region mRNA can encode only one of the T proteins.

When a cDNA clone of the polyoma virus middle T mRNA was joined to a transcriptional promoter within an appropriate vector, the resulting construct was able to transform an established rat cell line in culture, and the transformed cells were tumorigenic when injected into animals. At first, this result was thought to mean that polyoma middle T is the only viral gene required for transformation and tumorigenicity. Then, when the same experiment was repeated using primary cells instead of established cell lines, middle T was unable to induce transformation. However, when a cDNA clone of polyoma large T mRNA was introduced into the cells along with the middle T cDNA clone, the primary cells were transformed. Infection with SV40 or polyoma virus has long been known to establish primary rodent cells into permanent cell lines as well as to transform them. In fact, permanently growing transformants would not be obtained if this were not the case, since primary cells in culture are destined to die off within 10 to 20 doublings. Alone, the cDNA clone of polyoma large T antigen can establish primary cells into perpetually growing cell lines that require much lower concentrations of growth factors than normal cells to divide. These cells, however, are not transformed in appearance and are not tumorigenic: Middle T protein is needed to bring about the transformation of the established cells. These experiments demonstrated that in polyoma, different genes bring about different aspects of the transformed phenotype, and they provided one of the first insights into why multiple mutations may be needed to produce a cancer cell.

The cDNA clones of polyoma small T also affect cell behavior in culture, causing cells to adhere less firmly to plastic substrates without altering their morphology significantly. The role of small T in creating a fully cancerous phenotype is still unclear, however, since middle and large T proteins together seem capable of inducing full transformation in at least some cell types.

Polyoma large T protein is a phosphoprotein of about 100 kdal. It is located almost exclusively in the nucleus of both infected and transformed cells. Only the amino-terminal 40 percent or so of large T is required for immortalization. Like SV40 large T protein, it is a DNA binding protein that regulates several key steps in the virus replication cycle, involving early and late transcription and viral DNA replication (Chapter 24).

Polyoma middle T protein has a molecular weight of about 56 kdal. It is phosphorylated, and at least some of it is located at the inner surface of the plasma membrane. In fact, middle T may have to be anchored in the membrane to be active, since mutants defective in membrane attachment are defective in their ability to transform. Polyoma middle T coprecipitates and copurifies from cells with a cellular protein called $pp60^{c\text{-}src}$. This protein is a tyrosine-specific kinase, and recent studies indicate that interaction with middle T may alter its enzymatic activity. As discussed later in this chapter, *src* (pronounced "sarc") is a proto-oncogene that can be mutated to become a transforming gene, and now it seems possible that the transforming activity of middle T is the result of its interaction with the c-*src* gene product.

The biochemical properties of the polyoma small T protein are only poorly understood, and as already noted, the protein may not always be required for transformation. Small T seems to be found in the nucleus.

SV40 Large T Protein Both Immortalizes and Transforms Cells[42,43]

SV40 and polyoma DNAs do not cross-hybridize significantly, so until they were sequenced and found to encode related proteins, it was not realized how similar the two viruses are. Their early T proteins share extensive amounts of identical amino acids. SV40, however, has no middle T protein, and comparison of the DNA sequences of SV40 and polyoma shows that SV40 lacks the sequence that encodes the overlapping parts of polyoma small and middle T. Since polyoma middle T protein plays such a critical role in cell transformation, inducing both morphological alterations and anchorage-independent growth, in SV40 these functions must be taken on by SV40 large T (and possibly small T).

SV40 large T is a protein of about 90 kdal and most of it is located in the nucleus. The number of functions so far attributed to this one protein is remarkable: SV40 large T protein binds specifically to the SV40 genome (Chapter 24), playing a role in initiation of viral DNA replication and in control of early and late gene transcription; a C-terminal portion of the protein can serve a helper function for the translation of adenovirus late mRNAs in monkey cells that otherwise are nonpermissive for adenovirus growth; the protein has intrinsic ATPase activity; as already noted, SV40 large T induces the expression of host enzymes involved in DNA replication and causes a round of cellular DNA replication; and finally, a small amount of SV40 large T probably protrudes through the plasma membrane, where it serves as a tumor-specific transplantation antigen.

As with polyoma middle T, there is evidence that the transforming activity of SV40 large T protein might be exerted through its association with a cellular protein, in this case p53. A portion of large T purifies with the nuclear p53 protein. This association leads to increased stability of p53, thereby increasing the amount of p53 present in cells. At present, the normal role of p53 is unknown, and the protein has no known enzymatic function. There are, however, several lines of evidence that together suggest that the association with large T is important for transformation. First, the same cellular protein is also associated with another, completely different viral oncogene product—that of the adenovirus E1B gene, in adenovirus infected cells. Second, p53 is expressed at abnormally high levels in certain tumors. Third, under some conditions, the p53 gene can behave as an oncogene: When it is manipulated so that it is expressed at a high level, it can transform certain cell types.

Because there are so many functions in the SV40 large T polypeptide, as well as overlapping sequences coding for parts of little and large T, it is very difficult to use genetics to determine which of the activities of large T are responsible for cell transformation. Analysis of deletion mutants confirms that the ability of large T to induce S phase DNA synthesis in resting cells is a necessary but not sufficient activity for inducing transformation. It certainly makes sense that unregulated expression of a viral protein with the ability to drive a cell into S phase might be a transformation function, perhaps the basis for immortalization.

At present, there is no obvious function of SV40 that resembles the function of polyoma middle T. As with polyoma, SV40 small T is not required for transformation of many cell types.

SV40- and Polyoma-Transformed Cells Have Strong Tumor-Specific Transplantation Antigens[44,45,46]

The inability of SV40 and polyoma to induce tumors in adult animals is almost certainly because both viruses induce strong antigens on the surface of cells they transform. Some SV40 (or polyoma)-transformed cells can be serially passed in adult animals, but large numbers of cells must be used to induce progressive, fatal tumors. As expected, the SV40 transplantation antigens are distinct from those of polyoma but common to all SV40-transformed or tumor cells (likewise for polyoma). Large T protein is the SV40 TSTA, since injection of purified protein into an animal can induce immunity to SV40-transformed cells or to tumor induction by SV40 virus. By constructing deletions of the SV40 large T coding region, it has been shown that determinants of the SV40 TSTA are scattered along the large T protein. Why some of large T is stuck into the cell membrane and whether this is critical to cell transformation remain a mystery. In any case, its presence on the cell surface probably protects certain species of adult animals from what might otherwise be common and fatal tumors.

Only About 8 Percent of the Adenovirus DNA Genome Is Needed to Transform Cells[47,48]

Also intensively studied for their transforming ability are the larger DNA-containing adenoviruses. There are many isolates of human adenoviruses, and they normally multiply to produce common coldlike symptoms. But they can also transform nonpermissive or semipermissive rat and hamster fibroblasts and epithelial cells in culture, and some of them can induce tumors when injected into newborn hamsters. The adenovirus genome is a linear, double-stranded DNA molecule of about 35 kilobase pairs packaged in an icosahedral protein capsid. The molecular biology of adenovirus replication and gene expression was described in Chapter 24, and here we will focus only on the region of the genome involved in transformation and tumorigenicity.

Like SV40 and polyoma, adenovirus transformation involves stable integration of viral DNA into random locations in chromosomal DNA. Rat and hamster cells transformed by adenoviruses usually contain incomplete viral genomes. Since adenovirus lytic infection results in a complete takeover of host cell functions, including ultimately the shutoff of host protein synthesis and DNA replication, it may be that even in nonpermissive cells, a number of viral genes must be deleted to prevent killing the cells and thus allow stable cell transformation.

Adenovirus transformants always contain a segment of adenovirus DNA derived from the extreme left-hand end of the viral genome, often in multiple tandem arrays. Frequently, but not consistently,

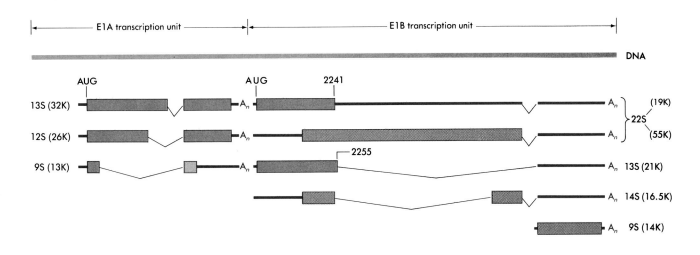

other parts of the genome are also present. The smallest restriction endonuclease fragment that can transform includes the left-hand 8 percent of the genome, about 3.2 kilobase pairs. This fragment contains two early genes called E1A and E1B (Figure 26-10).

There Are Multiple mRNAs from the Two Adenovirus Transforming Genes E1A and E1B[49–53]

E1A can be considered a pre–early gene because its expression is required for the full expression of all the other adenovirus early genes and so, in turn, the late genes. The effect of E1A on adenovirus early genes, including E1B, is at the level of transcription, so mutants defective in E1A transcribe only the E1A region efficiently. E1A is also responsible for stimulating host DNA synthesis, an effect seen transiently during lytic infection before cellular functions are shut down and an effect that may play an important role in cell transformation.

In a lytic infection, three polyadenylated mRNAs are made from the E1A region by differential splicing from a common precursor RNA (see Figure 26-10). The three mRNAs begin and end at the same nucleotide, but different splicing patterns give them different sizes and protein-coding potentials. The smallest is not an early mRNA, since it only appears after DNA replication, and it is not involved in transformation, since it is not expressed in transformed cells. The two larger E1A mRNAs are made early and are always present in transformed cells. They code for proteins that are identical in sequence except for an extra 43 amino acids in the middle of the larger one. Since cDNAs of both mRNAs have now been cloned, it is possible to test their biological activities independently. Remarkably, only the RNA encoding the larger protein can turn on early adenovirus genes. The role of the other in lytic infection is not yet known, but both mRNAs encode proteins that can play a role in cell transformation. The E1B region, like E1A, yields multiple mRNAs (see Figure 26-10). Two are present in transformed cells and encode 19 kdal and 55 kdal proteins.

Finding multiple mRNAs from the E1A and E1B regions made an understanding of adenovirus transformation seem more remote. As with polyoma and SV40, however, cDNA cloning now makes it possible to analyze the role of each gene product separately.

Figure 26-10

Region of the adenovirus genome (E1A and E1B) involved in transformation and transcripts derived from this region. The 9S RNA from E1A is made primarily late in a lytic infection and is often absent from transformed cells.

The E1A Gene Establishes Cells and Can Complement Polyoma Middle T for Transformation[54-59]

While DNA fragments making up the left-hand 8 percent of the adenovirus genome are needed to induce a fully transformed and oncogenic phenotype, DNA fragments that include just E1A, the left-hand 4.5 percent of the genome, can induce "atypical" transformation in some cells (in particular, primary rat cells). The atypical transformants are only partially transformed morphologically and are not tumorigenic, but like fully transformed cells, they are immortal and they are aneuploid (they have an abnormal number of chromosomes). Thus, E1A, like polyoma's large T gene, can rescue primary cells from senescence and cause them to grow indefinitely in culture, but E1A is not sufficient to induce full oncogenic transformation. Rather surprisingly, cDNAs for either of the large E1A mRNAs can induce prolonged growth in primary cells following their introduction by DNA transfection. Analysis of deletion mutants has localized this establishment function of E1A to within the amino-terminal 140 amino acids, a region common to the two proteins.

While the mechanism of action of the E1A gene products is unknown, it is now suspected that these nuclear proteins in some way alter the transcriptional machinery, so that expression of some genes is augmented while expression of other genes is repressed.

A DNA fragment carrying the E1B region cannot transform cells or immortalize them, even if it is first ligated to an SV40 promoter so that E1A is not needed for its transcription. However, E1B together with E1A induces complete transformation. Strikingly, a cDNA clone of polyoma middle T can also complement E1A to produce full transformation.

As yet, the role of the E1B proteins in transformation is poorly understood. Either the E1B 19K or 55K proteins seem sufficient, since mutations affecting just one or the other still allow complete morphological transformation in combination with E1A. The E1B 55K protein appears to be primarily nuclear and is associated with the same p53 protein as SV40 large T protein, while the 19K protein is located in the endoplasmic reticulum and also around, and probably within, the nucleus and may function to stimulate cell division following adenovirus infection. Studies on the roles of the adenovirus proteins in transformation are complicated by the dense packing of adenovirus genes and by the multiple uses to which single viral proteins are put to carefully regulate the complex lytic cycle. The ability of adenoviruses to transform cells is probably a small, almost inadvertent part of the viral life cycle, a consequence of their ability to turn quiescent cells into fertile environments for viral DNA replication.

Cells Transformed by a Highly Oncogenic Adenovirus May Escape Immune Surveillance by Eliminating Surface Histocompatibility Antigens[60]

Different human adenoviruses differ in their ability to induce tumors in animals. For example, adenovirus type 12 (Ad12) is strongly oncogenic, inducing tumors in rodents in 1 to 3 months, while Ad5 is nononcogenic. In vitro, however, both Ad12 and Ad5 viruses or their purified DNAs can transform rodent cells. But the tumorigenicity of

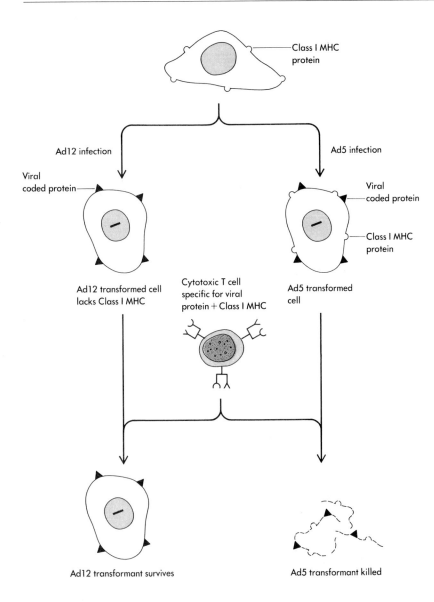

Figure 26-11
Highly oncogenic Ad12, but not weakly oncogenic Ad5, may suppress expression of class I MHC gene products on the cell surface, making transformed cells invisible to cytotoxic T cells and hence increasing their oncogenicity.

the resulting transformed cells reflects the oncogenicity of the transforming viruses. Thus, cells transformed by Ad12 will form tumors when injected into syngeneic rats, but cells transformed by Ad5 will not. This difference is also true of cells transformed by only the E1 regions of Ad12 or Ad5.

If the tumorigenicity of the Ad12 and Ad5 transformed cells is tested in immunodeficient animals, such as nude mice, then the Ad5 transformed cells fare better. They are still less tumorigenic than Ad12 transformants, but they can induce tumors in some animals. This last result suggests that the inability of Ad5 to induce tumors may in part be because the cells it transforms are immunologically rejected, while those transformed by Ad12 escape immune surveillance. Recently, a number of Ad12-transformed rat cells, but not Ad5 transformants, were shown to lack the rat class I major histocompatibility complex (MHC) gene product on their surface. Since cytotoxic T cells that kill virus-infected cells can only recognize viral antigens on the cell surface together with MHC class I antigens (Chapter 23), the absence of the MHC antigen would make the viral antigens, and hence the transformed cell, invisible to the killer T cells (Figure 26-11). The Ad12-transformed cells, but not the Ad5 transformants, lacked mRNA for the MHC class I heavy-chain gene. Furthermore, it is the

E1A region that must be derived from Ad12 to make adenovirus-transformed cells tumorigenic in immunocompetent animals. Hence, yet another function of the E1A region in transformation may be to negatively regulate expression of MHC genes and so help transformed cells to escape the host's immune system.

RNA TUMOR VIRUSES

Certain Features of Their Life Cycle Make Retroviruses Suitable Agents of Cell Transformation[61,62,63]

While the DNA tumor viruses encompass a variety of structurally diverse entities, those RNA viruses that have the capacity to cause cancer are all morphologically similar and probably are descended from a single ancestral virus. All belong to the retrovirus group. Unlike all other viruses, retroviruses replicate their RNA genome via a DNA intermediate that becomes permanently integrated into host cell DNA. Thus, like many other agents that cause cancer, oncogenic RNA viruses alter DNA.

A number of features of the retrovirus life cycle are central to the process of transformation (Figure 26-12; Chapter 24). Shortly after infection, the single-stranded viral RNA genome is copied into a linear molecule of double-stranded DNA. This DNA is longer than the viral RNA because of the presence of a specialized sequence, the long terminal repeat (LTR) of about 300 to 1000 base pairs found at each end. The LTR is derived from a combination of sequences present at the 3' end (U_3), the 5' end (U_5), or both ends (R) of the RNA genome and has the structure U_3–R–U_5. Following synthesis, the viral DNA is integrated into cellular DNA so that the ends of the LTR are directly joined to cellular sequences to form a stable structure, the provirus. Unlike DNA tumor virus integration, this is a highly regular and specific process essential for efficient replication of the virus. It is, however, random with respect to the site in cell DNA where the integration occurs.

Subsequent steps in viral replication are carried out by normal cell systems. Viral RNA is synthesized using cell RNA polymerase II, the enzyme responsible for cellular mRNA synthesis. Transcripts can serve either as mRNAs for virion proteins (in some cases after splicing) or as new genomes. Signals directing efficient transcription of the provirus are present in the U_3 portion of the LTR.

Several aspects of this cycle are of special importance to transformation by retroviruses. First, integration of viral DNA always occurs in an infected cell. Once integrated, the provirus is stable indefinitely. Second, infection of permissive cells does not kill them or noticeably impair their ability to divide. Thus, it is usual for a permissive cell to be transformed by a retrovirus and constantly release infectious virus. Third, defective viruses can be packaged with viral proteins supplied by a coinfecting virus and then can complete a new round of infection including provirus formation and expression in the absence of the helper virus. This means that all the viral genes (e.g., *gag, pol,* and *env*) can be missing from the genome, and yet, as long as its terminal sequences are intact, the genome can still be copied into viral DNA and the DNA can be integrated and expressed.

Table 26-3 Oncogene Families

Oncogene	Subcellular location of protein	Properties or normal function of protein
Class I: Protein kinases		
src	Plasma membrane	Tyrosine-specific protein kinase
yes	Plasma membrane	Tyrosine-specific protein kinase
fgr	?	Tyrosine-specific protein kinase
abl	Plasma membrane	Tyrosine-specific protein kinase
fps (fes)	Cytoplasm	Tyrosine-specific protein kinase
erbB	Plasma membrane (transmembrane)	EGF receptor/tyrosine-specific protein kinase
fms	Plasma membrane (transmembrane)	CSF-1 receptor/tyrosine-specific protein kinase
ros	Plasma membrane (transmembrane)	Tyrosine-specific protein kinase
kit	Plasma membrane	
mos	Cytoplasm	Serine/threonine protein kinase
raf (mil)	?	Serine/threonine protein kinase
Class II: GTP binding proteins		
H-ras	Plasma membrane	Guanine nucleotide binding protein with GTPase activity
K-ras	Plasma membrane	Guanine nucleotide binding protein with GTPase activity
Class III: Growth factors		
sis	Secreted	Derived from a gene that encodes PDGF
Class IV: Nuclear proteins		
myc	Nucleus	
myb	Nucleus	
fos	Nucleus	
ski	Nucleus	
Class V: Hormone receptor		
erbA	Cytoplasm	Thyroid hormone receptor
Unclassified:		
rel	?	
ets	?	

From this analysis, we can divide oncogenes into several distinct families of related protein sequences (Table 26-3). Oncogenes in the same family encode proteins with similar enzymatic activity or intracellular location, suggesting that their role in transformation may be similar.

src **Encodes a 60 kdal Phosphoprotein with Protein Kinase Activity**[86,91–95]

Although their existence had been known for many years, it was only in 1977 that the first oncogene protein, the v-src gene product, was detected. The crucial step was the development of antisera to the protein. When RSV induces tumors in rabbits, the rabbits produce antibodies to the src gene product as well as to viral structural gene products. The tumor sera could precipitate a 60 kdal protein from

Figure 26-19

Detection of the viral *src* gene product. Normal cells (chicken embryo fibroblasts), cells infected with Rous sarcoma virus, or cells infected with a transformation defective (td) mutant of Rous sarcoma virus that has a deletion of its *src* gene were grown in medium containing ^{35}S methionine so that their proteins would be radioactively labeled. Lysates of the cells were then mixed either with serum from normal rabbits, serum from rabbits that had been injected with Rous sarcoma virus and had developed tumors (TBR serum), or TBR serum that was mixed with disrupted virions to adsorb all the antibodies that can bind to viral proteins (the *src* gene product is not present in virions). Immune complexes were then precipitated, collected, and the proteins separated by electrophoresis on SDS polyacrylamide gels. The photo shown is of an autoradiogram of such a gel. When rabbits develop tumors induced by Rous sarcoma virus, they produce antibodies to all the viral coded proteins, even those that are only present inside the transformed cell (like the *src* gene product). Presumably this occurs because some tumor cells lyse and their contents are then seen by the immune system. T7: Molecular weight marker proteins. Lanes 1, 2, 3: normal cell proteins precipitated with (1) normal serum, (2) TBR serum, (3) TBR serum-blocked with viral proteins. Lanes 4, 5, 6: td sarcoma virus infected cell proteins precipitated with (4) normal serum, (5) TBR serum, and (6) TBR serum-blocked. Lanes 7, 8, 9: nondefective sarcoma virus transformed cell proteins precipitated with (7) normal serum, (8) TBR serum, and (9) TBR serum-blocked. In the autoradiogram shown, only the 60K protein band in lanes 8 and 9 fits the criteria for a product of the viral *src* gene: it is present in RSV transformed cells but not in cells infected by the td virus and its precipitation is not inhibited when viral proteins are used to block the TBR serum. Most of the other bands that are seen in viral infected cells are *gag, pol,* and *env* gene products. [From A. R. Purchio, E. Erikson, J. S. Brugge, R. L. Erikson, *Proc. Nat. Acad. Sci.* 75 (1978):1567–1571.]

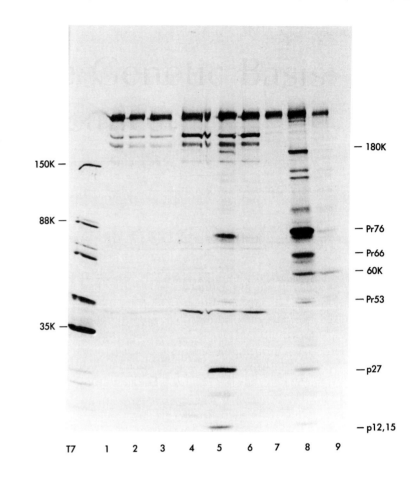

RSV-transformed cells (but in early experiments, not from uninfected or td RSV-infected cells), while antisera to viral structural proteins failed to precipitate this protein (Figure 26-19). Furthermore, the same protein could be identified as a product of in vitro translation from RNA fragments of the 3' third of the RSV genome.

The *src* gene product is a phosphoprotein and is designated pp60^{src}. Its amino terminus is modified by myristylation (covalent attachment of fatty acid), and in cells, the bulk of the protein is found attached to the inner surface of the plasma membrane. This attachment is essential for the oncogenic function of pp60^{src}, since mutants in the N terminus that cannot be myristylated do not attach well to the membrane and are defective in transformation. If cells transformed by RSV with a ts *src* mutation have their nucleus removed by exposure to the drug cytochalasin B, the remaining intact **cytoplasts** still have a temperature-sensitive transformed morphology. This remarkable result shows that at least some of the effects of *src* are not achieved via an effect on cellular gene expression or DNA replication, either directly or indirectly.

Early experiments failed to detect a c-*src* product in uninfected cells, but later, some tumor antisera detected low levels of pp60$^{c\text{-}src}$ in normal, uninfected cells. This is consistent with the presence of a few copies of c-*src* mRNA in virtually all uninfected cells. The pp60$^{c\text{-}src}$ differs from pp60$^{v\text{-}src}$ in just a few amino acids, mostly at its C terminus, and, like v-*src*, is localized at the inner surface of the plasma membrane.

Given a protein that can transform normal cells to cancer cells, it is, to say the least, not easy to guess what its function is. Soon after it

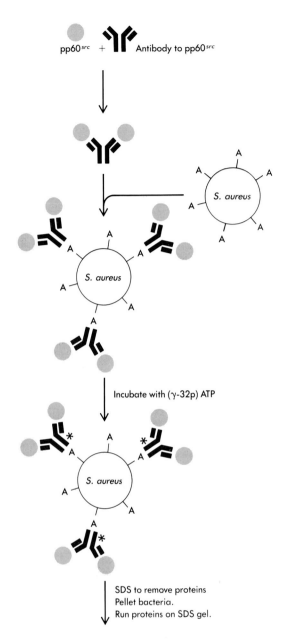

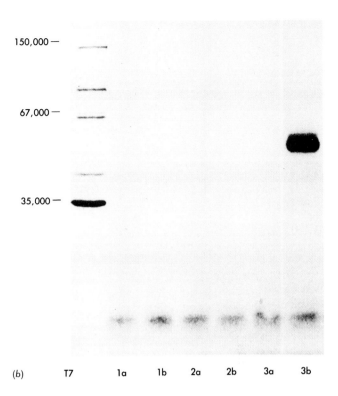

(a)

(b)

Figure 26-20
Demonstration of protein kinase activity in immunoprecipitates of pp60^{src}. Methodology is diagrammed in part (a), experimental results are shown in part (b). Cell extracts were prepared from chick embryo fibroblast cultures that were either uninfected [tracks 1a and 1b of part (b)], infected with avian leukosis viruses that lack a transforming gene [tracks 2], or infected with Rous sarcoma virus [tracks 3]. Each extract was mixed with either normal rabbit serum [a lanes in part (b)] or serum obtained from a rabbit bearing a Rous sarcoma virus induced tumor (TBR serum) (b lanes). Immune complexes were collected by binding to protein A–bearing *S. aureus* bacteria, washed, and the *S. aureus*–immune complexes suspended in a reaction buffer containing ^{32}P-ATP. After incubation to allow phosphorylation to occur, antibody and other proteins were separated from the bacteria, the bacteria were removed, and the remaining proteins were separated by gel electrophoresis. The heavy band in lane 3b is antibody to pp60^{src}, which has become radioactive because it was phosphorylated by pp60^{src}. Phage T7 virion proteins serve as molecular weight markers. [Part (b) is from M. S. Collett and R. L. Erikson, *Proc. Nat. Acad. Sci.* 75 (1978):2021–2024.]

was first detected, however, and long before it was purified, pp60^{v-src} was found to have protein kinase activity. By chance, a suitable in vitro substrate for pp60 is the immunoglobulin used to precipitate it. Thus, incubating immunoprecipitates with radioactive ATP causes the anti-pp60^{src} immunoglobulin to become radioactively labeled (Figure 26-20). Since then, many additional proteins, including pp60^{src} itself, have been found to be appropriate substrates for *src* kinase activity in vitro, and a plethora of normal cell proteins have been seen as targets in vivo.

All protein kinases that had been described before pp60^{src} phosphorylated either threonine or serine residues. Thus, researchers were quite surprised when analysis of the specificity of pp60^{src} showed that the protein catalyzes the transfer of the terminal phosphoryl group of ATP to the hydroxyl group of a tyrosine residue in its substrates. Then a careful search soon revealed a very low level of phosphotyrosine (P-tyrosine) in normal cells; RSV-transformed cells have about a tenfold higher level. Most important, in cells transformed by temperature-sensitive *src* mutants, both the transformed phenotype and the level of P-tyrosine in the cell shift rapidly in response to temperature shifts. Similarly, temperature-sensitive *src* mutants have reduced kinase activity in vitro. These results indicate that the kinase activity reflects, and may indeed be synonymous with, transforming ability.

Because v-*src* is under the transcriptional control of the viral LTR, it is expressed at much higher levels than c-*src*. As we noted, however, it is not just the higher level of v-*src* protein in cells that leads to transformation. When c-*src* is joined to an LTR (or other strong promoter) and introduced into cells, it is expressed at high levels but the cells are not transformed. Furthermore, the level of P-tyrosine in the cell is not elevated significantly. Thus, even though $pp60^{c\text{-}src}$ has kinase activity in vitro, apparently it has very little activity in vivo, in sharp contrast to $pp60^{v\text{-}src}$. Somehow, the mutations in v-*src* must enhance its kinase activity in vivo. These findings again point to the kinase activity of v-*src* as the means (or at least a correlate of the means) by which it transforms normal cells to cancerous ones.

Substrates for $pp60^{src}$ [96–101]

The finding that $pp60^{src}$ has a highly unusual protein kinase activity produced euphoria in the field. Protein phosphorylation is a major mechanism for inducing pleiotropic effects in cells, and so it seemed a most appropriate activity for $pp60^{src}$, a protein with the ability to induce the myriad changes associated with cancer. It seemed that by identifying cellular proteins phosphorylated on tyrosine residues after RSV infection, one would quickly begin to unravel the sequence of events that makes cells cancerous. However, while experimental results have come rapidly, insights have come more slowly. A major problem is that a very large number of different cellular proteins (Table 26-4) seem to be phosphorylated under the influence of pp60, and it is unlikely that more than a few of these are important to transformation. In most cases, either we do not know what the proteins do or we do not know why altering them would make a cell cancerous. A possible exception is vinculin. In normal cells, vinculin is located in adhesion plaques, the sites of cellular attachment to surfaces and sites where $pp60^{src}$ is also concentrated, and is thought to serve as a connecting anchor between actin cables and the membrane. In transformed cells, vinculin is found diffusely throughout the cell. Conceivably, phosphorylation causes vinculin redistribution, and this might help to bring about the transformed cell morphology.

A second problem in identifying important protein substrates for

Table 26-4   Some In Vivo Substrates for $pp60^{src}$

---

p36 (an abundant cell protein found on the inner surface of the plasma
   membrane; highest concentration in epithelial and endothelial cells,
   particularly the terminal web of the brush border of columnar epithelial
   cells of the intestine)
Vinculin
Enolase
Phosphoglycerate mutase
Lactate dehydrogenase

---

pp60 has been that many of the simplest technologies that can be applied to the problem (e.g., the use of two-dimensional protein gels) would only be able to detect relatively abundant substrates.

Recently, evidence has indicated that pp60^{src} may indirectly stimulate the inositol lipid pathway (Chapter 25). The phosphorylation of inositol lipids is significantly increased in RSV-transformed cells, although probably not directly by pp60^{src}. The synthesis and breakdown of inositol lipids is correlated with rapid cell proliferation. Furthermore, certain factors (e.g., vasopressin, prostaglandin F2, bombesin) that stimulate cell proliferation when they bind to specific cell surface receptors stimulate breakdown of the final product of the inositol biosynthetic pathway, phosphatidylinositol-4,5-biphosphate (see Figure 25-21). This breakdown releases diacylglycerol and inositol triphosphate. Both molecules are thought to be intracellular messengers. Diacylglycerol activates protein kinase C, thought to be a cellular target for tumor promoters. Inositol triphosphate may mobilize calcium ions (Ca^{2+}) from an intracellular store. Ca^{2+} is a well-known second messenger in cells (Chapter 25). Thus, according to this hypothesis, v-*src* may transform cells because it helps to drive the inositol lipid pathway, which in turn may generate natural signals for cell proliferation.

The study of *src* has run up against the poorly understood issue of how signals to proliferate are transmitted from the outer reaches of the cell (either the inner or outer surfaces of the plasma membrane) to the nucleus. While it was once thought that understanding transformation would require a prior understanding of the regulation of cell growth, the opposite now seems to be true. Analysis of oncogenes has in fact provided considerable new information on mechanisms regulating cell growth.

The *sis* Oncogene of Simian Sarcoma Virus Is Derived from the Cellular Gene for Platelet-Derived Growth Factor[102-106]

The fastest way to figure out what an oncogene does would be to identify the gene as one whose function is already known. Given that there are some 50 to 100,000 protein-coding genes in vertebrates and that we know the function of only a couple of thousand, this at first seemed like a remote possibility. Incredibly, though, it has already happened for several oncogenes.

The identity of the v-*sis* oncogene of simian sarcoma virus (SSV), with a cellular gene coding for platelet-derived growth factor (PDGF), was found quite by chance. PDGF is a protein secreted by platelets during clot formation and also by activated monocytes. It may promote wound healing by stimulating growth of nearby cells (Chapter 25). When the amino acid sequences of several fragments of human PDGF were published, one astute reader entered the sequences into a computer to look for related proteins. Out came the oncogene v-*sis*, whose nucleotide and predicted amino acid sequence had already been published, and with it a most important discovery in cancer research. Finding that a cellular gene known to induce normal cell proliferation was the progenitor of a viral oncogene with the capacity to induce abnormal cell proliferation was overwhelmingly reassuring to retrovirologists and growth factor researchers. They must indeed be on the right track (as they had known all along).

The v-*sis* oncogene is the oncogene of simian sarcoma virus, which was isolated from a fibrosarcoma of a pet woolly monkey. (The gene is also present as the oncogene of a feline sarcoma virus; see Table 26-2.) SSV induces fibrosarcomas when injected into monkeys, and it transforms fibroblasts in vitro. The v-*sis* oncogene consists of about 1 kilobase of woolly monkey cellular sequences recombined into the *env* gene of a helper virus to form SSV. The v-*sis* oncogene, which encodes a 28 kdal protein, p28^{sis}, replaces most of *env* and is fused to sequences encoding its amino terminus. Since these sequences include a signal involved in directing the *env* gene product to the cell surface, the resulting *env-sis* fusion protein is also directed to the surface and is secreted. However, fusion to *env* is not essential for the activation of the *sis* gene to a transforming oncogene. If a molecular clone of c-*sis* (which probably possesses its own amino-terminal signal sequence) is ligated to potent transcriptional signals (an LTR), it, too, can now transform fibroblasts in vitro.

PDGF is composed of dimers of two polypeptide chains, sometimes called A (or 1) and B (or 2), although it is not yet clear if PDGF is a heterodimer (AB) or a mixture of homodimers (AA + BB). Both A and B homodimers appear to possess biological activity. A and B are derived from different genes, but have related amino acid sequences, and v-*sis* is derived from the gene encoding the B polypeptide chain. Until the relationship of the A and B chains is better understood, the relationship of v-*sis* and c-*sis* will also remain somewhat confused.

Interestingly, v-*sis* cannot transform epithelial cells. It can only transform cells that possess PDGF receptors. Transformation by SSV seems to be an example of transformation by autocrine stimulation (Chapter 25): The cell that expresses both PDGF and PDGF receptors constitutively, continuously stimulates its own division.

Some Oncogenes (*erbB* and *fms*) Are Derived from Cellular Genes That Encode Receptors for Polypeptide Growth Factors[107,108]

Although there are numerous known growth factors, so far PDGF is the only one observed to have been converted to an oncogene. However, there appears to be a related mechanism for generating an oncogene, namely, mutation of a gene encoding a receptor for a polypeptide growth factor. Presumably, such mutations cause the encoded receptor to behave as though stimulated, even in the absence of its ligand.

The epidermal growth factor (EGF) receptor is a large glycoprotein with intrinsic kinase activity (Chapter 25). When EGF binds to the receptor, the kinase becomes active, and the receptor, as well as other cellular proteins, is rapidly phosphorylated. The discovery that this phosphorylation occurs on tyrosine residues immediately caused researchers to wonder whether growth factors and oncogenes might stimulate cell growth by similar mechanisms. Pursuing this line of thought, and spurred on by the finding that *sis* encodes PDGF, researchers generated antibodies to the EGF receptor, allowing enough of the protein to be purified from a human tumor cell line that makes unusually large amounts, and the receptor was partially sequenced. The remarkable discovery was immediately made (again by comparison with sequences in computer data banks) that this protein must be the product of the human analog of the *erbB* oncogene. The *erbB* oncogene, first found in the genome of avian erythroblastosis virus (AEV), is responsible for the ability of AEV to induce erythroleukemia in chickens and to transform both erythroblasts and fibroblasts in vitro.

The EGF receptor has at least three functional domains (see Figure 25-18): a large glycosylated portion outside the cell that binds EGF, a short transmembrane domain, and a cytoplasmic domain that has both kinase activity and the tyrosine sites of autophosphorylation. The v-*erbB* oncogene is missing most of the amino-terminal portion of the EGF receptor, a consequence of the recombination event that placed the gene in the retroviral genome and has also sustained mutation in its C-terminal coding portion. Somehow, the truncated, mutant EGF receptor of AEV must have lost not only its EGF binding capacity but also its normal controls, so that it constantly signals the cell to divide even in the absence of EGF.

Interestingly, the tyrosine kinase domain of *erbB* has amino acid sequence homology to the *src* gene product. However, as we have seen, in contrast to the EGF receptor, pp60*^{src}* is located entirely within the cell, at the inner face of the plasma membrane, and lacks any transmembrane or extracellular domains.

Analogous to the case of *erbB* is that of the oncogene *fms,* found in two cat sarcoma viruses. The *fms* oncogene is a transmembrane protein that turns out to be derived from the cellular gene for the receptor for CSF-1 (also known as M-CSF), a polypeptide growth factor that specifically stimulates the growth and differentiation of macrophages (Chapter 25). In contrast to *erbB*, v-*fms* does not differ from the authentic receptor by the loss of part of the extracellular domain. However, v-*fms* does possess an altered carboxyl terminus, including alteration of a conserved tyrosine residue (the tyrosine corresponding to Tyr-527 in pp60*^{src}*) that is an important site of autophosphorylation in this family of proteins. It is now thought that the loss of this site, and the resulting inability of the residue to be phosphorylated, is important in altering the regulation of activity of the protein, possibly leaving it turned on in the absence of CSF-1.

Two other oncogenes, *ros* (Table 26-3) and *neu* (Chapter 27) are also transmembrane proteins with tyrosine kinase domains and *kit* probably also shares these properties. All three may encode growth factor receptors, although the factors they bind have not yet been identified.

Members of the *ras* Gene Family Encode 21 kdal Proteins That Bind Guanine Nucleotides[109,110,111]

The *ras* oncogenes have received particularly intense attention because of their probable involvement in many human tumors (Chapter 27). The *ras* genes constitute a small multigene family with three functional members: H-*ras* (or c-H-*ras*, the cellular homolog of v-H-*ras*, the oncogene present in Harvey sarcoma virus); K-*ras* (or c-K-*ras*, present in Kirsten sarcoma virus as v-K-*ras*); and N-*ras* (which is not present on any known retrovirus genome and whose significance will be discussed in Chapter 27; see Figure 27-4). All three functional *ras* genes encode highly related proteins of 21 kdal called p21*^{ras}* or p21.

The *ras* proteins, both viral and cellular, specifically and tightly bind GTP as well as other guanosine nucleotides. In addition, it has recently been found that normal c-*ras*-coded proteins have GTPase activity. The GTPase activity is greatly reduced in transforming p21 variants. Other cellular proteins that bind guanine nucleotides and have GTPase activity include the G proteins that modulate activity of the adenylcyclase system. These proteins are active when they are bound to GTP, but become inactive when their GTPase converts the GTP to GDP. Although *ras* proteins do not correspond to any known G proteins, it is now thought that it is the loss of GTPase activity in activated *ras*-coded proteins that may be critical to their transforming activity (Chapter 27).

The virus-coded p21s, but not the normal cell proteins, have an autokinase activity in that they can transfer a phosphate group from GTP to a threonine residue at position 59. As it turns out, the c-*ras* proteins have an alanine residue at the corresponding position, and so they cannot have the autokinase activity. The presence of Thr at position 59, however, is not essential for converting a c-*ras* proto-oncogene to a transforming protein.

The *myc* Gene Product Is a Nuclear Protein[112–119]

Unlike the *src*, *sis*, *erbB*, and *ras* products found inside, outside, or within the plasma membrane, several oncogene products are found in the nucleus. The *myc* oncogene is the oncogene of several defective avian retroviruses that can induce myelocytomas (tumors of myeloid cells), sarcomas, and carcinomas and in vitro can transform chicken (but not readily rodent) fibroblasts, myeloid cells, and epithelial cells. In some viruses, v-*myc* is fused to *gag* and translated as a *gag-myc* fusion protein. In one virus, it is produced from a subgenomic mRNA and not fused to a viral protein (see Figure 26-15). Either way, the *myc*-coded protein is found in the nucleus. So far no evidence exists that it functions by binding to DNA (even though it can bind DNA in vitro). Its function remains unknown.

Like E1A, *myc* alone cannot transform rodent fibroblasts efficiently, but it can transform primary cells efficiently if introduced along with an activated *ras* oncogene (Chapter 27). In contrast to E1A and polyoma large T, *myc* is not very efficient at promoting the establishment of primary cells, except, as noted, when another oncogene is also present.

The c-*myc* oncogene almost certainly plays a role in cell division, and it is expressed in normal growing cells as well as in tumor cells. In cells whose growth has been arrested, *myc* mRNA is greatly reduced. But if such cells are stimulated by mitogens (e.g., PDGF in the case of fibroblasts), *myc* mRNA appears within 1 to 2 hours. Mitogens, like PDGF, set off a chain of events by inducing a cell's "competence" to respond to additional growth factors (Chapter 25). Since v-*sis* is related to PDGF, the finding that PDGF turns on *myc* expression implies that the two oncogenes, *sis* and *myc*, may lie on the same growth control pathway. Consistent with a regulatory role, both the viral and cellular *myc* proteins are unusually labile and disappear from the cell with a half-life of 20 minutes or so.

The cellular *myc* gene has three exons, the first of which is not translated into protein (see Figure 26-22). It has been suggested that the first exon may have a regulatory role in c-*myc* expression, perhaps in the stability of the RNA or perhaps in translational control. In any case, in retroviruses, v-*myc* has both lost its first exon and fallen under LTR control. These two disruptions of its normal regulation presumably make an oncogene out of c-*myc*. Unlike many other oncogenes discussed so far, mutational change in the *myc*-encoded protein does not seem to be important for its activities as an oncogene.

The c-*myc* gene is a member of a gene family. So far, it is the only member of the family that has been found in a retrovirus genome, but other members, L-*myc* and N-*myc*, are amplified in certain naturally occurring tumors (Chapter 27).

Another oncogene whose product is localized in the nucleus is *fos*, the oncogene of the mouse osteosarcoma virus. The function of the 55 kdal phosphoprotein pp55$^{c\text{-}fos}$ is unknown. c-*fos* expression is quickly stimulated when quiescent cells are exposed to mitogens such as PDGF, and within minutes of exposure, unstable c-*fos* mRNA accumulates. An enhancer-like element lying 5' of the *fos* gene is essential for this transcriptional activation in response to mitogens.

Oncogene Families: How Many Genes, How Many Functions?[89,120,121,122]

The fact that retroviruses isolated from different species have sometimes transduced homologous oncogenes was the first indication that the number of proto-oncogenes in cells might not be enormous. This possibility has received further support as cellular cancer genes detected in new ways (Chapter 27) have frequently turned out to be the homologs of known retrovirus oncogenes. Altogether, about 30 oncogenes are now known (see Table 26-3), and many people expect that the number will reach about 100. Fortunately, the known genes can be grouped by structure and function so that the number of oncogene classes is considerably less than 30. Moreover, now that several oncogenes have been tied to growth factor pathways, confidence that we can figure out their functions has soared.

About half the oncogenes in Table 26-3 turn out to resemble pp60^{src} in that they have protein kinase activity and share amino acid sequence homology with one another in the region that contains the active site for the kinase (Figure 26-21). In many cases, the homology at the amino acid level is so close that there is no doubt that the different oncogenes must have diverged from a common ancestor; and yet, the nucleotide sequences are very different, showing that the divergence of this family of oncogenes must have occurred early in eukaryotic evolution. For example, the products of *src* and *yes* have 82 percent amino acid homology, but the genes themselves have only 65 percent homology at the nucleotide level. Furthermore, *src* and *abl* are similarly related oncogenes, yet human c-*src* is more closely related to Drosophila c-*src* than either is to the corresponding c-*abl*. Thus, *src* and *abl* must have evolved from a common ancestor before the separation of vertebrates and insects. The full significance of this evolution into a highly conserved family will probably not become apparent until the normal functions of these proteins are better understood.

While most members of the protein kinase family of oncogenes are tyrosine-specific kinases, two, *mos* and *raf* (or *mil*) phosphorylate serine or threonine residues.

As already noted, in contrast to *erbB* and *fms* and also the more recently discovered *ros* and *neu*, and probably *kit* genes, most members of the tyrosine kinase family of oncogenes do not seem to encode proteins that protrude outside the cell and serve as receptors.

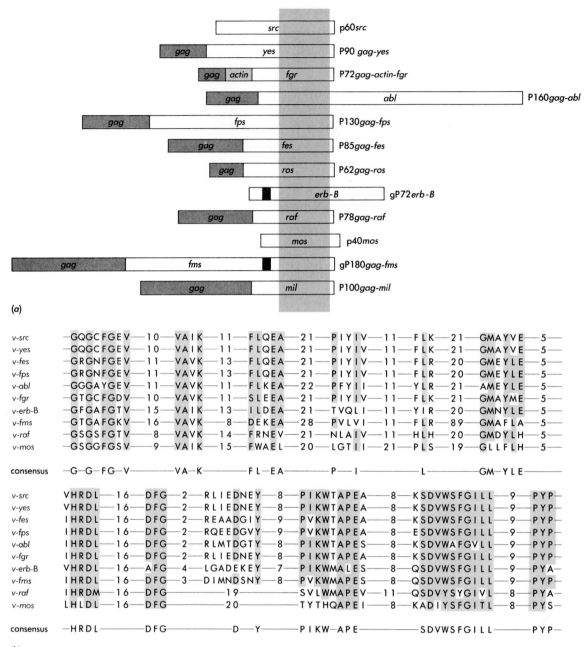

Figure 26-21

(a) Schematic representation of many of the proteins encoded by v-oncs belonging to the *src* family of oncogenes. Lengths are proportional to the number of amino acids in each protein. The NH₂ end of each chain is on the left. The colored strip represents a 250-amino-acid region of each protein that is similar in amino acid sequence [see part (b) of figure] and, by homology with pp60^src, is predicted to contain the protein kinase domain. Black bars represent the transmembrane domains of the *erb-B* and *fms* oncogenes, whose cellular homologs encode growth factor receptors. Note that most of the oncogenes are expressed as fusion proteins with proteins encoded by viral *gag* (and in one case *gag* plus a fragment of cellular actin). It is possible that in some cases, truncating the NH₂ end of the normal protein and/or joining it to a protein that is directed to the plasma membrane contributes to converting the proto-oncogene product into a transforming protein. [After T. Hunter, *Sci. Amer.* 251 (1984):70–79.] (b) Amino acid sequences from the regions of homology between members of the *src* oncogene family.

Nonetheless, the fact that *erbB* and *fms* encode growth factor receptors raises the possibility that the tyrosine kinase oncogene products that are entirely inside the cell may also be linked to extracellular signals, perhaps via interaction with other proteins that do have extracellular domains.

Another family of oncogenes, consisting of genes whose products are localized at the nucleus, is exemplified by *myc*. The *myc* gene product is remotely related in amino acid sequence to the *myb* and *ski* oncogene products, both of which are also nuclear proteins. As already noted, *fos* is also in the nucleus: Its function is totally unknown, and it is difficult to say whether it will continue to remain in the same gene family with *myc, myb,* and *ski.*

The *ras* genes constitute a third oncogene group, being localized at the inner surface of the plasma membrane, like many members of the *src* family, but binding guanine nucleotides and having GTPase activity. These properties are shared with the G proteins that interact with the adenylcyclase system to propagate signals initiated by polypeptide hormone-receptor interactions. However, *ras* proteins are different from any of the known G proteins, and their real function is still unknown.

The *erbA* gene may define a new group of oncogenes. Evidence indicates that this gene, which is present as a *gag–erbA* fusion in AEV, is derived from the cellular gene that encodes the receptor for thyroid hormone.

One of the v-oncs that did not originally fall into any of the classes just described was *sis,* but now its identification as a derivative of the gene encoding one chain of PDGF, as well as the connection of growth factors, their receptors, and their ability to turn on nuclear (*myc* and *fos*) oncogenes, provides the strongest link between the different classes of oncogenes. What acute transforming retroviruses have conveniently provided is a set of growth genes, or at least that subset of growth-controlling genes that can be altered so as to become oncogenes of retroviruses.

Some Retroviruses Have Two Oncogenes[123,124]

A striking difference between the DNA tumor viruses discussed earlier and many transforming retroviruses is that the former need two or three genes to cause cancer, whereas viruses like RSV seem to transform cells and cause tumors with just a single oncogene. While several exceptional retroviruses do have two different cell-derived genes (see Table 26-2), in these cases, related viruses that contain just one of the oncogenes are also capable of inducing tumors, so the significance of the extra gene is uncertain.

It is possible that retroviruses with a single transforming gene may only initiate tumors and that subsequent events may be needed to cause the infected cells to progress to a full-blown tumor. On the other hand, the rapid induction of tumors by acute transforming viruses (sometimes within a week or two), as well as the fact that the tumors often are not clonal but arise from many transformed cells, argues that a single transforming virus may indeed be able to transform cells in vivo. However, it must be noted that it is usually difficult to know whether additional mutations may have occurred in the tumor cells that contribute to their cancerous growth, and in most cases, such mutations have not been seriously looked for. Furthermore, many acute transforming retroviruses cause progressive fatal tumors only in certain highly susceptible, inbred strains of animals, so additional oncogenes might be necessary to allow the viruses to cause tumors in most animals. Clearly, the question of whether one oncogene is ever sufficient to fully transform a cell remains unanswered.

Slow Lymphoma-Inducing Retroviruses Activate Proto-Oncogenes in Vivo by Promoter or Enhancer Insertion[125-128]

Many retroviruses that can cause cancer lack oncogenes, are not defective, and encode only *gag, pol,* and *env.* These viruses, which are frequently found in chickens, mice, and cats, usually cause different forms of leukemia that typically take much longer to appear than tumors induced by transforming retroviruses. Because they never transform cells in culture, for a long time it was a mystery as to how these viruses induce tumors. The answer came from studying B cell leukemias induced by avian leukosis virus (ALV) in chickens.

First, every tumor induced by ALV has at least one provirus, suggesting the importance of viral DNA in maintaining the tumor. Second, mapping experiments using the Southern blot technique show that tumors contain proviruses integrated in preferred sites. Third, in some tumors, the ALV provirus is almost entirely deleted, although a viral LTR is always present, indicating that no viral gene product is necessary to maintain transformation, but the LTR is. And fourth, most ALV-induced tumors contain mRNA that can hybridize to cDNA complementary to the U_5 region of the LTR but not to other viral sequences. These facts came together to make spectacular sense when virtually all ALV-induced tumors were found to have ALV proviruses integrated near the cellular c-*myc* gene. Usually, integration occurs within the first c-*myc* exon or intron, and the virus is pointed in the same transcriptional orientation as c-*myc* (Figure 26-22). Furthermore, the virus usually has suffered a deletion, often near to or including its 5' LTR. As a result, the viral 3' LTR is often used to promote transcription, and an mRNA with U_5 sequences fused to c-*myc* sequences is made. Thus, c-*myc* is activated by **promoter insertion**. The level of the aberrant c-*myc* RNA in tumor cells can be 10 to 50 times higher than the normal level of c-*myc* RNA. This aberrant expression might reasonably be expected to disrupt growth control.

Sometimes, proviruses integrated near c-*myc* in retrovirus-induced tumors are oriented in the opposite transcriptional direction of c-*myc*. In these cases, *myc* mRNAs do not have viral sequences at their 5' ends, and activation of the gene is thought to result from the proximity of an enhancer element in the provirus. Enhancer elements can function in either orientation relative to the genes whose expression they affect.

It is important to remember that proviral integration is random in each infected cell. Thus, only a rare integration that by chance occurs near c-*myc* (in the case of ALV-induced B cell tumors of chickens) will

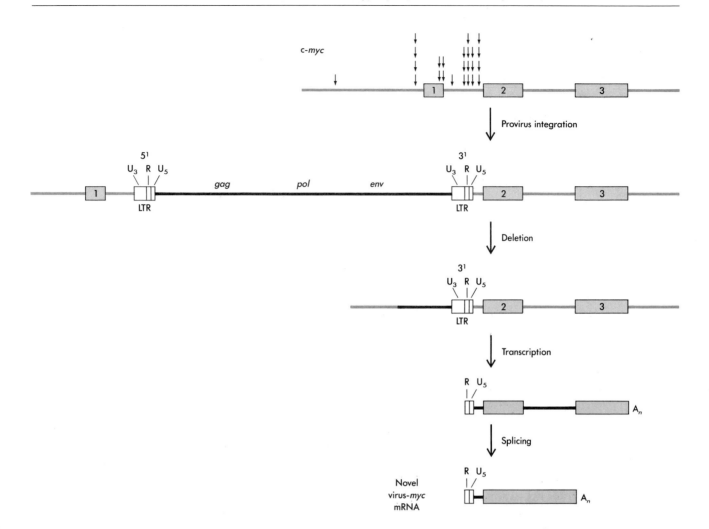

ultimately result in a tumor. The probability that any single infected cell will be transformed is very low, but since there are at least as many cells at risk as possible integration sites, it is highly probable that a tumor will arise in the infected animal if viral replication is extensive.

Both Viral and Cellular Genes Determine What Type of Leukemia Is Induced by Nondefective Retroviruses[129,130,131]

Nondefective C-type viruses that induce leukemias usually cause predominantly one type of disease, such as tumors of B cells, tumors of T cells, or tumors of cells of the erythroid lineage. Given the mechanism of proto-oncogene activation by provirus insertion, one wonders how this specificity is achieved. Furthermore, why is a particular oncogene, for example, c-*myc*, preferentially activated in a particular type of virus-induced tumor? While the answers are not completely known, it is clear that the age of the recipient animal, the availability of replicating target cells, and the genetic constitution of both the host and the virus determine the outcome of an infection. For example, although ALV causes primarily B cell leukemias in some lines of

Figure 26-22
Integration of an ALV provirus into the c-*myc* gene and activation of the gene by promoter insertion. The c-*myc* gene has three exons. The first exon is not translated but encodes the 5' untranslated portion of the *myc* mRNA. Arrows above the gene indicate actual sites of provirus integration that have been observed in different ALV-induced B cell tumors. In the example shown, the 5' LTR has been deleted. However, this is not usually the case, and it remains unclear why transcript that activate c-*myc* usually initiate in the 3' LTR rather than the 5' LTR. [Based on data from H. L. Robinson and G. C. Gagnon, *J. Virology* 57 (1986):28–36.]

chickens, in one particular strain it causes primarily erythroleukemias. Remarkably, in the latter, ALV proviruses are integrated near the c-*erbB* gene rather than near c-*myc*.

One clear-cut determinant of disease specificity in some retroviruses is the LTR. DNA copies of different retroviruses can be molecularly cloned, and fragments of each can be exchanged to construct recombinant viruses. When the U_3 portions of LTRs are exchanged between two mouse leukemia viruses, one of which causes T cell leukemias and the other erythroleukemias, the viruses switch their disease specificity. The exchange of LTRs causes a change in the preferred site of virus replication in injected animals, and this change in organ tropism probably underlies the change in the type of leukemia induced. It is probably different tissue specificities of the transcriptional enhancers in the LTRs of different viruses that account for the results.

Mouse Mammary Tumor Virus (MMTV) Activates Proto-Oncogenes During Tumor Induction[132,133,134]

Oncogenic mouse mammary tumor viruses (MMTVs), like the leukemia viruses already discussed, are replication-competent, lack cell-derived oncogenes, and induce tumors after a long latent period (4 to 9 months). Like the nondefective leukemia viruses, they, too, transform cells by integrating adjacent to cellular proto-oncogenes. Two cellular genes that lie on different mouse chromosomes have been identified as targets of MMTV activation. The two genes, called *int-1* and *int-2*, do not correspond to any of the oncogenes that have been discovered in retroviral genomes. Proviral integrations that activate *int-1* or *int-2* can occur over a region of about 30 kilobases, and the proviruses are positioned so that their transcriptional orientation is outward from the direction of transcription of the *int* genes (Figure 26-23). This positioning is consistent with activation of expression of the cellular gene by enhancer insertion rather than by promoter insertion, and indeed, *int* mRNAs in tumor cells do not carry MMTV sequences. Some tumors carry integrated MMTV proviruses near both *int-1* and *int-2* genes, indicating that the virus can contribute to multistep carcinogenesis.

The MMTV LTR probably plays a role in determining the target tissue for transformation. Supporting this idea are the results of experiments involving **transgenic mice.** Transgenic mice are generated by introducing cloned genes into an embryo at a very early stage in development. If the gene is taken up and integrated into cellular DNA, it becomes part of the animal's genetic makeup. If the gene is integrated into cells that differentiate to become germ cells, it can be passed on to future generations of mice. When an LTR from a molecular clone of MMTV was ligated to the c-*myc* gene and the resulting DNA was introduced into embryos, the resulting (female) transgenic mice developed breast tumors after pregnancy. The specificity of disease induction was presumably conferred by the MMTV LTR. Interestingly, only a few breast cells became cancerous, suggesting that aberrant expression of c-*myc* alone may not be sufficient to cause a tumor. Some additional genetic event may have to occur in a particular cell to cause it to lose growth control.

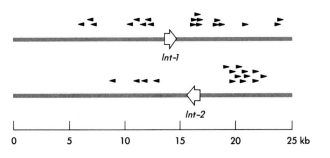

Figure 26-23
Integration sites of MMTV proviruses that have been observed in the vicinity of the *int-1* and *int-2* genes in breast tumors of mice. Each black arrowhead represents an integration in a separate tumor. Arrows indicate transcriptional orientations of proviruses (black arrows) or of the *int* genes (open arrows). The *int-1* transcript is about 2.6 kilobases; the *int-2* transcript, about 3.2 kilobases. [The *int-1* data are from R. Nusse and H. E. Varmus, *Cell* 31 (1982):99–109; and R. Nusse, A. van Ooyen, D. Westaway, Y. K. Fung, H. E. Varmus, and C. Moscovici, *Cancer Cells* 2 (1984):205–210. The *int-2* data are from G. Peters, S. Brookes, R. Smith, and C. Dickson, *Cell* 33 (1983):369–377; and C. Dickson, G. Peters, R. Smith, and S. Brookes, *Cancer Cells* 2 (1984):195–203. Figure modified from R. Weiss, N. Teich, H. Varmus, and J. Coffin, eds., *RNA Tumor Viruses*, 2nd ed. (Cold Spring Harbor, N.Y.: Cold Spring Harbor Laboratory, 1984), p. 212.]

Inherited Retroviruses Cause Cancer in High-Leukemic Inbred Mice[135–143]

The discovery in the early 1930s that certain inbred strains of mice develop leukemias and lymphomas between the age of 6 and 12 months argued strongly that cancer is a genetic disease in mice. But in the 1950s, it was found that cell-free extracts of spontaneous tumors of high leukemic inbred mice could induce leukemia when they were injected into newborn mice that normally would not develop the disease. The active agent was soon shown to be an RNA-containing virus. Since reverse transcriptase had not yet been discovered or even guessed at, and since chromosomal genes were known to be DNA, not RNA, the paradox of how leukemia could be both inherited like a gene and caused by RNA viruses seemed incomprehensible. The discovery of reverse transcriptase provided a resolution to the paradox: Perhaps the DNA proviruses of oncogenic RNA viruses could be inherited as genes. While initially surprising, it was soon shown that genes that cause certain strains of mice to develop lymphoma are indeed DNA copies of RNA tumor viruses, and these genes are inherited in a Mendelian fashion.

Since all mice inherit retrovirus proviruses, what distinguishes inherited viruses that cause lymphoma? The answer seems to be that these particular inherited viruses are able to replicate very efficiently in their hosts, so high-leukemic mouse strains exhibit high titers of replicating viruses in their tissues soon after birth. Even so, 6 to 12 months elapse before disease develops. Now we know that the inherited viruses so critical to lymphoma development in high-leukemic inbred mice do not usually cause disease directly. They undergo recombination with other inherited retrovirus sequences to generate novel viruses, called MCFs, whose *env* genes and LTRs are acquired during the recombination process. The MCF viruses then initiate tumors, usually of T cells. As already described for avian leukosis virus, they do this by integrating adjacent to one, or even more, of several different proto-oncogenes, including c-*myc* or a newly discovered gene called *pim-1*.

Viruses Also Cause Cancer in Animals Outside the Laboratory[144,145,146]

Many of the tumor viruses we have described cause cancer under highly unnatural circumstances. For example, SV40 is found in monkeys but causes tumors when injected into newborn hamsters; and almost all AKR mice are doomed to die from retrovirus-induced thymomas, but the strain of mice was selected and inbred for high cancer incidence. What about animals outside the lab? Are they, too, succumbing to virus-induced cancers? Next to laboratory animals, the most readily available animals for study are house pets and farm animals. It is clear that in certain species, at certain times and places, viruses are a significant cause of cancer. With a few exceptions, such as Marek's disease virus, a herpesvirus responsible for lymphoma induction in chickens, retroviruses seem to be involved in most of the natural virus-induced cancers.

A striking example of viruses causing cancer occurs in household cats. Feline leukemia viruses (FeLV) are spread horizontally from cat to cat, being secreted in large amounts in saliva and passed between animals by scratching and biting. FeLV causes many diseases besides lymphomas, and in fact, infected cats are more likely to die from anemia, immunosuppression, thymic atrophy and other degenerative diseases, probably caused by decimation of particular lymphoid or erythroid cell populations by FeLV.

Chicken flocks are frequently afflicted by retrovirus (avian leukemia virus) infections. These cause leukemias and can spread, through infection of the reproductive tract and the eggs, to newly hatched chicks, which show wasting and runting syndromes as well as later lymphomas. Having been infected so young, the chick's immune system sees the replicating retrovirus as self and does not reject it.

Mice trapped in the wild have shown high incidences of retrovirus-induced lymphomas and also of a retrovirus-induced neurological disease that resembles the devastating Lou Gehrig's disease. Mice live in small, closely packed groups called *demes*, and in some of these demes, retroviruses have become established as milk-borne agents and are transmitted to suckling mice.

Summary

Much cancer research in this century has been devoted to identifying agents that induce cancer in experimental animals and then to determining the mechanisms by which such agents act. Most such agents, called carcinogens, act either directly or indirectly on DNA. Agents that cause cancer are usually chemicals or radiation that can induce mutations in DNA, or they are viruses that alter the expression of cellular genes. These findings led to the realization that cancer is primarily a genetic disease that arises when specific mutations (or, in some cases, a viral genome) accumulate within a single cell and cause it to lose growth control.

Many compounds that can induce tumors in animals at first did not appear to be mutagens. But in the 1960s, it became apparent that these carcinogens are converted to powerful mutagens by enzymes in the body. This discovery led to the development of tests to assay the potential carcinogenicity of compounds in the environment. Most widely used is the Ames test, in which the compound to be tested is incubated with an extract of liver cells, and its mutagenicity then tested on bacteria that carry a mutation at the *his* locus. If the compound has mutagenic activity or has been converted to a strong mutagen by exposure to the liver enzymes, bacterial revertants arise, and these can grow on media that lack histidine.

Very important to the study of carcinogens, and particularly to the study of viruses that induce cancers, was the development of cell culture systems in which transformation of normal cells to malignant cells can be observed. Transformed cells are detected as foci of cells with altered morphology or as cells with altered growth requirements that can grow under conditions where normal cells cannot.

Several types of DNA viruses can cause tumors in animals. These include papovaviruses (SV40, polyoma, and papilloma viruses), adenoviruses, certain herpesviruses, poxviruses, and hepatitis B viruses. All tumorigenic RNA viruses belong to the retrovirus group. The best-studied tumor viruses are SV40, polyoma, adenoviruses, and retroviruses, although papilloma viruses and hepatitis B viruses are now receiving more attention as well. In all known cases, when viruses transform cells, at least a portion of their genome becomes stably associated with the cell. This usually involves integration of viral genes into cellular DNA. There are two known mechanisms of transformation. Either the virus encodes genes whose products alter cellular growth, or integration of the viral genome occurs in such a way as to alter the expression of a cellular gene. Viral or cellular genes whose protein products can induce cancerous changes in cell behavior are called oncogenes. Normal cellular genes with the capacity to become oncogenes when they suffer mutations or alterations in their expression are called proto-oncogenes.

SV40 and polyoma viruses usually transform only those cells that are not permissive for the complete lytic cycle of the virus. Only the early regions of the small circular viral genomes (about half the genome) are needed for transformation. In polyoma, this region encodes three proteins whose genes overlap: small, middle, and large T proteins. In SV40, the early region encodes two proteins: large and small T. The role of the different SV40/polyoma early genes in transformation has been studied by synthesizing cDNA copies of their mRNAs, joining these to transcriptional promoters, introducing the constructs into cells, and observing the effect of the gene on the growth and morphology of the cell. The studies reveal that polyoma large T can induce indefinite growth in primary cells and reduce their serum requirements for growth, while middle T is needed to induce the remainder of the transformed phenotype, including morphological transformation. If established cell lines are used instead of primary cells, middle T alone can transform them. Middle T may act by altering the activity of the cellular *src* gene product, a protein with tyrosine kinase activity. SV40 large T protein seems to induce many aspects of transformation single-handedly, although perhaps less efficiently than polyoma large T plus middle T. SV40 large T protein may act via its interaction with a cellular protein, p53, located in the nucleus, whose stability and thus amount increase in SV40-transformed cells. The role (or even need) of the SV40 and polyoma small T proteins in transformation remains unclear.

Only the left-hand 8 percent of the linear adenovirus genome is needed to transform cells. This region includes the E1A and E1B genes, each of which encodes overlapping mRNAs. Two E1A mRNAs and one or two E1B mRNAs encode proteins that contribute to the transformed phenotype. The E1A gene induces indefinite growth of cells and also induces certain other aspects of the transformed phenotype. Although either of its two transforming proteins can induce indefinite growth, both products are needed for cellular transformation. In lytic growth, E1A is needed for full expression of other adenovirus genes, and it is thought that E1A acts by altering gene expression. The role of the E1B gene products in transformation remains obscure. Different adenoviruses differ in their ability to induce tumors in rodents. One reason seems to be that at least some highly oncogenic adenoviruses suppress the expression of surface MHC proteins, so that the virus-infected cell is not seen and killed by cytotoxic T cells.

There are two types of cancer-causing retroviruses. One type is nondefective, lacks oncogenes, is unable to transform cells in culture, and induces primarily leukemias with a long latent period (3 to 12 months). With the exception of the Rous sarcoma virus, the second type is always defective, carries an oncogene, induces tumors in susceptible animals with a short latent period (days or a few weeks), and can usually transform cells in culture. These viruses are often called acute transforming retroviruses.

The oncogenes of acute transforming retroviruses are acquired from normal cells by recombination, and usually the acquisition involves both a loss of viral genetic material and an alteration to the protein encoded by the cellular gene. As a result, the virus is defective and requires a nondefective helper to replicate. However, a single defective viral genome can transform a cell.

The first retroviral oncogene to be studied in detail was *src*. It encodes a protein of 60 kdal with tyrosine protein kinase activity. This activity is probably responsible for the ability of the viral *src* gene to transform cells, but the putative substrates that are affected by phosphorylation and that thus bring about the transformed phenotype remain undiscovered. The cellular c-*src* gene encodes a protein that possesses protein kinase activity; however, it is now thought that the activity is regulated in some way that has been lost in the v-*src* gene product as a result of mutations.

About 50 different transforming retroviruses have been isolated from different animal species, and about 20 to 25 different oncogenes have been discovered in the genomes of these viruses. The oncogenes can be divided into families of related genes. (1) One family consists of genes that encode protein kinases and that have amino acid sequence homology to the *src* gene product. Like pp60^{src}, most of these proteins are located at the inner surface of the plasma membrane. However, at least two of them, *erbB* and *fms*, encode transmembrane proteins that are derived from cellular genes that encode receptors for polypeptide growth factors. The c-*erbB* gene encodes the receptor for epidermal growth factor (EGF), and c-*fms* encodes the receptor for the macrophage colony-stimulating factor (M-CSF or CSF-1). (2) Members of the *ras* gene family encode 21 kdal proteins with very similar amino acid sequence that are located at the inner surface of the plasma membrane, bind guanine nucleotides, and have GTPase activity. These properties make *ras* proteins analogous to the G proteins that mediate interactions between receptors and the cAMP system. (3) One oncogene, *sis*, is derived from a gene that encodes the B chain of the platelet-derived growth factor (PDGF). (4) Several oncogenes (*myc*, *fos*, and *myb*) encode proteins that are located in the nucleus and which are thought to alter the expression of other cellular genes, although there is little strong evidence to support this hypothesis. (5) One oncogene, *erbA*, encodes a protein with weak amino acid sequence homology to steroid (glucocorticoid and estrogen) receptors. Several more oncogenes remain unclassified.

Retroviruses that lack oncogenes and that induce leukemias and lymphomas do so by integrating adjacent to cellular proto-oncogenes. In the case of B cell lymphomas induced by avian leukosis viruses, proviruses almost invariably integrate within the first (noncoding) exon or the first intron of c-*myc*, undergo a deletion of a portion of the viral genome, and use their 3' LTR to initiate novel *myc* transcripts that have viral R and U$_5$ sequences covalently joined to their 5' ends. Sometimes, these viruses induce erythroleukemias. In these tumors, proviruses are found in the vicinity of the *erbB* gene. Mouse mammary tumor viruses induce breast tumors in mice by integrating adjacent to either (or both) the *int-1* or *int-2* gene, cellular genes not previously identified as oncogenes of retroviruses.

DNA copies of retroviruses are inherited in many animals. While most of these are probably harmless to their hosts, in some inbred strains of mice the inherited viruses cause cancer. This is usually because these particular proviruses are expressed and yield viruses that are able to replicate extensively in their hosts. Besides the inherited retroviruses of mice, retroviruses can cause cancer in animals outside the laboratory. In cats, retroviruses spread horizontally through biting and scratching and induce lymphomas, sarcomas, and immunosuppressive disorders. In chickens, the viruses can cause leukemias and be passed through the reproductive tract to the eggs and hence to newly hatched chicks.

Bibliography

General References

Bishop, J. M. 1985. "Viral Oncogenes." *Cell* 42:23–38.

Feramisco, J., B. Ozanne, and C. Stiles, eds. 1985. *Cancer Cells 3: Growth Factors and Transformation.* Cold Spring Harbor, N.Y.: Cold Spring Harbor Laboratory.

Gross, L. 1970. *Oncogenic Viruses.* 2nd ed. Elmsford, N.Y.: Pergamon Press.

Levine, A. J., G. F. Vande Woude, W. C. Topp, and J. D. Watson, eds. 1984. *Cancer Cells 1: The Transformed Phenotype.* Cold Spring Harbor, N.Y.: Cold Spring Harbor Laboratory.

Tooze, J., ed. 1980, 1981. *DNA Tumor Viruses. Molecular Biology of Tumor Viruses.* Rev. 2nd ed. Cold Spring Harbor, N.Y.: Cold Spring Harbor Laboratory.

Vande Woude, G. F., A. J. Levine, W. C. Topp, and J. D. Watson, eds. 1984. *Cancer Cells 2: Oncogenes and Viral Genes.* Cold Spring Harbor, N.Y.: Cold Spring Harbor Laboratory.

Varmus, H., and A. J. Levine, eds. 1983. *Readings in Tumor Virology.* Cold Spring Harbor, N.Y.: Cold Spring Harbor Laboratory.

Weiss, R., N. Teich, H. Varmus, and J. Coffin, eds. 1984. *RNA Tumor Viruses. Molecular Biology of Tumor Viruses.* 2nd ed. Cold Spring Harbor, N.Y.: Cold Spring Harbor Laboratory.

Weiss, R., N. Teich, H. Varmus, and J. Coffin, eds. 1985. *RNA Tumor Viruses. Molecular Biology of Tumor Viruses.* 2nd ed. Supplements and appendices. Cold Spring Harbor, N.Y.: Cold Spring Harbor Laboratory.

Cited References

1. Klein, G., U. Bregula, F. Wiener, and H. Harris. 1971. "The Analysis of Malignancy by Cell Fusion." *J. Cell Sci.* 8:659–672.
2. Stanbridge, E. J., C. J. Der, C-J. Doersen, R. Y. Nishimi, D. M. Peehl, B. E. Weissman, and J. E. Wilkinson. 1982. "Human Cell Hybrids: Analysis of Transformation and Tumorgenicity." *Science* 215:252–259.
3. Pereira-Smith, O. M., and J. R. Smith. 1983. "Evidence for the Recessive Nature of Cellular Immortality." *Science* 221:964–966.
4. Craig, R., and R. Sager. 1985. "Suppression of Tumorigenicity in Hybrids of Normal and Oncogene-Transformed CHEF Cells." *Proc. Nat. Acad. Sci.* 82:2062–2066.
5. Cairns, J. 1978. *Cancer, Science, and Society.* San Francisco: Freeman.
6. Upton, A. C. 1982. "Radiation Carcinogenesis." In *Cancer Medicine,* 2nd ed., ed. J. F. Holland and E. Frei. Philadelphia: Lea and Febiger, pp. 96–108.
7. Miller, E. C., and J. A. Miller. 1966. "Mechanisms of Chemical Carcinogenesis: Nature of Proximate Carcinogens and Interactions with Macromolecules." *Pharmacol. Rev.* 18:805.
8. Ames, B. N., W. E. Durston, E. Yamasaki, and F. D. Lee. 1973. "Carcinogens Are Mutagens: A Simple Test System Combining Liver Homogenates for Activation and Bacteria for Detection." *Proc. Nat. Acad. Sci.* 70:2281–2285.
9. Miller, E. C., and J. A. Miller. 1981. "Searches for Ultimate Chemical Carcinogens and Their Reactions with Cellular Macromolecules." *Cancer* 47:2327–2345.
10. Hecker, E., N. E. Fusenigne, W. Kunz, F. Marks, and H. W. Thielmann, eds. 1981. *Carcinogenesis: A Comprehensive Survey.* Vol. 7. *Cocarcinogenesis and Biological Effects of Tumor Promoters.* New York: Raven Press.
11. Tooze, J., ed. 1980, 1981. *DNA Tumor Viruses. Molecular Biology of Tumor Viruses.* Rev. 2nd ed. Cold Spring Harbor, N.Y.: Cold Spring Harbor Laboratory.
12. Weiss, R., N. Teich, H. Varmus, and J. Coffin, eds. 1984. *RNA Tumor Viruses. Molecular Biology of Tumor Viruses.* 2nd ed. Cold Spring Harbor, N.Y.: Cold Spring Harbor Laboratory.
13. Varmus, H., and A. J. Levine, eds. 1983. *Readings in Tumor Virology.* Cold Spring Harbor, N.Y.: Cold Spring Harbor Laboratory.
14. Temin, H. M., and H. Rubin. 1958. "Characteristics of an Assay for Rous Sarcoma Virus and Rous Sarcoma Cells in Tissue Culture." *Virology* 6:669–688.
15. Macpherson, I., and L. Montagnier. 1964. "Agar Suspension Culture for the Selective Assay of Cells Transformed by Polyoma Virus." *Virology* 23:291–294.
16. Gross, L. 1970. *Oncogenic Viruses.* 2nd ed. Elmsford, N.Y.: Pergamon Press.
17. Shin, S-I., V. H. Freedman, R. Risser, and R. Pollack. 1975. "Tumorigenicity of Virus-Transformed Cells in *Nude* Mice Is Correlated Specifically with Anchorage Independent Growth in Vitro." *Proc. Nat. Acad. Sci.* 72:4435–4439.
18. Bishop, J. M. 1985. "Viral Oncogenes." *Cell* 42:23–38.
19. Gross, L. 1953. "A Filterable Agent Recovered from Ak Leukemic Extracts Causing Salivary Gland Carcinomas in C3H Mice." *Proc. Soc. Exp. Biol. Med.* 83:414–421.
20. Sweet, B. H., and M. R. Hilleman. 1960. "The Vacuolating Virus, SV40." *Proc. Soc. Exp. Biol. Med.* 105:420–427.
21. Acheson, N. H. 1980, 1981. "Lytic Cycle of SV40 and Polyoma Virus in DNA Tumor Viruses." In *DNA Tumor Viruses. Molecular Biology of Tumor Viruses.* Rev. 2nd ed., ed. J. Tooze. Cold Spring Harbor, N.Y.: Cold Spring Harbor Laboratory, Chap. 3.
22. Topp, W. C., D. Lane, and R. Pollack. 1980. "Transformation by SV40 and Polyoma Virus." In *DNA Tumor Viruses. Molecular Biology of Tumor Viruses.* Rev. 2nd ed., ed. J. Tooze. Cold Spring Harbor, N.Y.: Cold Spring Harbor Laboratory.
23. Tegtmeyer, P., 1980, 1981. "Genetics of SV40 and Polyoma Virus." In *DNA Tumor Viruses. Molecular Biology of Tumor Viruses.* Rev. 2nd ed., ed. J. Tooze. Cold Spring Harbor, N.Y.: Cold Spring Harbor Laboratory, pp. 297–338d.
24. Graham, F. L. and A. J. Van der Eb. 1973. "A New Technique for the Assay of Infectivity of Human Adenovirus 5 DNA." *Virology* 52:456–467.
25. Kimura, G., and A. Itagaki. 1975. "Initiation and Maintenance of Cell Transformation by Simian Virus 40. A Viral Genetic Property." *Proc. Nat. Acad. Sci.* 72:673–677.
26. Fried, M. 1965. "Cell-Transforming Ability of a Temperature-Sensitive Mutant of Polyoma Virus." *Proc. Nat. Acad. Sci.* 53:486–491.
27. Benjamin, T. L. 1970. "Host Range Mutant of Polyoma Virus." *Proc. Nat. Acad. Sci.* 67:394–399.
28. Graham, F. G., P. J. Abrahams, C. Mudder, H. L. Heijneker, S. O. Warnoar, F. A. J. de Bries, W. Fiers, and A. J. Van der Eb. 1975. "Studies on in Vitro Transformation by DNA and DNA Fragments of Human Adenovirus and SV40." *Cold Spring Harbor Symp. Quant. Biol.* 39:637–650.
29. Sambrook, J., H. Westphal, P. R. Srinivasan, and R. Dulbecco. 1968. "The Integrated State of Viral DNA in SV40-Transformed Cells." *Proc. Nat. Acad. Sci.* 60:1288–1295.
30. Watkins, J. F., and R. Dulbecco. 1967. "Production of SV40 Virus in Heterokaryons of Transformed and Susceptible Cells." *Proc. Nat. Acad. Sci.* 58:1396–1403.
31. Botchan, M., J. Stringer, J. Mitchison, and J. Sambrook. 1980. "Integration and Excision of SV40 DNA from the Chromosome of a Transformed Cell." *Cell* 20:143–152.
32. Botchan, M. 1979. "Studies on SV40 Excision from Cellular Chromosomes." *Cold Spring Harbor Symp. Quant. Biol.* 43:709.

33. Hutchinson, M. A., T. Hunter, and W. Eckhart. 1978. "Characterization of T Antigens in Polyoma-Infected and Transformed Cells." *Cell* 15:65–80.

34. Paucha, E., A. Mellor, R. Harvey, A. E. Smith, R. M. Henwick, and M. Waterfield. 1978. "SV40 Large and Small Tumor Antigens from Simian Virus 40 Have Identical Amino Termini Mapping at 0.65 Map Units." *Proc. Nat. Acad. Sci.* 75:2165–2169.

35. Prives, C., Y. Gluzman, and E. Winocour. 1978. "Cellular and Cell-Free Synthesis of Simian Virus 40 T-Antigens in Permissive and Transformed Cells." *J. Virology* 25:587–595.

36. Staneloni, R. J., M. M. Fluck, and T. L. Benjamin. 1977. "Host-Range Selection of Transformation Defective 'HR-T' Mutants of Polyoma Virus." *Virology* 77:598–609.

37. Bouck, N., N. Beales, T. Shenk, P. Berg, and G. Di Mayorca. 1978. "New Region of the Simian Virus 40 Genome Required for Efficient Viral Transformation." *Proc. Nat. Acad. Sci.* 75:2473–2477.

38. Treisman, R., U. Novak, J. Favaloro, and R. Kamen. 1981. "Transformation of Rat Cells by an Altered Polyoma Virus Genome Expressing Only the Middle-T Protein." *Nature* 292:595–600.

39. Rassoulzadegan, M., A. Cowie, A. Carr, N. Glaichenhaus, R. Kamen, and F. Cuzin. 1982. "The Roles of Individual Polyoma Virus Early Proteins in Oncogenic Transformation." *Nature* 300:713–718.

40. Jat, P., and P. A. Sharp. 1986. "Large T Antigens of sV40 and Polyoma Virus Efficiently Establish Primary Fibroblasts." *J. Virology.* 59:745–750.

41. Courtneidge, S., and A. E. Smith. 1983. "Polyoma Virus Transforming Protein Associates with the Product of the c-*src* Cellular Gene." *Nature* 303:435–539.

42. Lane, D. P., and L. V. Crawford. 1979. "T Antigen Is Bound to a Host Protein in SV40-Transformed Cells." *Nature* 278:261–262.

43. Linzer, D. I. H., and A. J. Levine. 1979. "Characterization of a 54K Dalton Cellular SV40 Tumor Antigen Present in SV40-Transformed Cells and Uninfected Embryonal Carcinoma Cells." *Cell* 17:43–52.

44. Defendi, V. 1963. "Effects of SV40 Virus Immunization on Growth of Transplantable SV40 and Polyoma Virus Tumors in Hamsters." *Proc. Soc. Exp. Biol. Med.* 113:12–16.

45. Khera, K. S., A. Ashkenazi, F. Rapp, and J. L. Melnick. 1963. "Immunity in Hamsters to Cells Transformed in Vitro Among the Papovaviruses." *J. Immunol.* 91:604–613.

46. Tevethia, S. S., R. S. Greenfield, D. C. Flyer, and M. J. Tevethia. 1980. "SV40 Transplantation Antigen Relationship to SV40-Specific Proteins." *Cold Spring Harbor Symp. Quant. Biol.* 44:235–242.

47. Flint, S. J. 1980, 1981. "Transformation by Adenovirus." In *DNA Tumor Viruses. Molecular Biology of Tumor Viruses.* Rev. 2nd ed., ed. J. Tooze. Cold Spring Harbor, N.Y.: Cold Spring Harbor Laboratory, pp. 547–576e.

48. Trenton, J. J., Y. Yabe, and G. Taylor. 1962. "The Quest for Human Cancer Viruses." *Science* 137:835–841.

49. Berk, A. J., and P. A. Sharp. 1978. "Structure of the Adenovirus 2 Early mRNAs." *Cell* 14:695–711.

50. Chow, L. W., T. B. Broker, and J. B. Lewis. 1979. "Complex Splicing Patterns of RNAs from the Early Regions of Adenovirus-2." *J. Mol. Biol.* 134:265–303.

51. Perricaudet, M., G. Akusjarvi, A. Virtanen, and U. Pettersson. 1979. "Structure of Two Spliced mRNAs from the Transforming Region of Human Subgroup C Adenoviruses." *Nature* 281:694–696.

52. Virtanen, A., and U. Pettersson. 1983. "The Molecular Structure of the 9S mRNA from Early Region 1A of Adenovirus Serotype 2." *J. Mol. Biol.* 165:496–499.

53. Virtanen, A., and U. Pettersson. 1985. "Organization of Early Region 1B of Human Adenovirus Type 2: Identification of Four Differentially Spliced mRNAs." *J. Virology* 54:383–391.

54. Houweling, A., J. P. van den Elsen, and A. J. Van der Eb. 1980. "Partial transformation of primary rat cells by the leftmost 4.5% fragment of adenovirus DNA." *Virology* 105:537–550.

55. Van den Elsen, P., A. Houweling, and A. J. Van der Eb. 1983. "Expression of Region E1b of Human Adenovirus in the Absence of Region E1a Is Not Sufficient for Complete Transformation." *Virology* 128:377–390.

56. Land, H., L. P. Parada, and R. A. Weinberg. 1983. "Tumorigenic Conversion of Primary Embryo Fibroblasts Requires at Least Two Cooperating Oncogenes." *Nature* 304:596–602.

57. Ruley, H. E. 1983. "Adenovirus Early Region 1A Enables Viral and Cellular Transforming Genes to Transform Primary Cells in Culture." *Nature* 304:304–306.

58. Sarnow, P., Y. S. Ho, J. Williams, and A. J. Levine. 1982. "Adenovirus E1b-58kd Tumor Antigen and SV40 Large Tumor Antigen are Physically Associated with the Same 54kd Cellular Protein in Transformed Cells." *Cell* 28:387–394.

59. Kovesdi, I., R. Reichel, and J. R. Nevins. 1986. "Identification of a Cellular Transcription Factor Involved in E1A *Trans*-Activation." *Cell* 45:219–228.

60. Schrier, P. I., R. Bernards, R. T. M. J. Vaessen, A. Houweling, and A. J. Van der Eb. 1983. "Expression of Class I Major Histocompatibility Antigens Switched Off by Highly Oncogenic Adenovirus 12 in Transformed Rat Cells." *Nature* 305:771–775.

61. Coffin, J. M. 1985. "Structure of the Retroviral Genome." In *RNA Tumor Viruses. Molecular Biology of Tumor Viruses,* 2nd ed., ed. R. Weiss, N. Teich, H. Varmus, and J. Coffin. Cold Spring Harbor, N.Y.: Cold Spring Harbor Laboratory, pp. 261–368.

62. Coffin, J. M. 1985. "Genome Structure." In *RNA Tumor Viruses. Molecular Biology of Tumor Viruses,* 2nd ed., ed. R. Weiss, N. Teich, H. Varmus, and J. Coffin. Cold Spring Harbor, N.Y.: Cold Spring Harbor Laboratory, pp. 17–73.

63. Varmus, H. 1982. "Form and Function of Retroviral Proviruses." *Science* 216:812–821.

64. Bishop, J. M. 1983. "Cellular Oncogenes and Retroviruses." *Ann. Rev. Biochem.* 52:301–354.

65. Bishop, J. M., and H. E. Varmus. 1982. "Functions and Origins of Retroviral Transforming Genes." In *RNA Tumor Viruses. Molecular Biology of Tumor Viruses,* 2nd ed., ed. R. Weiss, N. Teich, H. Varmus, and J. Coffin. Cold Spring Harbor, N.Y.: Cold Spring Harbor Laboratory, pp. 999–1108.

66. Martin, G. S. 1970. "Rous Sarcoma Virus: A Function Required for the Maintenance of the Transformed State." *Nature* 227:1021–1023.

67. Duesberg, P., and P. Vogt. 1973. "RNA Species Obtained from Clonal Lines of Avian Sarcoma and from Avian Leukosis Virus." *Virology* 54:207–219.

68. Coffin, J. M., and M. A. Billeter. 1976. "A Physical Map of the Rous Sarcoma Virus Genome." *J. Mol. Biol.* 100:293–318.

69. Hayward, W. S. 1977. "Size and Genetic Content of Viral RNAs in Avian Oncovirus Infected Cells." *J. Virology* 24: 47–63.

70. Hanafusa, H., T. Hanafusa, and H. Rubin. 1963. "The Defectiveness of Rous Sarcoma Virus." *Proc. Nat. Acad. Sci.* 49:572–580.

71. Temin, H. 1963. "Separation of Morphological Conversion and Virus Production in Rous Sarcoma Virus Infection." *Cold Spring Harbor Symp. Quant. Biol.* 27:407–414.

72. Hartley, J. W., and W. P. Rowe. 1966. "Production of Altered Cell Foci in Tissue Culture by Defective Moloney Sarcoma Virus Particles." *Proc. Nat. Acad. Sci.* 55:780–786.

73. Vogt, P. K. 1971. "Spontaneous Segregation of Nontransforming Viruses from Cloned Sarcoma Viruses." *Virology* 46:939–946.

74. Stehelin, D., H. E. Varmus, J. M. Bishop, and P. K. Vogt. 1976. "DNA Related to the Transforming Gene(s) of Avian Sarcoma Viruses Is Present in Normal Avian DNA." *Nature* 260:170–173.

75. Scolnick, E. M., E. Rands, D. Williams, and W. P. Parks. 1973. "Studies on the Nucleic Acid Sequences of Kirsten Sarcoma Virus: A Model for Formation of a Mammalian RNA-Containing Sarcoma Virus." *J. Virology* 12:458–463.

76. Goldfarb, M., and R. A. Weinberg. 1981. "Generation of Novel, Biologically Active Harvey Sarcoma Viruses Via Apparent Illegitimate Recombination." *J. Virology* 38:136–150.

77. Swanstrom, R., R. C. Parker, H. E. Varmus, and J. M. Bishop. 1983. "Transduction of a Cellular Oncogene—The Genesis of Rous Sarcoma Virus." *Proc. Nat. Acad. Sci.* 80:2519–2523.

78. Takeya, T., and H. Hanafusa. 1983. "Structure and Sequence

of the Cellular Gene Homologous to the RSV *src* Gene and the Mechanism for Generating the Transforming Virus." *Cell* 32:881–890.

79. Spector, D. H., H. E. Varmus, and J. M. Bishop. 1978. "Nucleotide Sequences Related to the Transforming Gene of Avian Sarcoma Virus Are Present in DNA of Uninfected Vertebrates." *Proc. Nat. Acad. Sci.* 75:4102–4106.

80. Shilo, B., and R. A. Weinberg. 1981. "DNA Sequences Homologous to Vertebrate Oncogenes Are Conserved in *Drosophila melanogaster*." *Proc. Nat. Acad. Sci.* 78:6789–6792.

81. DeFeo-Jones, D., E. M. Scolnick, R. Koller, and R. Dhar. 1983. "*ras*-Related Gene Sequences Identified and Isolated from *Saccharomyces cerevisiae*." *Nature* 306:707–709.

82. Shibuya, M., T. Hanafusa, H. Hanafusa, and J. R. Stephenson. 1980. "Homology Exists Among the Transforming Sequences of Avian and Feline Sarcoma Viruses." *Proc. Nat. Acad. Sci.* 81:4697–4701.

83. Blair, D. G., M. Oskarsson, T. G. Wood, W. L. McClements, P. J. Fischinger, and G. F. Vande Woude. 1981. "Activation of the Transforming Potential of a Normal Cell Sequence: A Model for Oncogenesis." *Science* 212:941–943.

84. Chang, E. M., M. E. Furth, E. M. Scolnick, and D. R. Lowy. 1982. "Tumorigenic Transformation of Mammalian Cells Induced by a Normal Human Gene Homologous to the Oncogene of Harvey Murine Sarcoma Virus." *Nature* 297:479–483.

85. Iba, H., T. Takeya, F. R. Cross, T. Hanafusa, and H. Hanafusa. 1984. "Rous Sarcoma Virus Variants That Carry the Cellular *src* Gene Instead of the Viral *src* Gene Cannot Transform Chicken Embryo Fibroblasts." *Proc. Nat. Acad. Sci.* 81:4424–4429.

86. Parker, R. C., H. E. Varmus, and J. M. Bishop. 1984. "Expression of v-*src* and Chicken c-*src* in Rat Cells Demonstrates Qualitative Differences Between pp60^{v-src} and pp60^{c-src}." *Cell* 37:131–139.

87. Shalloway, D., P. M. Coussens, and P. Yaciuk. 1984. "Overexpression of the c-*src* Protein Does Not Induce Transformation of NIH 3T3 Cells." *Proc. Nat. Acad. Sci.* 81:7071–7075.

88. Miller, A. D., T. Curran, and I. M. Verma. 1984. "c-*fos* Protein Can Induce Cellular Transformation: A Novel Mechanism of Activation of a Cellular Oncogene." *Cell* 36:51–60.

89. Hunter, T. 1984. "The Proteins of Oncogenes." *Sci. Amer.* 251:70–79.

90. Kitamura, N., A. Kitamura, K. Toyoshima, Y. Hirayama, and M. Yoshida. 1982. "Avian Sarcoma Virus Y73 Genome Sequence and Structural Similarity of Its Transforming Gene Product to That of Rous Sarcoma Virus." *Nature* 297:205–207.

91. Brugge, J. S., and R. L. Erikson. 1977. "Identification of a Transformation-Specific Antigen Induced by an Avian Sarcoma Virus." *Nature* 269:346–348.

92. Collett, M. S., and R. L. Erikson. 1978. "Protein Kinase Activity Associated with the Avian Sarcoma Virus *src* Gene Product." *Proc. Nat. Acad. Sci.* 75:2021–2024.

93. Levinson, A. D., H. Oppermann, L. Levintow, H. E. Varmus, and J. M. Bishop. 1978. "Evidence That the Transforming Gene of Avian Sarcoma Virus Encodes a Protein Kinase Associated with a Phosphoprotein." *Cell* 15:561–572.

94. Hunter, T., and B. M. Sefton. 1980. "Transforming Gene Product of Rous Sarcoma Virus Phosphorylates Tyrosine." *Proc. Nat. Acad. Sci.* 77:1311–1315.

95. Cooper, J. A., K. L. Gould, C. A. Cartwright, and T. Hunter. 1986. "Tyr^{527} Is Phosphorylated in pp60^{c-src}: Implications for Regulation." *Science* 231:1431–1434.

96. Radke, K., and G. S. Martin. 1979. "Transformation by Rous Sarcoma Virus: Effects of *src* Gene Expression on the Synthesis and Phosphorylation of Cellular Polypeptides." *Proc. Nat. Acad. Sci.* 76:5212–5216.

97. Cooper, J. A., and T. Hunter. 1981. "Four Different Classes of Retroviruses Induce Phosphorylation of Tyrosine Present in Similar Cellular Proteins." *Mol. Cell Biol.* 1:394–407.

98. Sefton, B. M., T. Hunter, E. H. Ball, and S. J. Singer. 1981. "Vinculin: A Cytoskeletal Target of the Transforming Protein of Rous Sarcoma Virus." *Cell* 24:165–174.

99. Rohrschneider, L., M. Rosok, and K. Shriver. 1982. "Mechanism of Transformation by Rous Sarcoma Virus: Events Within Adhesion Plaques." *Cold Spring Harbor Symp. Quant. Biol.* 46:953–965.

100. Cross, F. R., E. A. Garber, and H. Hanafusa. 1985. "N-Terminal Deletions in Rous Sarcoma Virus p60^{src}: Effects on Tyrosine Kinase and Biological Activities and on Recombination in Tissue Culture with the Cellular *src* Gene." *Mol. Cell Biol.* 5:2789–2795.

101. Sugimoto, Y., M. Whitman, L. C. Cantley, and R. L. Erikson. 1984. "Evidence That the Rous Sarcoma Virus Transforming Gene Product Phosphorylates Phosphatidylinositol and Diacylglycerol." *Proc. Nat. Acad. Sci.* 81:2117–2122.

102. Doolittle, R. F., M. W. Hunkapiller, L. E. Hood, S. G. Devare, K. C. Robbins, S. A. Aaronson, and H. M. Antoniades. 1983. "Simian Sarcoma Virus Onc Gene, v-*sis*, Is Derived from the Gene (or Genes) Encoding a Platelet-Derived Growth Factor." *Science* 221:275–276.

103. Waterfield, M. D., G. J. Scrace, N. Whittle, P. Stroobant, A. Johnson, A. Wasteson, B. Westemark, C-H. Heldin, J. S. Huang, and T. F. Deuel. 1983. "Platelet-Derived Growth Factor is Structurally Related to the Putative Transforming Protein p28^{sis} of Simian Sarcoma Virus." *Nature* 304:35–39.

104. Gazit, A., H. Igarashi, I-M. Chiu, A. Srinivasan, A. Yaniv, S. R. Tronick, K. C. Robbins, and S. A. Aaronson. 1984. "Expression of the Normal Human *sis*/PDGF-2 Coding Sequence Induces Cellular Transformation." *Cell* 39:89–97.

105. Leal, F., L. T. Lewis, K. C. Robbins, and S. A. Aaronson. 1985. "Evidence That the v-*sis* Gene Product Transforms by Interaction with the Receptor for Platelet-Derived Growth Factor." *Science* 230:327–330.

106. C. Betsholz, A. Johnsson, C-H. Heldin, B. Westermark, P. Lind, M. S. Urdea, R. Eddy, T. B. Shows, K. Philpott, A. L. Mellor, T. J. Knott, and J. Scott. 1986. "cDNA Sequence and Chromosomal Localization of Human Platelet-Derived Growth Factor A-Chain and Its Expression in Tumor Cell Lines." *Nature* 320:695–699.

107. Downward, J., Y. Yarden, E. Mayes, G. Scrace, N. Totty, P. Stockwell, A. Ullrich, J. Schlessinger, and M. D. Waterfield. 1984. "Close Similarity of Epidermal Growth Factor Receptor and v-*erb*-B Oncogene Protein Sequences." *Nature* 307:521–527.

108. Sherr, C. J., C. W. Rettenmier, R. Sacca, M. F. Roussel, A. T. Look, and E. R. Stanley. 1985. "The c-*fms* Proto-Oncogene Product Is Related to the Receptor for the Mononuclear Phagocyte Growth Factor, CSF-1." *Cell* 41:665–676.

109. Ellis, R. W., D. DeFeo, T. Y. Shih, M. A. Gonda, H. A. Young, N. Tsvchida, D. R. Lowy, and E. M. Scolnick. 1981. "The p21 *src* Genes of Harvey and Kirsten Sarcoma Viruses Originate from Divergent Members of a Family of Normal Vertebrate Genes." *Nature* 292:506–511.

110. Shih, T. Y., A. G. Papageorge, P. E. Stokes, M. O. Weeks, and E. M. Scolnick. 1980. "Guanine Nucleotide-Binding and Autophosphorylating Activities Associated with the p21 Protein of Harvey Murine Sarcoma Virus." *Nature* 287:686–691.

111. Gibbs, J. M., I. S. Sigal, M. Poe, and E. M. Scolnick. 1984. "Intrinsic GPTase Activity Distinguishes Normal and Oncogenic *ras* p21 Molecules." *Proc. Nat. Acad. Sci.* 81:5704–5708.

112. Donner, P., I. Greiser-Wilke, and K. Moelling. 1982. "Nuclear Localization and DNA Binding of the Transforming Gene Product of Avian Myelocytomatosis Virus." *Nature* 296:262–265.

113. Hann, S. R., H. D. Abrams, L. R. Rohrschneider, and R. N. Eisenman. 1983. "Proteins Encoded by the v-*myc* and c-*myc* Oncogenes: Identification and Localization in Acute Leukemia Virus Transformants and Bursal Lymphoma Cell Line." *Cell* 34:789–798.

114. Eisenman, R. N., C. Y. Tachibana, H. D. Abrams, and S. R. Hann. 1985. "v-*myc* and c-*myc* Encoded Proteins Are Associated with the Nuclear Matrix." *Mol. Cell Biol.* 5:114–126.

115. Ralston, R., and J. M. Bishop. 1983. "The Protein Products of the Oncogene *myc*, *myb* and Adenovirus E1a are Structurally Related." *Nature* 306:803–806.

116. Kelly, K., B. H. Cochran, C. D. Stiles, and P. Leder. 1983. "Cell-Specific Regulation of the c-*myc* Gene by Lymphocyte Mitogens and Platelet-Derived Growth Factor." *Cell* 35:603–610.

117. Van Beveren, C., F. van Straaten, T. Curran, R. Muller, and I. M. Verma. 1983. "Analysis of FBJ-MuSV Provirus and c-*fos* (Mouse) Gene Reveals That Viral and Cellular *fos* Gene Products Have Different Carboxy Termini." *Cell* 32:1241–1255.

118. Cochran, B. H., J. Zullo, I. M. Verma, and C. D. Stiles. 1984. "Expression of the c-*fos* Gene and of a *fos*-Related Gene Is Stimulated by Platelet-Derived Growth Factor." *Science* 226:1080–1082.

119. Treisman, R. 1985. "Transient Accumulation of c-*fos* RNA Following Serum Stimulation Requires a Conserved 5' Element and c-*fos* 3' Sequences." *Cell* 42:889–902.

120. Weinberger, C., S. M. Hollenberg, M. G. Rosenfeld, and R. M. Evans. 1985. "Domain Structure of Human Glucocorticoid Receptor and Its Relationship to the v-*erb-A* Oncogene Product." *Nature* 318:670–672.

121. Greene, G. L., P. Gilna, M. Waterfield, A. Baker, Y. Hort, and J. Shine. 1986. "Sequence and Expression of Human Estrogen Receptor Complementary DNA." *Science* 231:1150–1154.

122. Green, S., P. Walter, V. Kumar, A. Krust, J-M. Bornet, P. Argos, and P. Chambon. 1986. "Human Oestrogen Receptor cDNA: Sequence, Expression and Homology to v-*erb-A*." *Nature* 320:134–139.

123. Anderson, S. M., W. S. Hayward, B. G. Neel, and H. Hanafusa. 1980. "Avian Erythroblastosis Virus Produces Two mRNAs." *J. Virology* 36:676–683.

124. Mushinski, J. F., M. Potter, S. R. Bauer, and E. P. Reddy. 1983. "DNA Rearrangement and Altered RNA Expression of the c-*myb* Oncogene in Mouse Plasmacytoid Lymphosarcomas." *Science* 220:795–798.

125. Hayward, W. S., B. G. Neel, and S. M. Astrin. 1981. "Activation of a Cellular Onc Gene by Promoter Insertion in ALV-Induced Lymphoid Leukosis." *Nature* 290:475–480.

126. Payne, G. S., J. M. Bishop, and H. E. Varmus. 1982. "Multiple Arrangements of Viral DNA and an Activated Host Oncogene in Bursal Lymphomas." *Nature* 295:209–214.

127. Tsichlis, P. N., and J. M. Coffin. 1980. "Recombinants Between Endogenous and Exogenous Avian Tumor Viruses: Role of the C Region and Other Portions of the Genome in the Control of Replication and Transformation." *J. Virology* 33:238–249.

128. Robinson, H. L., and G. C. Gagnon. 1986. "Patterns of Proviral Insertion in Avian Leukosis Virus-Induced Lymphomas." *J. Virology* 57:28–36.

129. Nilsen, T. W., P. A. Maroney, R. G. Goodwin, F. M. Rottman, L. B. Crittenden, M. A. Raines, and H. Kung. 1985. "c-*erb-B* Activation in ALV-Induced Erythroblastosis: Novel RNA Processing and Promoter Insertion Result in Expression of an Amino-Truncated EGF Receptor." *Cell* 41:719–726.

130. Chatis, P. A., C. A. Holland, J. W. Hartley, W. D. Rowe, and N. Hopkins. 1983. "Role for the 3' End of the Genome in Determining Disease Specificity of Friend and Moloney Murine Leukemia Viruses." *Proc. Nat. Acad. Sci.* 80:4408–4411.

131. DesGroseillers, L., E. Rassart, and P. Jolicoeur. 1983. "Thymotropism of Murine Leukemia Virus Is Conferred by Its Long Terminal Repeat." *Proc. Nat. Acad. Sci.* 80:4203–4207.

132. Nusse, R., A. Van Ooyen, D. Cox, Y. K. T. Fung, and H. Varmus. 1984. "Mode of Proviral Activation of a Putative Mammary Oncogene (*int-1*) on Mouse Chromosome 15." *Nature* 307:131–136.

133. Peters, G., S. Brookes, R. Smith, and C. Dickson. 1983. "Tumorigenesis by Mouse Mammary Tumor Virus: Evidence for a Common Region for Provirus Integration in Mammary Tumors." *Cell* 33:369–377.

134. Stewart, T. A., P. K. Pattengale, and P. Leder. 1984. "Spontaneous Mammary Adenocarcinomas in Transgenic Mice That Carry and Express MMTV/*myc* Fusion Genes." *Cell* 38:627–637.

135. Furth, J., H. R. Seibold, and R. R. Rathbone. 1933. "Experimental Studies on Lymphomatosis of Mice." *Amer. J. Cancer* 19:521–604.

136. Gross, L., 1951. "Spontaneous Leukemia Developing in C3H Mice Following Inoculation, in Infancy, with AK-Leukemic Extracts, or AK-Embryos." *Proc. Soc. Exp. Biol. Med.* 76:27–32.

137. Rowe, W. P. 1973. "Genetic Factors in the Natural History of Murine Leukemia Virus Infection: G. H. Clowes Memorial Lecture." *Cancer Res.* 33:3061–3068.

138. Chattopadhyay, S. K., W. P. Rowe, N. M. Teich, and D. R. Lowy. 1975. "Definitive Evidence That the Murine C-Type Virus Inducing Locus *Akv-1* Is Viral Genetic Material." *Proc. Nat. Acad. Sci.* 72:906–910.

139. Hartley, J. W., N. K. Wolford, L. J. Old, and W. P. Rowe. 1977. "A New Class of Murine Leukemia Virus Associated with Development of Spontaneous Lymphomas." *Proc. Nat. Acad. Sci.* 74:789–792.

140. Cuypers, H. T., G. Selten, W. Quint, M. Zijlstra, E. R. Maandag, W. Boelens, P. Van Wezenbeek, C. Melief, and A. Berns. 1984. "Murine Leukemia Virus-Induced T-Cell Lymphomagenesis: Integration of Proviruses in a Distinct Chromosomal Region." *Cell* 37:141–150.

141. Tsichlis, P. N., P. G. Strauss, and L. F. Hu. 1983. "A Common Region for Proviral DNA Integration in Mo MuLV-Induced Rat Thymic Lymphomas." *Nature* 302:445–448.

142. Li, Y., C. A. Holland, J. W. Hartley, and N. Hopkins. 1984. "Viral Integration Near c-*myc* in 10–20% of MCF 247 Induced AKR Lymphomas." *Proc. Nat. Acad. Sci.* 81:6808–6811.

143. Steffen, D. 1984. "Proviruses Are Adjacent to c-*myc* in Some Murine Leukemia Virus-Induced Lymphomas." *Proc. Nat. Acad. Sci.* 81:2097–2101.

144. Hardy, W. D., Jr., P. W. Hess, E. G. MacEwen, A. J. McClelland, E. E. Zuckerman, M. Essex, S. M. Cotter, and O. Jarrett. 1976. "Biology of Feline Leukemia Virus in the Natural Environment." *Cancer Res.* 36:582–588.

145. Rubin, H., L. Fanshier, A. Cornelius, and W. F. Hughes. 1962. "Tolerance and Immunity in Chickens After Congenital and Contact Infection with an Avian Leukosis Virus." *Virology* 17:143–156.

146. Gardner, M. B., V. Klement, R. R. Rongey, P. McConahey, J. D. Estes, and R. J. Huebner. 1976. "Type C Virus Expression in Lymphoma-Paralysis-Prone Wild Mice." *J. Nat. Cancer Inst.* 57:585–590.

CHAPTER 27

The Origins of Human Cancer

During the past 30 years it has become increasingly clear that cancer is a genetic disease that results when multiple mutations (genetic events) accumulate within the DNA of a single somatic cell, causing the cell to lose growth control. But the question was, How could we identify the cancer-causing genes in human tumors? Furthermore, even if we could identify the genes, could we understand their function or discover how they had gone astray in the cancer cell? In the past few years, we have begun to obtain such convincing answers for several cancers that we can begin to use this knowledge to search for means both to prevent and cure these diseases. This chapter describes the types of research that suddenly came together in the 1970s and '80s to begin revealing the genetic basis of human cancer.

THE GENETIC BASIS OF HUMAN CANCER

Cancer Is Clonal and Arises More Frequently in Cell Types That Undergo Frequent Division[1]

The human body consists of about 10^{13} cells, and these are differentiated into the many different types that form the different organs: skin, liver, kidney, blood cells, and so on. In the mature body, some cell types divide rarely or even never (e.g., nerve cells), while others, like skin and the progenitors of the circulating blood cells, must divide throughout life in order to replace the billions of cells that die every day. Clearly, a very, very carefully controlled program must determine the growth of every type of cell in the body, since a few extra doublings or insufficient divisions would soon produce chaos. Tumors arise when a single cell loses its sense of these controls.

Evidence that most tumors arise from a single cell came first from analyzing tumors in females who inherit different isozymes of the X-chromosome-coded enzyme *glucose-6-phosphate dehydrogenase (G6PD)*. The human female has two X chromosomes, but only one of the two is active in each cell. Early in embryonic life, one X chromosome in each cell is modified so that it is transcriptionally silent. However, the particular chromosome (paternal or maternal) chosen for inactivation is different in different cells. Thus, females are **mosaics**: Most of their tissues consist of a mixture of cells, some using the paternal and some the maternal X chromosome. If a tumor arose in many cells, then different cells of the same tumor should express different isozymes of G6PD. However, tumors turn out to express only one isozyme, indicating their origin from a single cell.

There are two types of tumors, only one of which is cancerous. If a cell divides when it should not, but nonetheless stays within its normal location, then its growth produces a mass of cells called a *benign tumor*, which is not a cancer. Benign tumors can become very large so that they require surgical removal, but they are not usually life threatening. In contrast, some cells that begin to divide when they should not also acquire the ability to invade surrounding tissues and to move to alien sites in the body (i.e., **metastasize**). Thus, a tumor may begin, for example, in a single lung cell; but at some point after a mass of cells has formed, tumorous lung cells from the mass break off and migrate to the brain or other organs where they seed new tumor growths, called *metastases*. Metastatic cells, still identifiable as lung cells in the above example, multiply and destroy the architecture and ultimately the function of the invaded organ. Tumors that are highly invasive of surrounding tissues and that can metastasize are called *malignant tumors* or *cancers*. It is their ability to spread to vital organs and so escape the surgeon's knife that makes them deadly.

Cancers can arise in many types of differentiated cells and since cancers are classified according to the cell type in which they arise, there are over 200 different forms of human cancer, each named for the cell type of origin. For example, gliomas are cancers of glial cells; hepatomas, cancers of liver cells (hepatocytes); melanomas, cancers of the pigment-producing melanocytes that are particularly densely located in moles. The many different forms of human cancer are classified into one of three groups. **Carcinomas** include all the cancers that arise in cells of the sheets or epithelia that cover our surfaces (skin, gut, etc.). **Sarcomas** include cancers of the supporting tissues (bone, muscle, blood vessels, fibroblasts). **Leukemias** and **lymphomas** include cancers of the cells that produce the circulating cells of the blood and the immune system. More than 90 percent of human cancers are carcinomas. Sarcomas and leukemias together account for only about 8 percent of human tumors. Furthermore, among the carcinomas, cancers of the lung, breast, and colon are the most common fatal tumors in Western countries and account for more than half of cancer deaths, while liver cancer is most common in Asia and Africa. Although not perfectly correlated, cancers tend to arise in cells that continue to proliferate in adult life. In contrast, nerve cells, for example, never become cancerous in adults, although they can do so in infants, where they are still undergoing division. In addition, major sites of cancer are the external epithelia that are in contact with the environment: the air we breathe and the food and water we consume.

Human Cancers Are the
Result of Multiple Mutations[2-5]

Although a few forms of cancer are associated with childhood, for most cancers incidence increases sharply with age. For example, the death rate from cancer of the large intestine increases more than a thousandfold between the ages of 30 and 80 (Figure 27-1), and this is quite typical of many cancers. The model that has been used to explain curves like the one in Figure 27-1 is that cancer arises when a cell

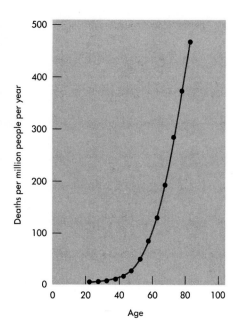

Figure 27-1
Annual death rate in relation to age due to cancer of the large intestine. Statistics are for the United States. The data are plotted on a linear scale. The death rate from this disease increases about a thousandfold between the ages of 30 and 80.

accumulates several specific mutations (genetic events). Since mutations can occur at any time, the probability that any particular cell will acquire a particular mutation increases in direct proportion to age. The probability that the cell has acquired mutations in all N of the genes needed to produce a cancer cell rises as the Nth power of age. Although it is very difficult to arrive at reliable values for N, extensive analysis of cancer incidence data has led to estimates of 2 to 7. Most leukemias are believed to result from the accumulation of two to four specific mutations (genetic events) within a single cell, while carcinomas may require anywhere from two to six or seven.

Other evidence that cancer results from multiple mutations comes from studying certain individuals who inherit a very high probability of developing a specific form of cancer, for example, a tumor called retinoblastoma. These people appear to have inherited one of the mutations that can lead to the cancer. However, only when a particular cell acquires one or more additional changes does the disease develop.

Epidemiology Reveals That Environmental Factors Cause Most Cancers[6]

Until about 1940, it was widely believed that cancer was an inevitable consequence of aging, that somehow the very process of living included the inevitable creation of cancer cells at some significant frequency. Since cell division requires DNA replication, and DNA replication produces mutations at a low rate, even today this remains an entirely reasonable hypothesis. Fortunately, however, it does not seem to be true for the majority of cancers that kill us. We know this because of epidemiological studies that have analyzed statistics of cancer deaths in relation to country, social group, occupation, living habits, and different time periods. The major finding from these studies is that there is great variation in the incidence of most cancers from place to place, from time to time, and depending on people's activities (Table 27-1). Furthermore, when individuals migrate to a new region, usually, within one or two generations, they acquire the cancer incidence profile of their new home, even when they continue to intermarry with genetically similar individuals. Thus, environmental factors rather than intrinsic, inevitable aging processes must play a large role in determining the probability of getting cancer.

Table 27-1 Geographic Variation in Incidence of Some Common Cancers

Type of Cancer	Region of Highest Incidence	Range of Variation*	Region of Lowest Incidence
Lung	Great Britain	35	Nigeria
Stomach	Japan	25	Uganda
Liver	Mozambique	70	Norway
Skin	Queensland	>200	Bombay
Colon	United States	10	Nigeria
Breast	United States	15	Uganda

*Range of variation is the highest incidence observed in any country divided by the lowest incidence.
SOURCE: Data from R. Doll, "Strategy for Detection of Cancer Hazards to Man," *Nature* 265 (1977): 589–596. Except for breast cancer, the data are for men.

It is not always easy to identify the factors in the environment, diet, or personal habits that cause cancer. However, clear-cut examples now include cigarette smoking, which probably causes 85 to 90 percent of all lung cancers; hepatitis B virus infection, which in some unknown way is thought to lead to 80 to 90 percent of liver cancers; asbestos, which causes mesotheliomas and, in combination with cigarette smoking, greatly increases the incidence of lung cancer: and sunlight, which induces skin cancers.

Mutagens and Certain Viruses Are Carcinogens for Humans[7, 8, 9]

The fact that the cancerous phenotype of a cell is stably inherited from generation to generation had suggested many years ago that genetic changes in DNA might underlie the loss of cellular growth control seen in cancers. However, this evidence alone was not at all compelling, since specific patterns of gene expression, as in differentiated cell types, can be stably inherited in the absence of changes to DNA base sequences. It was the realization that the majority of carcinogens are mutagens for DNA that first led to general acceptance of the idea that many cancers arise from changes at the DNA level. In addition, two lines of evidence had previously indicated that at least some human cancers involve very specific alterations in DNA. First, as already mentioned, a high probability of contracting certain very rare forms of cancer is inherited; and in some of these cases, specific chromosomal defects are found to be the basis for the inherited tendency. Second, malignant tumors frequently have specific chromosome abnormalities that are lacking in the normal cells of the same individual. As we shall see, advances in molecular cloning technologies have made it possible to isolate at least some of the specific genes that are altered to produce cancer cells.

Frequently in conflict with the notion of a mutational origin of human cancer has been the belief that viruses are a significant cause of human cancer. Now there is little doubt that both viruses (e.g., hepatitis B) and other types of carcinogens cause human cancers. Furthermore, some viruses may cause cancer precisely because they can cause mutations in cellular DNA. The only question is the extent to which viruses versus carcinogens or other agents contribute to the disease.

HUMAN CANCER GENES DETECTED BY DNA TRANSFECTION

Chromosomal DNA from Human Tumors Can Induce Transformed Foci in 3T3 Cells[10–15]

One type of experiment that led to the isolation of cellular genes involved in human cancer relies on the technique of DNA transfection. Many types of cells, from bacterial to mammalian cells, can take up DNA and, at a low frequency, integrate some of the ingested DNA into their own chromosomes. Furthermore, if the donor DNA encodes a genetic trait that the recipient cell lacks, the recipient cell can acquire that trait. In bacteria, the process is called **transformation**

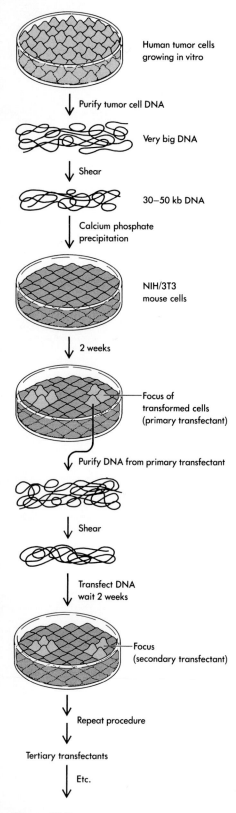

Figure 27-2
Assay for cellular oncogenes using transfection transformation of mouse 3T3 cells.

(confusingly, the same word used to describe the cancerous alteration of animal cells growing in tissue culture); in mammalian cells, it is usually called **transfection**. Although the frequency of DNA uptake is extremely low, it can be greatly increased through special techniques. In mammalian cells, transfection frequency increases dramatically if the DNA is fed to the recipient cells as a microscopic-size DNA–calcium phosphate coprecipitate.

If human cancer cells have altered genes that cause the cells to become cancerous, then, it was reasoned, DNA from cancer cells might be able to confer the cancerous phenotype on noncancerous recipient cells. To look for the hypothetical human cancer genes, DNA from human tumor cell lines was purified, sheared mechanically to break it into manageable pieces of about 30 to 50 kilobase pairs, dissolved in phosphate buffer, and then precipitated by the addition of calcium chloride (Figure 27-2). The solution was poured onto a monolayer of mouse 3T3 cells and the fine precipitate allowed to settle down onto the cells. 3T3 cells were selected as recipients because they are highly efficient at DNA uptake and also because they had been used for decades to assay cancer-causing genes of tumor viruses. Remarkably, DNA from human tumor cell lines was found to induce foci of transformed cells in the 3T3 monolayer, and these transformed cells could induce tumors in animals. Importantly, DNAs isolated from normal cells and used side by side as a negative control did not induce foci. This experiment indicated that human cancer cells have indeed sustained genetic alterations that can be assayed by DNA transfection. The number of foci obtained in this type of experiment increases linearly with the amount of DNA added, in-

Table 27-2   Human *ras* Oncogenes Detected by DNA Transfection

	Cell Lines (Positive/Tested)	Tumors	*ras* Oncogenes H	K	N
Carcinomas					
Bladder	2/7	2/25	3	1	—
Breast	1/19	0/18	1	—	—
Colon	1/2	2/2	—	2	1
Liver	1/4	1/8	—	—	2
Lung	6/12	1/5	1	5	1
Ovary	0/1	1/5	—	1	—
Pancreas	1/1	1/2	—	2	—
Sarcomas					
Fibrosarcoma	1/4	0/4	—	—	1
Rhabdomyosarcoma	1/3	1/2	—	1	1
Leukemias					
ALL	7/12	—	—	1	6
AML	1/2	—	—	—	1
APL	1/1	0/1	—	—	1
CML	1/2	0/3	—	—	1
Total:	24/70*	9/75	5	13	15

*The data are compiled from many published studies. In some reports, only cell lines giving positive results are described. Thus, in this table, the ratio of positive cell lines to those tested is an overestimate. Usually, about 10 to 20 percent of cell lines are positive for transfectable *ras* genes; the number varies considerably, depending on tumor type.
SOURCE: M. Barbacid, "Human Oncogenes," in *Important Advances in Oncology*, ed. V. DeVita, S. Hellman, and S. Rosenberg (Philadelphia: Lippincott, 1986).

dicating that foci are induced by a single gene that must act in a dominant fashion.

Foci of 3T3 cells transformed by human tumor DNA are called primary transfectants. When DNA from these cells is purified, it, too, can pass the transformed phenotype on to fresh 3T3 cells, inducing foci called secondary transfectants. In contrast, DNA from normal 3T3 cells cannot induce foci. The procedure can be repeated serially over and over again, generating tertiary transfectants, and so on.

The first human tumor cells whose DNA was shown to induce foci in the 3T3 transfection transformation assay were certain established cell lines derived from tumors of the bladder, lung, colon, lymphoid cells, and a neuroblastoma. Of the many human tumor cell lines that have now been tested by this method, approximately 15 to 20 percent possess detectable transforming genes (Table 27-2). As soon as these results were obtained, it was immediately of great interest to know whether DNA isolated from primary human tumor material fresh from a patient rather than from established cell lines would also possess transforming activity in the 3T3 focus assay. Once again, about 15 percent of primary human tumor DNAs have been found to possess focus-inducing activity (see Table 27-2). Together these results show that at least some human tumors of unidentified cause possess genetic alterations that result in dominant transforming genes. Furthermore, since DNA from normal human cells lacks such transforming genes, the alterations must be related to the cancerous phenotype of the tumor cells.

Molecular Cloning of Transforming Genes from Human Tumors[16, 17]

To begin characterizing the human transforming genes biochemically, tumor DNAs were digested with various restriction endonucleases that cleave at specific 6-base sequences; the cleaved DNAs then were tested to see if they retained focus-inducing ability. We would expect a gene to possess cleavage sites for some enzymes but not others, and thus we would expect that some enzymes but not others would destroy its biological activity. This is precisely what was found for the human transforming genes. Each tumor DNA possesses a characteristic pattern of inactivation by restriction endonucleases, and furthermore, this pattern is retained through serial transfer from the tumor DNA to the DNA of all primary 3T3 foci, to secondary foci, and so on.

A most surprising result that soon emerged was that the same enzymes would destroy the transforming activity of DNAs from certain different human tumor cell lines, and in these cases, enzymes that failed to destroy the transforming gene of one line would also fail to destroy the transforming gene of the other lines. This result implied that the same genes were frequently involved in causing different cancers, a result that was soon verified by additional studies. In some cases tumors of the same cell type are somewhat more likely to harbor the same transforming gene, however, human tumors of different cell types can involve the same activated oncogene (see Table 27-2).

The next step in analyzing the human transforming genes was to clone them. The simplest molecular cloning procedures rely on having a hybridization probe for the desired gene. At first, since the identity of the transforming genes was unknown, it seemed that no such probes existed. Then an extremely useful observation supplied a hybridization probe that made it relatively easy to clone the human

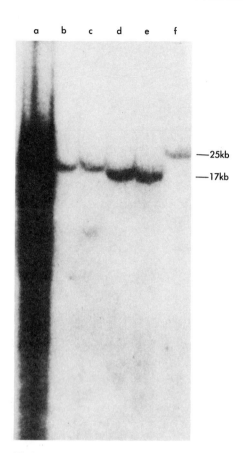

—25kb

—17kb

Figure 27-3
Detection of human *Alu* containing DNA in mouse NIH/3T3 transfectants. Primary foci were induced in NIH/3T3 cells with DNA isolated from the human EJ bladder tumor cell line. DNA from a primary focus was used to induce secondary foci. DNAs from a primary focus and from five secondary foci were digested with a restriction endonuclease (*Eco* RI) and subjected to Southern transfer. Blots were probed with a human repetitive sequence DNA. Lane a: DNA from a primary focus. Lanes b–f: DNAs from five secondary foci. The differences in size of some *Alu*-containing bands in different foci is due to differences in degradation and integration of the EJ transforming gene during the transfection procedure. [Courtesy of C. Shih and R. A. Weinberg, *Cell* 29 (1982):161–169.]

transforming genes. Vertebrates possess short, highly reiterated sequences scattered throughout their genomes. Although these sequences are shared by all members of a species, they have diverged from species to species. Thus, these sequences can be used as hybridization probes to distinguish the DNA of, for example, humans and mice. In humans, the highly reiterated set of sequences is called the *Alu* family. Because approximately 300,000 copies of the *Alu* family sequence are scattered throughout the genome, there is likely to be such a sequence near any particular gene of interest, including, for example, the human cancer genes.

When a mouse cell transformed by human tumor cell DNA is digested with a restriction enzyme and then subjected to gel electrophoresis and Southern transfer of the DNA to nitrocellulose paper, the location of human DNA in the Southern blot can be seen with a radioactive probe for human *Alu* sequences, while the mouse DNA remains invisible (Figure 27-3). It is apparent from Figure 27-3 that primary 3T3 transfectants take up and integrate a great number of human genes. Usually, as much as 0.1 percent of the donor cell genome, the equivalent of about 3×10^3 kilobase pairs or, in this experiment, about 100 fragments of 30- to 50-kilobase-pair DNA, is incorporated. However, when DNA from the primary transfectant is used to induce foci a second time, even though once again about 100 pieces of donor DNA are taken up, since there is about 10^4 to 10^5 times as much mouse as human DNA in a primary 3T3 transfectant, most of the DNA taken up in the secondary transfectant is mouse DNA. The only human DNA that will tend to be found in the secondary transformant and in all subsequent transfectants, and thus the only DNA seen by the *Alu* probe, is that surrounding the human gene being selected because of its ability to induce transformed foci. This is what is found in Southern blots of DNA from secondary and tertiary 3T3 transfectants when probed with the human *Alu* probe. DNA from a secondary or tertiary 3T3 transfectant transformed by a human tumor gene contains only a few (or, depending on the gene and the restriction endonuclease used in the analysis, sometimes only one) *Alu* hybridizable bands (see Figure 27-3).

It is possible to clone many human transforming genes by constructing a λ library of chromosomal DNA isolated from a secondary or tertiary 3T3 transformant. The library is screened with the human *Alu* probe, and clones that contain this sequence are then tested for their ability to transform 3T3 cells. This approach was first used successfully to isolate the transforming gene of a human bladder carcinoma cell line, the so-called EJ cell line.

So Far, Most 3T3-Transforming Genes from Human Tumors Are H-*ras*, K-*ras*, or N-*ras* Genes[18-28]

As soon as the human EJ bladder oncogene was obtained, researchers wondered where this gene came from: Was it viral or, as expected, cellular in origin? A radioactive probe was prepared from a fragment of the gene (a fragment lacking repetitive *Alu* sequences) and used to analyze Southern blots of normal human DNA and of DNA from the EJ bladder tumor cell line. It was apparent that this gene was a normal cellular gene present in both DNAs. Furthermore, using a variety of restriction enzymes, no difference could be seen between the arrangement of the gene found in normal DNA and that found in bladder tumor cell DNA. Since normal human cell DNA is unable to transform 3T3 cells, this result indicated that whatever genetic change had activated the EJ bladder oncogene, it did not involve a major structural alteration of the gene.

Shortly after the human bladder oncogene was cloned, the startling finding was made that the gene was none other than the human H-*ras-1* gene, the human cellular homolog of the Harvey *ras* gene, already well known as the transforming gene of the highly oncogenic rodent retrovirus Harvey sarcoma virus (Chapter 26). At the same time, the transforming genes of a human lung tumor and of a colon cancer were found to be the human K-*ras-2* gene, the homolog of the transforming gene of Kirsten sarcoma virus (see Table 27-2). Although only the H- and K-*ras* genes were known from the study of transforming retroviruses, a new member of the family, N-*ras*, was soon found when a transforming gene from a human neuroblastoma was cloned. It turned out to have only very weak nucleotide homology to H- or K-*ras* genes, but when portions of the gene were sequenced, it was clear that it encoded a very similar protein.

As discussed in Chapter 26, vertebrate *ras* genes constitute a small multigene family. In humans, five *ras* genes have been found. H-*ras-1* and -2 are highly homologous in nucleotide sequence, but H-*ras-2* is a nonfunctional pseudogene and lacks introns. K-*ras-1* and -2 are highly homologous to each other and are only weakly homologous to the H-*ras* genes in nucleotide sequence; and K-*ras-1* is a nonfunctional pseudogene. Finally, as already noted, N-*ras* is only weakly homologous to the other *ras* genes. Functional *ras* genes (H-*ras-1*, K-*ras-2*, and N-*ras*) vary greatly in size (from about 5 to 45 kilobase pairs) owing to differences in intron size, but they have similar exon structures (Figure 27-4a). The K-*ras-2* gene differs from the other two, however, in having two fourth protein-coding exons (4a and 4b) that permit the formation of two proteins with different amino acid sequences near their carboxyl termini. Their extreme C-termini, however, share the structure found in all *ras* proteins (Cys-A-A-X, where A is a aliphatib amino acid; Figure 27-4b).

Despite differences in nucleotide sequence, the three functional *ras* genes encode highly related proteins, called p21s, that consist of 189 amino acids (188 when the more usual 4b exon of K-*ras-2* is used). The sequence of the first 86 amino acids is virtually identical among the three *ras* gene products; the next 79 amino acids are about 85 percent homologous between any two; and the carboxyl-terminal 24 amino acids are strikingly different in sequence between the three except for the final four residues (see Figure 27-4b). The *ras* proteins are post-translationally modified by covalent attachment of palmitic acid to the Cys residue present near their carboxyl termini. After modification, and probably as a result of it, they become associated with the inner surface of the plasma membrane. Although they are known to bind guanine nucleotides (GDP and GTP) and to possess GTPase activity, their biological function is unknown.

Since *ras* genes were already known to function as oncogenes in certain acute transforming retroviruses, the stunning finding that they could also become oncogenes in spontaneous human tumors provided strong circumstantial evidence that the human oncogenes detected by DNA transfection had indeed played a role in the development of their respective tumors. In addition, it provided unity between viral and spontaneous carcinogenesis.

It is not known why activated *ras* genes are the most common type of transfectable human cancer gene. In part, it may be that these

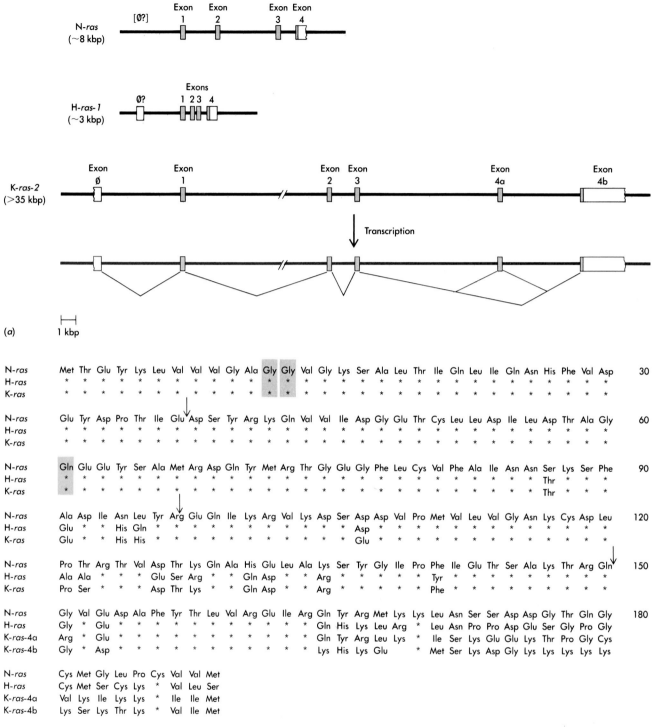

(a)

(b)

Figure 27-4

Structure of the human N-*ras*, H-*ras-1*, and K-*ras-2* genes and amino acid sequences of the polypeptides (p21s) they encode. (a) Exons that encode protein are shown as solid boxes; untranslated exons are white boxes. Whether there is a fifth untranslated exon (exon 0) for the N-*ras* gene is not yet known. Primary transcript of the K-*ras-2* gene and the two ways it can be spliced are indicated below the gene. The 4a exon can be skipped over and the 4b exon joined to exon 3. (b) Amino acids shared by all three proteins are indicated by asterisks. Arrows indicate boundaries of coding exons: these are identical in all three genes. Amino acids (12 and 61) that have been found to be altered and to lead to activated *ras* genes in human tumors are shaded. All *ras* proteins that have been studied terminate in the sequence cysAAX, where A stands for an aliphatic amino acid and X is any carboxyl terminal residue.

genes are more frequently involved in human cancers, possibly because of the simple mechanism by which they can become activated oncogenes. It is also possible that the 3T3 transformation assay is particularly sensitive to detecting this class of oncogene, while 3T3 cells are known to be relatively insensitive to morphological transformation by certain other oncogenes, for example, *myc* and *myb*.

Although the vast majority of transfectable human transforming genes have turned out to be one of the *ras* genes, a few other genes have been detected as well. A gene called *met* has been found in a chemically transformed human tumor cell line; and an activated *raf* gene (Chapter 26) has been detected in several human carcinomas. How many different human oncogenes will ultimately be detected by the 3T3 transfection transformation assay is hard to say, but now most researchers expect no more than about a dozen.

Activated *ras* Genes Result from a Single-Base Change That Leads to a Single-Amino-Acid Change[29-34]

The molecularly cloned normal human H-*ras-1* gene cannot transform 3T3 cells, although it can become a transforming gene if it is joined to a very strong transcriptional promoter so that abnormally high levels of p21 are synthesized in transfected cells. In contrast to these findings, the molecularly cloned EJ bladder oncogene can, of course, transform 3T3 cells efficiently without being joined to a new viral promoter. Thus, the EJ gene must carry a mutation relative to its normal counterpart.

With molecular clones of the normal and EJ H-*ras-1* genes available, it was easy to find the difference between them. The two genes were digested with restriction endonucleases; then fragments from each were rejoined in order to identify the region of the EJ gene that gave it transforming activity (Figure 27-5). A small fragment responsible for the property was soon found. When this fragment was sequenced, it turned out to differ from the normal gene by just one nucleotide; a G in the twelfth codon of the normal gene had become a T. The nucleotide difference leads to a change in the twelfth amino acid of the *ras* protein from glycine to valine. Analysis of other human transforming genes revealed that a single nucleotide change in the twelfth codon is a common mechanism by which a K-, H-, or N-*ras* gene is activated to become an oncogene. Single-base changes leading to an altered amino acid at position 61 or 13 have also been found (Table 27-3). *ras* proteins activated by a single-amino-acid change at position 12 often have altered electrophoretic mobility on SDS gels.

The technique of in vitro mutagenesis has been used to introduce mutations into many different sites of a molecularly cloned *ras* gene. The experiments show that in addition to amino acids 12 and 61, mutations that alter residues 13, 59, or 63 can also turn *ras* genes into transforming genes. In addition, in vitro mutagenesis has been used to introduce every possible amino acid change into position 12 of the H-*ras-1* gene. Replacement of glycine with any amino acid except proline results in an activated gene. Interestingly, *ras* proteins have homology to the guanine nucleotide binding, protein synthesis elongation factor, EF-Tu. The structure of EF-Tu is known, and it turns out that regions of good homology between *ras* and EF-Tu are in areas of

Figure 27-5

3T3 transforming ability of DNA molecules constructed from fragments of the normal human H-*ras-1* gene and the activated H-*ras-1* gene from the EJ bladder carcinoma. The H-*ras-1* gene with its four protein-coding exons is shown at the top of the figure. Each line below this represents a recombinant molecule. Portions of each molecule derived from the EJ H-*ras-1* gene are shown as black lines; portions derived from the normal H-*ras-1* gene are shown as gray lines. The restriction enzyme sites used to generate the recombinants are shown. The ability of each to induce transformed foci in 3T3 cells is shown at the right. For each pair of reciprocal recombinants (designated 1 and 2), only one member of the pair is active. The results identify the region important to oncogene activation as lying between the Xma I and the KpnI restriction endonuclease sites and encompassing the first protein-coding exon of the gene. [After C. J. Tabin et al., *Nature* 300 (1982):143–149.]

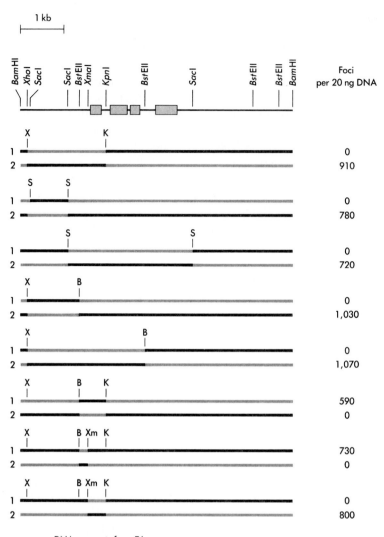

EF-Tu that bind guanine nucleotides. The amino acid of EF-Tu that corresponds to residue 12 in *ras* proteins lies on a loop, located between an α helix and a β strand, that includes the region that interacts with the phosphate group of the guanine nucleotides.

As soon as it was learned that single-amino-acid substitutions could activate *ras* proteins, viral *ras* genes were reexamined to see whether their transforming activity might be a result of mutations in the protein-coding region as well as a result of high-level expression of p21 due to the influence of potent viral transcriptional signals. Indeed, viral *ras* genes turn out to carry several mutations relative to cellular *ras* genes, and these include mutations at codons 12 and 59 (see Table 27-3). Thus, viral *ras* genes are altered in three ways, any one of which could give them transforming activity. Interestingly, the alteration at codon 59 introduces a threonine residue in both the *v*-H- and *v*-K-*ras* p21s, and this residue is phosphorylated in both proteins. A kinase activity latent in p21 itself is responsible for the phosphorylation. Normal *ras* gene products lack this residue and are not phosphorylated.

Table 27-3 Comparison of Amino Acids That Distinguish Normal and Transforming *ras* Proteins

ras Gene	Amino Acid Positions			
	12	13	59	61
H-ras-1-Derived Genes				
Human *c*-H-*ras*-1	Gly	Gly	Ala	Gln
Human EJ bladder tumor	Val			
Human HS242 breast tumor				Leu
Harvey virus *v*-H-*ras*	Arg		Thr	
K-ras-2-Derived Genes				
Human *c*-K-*ras*-2	Gly	Gly	Ala	Gln
Human Calu lung tumor	Lys			
Human SW480 colon tumor	Val			
Kirsten virus *v*-K-*ras*	Ser		Thr	
N-ras-Derived Genes				
Human N-*ras*	Gly	Gly	Ala	Gln
Human SK-N-SH neuroblastoma				Lys
Human HL60 promyelocytic leukemia				Lys
Human HT-1080 fibrosarcoma				Lys
Human AML 33		Asp		

Activated *ras* Genes Have Decreased GTPase Activity[35–42]

How does a single-amino-acid change in p21 have such a dramatic effect on the biological function of the molecule? Until we know the normal cellular function of the *ras*-coded proteins, we cannot have a satisfactory answer to this question. However, one clue is that alterations at amino acid 12 reduce or abolish the GTPase activity inherent in p21. The mutations do not alter the ability of the protein to bind guanine nucleotides, however. How might a change in GTPase activity of p21 lead to alterations in growth control?

At the moment, our best clue to what the normal function of *ras* gene products may be is the knowledge that *ras* proteins bind guanine nucleotides and have GTPase activity. There are other classes of proteins with these properties, for example, the G proteins that interact with adenylate cyclase to alter cAMP levels in cells (Chapter 25). Normally, these proteins, which are located at the plasma membrane, are inactive; but when cells are stimulated by certain hormone-receptor interactions, the G proteins can become activated and bind GTP. In this form, they can stimulate or depress adenylate cyclase, thus changing the intracellular level of cAMP and altering cellular behavior. Then a GTPase activity present in the G protein hydrolyzes GTP to GDP, thereby returning the protein to its inactive form. Although *ras*-coded p21s do not correspond to any of the three subunits of known G proteins, this model provides a useful way of understanding how a mutation that destroyed the GTPase activity of a protein might cause the protein to remain in an active form and so to send signals relentlessly that affect cell behavior (Figure 27-6).

New clues to the function of *ras*-coded proteins in mammalian cells may come from studying *ras* gene products in yeast. In yeast, there

Figure 27-6
Highly schematized view of the inactive form of a *ras* protein binding GDP while the activated form binds GTP. If a mutation destroys the GTPase present in the molecule, then the molecule may remain in the active state.

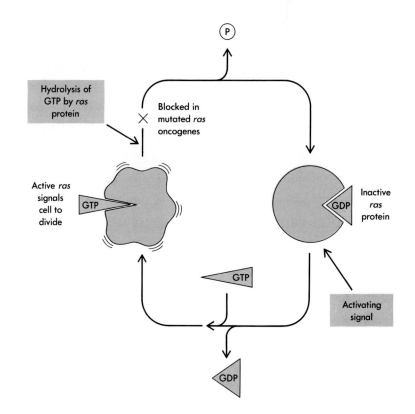

are two *ras* genes, *RAS1* and *RAS2*. The proteins they encode are considerably larger than mammalian p21s (309 and 322 amino acids), but they have a surprising degree of amino acid sequence homology to p21s (90 percent in the first 80 amino acids), including the same amino acids at the critical positions 12, 13, 59, 61, and 63, of mammalian *ras* gene products (Figure 27-7). While yeast cells can survive with just one of their two *RAS* genes, if both are destroyed by mutations, cells die. Remarkably, when the human H-*ras-1* gene is introduced into a yeast *ras1⁻ras2⁻* cell, it complements the defect, rescuing the cell and permitting vegetative growth. In addition, a modified yeast *RAS*-encoded protein with a mutation at amino acid 12 and a portion of its carboxyl terminus removed can induce transformed foci in 3T3 cells. These experiments provide evidence that yeast and mammalian *ras* proteins must serve at least some analogous functions in their respective cells, and so they encourage us to believe that powerful yeast genetics can be used to get at the normal cellular roles of mammalian as well as yeast *ras* genes.

Preliminary evidence that *ras* proteins play an essential role in controlling the mammalian cell cycle comes from studies using the technique of microinjection. Using very fine glass needles, purified p21 can be microinjected into mouse fibroblasts growing in vitro. The activated form of the protein is particularly effective in triggering cellular DNA synthesis. Furthermore, microinjection of monoclonal antibodies against p21 leads to G_1-induced arrest of cell growth. Resting 3T3 cells injected with anti-*ras* antibodies are unable to enter S phase when they are subsequently stimulated to divide, while dividing 3T3 cells injected with the antibody complete ongoing rounds of DNA synthesis but fail to initiate new rounds.

```
Ras 2   Met Pro Leu Asn Lys Ser Asn Ile Arg Glu Tyr Lys Leu Val Val Val Gly Gly Gly Gly Val Gly Lys Ser Ala Leu Thr Ile Gln Leu   30
Ras 1    *  Gln Gly  *   *   *  Thr Ile Arg  *   *   *  Ile  *   *   *  Gly  *   *   *   *   *   *   *   *   *   *   *   *  Phe
H-ras                           Met Thr  *   *   *  Leu  *   *   *  Ala  *   *   *   *   *   *   *   *   *   *   *  Leu

Ras 2   Thr Gln Ser His Phe Val Asp Glu Tyr Asp Pro Thr Ile Glu Asp Ser Tyr Arg Lys Gln Val Val Ile Asp Asp Glu Val Ser Ile Leu   60
Ras 1   Ile  *  Ser Tyr  *   *   *  Gly  *   *   *   *   *   *   *   *   *   *   *   *   *   *   *   *  Asp Lys Val Ser Ile  *
H-ras   Ile  *  Asn His  *   *   *  Glu  *   *   *   *   *   *   *   *   *   *   *   *   *   *   *   *  Gly Glu Thr Cys Leu  *

Ras 2   Asp Ile Leu Asp Thr Ala Gly Gln Glu Glu Tyr Ser Ala Met Arg Glu Gln Tyr Met Arg Asn Gly Glu Gly Phe Leu Leu Val Tyr Ser   90
Ras 1    *   *   *   *   *   *   *   *   *   *   *   *   *   *   *  Glu  *   *   *   *  Thr  *   *   *   *   *  Leu  *  Tyr Ser
H-ras    *   *   *   *   *   *   *   *   *   *   *   *   *   *   *  Asp  *   *   *   *  Thr  *   *   *   *   *  Cys  *  Phe Ala

Ras 2   Ile Thr Ser Lys Ser Ser Leu Asp Glu Leu Met Thr Tyr Tyr Gln Gln Ile Leu Arg Val Lys Asp Thr Asp Tyr Val Pro Ile Val Val  120
Ras 1   Val Thr Ser Arg Asn  *  Phe Asp Glu Leu Leu Ser  *  Tyr Gln  *   *  Gln  *   *   *   *  Ser  *  Tyr Ile  *  Val  *  Val
H-ras   Ile Asn Asn Thr Lys  *  Phe Glu Asp Ile His Gln  *  Arg Glu  *   *  Lys  *   *   *   *  Ser  *  Asp Val  *  Met  *  Leu

Ras 2   Val Gly Asn Lys Ser Asp Leu Glu Asn Glu Lys Gln Val Ser Tyr Gln Asp Gly Leu Asn Met Ala Lys Gln Met Asn Ala Pro Phe Leu  150
Ras 1    *   *   *   *  Leu  *   *  Glu Asn Glu Arg Gln Val  *  Tyr Glu Asp Gly Leu Arg Leu  *  Lys Gln Leu Asn Ala  *  Phe Leu
H-ras    *   *   *   *  Cys  *   *  Ala Ala Arg Thr Val Glu  *  Arg Gln     Ala Gln Asp Leu  *  Arg Ser Tyr Gly Ile  *  Tyr Ile

Ras 2   Glu Thr Ser Ala Lys Gln Ala Ile Asn Val Glu Glu Ala Phe Tyr Thr Leu Ala Arg Leu Val Arg Asp Glu Gly Gly Lys Tyr Asn Lys  180
Ras 1    *   *   *   *   *  Gln Ala Ile Asn  *  Asp Glu  *   *   *  Ser  *  Ile  *  Leu Val  *  Asp Asp Gly Gly Lys Tyr Asn Ser
H-ras    *   *   *   *   *  Thr Arg Gln Gly  *  Glu Asp  *   *   *  Thr  *  Val  *  Glu Ile  *  Gln His Lys Leu Arg Lys Leu Asn

Ras 2   Thr Leu Thr Glu Asn Asp Asn Ser Lys Gln Thr Ser Gln Asp Thr Lys Gly Ser Gly Ala Asn Ser Val Pro Arg Asn Ser Gly Gly His  210
Ras 1   Met Asn Arg Gln Leu Asp Asn Thr Asn Glu Ile Arg Asp Ser Glu Leu Thr  *  Ser  *  Thr Ala Asp Ile Glu Lys Lys Asn Asn Gly
H-ras   Pro Pro Asp Glu Ser Gly Pro Gly Cys Met Ser Cys Lys

Ras 2   Arg Lys Met Ser Asn Ala Ala Asn Gly Lys Asn Val Asn Ser Ser Thr Thr Val Val Asn Ala Arg Asn Ala Ser Ile Glu Ser Lys Thr  240
Ras 1   Ser Tyr Val Leu Asp Asn Ser Leu Thr Asn Ala Gly Thr Gly  *  Ser Ser Lys Ser Ala Val Asn His Asn Gly Glu Thr Thr  *  Arg
H-ras

Ras 2   Gly Leu Ala Gly Asn Gln Ala Thr Asn Gly Lys Thr Gln Thr Val Arg Thr Asn Ile Asp Asn Ser Thr Gly Gln Ala Gly Gln Ala Asn  270
Ras 1   Thr Asp Glu Lys  *  Tyr Val Asn Gln Asn Asn Asn Asn Glu Gly Asn  *  Lys Tyr Ser Ser Asn Gly Asn Gly Asn Arg Ser Asp Ile
H-ras

Ras 2   Ala Gln Ser Ala Asn Thr Val Asn Asn Arg Val Asn Asn Asn Ser Lys Ala Gly Gln Val Ser Asn Ala Lys Gln Ala Arg Lys Gln Gln  300
Ras 1   Ser Arg Gly Asn Gln Asn Asn Ala Leu Asn Ser Arg Ser Lys Gln Ser  *  Glu Pro Gln Lys  *  Ser Ser Ala Asn Ala Arg Lys Glu
H-ras

Ras 2   Ala Ala Pro Gly Gly Asn Thr Ser Glu Ala Ser Lys Ser Gly Ser Gly Gly Cys Cys Ile Ile Ser
Ras 1   Tyr                                            *   *   *   *   *  Ile Ile Cys
H-ras                                                  *  Val Leu Ser
```

Figure 27-7

Comparison of the amino acid sequences of proteins encoded by the yeast *RAS1* and *RAS2* genes and the human H-*ras-1* gene. The sequences are aligned for maximum homology. Amino acids that are common to all three proteins are indicated by an asterisk. Note that quite a few of the amino acid differences between the proteins are conservative amino acid changes. Amino acids that can be altered to produce activated mammalian *ras* oncogenes are shaded (positions corresponding to residues 12, 13, 59, 61, and 63 of the H-*ras-1* protein. [After S. Powers et al., *Cell* 36 (1984):607–612.]

Activated *ras* Genes in Carcinogen-Induced Rodent Tumors[43–46]

Even before human tumor DNAs were shown to harbor genes capable of transforming 3T3 cells, DNAs from certain mouse and rat cells transformed by chemical carcinogens had been shown to possess transfectable transforming genes. These results provided evidence that chemical carcinogens, most of which are potent mutagens for DNA, may work by activating oncogenes directly.

One striking example of oncogene activation by chemical carcinogens involves mammary tumors induced in rats by the carcinogen N-*nitroso*-N-*methylurea (NMU).* This carcinogen is known to induce G to A transitions in DNA by methylation at the oxygen 6 position. If 50-day-old rats, young enough to still be undergoing sexual development, are injected with a single dose of NMU, 6 to 12 months later they develop mammary tumors. More than 80 percent of these tumors contain an activated H-*ras-1* gene. Furthermore, 100 percent of these result from a G to A mutation in the second nucleotide of the twelfth codon, a change that causes the normal glycine to be replaced

Table 27-4 Transforming Genes Detected in Chemically Induced Rodent Tumors or Chemically Transformed Cell Lines

Tumor or Cell Line	Gene
NMU-induced mammary carcinomas (rat) DMBA-induced mammary carcinomas (rat) Benign and malignant DMBA/TPA-induced skin papillomas (mouse) MC-, BP-, and MNNG-transformed primary cells (guinea pig)	c-H-*ras*-1
MC-induced fibrosarcomas and MC-transformed fibroblasts (mouse) MC-induced thymic lymphoma (mouse) BP-induced fibrosarcoma (mouse)	c-K-*ras*-2
ENU-induced neuroblastoma cell lines (rat)	*neu*
MNNG-"retransformed" human osteosarcoma cell line	*met*

by a glutamic acid residue. NMU is unstable in vivo, and the total exposure time to the carcinogen may be no more than a few hours. Thus, within this time, random G to A mutations must occur. Presumably, some of these lie in H-*ras-1* genes of breast cells and so put these cells at risk to become tumors. Mammary tumors induced in rats by the carcinogen *DMBA (7,12-dimethylbenz(a)anthracene)* (which forms adducts with deoxyguanosine and deoxyadenosine that lead to subsequent excision repair) have a lower incidence of activated H-*ras-1* genes than NMU-induced tumors, and those that have been examined have mutations in codon 61.

Activated *ras* genes have been detected in DMBA-induced mouse skin carcinomas, in methylcholanthrene-transformed mouse fibroblasts, and in mouse thymomas induced by NMU and X-rays (Table 27-4). Interestingly, a different transforming gene is activated in rat neuroblastomas induced by exposing rat embryos in utero to *nitrosoethylurea*. This oncogene, called *neu*, is distinct from but related to the gene encoding the EGF receptor and may encode a receptor for a different polypeptide growth factor.

Activated *ras* Genes Alone May Not Cause Cancer[47]

If cancer results from the accumulation of several genetic alterations within a single cell, then we might expect that an activated *ras* gene alone would not be sufficient to induce cancer. Now numerous lines of evidence support this view.

Skin tumors can be induced experimentally in mice by sequential treatments with an initiator such as DMBA followed by a tumor promoter such as *12-O-tetradecanoyl-phorbol-13-acetate (TPA)*. In this system, it is possible to obtain both premalignant, benign papillomas as well as malignant carcinomas. When DNA is isolated from either the precancerous lesions or a transplantable carcinoma and analyzed for the presence of activated *ras* genes, often they are found to be present. The fact that these genes are present in the benign papillomas argues that the presence of an activated *ras* gene alone does not make a cell cancerous. Thus, additional events must be required for the *ras* oncogene–containing cell to become a full cancer. The result also suggests that *ras* activation may be an early step in the development of these mouse skin tumors.

Direct Evidence That Two or More Oncogenes Cooperate to Produce Cancer[48–52]

If multiple mutations are required to cause cancer, then how can we explain the ability of a single activated *ras* gene to transform 3T3 cells to cancer cells? Insight into this paradox came from experiments in which attempts were made to use activated *ras* genes to transform primary rodent cells rather than established 3T3 cell lines. The *ras* genes were essentially unable to transform the primary cells.

As discussed in Chapter 25, although 3T3 cell lines are not tumorigenic, they nonetheless have enhanced growth potential relative to normal cells: They are established (immortal) and grow readily at low cell density. Previous studies with DNA tumor viruses, polyoma and adenovirus, had shown that although a single viral gene can induce transformation of established 3T3-like cell lines, two, and sometimes three, viral genes are required to transform primary cells. For example, polyoma large T protein alone induces primary cells to grow like established cells, but polyoma middle T protein is needed in addition to induce the rounded morphology and in vivo growth properties typical of a fully transformed cell. In at least some established cell lines, the need for the former gene is bypassed, and polyoma middle T alone can transform the cells. A similar situation may apply to cellular transformation by activated *ras* genes.

Activated *ras* genes alone are highly inefficient at transforming at least some types of primary rodent cells. However, if a second oncogene is present, then they readily induce transformed foci (Figure 27-8). Oncogenes that can complement *ras* include *myc*, the E1A gene of adenovirus, *myb*, and SV40 and polyoma large T (Chapter 26). Interestingly, these genes all share the ability to induce establishment

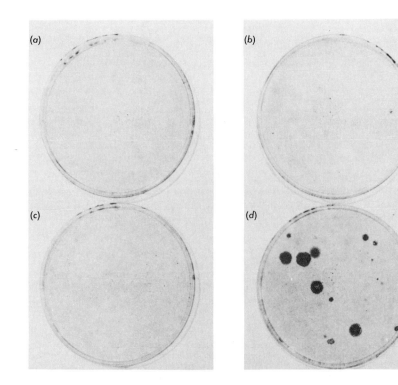

Figure 27-8
The activated H-*ras-1* bladder oncogene alone cannot induce visible foci of transformed cells in primary baby rat kidney cells. Cells were transfected with appropriate DNAs and fixed and stained about 3 weeks later. (a) Negative control (no oncogenes). (b) *E1a* gene of adenovirus. (c) Activated H-*ras-1* gene alone. (d) *E1a* plus activated H-*ras-1* genes. [Courtesy of H. E. Ruley. Reprinted with permission from *Nature* 304 (1983):602-606.]

or immortalization of primary cells, and they all encode proteins that are localized primarily in the cell nucleus.

Now some evidence indicates that even two oncogenes may not be sufficient to induce the fully cancerous phenotype. Some primary cells transformed by both *myc* and *ras* oncogenes fail to grow as progressive tumors, suggesting that additional oncogenes may also be needed.

If multiple oncogenes are needed to induce cancers, then how can the Harvey and Kirsten sarcoma viruses, which carry just a single activated *ras* oncogene, induce tumors so efficiently in mice? The answer to this question is not clear. One finding is that extremely high levels of activated *ras* genes may be sufficient to transform some types of primary cells in culture. As already noted, the *ras* genes of Harvey and Kirsten sarcoma viruses are activated in three ways: by point mutations in codons 12 and 59 and by the proximity to the *ras* gene of the strong retrovirus transcriptional signals. Conceivably, this makes the viral *ras* genes potent enough to transform cells single-handedly. Yet a second possibility is that cells infected by these viruses are at high risk to acquire the additional changes necessary to become cancer cells, since they already have one cancerous change. Or maybe only rare cells that already have aberrant gene expression are transformed by the virus.

If activated *ras* genes represent only one type of mutant gene that may contribute to the formation of human tumors, then clearly other genes remain to be found. Some of these genes are discussed in the following sections.

CHROMOSOME ABNORMALITIES ASSOCIATED WITH HUMAN TUMORS

Chromosome Abnormalities Are Associated with Many Forms of Human Cancer[53, 54]

Many human tumors have chromosome abnormalities, and frequently specific abnormalities are associated with specific forms of cancer. These findings have emerged gradually over the past 30 years as better and better techniques have become available for identifying and analyzing human chromosomes and for growing primary human tumor cells in vitro.

To visualize chromosomes, cells are usually captured in the metaphase stage, where the chromosomes are most condensed. A larger number of such cells can be obtained if the cells can be grown in vitro for a short time so that colchicine can be added to accumulate cells at mitosis. Next the cells are immersed in a hypotonic solution. This causes them to swell so that the chromosomes are farther apart and thus less likely to fall on top of one another when the cell is opened (see Figure 25-1). The cells are allowed to settle onto a slide, air-dried, fixed, and stained. Often, they are photographed and the 46 chromosomes cut out and aligned as shown in Figure 27-9, with the 22 pairs of autosomes arranged by size and followed by the two sex chromosomes.

Early chromosome preparations did not permit all the human chromosomes to be identified, since the only distinguishing characteris-

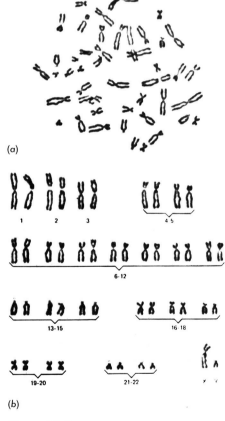

(a)

(b)

Figure 27–9
Typical preparation of human metaphase chromosomes before the discovery of banding techniques. (a) Metaphase spread from a lymphocyte of a normal male. (b) Arrangement of chromosomes in (a) into a standard karyotype. [Photo by J. J. Biesele, courtesy of H. E. Sutton, *An Introduction to Human Genetics,* 3rd ed. (Philadelphia: Saunders, 1980), p. 65.]

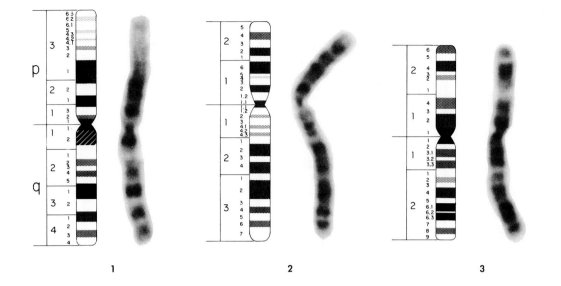

tics available were length and centromere location. Then, starting in 1970, **banding** techniques were developed that allowed every chromosome and even regions within chromosomes to be distinguished. The first method used fluorescent derivatives of quinacrine, which, for unknown reasons, bind preferentially to some regions of chromosomes. Later, the *modified Giemsa stain* was developed, which results in banding patterns (G bands) similar to those produced by the quinacrine method (Q bands) but is simpler to perform. These methods have led to diagrammatic representations of human chromosomes and to the nomenclature for specific chromosomal regions described in Figure 27-10. Now it is also possible to apply the banding techniques to chromosomes at the less condensed prophase stage, where even more bands (as many as 3000) can be identified.

Chromosome abnormalities in human tumors can involve both the number and, more frequently, the structure of chromosomes. During mitosis, one or more chromosomes may fail to migrate properly at anaphase. As a result of this **nondisjunction**, both daughter cells have an abnormal chromosome number: One daughter cell receives both chromosomes of a pair and hence has a **trisomy** of that particular chromosome, while the other receives neither and so is **monosomic**. Structural abnormalities in chromosomes arise through errors in breakage and reunion and can involve deletions (del), inversions (inv), and reciprocal translocations (t). In addition, some cancer cells have enormous amplifications of specific DNA sequences. These amplifications may be located within a chromosome, where they appear as **homogeneously staining regions** (HSRs), or they may be present on tiny chromosomes called **double minutes** (DMs).

Since a specific chromosome abnormality may be a primary cause of a certain type of tumor, it may be present in the tumor from its inception. But in addition, a tumor that at first displays just one chromosome abnormality may acquire additional changes with time. In at least some cases, these changes go hand in hand with an increasingly poor prognosis for the patient's survival. Once under way, tumors seem to give rise to variants with less and less controlled growth, and at least some of these changes in behavior may result from changes in chromosomes.

Figure 27-10

Examples of G banding and the nomenclature for chromosomes that has arisen from this methodology (400-band stage). The short arm of a chromosome is designated p and the long arm q. The arms are divided into one to three regions by landmarks. Landmarks are consistent features useful for identification such as the ends of the arm, the centromere, and certain very prominent bands. A region is the space between adjacent landmarks. Regions are numbered consecutively from the centromere, and bands are numbered consecutively within each region. Thus, for example, band 2q23 refers to chromosome 2, the long arm, region 2, band 3. [Courtesy of Dr. Jorge J. Yunis.]

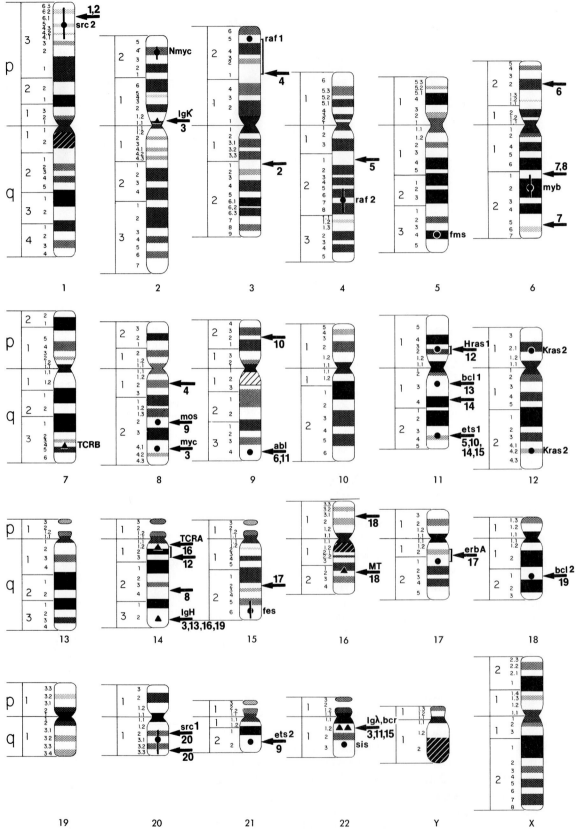

Figure 27-11

Human chromosome map showing 400 bands defined by Giemsa staining. The locations of 20 oncogenes (dot), 6 other cellular genes (triangle), and 2 breakpoints involved in specific chromosome rearrangements seen in cancers are marked. For oncogene abbreviations see Table 26-3. Other cellular genes are IgH, IgK, and Igλ: antibody genes for heavy chains, and for K and λ light chains, respectively. TCRA and TCRB: genes for the α and β chains of the T cell receptor. MT: metallothinein gene cluster. bcr: breakpoint cluster region. Rearrangements are numbered as follows: 1 = del 1p in neuroblastoma; 2 = t(1;3) in myelodysplasia (MDS); 3 = t(2;8), t(8;14), or t(8;22) in Burkitt's lymphoma; 4 = t(3;8) in mixed parotid gland tumor; 5 = t(4;11) in MDS; 6 = t(6;9) in ANLL; 7 = del 6q in nonHodgkin's lymphoma (NHL); 8 = t(6;14) in cystadenocarcinoma of ovary; 9 = t(8;21) in ANLL; 10 = t(9;11) in ANLL; 11 = t(9;22) in chronic myelogenous leukemia; 12 = t(11;14)(p13-14.1;q11.2-13) in T-cell acute lymphoblastic leukemia; 13 = t(11;14) × (q13.3;q32.3) in NHL; 14 = del 11q in NHL; 15 = t(11;22) in Ewing's sarcoma; 16 = inv(14) in T-cell chronic lymphocytic leukemia; 17 = t(15;17) in ANLL; 18 = inv(16) in ANLL; 19 = t(14;18) in NHL; 20 = del 20q in MDS. [Courtesy of J. J. Yunis. Reproduced with permission from *Important Advances in Oncology*, eds. V. deVita, S. Hellman, and S. Rosenberg (Philadelphia: Lippincott, 1986.)]

Some years ago, it was suggested that chromosome abnormalities might cause cancer by altering the expression of specific genes. In the past few years, specific translocations associated with several human tumors have been found to involve activation of the proto-oncogenes *c-myc* and *c-abl* (Chapter 26). In addition, amplified genes in HSRs and DMs of certain tumors are members of the *myc* gene family. Using molecular cloning technologies, the diverse mechanisms by which chromosome abnormalities lead to oncogene activation are rapidly being elucidated.

Some specific chromosome abnormalities associated with specific forms of cancer are summarized in Figure 27-11 and Table 27-5. It is easiest to analyze the chromosomes of blood cells and hence of leukemias. It is more difficult to separate the cells of solid tumors and to grow them in culture in order to accumulate mitotic cells. Hence, chromosome abnormalities have been recognized most often in various forms of leukemia, and only recently have specific abnormalities begun to be seen as common features of many solid tumors as well.

The Philadelphia Chromosome and *abl* Proto-Oncogene Activation in Chronic Myelogenous Leukemia[55–62]

The first specific chromosome abnormality to be associated with cancer was the *Philadelphia chromosome (Ph¹)*, named for the city in which it was discovered. This small chromosome is present in the leukemic cells of at least 90 percent of patients with *chronic myelogenous leukemia (CML)*, an invariably fatal cancer involving uncontrolled multiplication of myeloid stem cells (Chapter 25). The abnormality is also seen in some patients with ANLL (acute nonlymphocytic leukemia) and ALL (acute lymphocytic leukemia). Ph¹ is derived from a chromosome 22 by a *reciprocal translocation* involving chromosome 9. (A reciprocal translocation is one in which no genetic material is lost when fragments of the chromosomes are exchanged.) A portion of the long arm of 22 is translocated to 9 while a small fragment from the tip of the long arm of 9 is translocated to 22 (Figure 27-12). Thus, there are actually two abnormal chromosomes present in CML cells: the short-

Table 27-5 Some of the Neoplasms with a Known Consistent Chromosome Defect*

Disease	Chromosome Defect	Breakpoints or Deletion
Leukemias		
Chronic myelogenous leukemia	t(9;22)	9q34.1 and 22q11.21
Acute nonlymphocytic leukemia		
M1	t(9;22)	9q34.1 and 22q11.21
M2	t(8;21)	8q22.1 and 21q22.3
M3	t(15;17)	15q22 and 17q11.2
M4[†]	inv 16[‡]	p13.2 and q22
M4[†], M5[†]	t(9[§];11)	9p22 and 11q23
M1, M2, M4, M5, M6	del 5q	5q22q23
	del 7q	7q33q36
	+8	
Chronic lymphocytic leukemia	+12	
	t(11;14)[‡]	11q13 and 14q32
Acute lymphocytic leukemia		
L1-L2	t(9;22)	9q34.1 and 22q11.21
L2[‖]	t(4;11)	4q21 and 11q23
L3	t(8;14)	8q24.13 and 14q32.33
Lymphomas		
Burkitt's, small noncleaved cell (non-Burkitt),[†] large-cell immunoblastic[†‡]	t(8;14)	8q24.13 and 14q32.33
Follicular small cleaved,[†] follicular mixed,[†] and follicular large cell[†]	t(14;18)	14q32.3 and 18q21.3
Small-cell lymphocytic[†]	+12	
Small-cell lymphocytic, transformed to diffuse large cell[†]	t(11;14)[‡]	11q13 and 14q32
Carcinomas		
Neuroblastoma, disseminated	del 1p	1p31p36
Small-cell lung carcinoma	del 3p	3p14p23
Papillary cystadenocarcinoma of ovary	t(6;14)	6q21 and 14q24
Constitutional retinoblastoma[†]	del 13q	13q14.13
Retinoblastoma[‡]	del 13q	13q14
Aniridia-Wilms' tumor[†]	del 11p	11p13
Wilms' tumor[‡]	del 11p	11p13
Benign Solid Tumors		
Mixed parotid gland tumor	t(3;8)[‡]	3p25 and 8q21
Meningioma	−22	22

*Subbands, denoted by a decimal-digit system, were defined by using high-resolution banding techniques. M1, M2, etc., and L1, L2, etc., represent distinct forms of nonlymphocytic leukemia and acute lymphocytic leukemia, respectively.
[†]Consistent chromosome defects revealed by high-resolution banding techniques.
[‡]Few cases described.
[§]Chromosomes 6, 10, 17, and 19 may serve as alternative receptor chromosomes.
[‖]Recently suspected to represent an undifferentiated form of ANLL-M4 leukemia, with the same breakpoint 11q23 as the 9;11 translocation.
SOURCE: Table compiled by J. J. Yunis and reprinted with modification from *Science* 221 (1983):227–236.

ened 22 (Ph[1]) and an abnormal chromosome 9 called 9q[+]. The translocation breakpoints involved in generating Ph[1] and 9q[+] always occur within the same bands: q34 on chromosome 9 and q11 on chromosome 22.

The first hint that the Philadelphia chromosome might lead to activation of the *abl* proto-oncogene came when somatic cell hybrids were used to map *c-abl* and it was found to lie on chromosome 9 (see Figure 27-13). Next, *c-abl* was shown to be present on the Ph[1] chromosome, indicating that it lay in the portion that is translocated to 22 to generate the aberrant Ph[1]. Then an abnormally large *abl* mRNA was found in CML cells, suggesting that a structural alteration had indeed occurred in the immediate vicinity of the gene. Now, molecular cloning techniques have made it possible to analyze the 9-22 breakpoints in detail and have shown how *c-abl* expression is altered as a result of the translocation that generates the Philadelphia chromosome.

Two chromosome breaks are required to generate Ph[1] and 9q[+]. One occurs within a 6-kilobase-pair region on chromosome 22 called the *breakpoint cluster region* (*bcr*). The second break occurs on chromosome 9 at the 5' side of the *abl* gene. Apparently, this break can lie anywhere within a region of about 50 kilobase pairs. The bcr itself lies within a large gene, now called the *bcr* gene. When chromosomes 9 and 22 fuse, the 5' half of the *bcr* gene ends up on the 5' side of *abl*, with the two genes lying in the same transcriptional orientation (Figure 27-13). A large precursor RNA encompassing both genes is spliced so that the 5' exons of the *bcr* gene are joined to a specific exon in the middle of *c-abl* (see Figure 27-13). This explains why most Philadelphia chromosomes, regardless of their precise breakpoints, generate novel *abl* transcripts of about the same size. These 8-kilobase-pair transcripts are translated to yield polypeptides of about 210 kdal (P210) instead of the normal 145 kdal. In P210, about 25 amino acids normally present at the amino terminus of *c-abl* are replaced with about 600 to 700 residues of the *bcr* protein.

The *abl* gene was first discovered as the transforming gene of the highly oncogenic Abelson murine leukemia virus. In the virus, *v-abl* is expressed as a fusion protein with the viral *gag* gene product (Chapter 26). The *abl* gene has amino acid sequence homology to the tyrosine kinase family of oncogenes, and the viral *gag-abl* fusion protein has detectable tyrosine protein kinase activity. In contrast, the tyrosine kinase activity presumed to be present in the product of the normal proto-oncogene, *c-abl*, is not detectable. P210, the *bcr-abl* fusion protein found in CML cells, resembles the viral *gag-abl* in having detectable tyrosine specific kinase activity. Although *c-abl* probably acts as a kinase in some appropriate circumstances in vivo, it seems likely that alterations of its amino terminus remove a regulatory mechanism that normally controls its activity and allow the kinase activity to be monitored experimentally. Presumably this is related to why the *bcr-abl* and *gag-abl* fusions are oncogenic.

A question of interest is the specific role of the *bcr* gene in activating *abl*. Perhaps the *bcr* portion of the *bcr-abl* fusion protein modulates the activity of the *abl* portion in such a way that the hybrid protein specifically acquires the ability to transform myeloid stem cells (recall that Abelson virus with its *gag-abl* protein primarily induces tumors of B cells). Alternatively, it is possible that the *bcr* gene is expressed specifically in stem cells. Thus, fusion to this gene might ensure that the aberrant oncogenic *abl* fusion protein is expressed efficiently in this

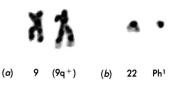

(a) 9 (9q[+]) *(b)* 22 Ph[1]

Figure 27-12
The Philadelphia chromosome in CML. (b) Appearance of normal and aberrant chromosomes 9 and 22 from a CML patient.

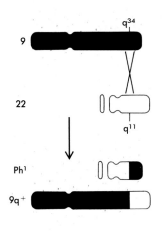

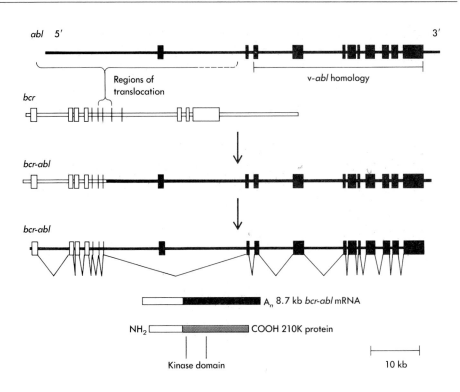

Figure 27-13
Diagram showing how the Philadelphia chromosome arises by a reciprocal translocation involving chromosomes 9 and 22 and how this event joins the *bcr* and *abl* genes. The *abl* gene is located on band q34 of chromosome 9; *bcr* is at q11 of chromosome 22. In the diagrams of the genes, boxes denote exons (these appear as vertical lines for small exons), and horizontal lines represent DNA. Not all the exons of *bcr* are defined yet, but breakpoints cluster within the bracketed region. The 5′ exons of the very large *abl* gene are not yet completely defined. Thus, it is not clear whether the breakpoints on chromosome 9 fall within the gene or can occur 5′ of the gene. The former possibility is more likely. Following translocation, a transcript of the 5′ exons of *bcr* is apparently spliced to a specific internal exon of the *abl* gene. The resulting transcript is translated to yield the aberrant *bcr-abl* fusion protein.
[After J. Adams, *Nature* 315 (1985):542.]

cell type and hence can participate in the cancerous transformation of such cells.

Translocation of the *myc* Proto-Oncogene to Immunoglobulin Gene Loci in B Cell Tumors of Mice and Humans[63–70]

Burkitt's lymphoma is a tumor of B cells that is common in Africa. It probably arises as a result of viral infection. Tumor cells generally secrete antibody, and they possess highly characteristic chromosome translocations. The translocations invariably involve chromosome 8 and one of the three chromosomes that carry the antibody light- or heavy-chain genes: chromosomes 14 (heavy-chain genes), 2 (λ light-chain genes), or 22 (κ light-chain genes). Furthermore, the translocation breakpoints are very specific, occurring at the same bands in different tumors (Figure 27-14). Translocations involving chromosomes 8 and 14 [designated t(8:14)] are by far the most common type in Burkitt's lymphoma and are present in 90 percent of the tumors.

The fact that antibody genes, which are so active in B cells, map to the same chromosomal regions involved in the specific translocations in Burkitt's tumors led to the suggestion that a specific oncogene on chromosome 8 might be activated if it were placed near to these genes as a result of a translocation. Considerable evidence now indicates that this hypothesis was fundamentally correct and that the proto-oncogene involved is *c-myc*.

c-myc is the cellular homolog of the transforming gene of the chicken retrovirus MC29 (Chapter 26). In humans, *c-myc* lies on the long arm of chromosome 8, at band q24, and spans about 5 kilobase pairs. The gene has three exons: the first specifies a long (550 base pairs) untranslated region of the mRNA, while the second two en-

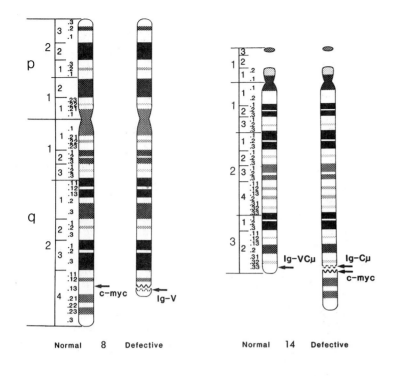

Figure 27-14
The common t(8:14) translocation seen in Burkitt's lymphoma. Diagrams of normal and aberrant chromosomes are represented at the 1200 Giemsa band stage. The approximate locations of the c-*myc* and heavy chain genes are shown. [Courtesy of J. J. Yunis, *Science* 221 (1983):227–236.]

code the *myc* protein (439 amino acids). It is a phosphoprotein that is present in the nucleus. Two forms are detected in cells, and they migrate in SDS gels with molecular weights of 64 and 67 kdal. The normal function of *c-myc* is unknown. However, quiescent cells that are stimulated to divide by addition of growth factors or nonphysiological mitogens show a rapid, sharp increase in *myc* mRNA, suggesting a role for this gene in cell proliferation. In dividing cells, levels of *myc* mRNA seem to be fairly constant, suggesting that the gene does not function in a cell-cycle-dependent manner.

In translocations that involve chromosome 14, *c-myc*, along with the distal portion of chromosome 8, moves adjacent to DNA encoding the heavy-chain constant regions. The two loci are joined head to head in opposite transcriptional orientations. As discussed in Chapter 23, antibody genes show allelic exclusion, meaning that heavy-chain genes on only one of the two chromosomes are used. The *myc* gene is translocated to the nonfunctional heavy-chain locus, which explains the ability of the cells to synthesize antibodies. In the less frequent translocations involving antibody light-chain loci, *myc* remains on chromosome 8, and fragments of either 2 or 22 are brought nearby. In these cases, the antibody genes and *myc* are joined tail to head, in the same transcriptional orientation, with the antibody gene lying to the 3′ side of *myc*.

It is now widely believed that translocations involving c-*myc* lead to altered transcription of the gene and that this is how they contribute to malignant transformation. One piece of evidence that indirectly supports this view is that the translocations involving *c-myc* do not alter the protein-coding portion of the gene. Translocated genes that have been completely sequenced usually have not sustained even a single-base change that might alter the activity of the encoded *myc* protein. It has proved difficult to demonstrate that *myc* mRNA is consistently elevated in different Burkitt's tumors. Part of the problem is that it is difficult to decide what a normal level of *myc* mRNA should be. If the B cell is supposed to be quiescent, awaiting antigen stimulation, then the presence of any *myc* mRNA might be sufficient to stimulate inappropriate cell division, so any disturbance of carefully controlled levels of *myc* expression might contribute to transformation.

How does the proximity of antibody genes change the transcription of a translocated *myc* gene? The answer to this question appears to be complex and may involve different mechanisms, depending on the precise breakpoints involved in the translocation. The relative orientation of the antibody genes and *c-myc* in the translocations tells us that activation cannot involve use of the antibody gene promoters to drive transcription of the *myc* gene. In some cases, the *myc* gene may fall under the influence of a powerful tissue-specific enhancer present near the antibody constant-region genes. In other cases, sequences 5' of *c-myc* may be lost that normally allow the gene to be negatively regulated. In still other cases, enhancer elements that have not yet been identified may turn out to lie near the rearranged *myc* gene.

In the case of translocations to the heavy-chain locus, most breakpoints on chromosome 14 occur within the switch regions lying in front of sequences that encode the heavy-chain μ, α, or γ constant regions, while the breakpoints near *c-myc* are widely scattered. The latter can occur from somewhere within the first *myc* intron to more than 25 kilobase pairs 5' of *c-myc*. In the case of the less frequent translocations involving antibody light-chain genes, *myc* and the antibody gene may be as much as 100 kilobase pairs apart.

Mouse plasmacytomas (tumors of mature B cells) also possess chromosome translocations that involve *c-myc* and one or another of the immunoglobulin loci. Except for some differences in the position of preferred breakpoints, many of the observations already described apply equally well to these translocations and in many cases were discovered first in the mouse system. Some other types of B cell tumors in humans also involve translocations of *c-myc* and antibody gene loci, revealing that these specific translocations are not limited to a single tumor type. B cell lymphomas induced in chickens by avian retroviruses that lack oncogenes usually involve activation of the *c-myc* gene by proviral insertion (Chapter 26).

Why is *c-myc* so frequently involved as an oncogene in these particular tumors, and why does the activation mechanism so frequently involve translocation to the antibody gene loci? So far, there is no adequate explanation for why a particular oncogene is frequently activated in a particular type of tumor. Furthermore, it should be noted that retroviruses that carry *myc* as an oncogene can induce a variety of tumors (Chapter 26), and there is evidence that *c-myc* can serve as an oncogene in human tumors of cell types other than B cells. As for the use of the antibody gene loci, there are two plausible reasons for their involvement. As originally proposed, the fact that they are transcriptionally active in B cells may be an essential aspect of their role in activating *c-myc*. Second, it seems likely that at least some, and possibly most, of the translocations in B cell tumors result when the recombination mechanisms involved in constructing functional antibody genes make a mistake and, instead of joining two segments of an antibody gene, join a proto-oncogene to an antibody gene. This explanation could account for the frequent cases where *c-myc* is introduced into a switch region of the heavy-chain locus.

As in the case of tumors with activated *ras* genes, more than one oncogene is probably involved in generating Burkitt's lymphomas. As we will discuss in detail shortly, viral infection may be an important first step in launching these tumors. Furthermore, some Burkitt's tumor cell lines have activated *ras* genes that can be detected in the 3T3 transfection assay, in addition to specific chromosome translocations.

Some B cell tumors have translocations that involve the antibody loci but oncogenes other than *c-myc*. Recently, two such genes have been molecularly cloned based on the assumption that sequences lying near the translocation breakpoints in the antibody loci might encode oncogenes. These new candidate oncogenes are designated *bcl-1* and *bcl-2*.

Retinoblastoma Involves the Loss of Both Copies of a Gene (*Rb-1*) on Band q14 of Human Chromosome 13[71,72]

Retinoblastoma is an eye tumor of children and is one of several heritable forms of childhood cancer. About half the cases of retinoblastoma are hereditary, while the remainder, called sporadic, occur in genetically normal individuals. Retinoblastoma is inherited as a highly penetrant autosomal dominant trait, and a child who inherits the single gene predisposing to retinoblastoma will almost certainly develop the tumor. These individuals often develop tumors in both eyes, while sporadic cases of retinoblastoma occur in only one eye. These findings led to the proposal that retinoblastoma results from the accumulation of at least two mutations within a single cell and that people with a high probability of contracting the disease inherit one of the necessary mutations.

Probably all forms of retinoblastoma involve abnormalities in band q14 on chromosome 13, and in some cases these are visible microscopically as a deletion of the q14 band. Tumor cells lack a normal chromosome 13 and are homozygous for the deleted version of chromosome 13. In addition, normal cells from individuals with a high tendency to the disease can sometimes be seen to carry one normal chromosome 13 and one with a deletion of band q14, while their tumor cells are homozygous for the abnormal chromosome. The deletions of 13q14 seen in different cases of retinoblastoma can vary at their endpoints, indicating that the deletion does not activate an oncogene by introducing breaks into a specific region of DNA, but rather that its effect may arise through the removal of a gene. Thus, retinoblastoma seems to be the result of a recessive mutation. Mutations (deletions) at the *Rb-1* locus are recessive to a normal gene and only contribute to tumorigenesis when the normal allele is inactivated. As shown in Figure 27-15, this can occur through errors of mitosis or through recombination events during mitosis that lead to loss of the normal chromosome, as well as through mutations of the normal allele.

Many years ago, experiments were performed to determine whether cancer is dominant or recessive. Normal cells and cancer cells were fused and the resulting cell hybrids tested for their tumorigenicity. Years of such experimentation have shown that hybrids can be either cancerous or normal, depending on the particular pair of cells involved in the fusion. These results suggest that at least some cancers are the result of recessive mutations. The study of retinoblastoma provides additional support for this idea. Clearly, it is important now to clone the *Rb-1* gene, since it may represent a new class of oncogene distinct from the dominant transforming genes such as acti-

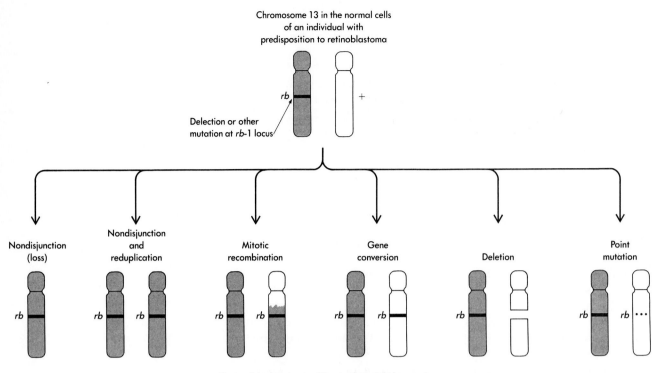

Chromosome 13 in the normal cells
of an individual with
predisposition to retinoblastoma

Deletion or other
mutation at *rb*-1 locus

Nondisjunction
(loss)

Nondisjunction
and
reduplication

Mitotic
recombination

Gene
conversion

Deletion

Point
mutation

Six possible chromosome 13 pairs that might be seen in
a retinoblastoma tumor arising in the individual shown above

Figure 27-15
Schematic representation of the chromosome 13s present in an individual with inherited predisposition to retinoblastoma, and the mechanisms that could lead to homozygosity of the recessive defect and hence to a tumor. [After W. K. Cavanee et al., *Nature* 305 (1983):779–784.]

vated *ras*. Conceivably, recessive cancer genes could turn out to be negative regulators of oncogene expression, but many other possibilities can also be imagined.

Mutations at *Rb-1* predispose to other forms of cancer as well as retinoblastoma. A fairly high percentage of patients that recover from heritable retinoblastoma (but not sporadic retinoblastoma patients) develop osteosarcoma. As in retinoblastoma, homozygosity of an abnormal chromosome 13 with a deletion at q14 appears to be involved. This result indicates that recessive mutations that predispose to cancer may be common to many different forms of disease, and a relatively small number of such genes may turn out to underlie many types of tumors.

Amplified Oncogenes in Double Minutes and Homogeneously Staining Regions in Certain Human Tumors[73–80]

Double minute chromosomes (DMs) and homogeneously staining regions (HSRs) within chromosomes are related karyotypic abnormalities seen in tumor cells. As discussed in Chapter 20, these abnormalities are also seen in cells following selection for drug resistance (see Figure 20-28). They result from the amplification of cellular genes. The precise structure of the amplified fragment within a particular DM or HSR is difficult to establish, since the fragments involved are enormous, ranging in size from 100 to more than 1000 kilobase pairs. DMs and HSRs apparently represent interchangeable forms of an amplified sequence, and a single tumor can give rise to cell lines that contain one or the other structure. In addition, HSRs can vary in size and number and can move to different chromosomal locations.

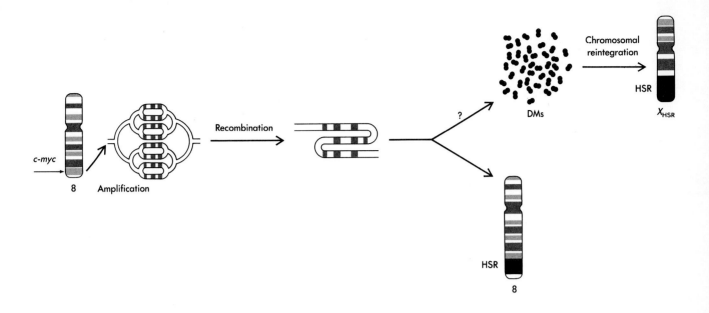

Figure 27-16
Model for how HSRs and DMs may arise and move from one chromosome to another. DNA encompassing an oncogene may be replicated excessively owing to reinitiations at a particular origin of replication (the onionskin model). Recombination within the onionskin structure generates tandem arrays of the sequence. These might leave the chromosome to become double minutes, which in turn might be able to reintegrate into new chromosomal locations, for example, the X chromosome in the case shown, giving rise to HSRs. [After K. Alitalo, et al., *Genes and Cancer* (Glen R. Lis, Inc., 1984) pp. 383–397.]

Although the sequence of events that underlies these alterations is not known, plausible scenarios are illustrated in Figure 27-16.

It is highly likely that DMs and HSRs contribute to the transformed phenotype of tumors that carry them, probably conferring a growth advantage. Otherwise, it is difficult to see why these structures would be retained. It was thought that the sequences amplified in DMs and HSRs might encompass cellular proto-oncogenes, since alteration in the level of expression, particularly increased transcription, was known to be one mechanism for activating some of these genes (e.g., *ras*, *mos*, and *myc*). This hypothesis has turned out to be correct. Several different oncogenes have been identified in the DMs and HSRs of different human tumors (Table 27-6). It is most unlikely that this is a coincidence. Rather, the increased expression of the oncogene resulting from its amplification almost certainly contributes to tumorigenicity and explains the persistence of DMs and HSRs in tumors.

Human neuroblastomas were one of the first tumors shown to harbor DMs and HSRs. When DNA from neuroblastoma cells was hybridized with radioactive viral oncogene probes, dark bands were observed in Southern blots when *myc* probes were used. When the homologous human sequences were cloned, they turned out to be a new member of the *myc* gene family, a gene now called N-*myc*. N-*myc* sequences can be amplified as much as a thousandfold in some neuroblastomas (Figure 27-17). Neuroblastomas, like many tumors, can undergo progression with time from less malignant to more malignant forms. Interestingly, N-*myc* amplifications are usually seen in the later stage III and IV tumors. It is these that also can be most readily established into cell lines in tissue culture.

Some small-cell lung carcinomas also have amplified *myc* genes in DMs and HSRs. The amplifications are found in late stages of tumor evolution, in the so-called variant cells that display a more malignant behavior. Amplifications can involve either *c-myc*, N-*myc*, or a new member of the *myc* gene family now designated L-*myc*.

Additional evidence that elevated levels of *myc* gene expression resulting from gene amplification are related to growth properties of tumor cells comes from studies with the HL60 cell line. HL60 is derived from a human promyelocytic leukemia. The cells have an ampli-

Table 27-6 Examples of Amplification of Cellular Oncogenes in Human Tumors

Amplified Gene	Tumor	Degree of Amplification	DM or HSR Present
c-*myc*	Promyelocytic leukemia cell line, HL60	20×	+
	Small-cell lung carcinoma cell lines	5–30×	?
N-*myc*	Primary neuroblastomas (stages III and IV) and neuroblastoma cell lines	5–1000×	+
	Retinoblastoma cell line and primary tumors	10–200×	+
	Small-cell lung carcinoma cell lines and tumors	50×	+
L-*myc*	Small-cell lung carcinoma cell lines and tumors	10–20×	?
c-*myb*	Acute myeloid leukemia	5–10×	?
	Colon carcinoma cell lines	10×	?
c-*erbB*	Epidermoid carcinoma cell line	30×	?
	Primary gliomas		?
c-K-*ras*-2	Primary carcinomas of lung, colon, bladder, and rectum	4–20×	?
N-*ras*	Mammary carcinoma cell line	5–10×	?

SOURCE: Modified from H. E. Varmus, *Ann. Rev. Genetics* 18 (1984):553–612.

fied c-*myc* gene and an activated N-*ras* gene that can be detected by transfection transformation of 3T3 cells. Addition of retinoic acid to HL60 cells induces them to differentiate into granulocyte-like cells that have only a very limited ability to proliferate. When this occurs, c-*myc* expression drops dramatically.

VIRUSES AND HUMAN CANCER

The Long Search for the Human Tumor Viruses

In the past 25 years, two discoveries were particularly important in fueling support for the viral origins of human cancer. One was the ability of certain DNA viruses, particularly the papovaviruses and adenoviruses, to transform cells in vitro and to induce tumors in rodents. The second was the discovery of reverse transcriptase and the subsequent explosion of the retrovirus field. Adenoviruses are extremely common in humans, where they cause colds and other types of infections, so it seemed possible that they might also be human tumor viruses. The realization that many forms of cancer in animals are caused by retroviruses and that these can be both inherited and horizontally spread made it seem very inevitable that retroviruses would be found to be a cause of one or more forms of human cancer.

Many human tumors were thus screened for the presence of integrated papovavirus and adenovirus DNA with almost entirely nega-

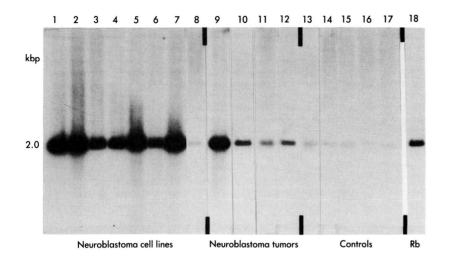

Neuroblastoma cell lines   Neuroblastoma tumors   Controls   Rb

Figure 27-17

A Southern blot showing amplification of N-*myc* in neuroblastoma cells and in a retinoblastoma. DNAs from cell lines or primary tumors were digested with a restriction endonuclease, and subjected to electrophoresis and transfer to nitrocellulose paper; the paper was hybridized to a radioactive probe prepared from a molecular clone of the N-*myc* gene. DNAs in lanes 1–8 were isolated from cell lines derived from neuroblastomas; lanes 9–12, from primary neuroblastomas; lanes 13–16, from other types of human tumors that also possess HSRs but which do not involve N-*myc*; lane 17, from normal human skin fibroblasts; and lane 18, from a human retinoblastoma cell line. Amplifications of N-*myc* range from about 5- to 8-fold (lane 8) to 20- to 25-fold (lanes 3 and 4) to 80- to 140-fold (lanes 1, 2, 5, 7, and 9). Lanes 13–17 show the appearance of hybridization to N-*myc* that is not amplified. [Courtesy of M. Schwab. From M. Schwab et al., *Nature* 305 (1983):245–248.]

tive results. Then more than a decade was spent trying to isolate the putative human retroviruses from tumors. Only one such virus, one that causes a rather rare form of T cell leukemia, has been found so far.

The several virus-cancer relationships that have emerged in the past decade have revealed just how difficult it can be to prove a causal relationship between virus infection and tumor development. No known human cancer appears to be caused by a virus that rapidly and single-handedly induces tumors. Rather, viruses that cause human cancers, like the carcinogens suspected of causing human cancers, act over long periods of time and probably only in combination with other events. It is for this reason that it is still difficult to know how frequently viruses may be involved in causing many types of human tumors.

Chronic Hepatitis B Virus Infection Leads to Liver Cancer[81–84]

Hepatitis B is a tiny DNA virus that infects humans and certain nonhuman primates and has a strong tropism for liver cells. When a nonimmune person becomes infected with hepatitis B, viral replication occurs in the liver. In about 95 percent of the cases, replication terminates when neutralizing antibodies develop and the individual becomes immune to reinfection. However, in about 5 percent of the cases, persistent infection of the liver is established and continues for life. Often, persistent infections are not associated with liver damage; but in some cases, symptomatic and progressive forms of liver damage occur. The latter condition is known as *chronic hepatitis B.* Persistent hepatitis B virus infections, particularly those associated with chronic hepatitis and cirrhosis, are believed to help cause 80 to 90 percent of primary hepatocellular carcinomas (also called hepatoma or liver cancer). Hepatitis B virus infection is estimated to cause about 500,000 cancer deaths a year worldwide.

The association of hepatitis B and liver cancer was discovered when the worldwide distribution of chronic hepatitis was determined and compared with the worldwide distribution of hepatoma. As shown in Figure 27-18, the two diseases show a striking overlap in geographic distribution. Further studies have shown that patients with hepatoma

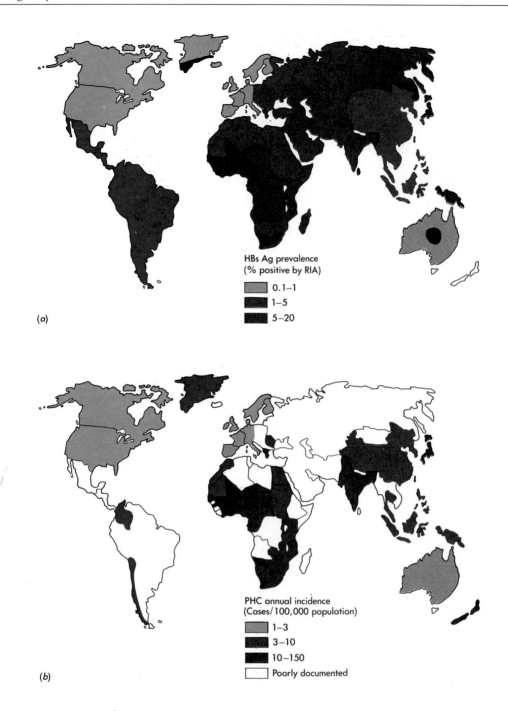

HBs Ag prevalence
(% positive by RIA)

- 0.1–1
- 1–5
- 5–20

(a)

PHC annual incidence
(Cases/100,000 population)

- 1–3
- 3–10
- 10–150
- Poorly documented

(b)

Figure 27-18
Maps showing the similarity in world-wide distribution of (a) hepatitis B virus infection and (b) annual incidence of primary hepatocellular carcinoma (PHC). Evidence for virus infection is presence of the hepatitis B surface antigen (HBsAg) in serum as determined by radioimmunoassay (RIA). [After P. Maupos and J. L. Melnick, *Prog. Med. Virol.* 27 (1981):1–5.]

are 5 to 70 times more likely to show evidence of hepatitis B infection than similar individuals without hepatoma. Although most hepatitis B virus carriers never develop hepatoma, the overwhelming majority of hepatoma patients show evidence of infection. Chronic hepatitis and cirrhosis combined with infection early in life create the greatest risk of developing hepatoma, a risk that can be 250-fold that of noncarriers. Primary hepatitis B infection usually precedes the onset of liver cancer by about 25 to 30 years.

Until recently, it was difficult to study the life cycle of hepatitis B virus because it could not be grown in tissue culture. However, molecular cloning of the genome and the discovery of hepatitis B–like viruses that induce liver cancer in woodchucks have allowed some important advances (Chapter 24).

What is the mechanism by which hepatitis B virus induces cancer?

Based on our knowledge of tumor viruses, we might suppose that hepatitis B carries a transforming gene, or that the virus integrates into cellular DNA and in the process activates a cellular proto-oncogene. So far, there is no evidence that hepatitis B virus possesses any type of transforming gene. Whether the virus activates cellular oncogenes is still an open question. Consistent with this possibility are the facts that hepatomas are clonal and viral DNA is found to be integrated into cellular DNA in tumor cells. However, no evidence for integrations near known proto-oncogenes has yet been obtained. Nonetheless, the fact that viral integrations occur leaves the possibility of oncogene activation viable. In addition, deletions at the sites of viral integrations have been observed in some tumor lines, so conceivably hepatitis B acts as a mutagen. In contrast to these models, some researchers now believe that hepatitis B may contribute to liver cancer by a different mechanism. Chronic infection is accompanied by chronic hepatic injury and regeneration. Perhaps this long-term trauma, like chronic alcohol abuse, which is also associated with liver cancer, in some way promotes the induction and growth of malignant cells.

Epstein-Barr Virus (EBV) and Its Relationship to Burkitt's Lymphoma, Nasopharyngeal Carcinoma, and Mononucleosis[85,86,87]

Burkitt's lymphoma, a tumor of mature B cells, is found throughout the world, but is most frequent among children in certain parts of East Africa. In the early 1960s, the physician Burkitt suggested that the tumor might be caused by a virus. Examination of tumor specimens soon led to the discovery of a herpes virus called **Epstein-Barr virus (EBV)**. EBV infects B cells in vitro and immortalizes them. It establishes a latent state in which just some of its genes are expressed in most cells, although a rare cell expresses all the viral genes and produces infectious virus.

It has been difficult to obtain conclusive evidence that EBV causes Burkitt's lymphoma, because infection by EBV is not limited to the people or areas where the tumors are found but is present throughout the world. However, areas with a high incidence of the tumor have a higher incidence of infection, and infection occurs at an earlier age. For example, in some tropical areas where the tumor is frequent, more than 90 percent of the children are infected by six years of age, while in the United States, only 30 to 40 percent show evidence of infection by this age. In addition, more than 90 percent of the Burkitt's tumors that arise in Africa are associated with EBV, whereas other types of leukemias do not show such an association. But if the virus is so widespread, why is the tumor so much more frequent in certain tropical areas of the world? There is a good correlation between the affected areas and a high incidence of malaria, and many researchers think that it is the combination of EBV and malaria that produces a high incidence of Burkitt's lymphoma.

In contrast to the difficulty in proving an association of EBV with Burkitt's lymphoma, it was soon shown conclusively that EBV causes mononucleosis. This disease, probably more common in the United States than in Africa, results when individuals are infected by EBV at a later age. Mononucleosis is communicated through kissing or through use of saliva-contaminated utensils, since EBV replicates readily in the oropharyngeal region. Mononucleosis is a disease of excessive lymphoid cell proliferation. Many of the symptoms are probably caused by the immune system's response to EBV-infected cells.

As discussed earlier in this chapter, Burkitt's lymphomas possess specific chromosome translocations involving the *c-myc* proto-oncogene, and these are almost certainly involved in tumor induction. Thus, EBV may serve as just one factor in tumor formation. The role of the virus in the cancerous transformation is not known, but presumably, the viral genes that can immortalize B cells are involved. Whether the virus helps to maintain the tumor or simply helps it to get started is also unknown. That EBV can be a potent tumor virus is indicated by the fact that individuals who are immunosuppressed can develop lymphomas (as well as fatal mononucleosis) following EBV infection. These tumors lack the characteristic chromosome abnormalities seen in Burkitt's lymphoma and are polyclonal instead of monoclonal like the Burkitt's tumors. Thus, the immune system probably saves us from a high probability of getting EBV-induced B cell cancers.

Why do children without obvious immune deficiency get Burkitt's lymphoma? This is unclear, since their tumor cells often appear to be highly antigenic owing to the expression of viral antigens on their surface. A hypothesis is that simultaneous infection with malaria somehow interferes with immune rejection of the tumor cells.

EBV is also implicated as a causal agent in nasopharyngeal carcinoma (NPC), a tumor of adults that is quite rare except among the Chinese. There seems to be a genetic susceptibility factor involved, since Chinese people retain an elevated incidence of NCP irrespective of where they live. In addition, environmental factors such as diet (particularly cured fish), smoke, and chemicals may be involved in contributing to the onset of disease. EBV genomes and their transcripts are found in almost all undifferentiated NPC biopsies, regardless of the country or the genetic constitution of the host.

A Retrovirus Called HTLV I Causes Adult T Cell Leukemia[88–91]

So far, the only retrovirus that has been identified as the cause of a human cancer is the human *T cell leukemia/lymphoma virus type I* (*HTLV I*), which causes a form of T cell cancer called *adult T cell leukemia-lymphoma (ATLL)*. This is a fatal tumor that is quite common in parts of Japan and the Caribbean. HTLV I was first detected in a continuously proliferating T cell line derived from a patient in the United States. Subsequently, the virus was found in T cell tumors from patients in Japan. An important advance was the ability to transmit the virus to fresh human cord-blood T cells in vitro by cocultivation with HTLV I–producing cells.

Several lines of evidence strongly implicate HTLV I as the cause of ATLL. First, the geographic distribution of the disease corresponds closely to areas where virus infection is frequent (the latter is determined by the presence of antibodies to the virus in serum). Second, all ATLL patients are infected with HTLV I. Third, the proviral DNA is always present in tumor cells, but not usually in other cells from the same patient. Finally, HTLV I can transform human T cells in vitro.

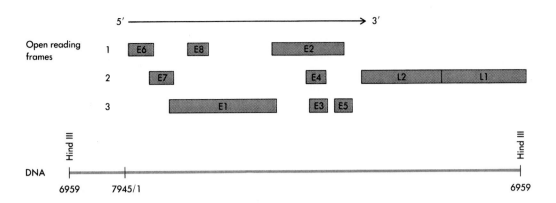

Figure 27-19
Genomic organization of bovine papillomavirus DNA. The circular 7945-base-pair genome is represented as a linear molecule. Open boxes denote regions of open reading frames (ORFs) and hence potential protein-coding segments. The segment of the genome that is transcribed in transformed cells and the direction of transcription are indicated by the arrow at the top of the figure. ORFs within this region are designated E1-E8. [After N. Sarver, M. S. Rabson, Y.-C. Yang, J. C. Byrne, and P. M. Howley, *J. Virology* 52 (1984):377–388.]

As in the case of the specific cancers associated with hepatitis and EBV infections, the majority of individuals who carry HTLV I virus do not develop T cell leukemia; and, as in the case of these viruses, incubation times between infection and disease are long.

Although integrated HTLV I proviral DNA is always present in tumor cells, the virus does not integrate into specific sites in chromosomal DNA or near proto-oncogenes. The HTLV I provirus has been completely sequenced, and the virus does not have a cell-derived oncogene. However, the virus can transform (immortalize) T cells in vitro, suggesting that it carries a virus-coded transforming gene. Thus, in terms of its transforming ability, HTLV I more closely resembles oncogenic DNA viruses than replication-competent retroviruses. The candidate transforming gene is encoded at the 3' end of the virus by a gene called *pX* or *lor*. This gene may encode a protein that functions as a *trans*-acting factor to increase viral transcription (Chapter 24). Conceivably, the *pX* gene product also activates transcription of specific cellular genes. For example, overexpression of genes that could stimulate excessive T cell division might put the cells at risk to accumulate additional cancer-causing mutations.

Papillomaviruses and Cervical Carcinoma[92–97]

Certain human genital cancers show epidemiological features characteristic of infectious disease; for example, cervical carcinoma is most common among women with multiple sex partners. Now evidence is accumulating that papillomaviruses may play a role in causing cervical carcinoma and also penile cancer, as well as benign genital warts.

Papillomaviruses are small DNA viruses belonging to the papovavirus group. They are found in a variety of animals, and isolates from rabbits (Shope papillomavirus) and cows (bovine papillomavirus, or BPV) have been particularly well studied. Papillomaviruses cause benign tumors (warts or papillomas) of the cutaneous and mucosal epithelia, including many types of warts common in humans. In some cases, certain papillomas can undergo malignant transformation. Furthermore, some papillomaviruses can transform cells in vitro. Papillomaviruses replicate as extrachromosomal circular DNA, maintaining a fairly constant number of genomes per infected cell. In warts, DNA is stably maintained as a plasmid in basal cells, and a complete viral replication cycle resulting in the production of progeny genomes occurs only in the outer layers of skin as cells undergo terminal differentiation.

Different papillomaviruses have similar genome organization (Figure 27-19). At least in some viruses, two regions of the genome, one encoding open reading frame (ORF) E6 and the other the E2 to E5 ORFS, have transforming activity.

There are now at least 30 distinct but related human papillomavirus isolates. Many were isolated using known viral DNAs as hybridization probes and conditions of reduced stringency to detect new members of the family. Most interesting has been the finding that two particular isolates, human papillomavirus type 16 (HPV-16) and type 18 (HPV-18) are particularly common in cervical carcinomas: Over 80 percent of invasive cervical cancers harbor one of these viruses. Even the well-known human cell line HeLa, which originated from a cervical carcinoma, harbors HPV-18. Interestingly, in tumors, viral DNA is integrated into cellular DNA and is usually transcriptionally active. At least some tumors have been found to express the E6 open reading frame.

It remains unproved that HPV causes cervical carcinoma, and as usual, it is difficult to rule out the possibility that the association of the virus with the tumor is just a result of an ability of HPV to replicate in tumor cells plus a high probability that sexually active individuals may harbor these viruses. Nonetheless, the circumstantial evidence implicating HPV as a cause of cervical cancer is compelling.

Summary

Most cancers probably arise when several mutations accumulate within the DNA of a single cell, causing it to lose growth control. Many lines of evidence support a genetic basis for cancer: The cancerous phenotype of a cell is stably inherited; the majority of agents that induce cancer in animals or transform cells in vitro are mutagens that act on DNA; and specific chromosomal abnormalities predispose to cancer or are present in tumors. Evidence that more than one genetic event is required to induce a tumor has come from analyzing cancer incidence as a function of age and from the study of inherited forms of cancer. These studies suggest that at least two and possibly six or seven genetic events are required to produce a cancer cell.

Carcinogens in the environment probably cause the majority of human cancers. While the particular substances remain to be identified in most cases, some clear-cut examples include tobacco (lung cancer), asbestos (mesothelioma), and sunlight (skin cancer).

Viruses are another important cause of some human cancers. They probably act in combination with other genetic events to bring about the cancerous transformation of cells. A still-unanswered question is the percent of human cancers that have a viral etiology, but now the best guess is probably about 10 to 20 percent.

For some time, a major goal of molecular biologists has been to identify the cellular genes involved in human tumors, the so-called human oncogenes. Several lines of research converged in the 1980s to make this goal achievable. Absolutely critical was the development of molecular cloning technologies. In addition, the analysis of oncogenes carried by tumor viruses (particularly the cell-derived oncogenes carried by retroviruses), the analysis of specific chromosome abnormalities in tumors, and the ability to assay for single-copy cellular genes using the technique of DNA transfection all played essential roles.

DNA isolated from about 15 to 20 percent of human tumors can induce transformed foci in mouse 3T3 cells, while DNA from normal cells does not induce foci under the same experimental conditions. The transformed phenotype can be serially passed via DNA from the primary transfectants to fresh 3T3 cells, then from the resulting secondary transfectants to additional cells, and so on. A single human gene that acts in a dominant fashion must be responsible for the transforming ability of the tumor DNA. These human oncogenes can be molecularly cloned from the DNA of secondary or tertiary 3T3 transfectants by using a human species-specific repeated DNA sequence, the *Alu* repeat, as a hybridization probe.

Remarkably, the majority of the human oncogenes that have been detected by DNA transfection of 3T3 cells turn out to be activated human *ras* genes, homologs of the genes first identified as the oncogenes of the Harvey and Kirsten mouse sarcoma retroviruses. There are three functional *ras* genes in human cells, H-*ras*-1, K-*ras*-2, and N-*ras*. Any one of these can become an oncogene as a result of mutation. Frequently a single-base change that alters the twelfth codon of the gene and consequently changes the twelfth amino acid from the normal glycine converts the normal cellular gene to an oncogene. Alterations of the twelfth codon can also activate N- and K-*ras* genes. Some tumors have *ras* genes that are activated as a result of mutations of the thirteenth or sixty-first codon.

Since the normal cellular function of *ras* genes is not known, it is difficult to say why a single-base change converts the gene to an oncogene. However, *ras* genes encode 21 kdal proteins, located at the inner surface of the plasma membrane, that bind guanine nucleotides and possess GTPase activity. Although activated *ras* genes have guanine nucleotide binding activity, they have sharply decreased GTPase activity. By analogy to other guanine nucleotide binding proteins, such as the G proteins that regulate cAMP levels, it is postulated that loss of GTPase activity leads to loss of a regulatory mechanism and results in the *ras* protein remaining in an active form much of the time.

A few other cellular oncogenes have been detected by DNA transfection. One, the *neu* gene (in carcinogen-induced neuroblastomas and as glioblastomas) is distinct from but related to the gene encoding the EGF receptor. It is not known why *ras* genes are the predominant type detected by the assay. It may reflect either their preferential activation in human tumors or the greater ability of *ras* oncogenes to be seen in the 3T3 transformation assay.

It is difficult to prove that activated *ras* genes detected by transfection were involved in causing the human tumors that harbor them. However, it seems likely that this is the case. Using carcinogen-induced breast tumors of rats as a model system, it has been shown that a single brief exposure to a carcinogen can induce tumors almost all of which have an alteration at codon 12 of the rat H-*ras*-1 gene.

Although activated *ras* genes can transform 3T3 cells, they are unable to transform some types of primary rodent cells efficiently. However, if they are transfected along with a second oncogene such as *myc* or the E1A gene of adenovirus, then transformants are obtained. This result provides direct evidence that more than one oncogene is needed to cause cancerous transformation.

It has been known for some time that human tumors possess chromosome abnormalities, but the field has exploded as better and better techniques have become available for preparing and analyzing chromosomes. Most of the work has been done with leukemias, but solid tumors, frequently more difficult to study, are now beginning to be analyzed in detail as well. Abnormalities can involve both chromosome number and structure. So far, most effort has gone to identifying specific abnormalities that are present in particular forms of cancer. However, superimposed on these almost invariable abnormalities are often numerous other changes, some common but not ubiquitous for a specific form of cancer. There is evidence that as tumors progress, often their chromosomes become increasingly aberrant, both in number and structure.

It was suggested that specific chromosome abnormalities might involve specific oncogene activation. Remarkably, several examples of this have now been found. The Philadelphia chromosome in chronic myelogenous leukemia, the first consistent abnormality associated with a human cancer, results from a translocation between chromosomes 9 and 22 that joins the 5' half of a gene called *bcr* to the 5' side of the *abl* oncogene. As a result, a long *bcr-abl* primary transcript is made. After splicing, this *bcr-abl* RNA is translated to yield a *bcr-abl* fusion protein that is probably involved in causing CML.

Burkitt's lymphomas, as well as certain other B cell lymphomas, have translocations that involve the immunoglobulin gene loci and the *c-myc* oncogene. Most common is a t(8;14) translocation in which the immunoglobulin heavy-chain constant region is fused head to head with *c-myc*. While the way in which this event activates *c-myc* is not yet completely clear, it is suspected that the fusions in some way alter c-*myc* gene expression and that this, in turn, contributes to the development of the tumor.

Certain tumors possess tiny chromosomes, called double minutes (DMs), that lack centromeres and can vary in number from cell to cell. These tumors frequently also have homogeneously staining regions (HSRs) within certain chromosomes. The two structures appear to be interchangeable forms of an amplified DNA sequence. Now it has been shown that the DMs and HSRs of neuroblastomas involve the amplification of a fragment of DNA carrying a gene called N-*myc*, while variants of small-cell lung tumors have amplifications of either c-*myc*, N-*myc*, or L-*myc*. In both cases, amplification appears to occur with tumor progression, and is present only in later stages of tumor development. The normal cellular functions of members of the *myc* gene family are not known, but the genes encode proteins that are localized to the nucleus and may be involved in regulating the expression of other cellular genes.

Deletion of a specific chromosomal band (band 14 on chromosome 13) is associated with a tumor called retinoblastoma. Tumors are thought to arise when both copies of a gene at this location, a gene called *Rb-1*, are deleted or mutated. Thus, *Rb-1* may represent a new form of oncogene that is recessive rather than dominant. The homozygous deletion that usually gives rise to retinoblastoma can probably cause other types of tumors as well. Other specific band deletions are associated with yet other forms of cancer.

Most searches for human tumor viruses have ended in frustration. However, there are now several cases where a viral cause of cancer seems either clear, probable, or at least possible. These include hepatitis B virus as a cause of liver cancer, a retrovirus called HTLV I as the cause of adult T cell lymphoma, papilloma viruses as a cause of cervical cancer, and possibly the herpes virus Epstein-Barr as a factor in causing nasopharyngeal carcinoma and Burkitt's lymphoma. In none of these cases is the mechanism by which the virus causes the disease known with certainty. In all cases, the majority of people infected with the virus do not contract cancer. There are many examples where immunosuppression is associated with rare forms of cancer, and these cancers may have a viral origin. Our immune system probably protects us from most viruses that are capable of causing cancer.

The next decade will surely see a revolution in our understanding of the molecular basis of cancer. When or how this information will allow us to control the disease is difficult to say. For now, the best approach is still prevention. Elimination of cigarette smoking and vaccination against hepatitis B virus are probably the major new "cancer cures" presently available.

Bibliography

General References

Bishop, J. M. 1982. "Oncogenes." *Sci. Amer.* 246:80–92.

Cairns, J. 1978. Cancer, Science and Society. San Francisco: Freeman.

DeVita, V., S. Hellman, and S. Rosenberg, eds. 1986. *Important Advances in Oncology*. Philadelphia: Lippincott.

Sendberg, A. 1980. *The Chromosomes in Human Cancer and Leukemia*. North Holland and Amsterdam: Elsevier.

Weinberg, R. A. 1983. "A Molecular Basis of Cancer." *Sci. Amer.* 249:126–142.

Cited References

1. Fialkow, P. J. 1974. "The Origin and Development of Human Tumors Studied with Cell Markers." *New Eng. J. Med.* 291:26–35.
2. Farber, E., and R. Cameron. 1980. "The Sequential Analysis of Cancer Development." *Adv. Cancer Res.* 31:125–226.
3. Knudson, A. G., H. W. Hethcote, and B. W. Brown. 1975. "Mutation and Childhood Cancer: A Probabilistic Model for the Incidence of Retinoblastoma." *Proc. Nat. Acad. Sci.* 72:5116–5120.
4. Foulds, L. 1975. *Neoplastic Development*. Vol. 2. New York: Academic Press.
5. Klein, G., and E. Klein. 1985. "Evolution of Tumors and the Impact of Molecular Oncology." *Nature* 315:190–195.
6. Hiatt, H. H., J. D. Watson, J. A. Winsten, eds. 1977. *Origins of Human Cancer. Cold Spring Harbor Conferences on Cell Proliferation.* Vol. 4. Cold Spring Harbor, N.Y.: Cold Spring Harbor Laboratory.
7. Ames, B. N., W. E. Durston, E. Yamasaki, and F. D. Lee. 1973. "Carcinogens Are Mutagens: A Simple Test System Combining Liver Homogenates for Activation and Bacteria for Detection." *Proc. Nat. Acad. Sci.* 70:2281–2285.
8. Yunis, J. J. 1983. "The Chromosomal Basis of Human Neoplasia." *Science* 221:227–236.
9. Essex, M., G. Todaro, H. zur Hausen, eds. 1980. *Viruses in Naturally Occurring Cancer. Cold Spring Harbor Conferences on Cell Proliferation.* Vol. 7. Cold Spring Harbor, N.Y.: Cold Spring Harbor Laboratory.
10. Hill, M., and J. Hillova. 1971. "Production Virale dans les Fibroblasts de Poule Traites per l'Acide Desoxyribonucleique de Cellules XC de Rat Transformees par le Virus de Rous." *Compt. Rend. Acad. Sci.* 272:3094–3097.
11. Graham, F. L., and A. J. van der Eb. 1973. "A New Technique for the Assay of Infectivity of Human Adenovirus 5 DNA." *Virology* 52:456–467.
12. Pellicer, A., D. Robins, B. Wold, R. Sweet, J. Jackson, I. Lowy, J. M. Roberts, G. K. Sim, S. Silverstein, and R. Axel. 1980. "Altering Genotype and Phenotype by DNA-Mediated Transfer." *Science* 209:1414–1422.
13. Cooper, G. M. 1982. "Cellular Transforming Genes." *Science* 217:801–806.
14. Weinberg, R. A. 1982. "Oncogenes of Spontaneous and Chemically Induced Tumors." *Adv. Cancer Res.* 36:149–163.
15. Barbacid, M. 1986. "Human Oncogenes." In *Important Advances in Oncology*, ed. V. DeVita, S. Hellman, and S. Rosenberg. eds. Philadelphia: Lippincott.
16. Gusella, J. F., C. Keys, A. Varsanyi-Breiner, F. Kao, C. Jones, T. T. Puck, and D. Housman. 1980. "Isolation and Localization of DNA Segments from Specific Human Chromosomes." *Proc. Nat. Acad. Sci.* 77:2829–2833.
17. Shih, C., and R. A. Weinberg. 1982. "Isolation of a Transforming Sequence from a Human Bladder Carcinoma Cell Line." *Cell* 29:161–169.
18. Ellis, R. W., D. De Feo, T. Y. Shih, M. A. Gonda, H. A. Young, N. Tsuchida, D. R. Lowy, and E. M. Scolnick. 1981. "The p21 src Genes of Harvey and Kirsten Sarcoma Viruses Originate from Divergent Members of a Family of Normal Vertebrate Genes." *Nature* 292:506–511.

19. Der, C., T. Krontiris, and G. M. Cooper. 1982. "Transforming Genes of Human Bladder and Lung Carcinoma Cells Are Homologous to the ras Genes of Harvey and Kirsten Sarcoma Viruses." *Proc. Nat. Acad. Sci.* 79:3637–3640.
20. Parada, L. F., C. J. Tabin, C. Shih, and R. A. Weinberg. 1982. "Human EJ Bladder Carcinoma Oncogene Is Homologue of Harvey Sarcoma Virus ras Gene." *Nature* 297:474–478.
21. Santos, E., S. R. Tronick, S. A. Aaronson, S. Pulciani, and M. Barbacid. 1982. "T24 Human Bladder Carcinoma Oncogene Is an Activated Form of the Normal Human Homologue of BALB- and Harvey-MSV Transforming Genes." *Nature* 298:343–347.
22. Capon, D., Y. Ellson, A. Levinson, P. Seeburg, and D. Goeddel. 1983. "Complete Nucleotide Sequences of the T24 Human Bladder Carcinoma Oncogene and Its Normal Homologue." *Nature* 302:33–37.
23. Shimizu, K., D. Birnbaum, M. Ruley, O. Fasano, Y. Suard, L. Edlund, E. Taparowsky, M. Goldfarb, and M. Wigler. 1983. "The Structure of the K-ras Gene of the Human Lung Carcinoma Cell Line Calu-1." *Nature* 304:497–500.
24. McGrath, J. P., D. J. Capon, D. H. Smith, E. Y. Chen, P. H. Seeburg, D. V. Goeddel, and A. D. Levinson. 1983. "Structure and Organization of the Human Ki-ras Proto-Oncogene and a Related Processed Pseudogene." *Nature* 304:501–506.
25. Taparowsky, E., K. Shimizu, M. Goldfarb, and M. Wigler. 1983. "Structure and Activation of the Human N-ras Gene." *Cell* 34:581–586.
26. Shih, T., A. Papageorge, P. Stokes, M. Weeks, and E. Scolnick. 1980. "Guanine-Nucleotide Binding and Autophosphorylating Activities Associated with the p21^{src} Protein of Harvey Murine Sarcoma Virus." *Nature* 287:686–691.
27. Aaronson, S. A., and S. R. Tronick. 1986. "The Role of Oncogenes in Human Neoplasia." In *Important Advances in Oncology*, ed. V. DeVita, S. Hellman, and S. Rosenberg. Philadelphia: Lippincott.
28. Shimizu, K., Y. Nakatsu, M. Sekiguchi, K. Hokamura, K. Tanaka, M. Terada, and T. Sugimura. 1985. "Molecular Cloning of an Activated Human Oncogene, Homologous to v-raf, from Primary Stomach Cancer." *Proc. Nat. Acad. Sci.* 82:5641–5645.
29. Tabin, C. J., S. M. Bradley, C. I. Bargmann, R. A. Weinberg, A. G. Papageorge, E. M. Scolnick, R. Dhar, D. R. Lowy, and E. H. Chang. 1982. "Mechanism of Activation of a Human Oncogene." *Nature* 300:143–149.
30. Reddy, E. P., R. K. Reynold, E. Santos, and M. Barbacid. 1982. "A Point Mutation Is Responsible for the Acquisition of Transforming Properties of the T24 Human Bladder Carcinoma Oncogene." *Nature* 300:149–152.
31. Fasano, O., T. Aldrich, F. Tamanoi, E. Taparowsky, M. Furth, and M. Wigler. 1984. "Analysis of the Transforming Potential of the Human H-ras Gene byRandom Mutagenesis." *Proc. Nat. Acad. Sci.* 81:4008–4012.
32. Seeburg, P. H., W. W. Colby, D. J. Capon, D. V. Goedell, and A. D. Levinson. 1984. "Biological Properties of Human c-Ha-ras Genes Mutated at Codon 12." *Nature* 312:71–75.
33. Jurnak, F., 1985. "Structure of the GDP Domain of EF-Tu and Location of Amino Acids Homologous to ras Oncogene Proteins." *Science* 230:32–37.
34. McCormick, F., B. F. C. Clark, T. F. M. la Cour, M. Kjeldgaard, L. Norskov-Lauritsen, and J. Nyborg. 1985. "A Model for the Tertiary Structure of p21, the Product of the ras Oncogene." *Science* 230:78–82.
35. Gibbs, J. B., I. S. Sigal, M. Poe, and E. M. Scolnick. 1984. "Intrinsic GTPase Activity Distinguishes Normal and Oncogenic ras p21 Molecules." *Proc. Nat. Acad. Sci.* 81:5704–5708.
36. McGrath, J., D. Capon, D. Goeddel, and A. Levinson. 1984. "Comparative Biochemical Properties of Normal and Activated Human ras p21 Protein." *Nature* 310:644–655.
37. Dhar, R., A. Nieto, R. Koller, D. DeFeo-Jones, P. Robinson, G. Temeles, and E. M. Scolnick. 1984. "Nucleotide Sequence of Two H-ras-Related Genes Isolated from the Yeast *Saccharomyces cerevisiae*." *Nucleic Acid Res.* 12:3611–3618.

38. Powers, S., T. Kataoka, O. Fasano, M. Goldfarb, J. Strathern, J. Broach, and M. Wigler. 1984. "Genes in *S. cerevisiae* Encoding Proteins with Domains Homologous to the Mammalian *ras* Proteins." *Cell* 36:607–612.

39. Kataoka, T., S. Powers, S. Cameron, O. Fasano, M. Goldfarb, J. Broach, and M. Wigler. 1985. "Functional Homology of Mammalian and Yeast *RAS* Genes." *Cell* 40:19–26.

40. Hurley, J., M. Simon, D. Teplow, J. Robishaw, and A. Gilman. 1984. "Homologies Between Signal Transducing G Proteins and *ras* Gene Products." *Science* 226:860–862.

41. Feramisco, J. R., M. Gross, T. Kamata, M. Rosenberg, and R. W. Sweet. 1984. "Microinjection of the Oncogene Form of the Human H-*ras* (T24) Protein Results in Rapid Proliferation of Quiescent Cells." *Cell* 38:109–117.

42. Mulcahy, L. S., M. R. Smith, and D. W. Stacey. 1985. "Requirement for *ras* Proto-Oncogene Function During Serum-Stimulated Growth of NIH/3T3 Cells." *Nature* 313:241–243.

43. Shih, C., B. Z. Shilo, M. P. Goldfarb, A. Dannenberg, and R. A. Weinberg. 1979. "Passage of Phenotypes of Chemically Transformed Cells Via Transfection of DNA and Chromatin." *Proc. Nat. Acad. Sci.* 76:5714–5718.

44. Zarbl, H., S. Sukamar, A. V. Arthur, D. Martin-Zanca, and M. Barbacid. 1985. "Direct Mutagenesis of H-*ras*-1 Oncogenes by Nitroso-Methyl-Urea During Initiation of Mammary Carcinogenesis in Rats." *Nature* 315:382–385.

45. Guerrero, I., A. Villasante, A. Mayer, and A. Pellicer. 1984. "Carcinogen and Radiation-Induced Mouse Lymphomas Contain an Activated *c-ras* Oncogene." In *Cancer Cells: Oncogenes and Viral Genes*, ed. G. F. Vande Woude, A. J. Levine, W. C. Topp, and J. D. Watson. Cold Spring Harbor, N.Y.: Cold Spring Harbor Laboratory.

46. Schechter, A. L., D. F. Stern, L. Vaidyanathan, S. J. Decker, J. A. Drebin, M. J. Greene, and R. A. Weinberg. 1984. "The *neu* Oncogene: An *erbB*-Related Gene Encoding a 185,000-Mr Tumor Antigen." *Nature* 312:513–516.

47. Balmain, A., M. Ramsden, G. T. Bowden, and J. Smith. 1984. "Activation of the Mouse Cellular Harvey-*ras* Gene in Chemically Induced Benign Skin Papillomas." *Nature* 307:658–660.

48. Van der Eb, A. J., H. van Ormondt, P. I. Schier, J. H. Lupker, H. Jochemsen, P. J. van der Elsen, R. J. De Leys, J. Maat, C. P. van Beveren, R. Dijkeme, and A. de Waard. 1980. "Structure and Function of the Transforming Genes of Human Adenoviruses and SV40." *Cold Spring Harbor Symp. Quant. Biol.* 44:383–399.

49. Rassoulzadegan, M., Z. Naghashfar, A. Cowie, A. Carr, M. Grisoni, R. Kamen, and F. Cuzin. 1983. "Expression of the Large T Protein of Polyoma Virus Promotes the Establishment in Culture of 'Normal' Rodent Fibroblast Cell Lines." *Proc. Nat. Acad. Sci.* 80:4354–4358.

50. Land, H., L. F. Parada, and R. A. Weinberg. 1983. "Tumorigenic Conversion of Primary Embryo Fibroblasts Requires at Least Two Cooperating Oncogenes." *Nature* 304:596–602.

51. Ruley, H. E. 1983. "Adenovirus Early Region 1A Enables Viral and Cellular Transforming Genes To Transform Primary Cells in Culture." *Nature* 304:602–606.

52. Spandidos, D. A., and N. M. Wilkie. 1984. "Malignant Transformation of Early Passage Rodent Cells by a Single Mutated Human Oncogene." *Nature* 310:469–475.

53. Rowley, J. D. 1982. "Identification of the Constant Chromosome Regions Involved in Human Hematologic Malignant Disease." *Science* 216:749–755.

54. Yunis, J. J. 1986. "Chromosomal Rearrangements, Genes, and Fragile Sites in Cancer: Clinical and Biological Implications." In *Important Advances in Oncology*, ed. V. DeVita, S. Hellman, and S. Rosenberg. Philadelphia: Lippincott.

55. Nowell, P. C., and D. A. Hungerford. 1960. "A Minute Chromosome in Human Granulocytic Leukemia." *Science* 132:125–132.

56. Rowley, J. D. 1973. "A New Consistent Chromosomal Abnormality in Chronic Myelogenous Leukemia Identified by Quinacrine Fluoresence and Giemsa Staining." *Nature* 243:290–293.

57. De Klein, A., A. G. van Kessel, G. Grosveld, C. R. Bartram, A. Hagemeijer, D. Bootsma, N. K. Spurr, N. Heisterkamp, J. Groffen, and J. R. Stephenson. 1982. "A Cellular Oncogene Is Translocated to the Philadelphia Chromosome in Chronic Myelocytic Leukemia." *Nature* 300:765–767.

58. Gale, R. P., and E. Canaani. 1984. "An 8-Kilobase *abl* RNA Transcript in Chronic Myelogenous Leukemia." *Proc. Nat. Acad. Sci.* 81:5648–5652.

59. Groffen, J., J. R. Stephenson, N. Heisterkamp, A. de Klein, C. R. Bartram, and G. Grosveld. 1984. "Philadelphia Chromosomal Breakpoints Are Clustered Within a Limited Region, *bcr*, on Chromosome 22." *Cell* 36:93–99.

60. Konopka, J. B., S. M. Watanabe, and O. N. Witte. 1984. "An Alteration of the Human *c-abl* Protein in K562 Leukemia Cells Unmasks Associated Tyrosine Kinase Activity." *Cell* 37:1035–1042.

61. Shtivelman, E., B. Lifchitz, R. Gale, and E. Canaani. 1985. "Fused Transcript of *abl* and *bcr* Genes in Chronic Myelogenous Leukemia." *Nature* 315:550–554.

62. Adams, J. 1985. "Oncogene Activation by Fusion of Chromosomes in Leukemia." *Nature* 315:542–543.

63. Ohno, S., M. Babonits, F. Weiner, J. Spira, G. Klein, and M. Potter. 1979. "Non-Random Chromosome Changes Involving Ig-Gene Chromosomes (Nos. 12 and 6) in Pristane Induced Mouse Plasmacytomas." *Cell* 18:1001–1008.

64. Shen-Ong, G. L. C., E. J. Keath, S. P. Piccoli, and M. D. Cole. 1982. "Novel *myc* Oncogene RNA from Abortive Immunoglobulin Gene Recombination in Mouse Plasmacytomas." *Cell* 31:443–452.

65. Dalla-Favera, R., S. Martinotti, R. Gallo, J. Erikson, and C. Croce. 1983. "Translocation and Rearrangements of the *c-myc* Oncogene Locus in Human Undifferentiated B-Cell Lymphomas." *Science* 219:963–997.

66. Leder, P., J. Battey, G. Lenoir, C. Moulding, W. Murphy, H. Potter, T. Stewart, and R. Taub. 1983. "Translocations Among Antibody Genes in Human Cancer." *Science* 222:765–771.

67. Corcoran, L. M., S. Cory, and J. M. Adams. 1985. "Transposition of the Immunoglobulin Heavy Chain Enhancer to the *myc* Oncogene in a Murine Plasmacytoma." *Cell* 40:71–79.

68. Hamlyn, P. H., and T. H. Rabbitts. 1983. "Translocation Joins *c-myc* and Immunoglobulin Gamma-1 Genes in a Burkitt Lymphoma Revealing a Third Exon in the *c-myc* Gene." *Nature* 304:135–139.

69. Seibenlist, U., L. Hennighausen, J. Battey, and P. Leder. 1984. "Chromatin Structure and Protein Binding in the Putative Regulatory Region of the *c-myc* Gene in a Burkitt Lymphoma." *Cell* 37:381–391.

70. Kelly, K., B. A. Cochran, C. D. Stiles, and P. Leder. 1983. "Cell-Specific Regulation of the *c-myc* Gene by Lymphocyte Mitogens and Platelet-Derived Growth Factor." *Cell* 35:603–610.

71. Knudson, A. G. 1985. "Hereditary Cancer, Oncogenes, and Antioncogenes." *Cancer Res.* 45:1437–1443.

72. Cavanee, W. K., T. P. Dryja, R. A. Phillips, W. F. Benedict, R. Godbout, B. L. Gallie, A. L. Murphree, L. C. Strong, and R. L. White. 1983. "Expression of Recessive Alleles by Chromosomal Mechanisms in Retinoblastoma." *Nature* 305:779–784.

73. Biedler, J. L. and B. A. Spengler. 1976. A Novel Chromosome Abnormality in Human Neuroblastoma and Antifolate-Resistant Chinese Hamster Cell Lines in Culture. *J. Nat. Cancer Inst.* 57:683–695.

74. Schimke, R. T., ed. 1982. *Gene Amplification*. Cold Spring Harbor, N.Y.: Cold Spring Harbor Laboratory.

75. Varmus, H. E. 1984. "The Molecular Genetics of Cellular Oncogenes." *Ann. Rev. Genetics* 18:553–612.

76. Collins, S., and M. Groudine. 1982. "Amplification of Endogenous *myc*-Related DNA Sequences in a Human Myeloid Leukemia Cell Line." *Nature* 298:679–681.

77. Schwab, M., K. Alitalo, K. Klemphauer, H. E. Varmus, J. M. Bishop, F. Gilbert, G. Brodeur, M. Goldstein, and J. Trent. 1983. "Amplified DNA with Limited Homology to *myc* Cellular Oncogene Is Shared by Human Neuroblastoma Tumor." *Nature* 305:245–248.

78. Brodeur, G. M., R. C. Seeger, M. Schwab, H. E. Varmus, and J. M. Bishop. 1984. "Amplification of N-*myc* in Untreated Human Neuroblastomas Correlates with Advanced Disease Stage." *Science* 224:1121–1124.

79. Nau, M. M., B. J. Brooks, J. Battey, E. Sausville, A. F. Gazdar, I. R. Kirsch, O. W. McBride, V. Bertness, G. F. Hollis, and J. D. Minna. 1985. "L-*myc*, a New *myc*-Related Gene Amplified and Expressed in Human Small Cell Lung Cancer." *Nature* 318:69–73.

80. Ullrich, A. L., L. Coussens, J. S. Hayflick, T. J. Dull, A. Gray, A. W. Tam, J. Lee, Y. Yarden, T. A. Libermann, J. Schlessinger, J. Downward, E. L. V. Mayes, N. Whittle, M. D. Waterfield, and P. H. Seeburg. 1984. "Human Epidermal Growth Factor Receptor cDNA Sequence and Aberrant Expression of the Amplified Gene in A431 Epidermoid Carcinoma Cells." *Nature* 309:418–423.

81. Szmuness, W. 1978. "Hepatocellular Carcinoma and the Hepatitis B Virus: Evidence for a Causal Association." *Prog. Med. Virol.* 24:40–69.

82. Summers, J., J. M. Smolec, and R. Snyder. 1978. "A Virus Similar to Human Hepatitis B Virus Associated with Hepatitis and Hepatoma in Woodchucks." *Proc. Nat. Acad. Sci.* 75:4533–4537.

83. Tiollais, P., C. Pourcel, and A. Dejean. 1985. "The Hepatitis B Virus." *Nature* 317:489–495.

84. Melnick, J. L. 1983. "Hepatitis B Virus and Liver Cancer." In *Viruses Associated with Human Cancer*, ed. L. A. Phillips. New York: Dekker, pp. 337–367.

85. Epstein, M. A., and B. G. Anchong. 1979. *The Epstein-Barr Virus.* New York: Springer-Verlag.

86. De The, G., A. Geser, N. E. Day, P. M. Tukei, E. H. Williams, D. P. Beri, P. G. Smith, A. G. Dean, G. W. Born Kamm, P. Feorino, and W. Henle. 1978. "Epidemiological Evidence for Causal Relationship Between Epstein-Barr Virus and Burkitt's Lymphoma: Results of the Ugandan Prospective Study." *Nature* 272:756–761.

87. Miller G. 1985. "Epstein-Barr Virus." In *Virology*, ed. B. N. Fields, D. M. Knipe, R. M. Chanock, J. L. Melnick, B. Roizman, and R. E. Shope. New York: Raven Press, pp. 563–589.

88. Poiesz, B. J., F. W. Ruscetti, A. F. Gazdar, P. A. Bunn, J. D. Minna, and R. C. Gallo. 1980. "Detection and Isolation of Type C Retrovirus Particles from Fresh and Cultured Lymphocytes of a Patient with Cutaneous T-Cell Lymphoma." *Proc. Nat. Acad. Sci.* 77:7415–7419.

89. Poiesz, B. J., F. W. Ruscetti, M. S. Reitz, V. S. Kalyanaraman, and R. C. Gallo. 1981. "Isolation of a New Type C Retrovirus (HTLV) in Primary Uncultured Cells of a Patient with Sezary T-Cell Leukemia." *Nature* 294:268–271.

90. Miyoshi, I., I. Kubonishi, S. Yoshimoto, T. Akagi, Y. Ohtsuki, Y. Shiraishi, K. Nagata, and Y. Hinuma. 1981. "Type C Virus Particles in a Cord Blood T-Cell Line Derived from Cocultivating Normal Human Cord Leukocytes and Human Leukaemic T Cells." *Nature* 294:770–771.

91. Wong-Staal, F., L. Ratner, G. Shaw, B. Hahn, M. Harper, G. Francini, and R. Gallo. 1985. "Molecular Biology of Human T-Lymphotropic Retroviruses." *Cancer Res.* 45:4539–4544.

92. Orth, G., F. Breitburd, M. Faure, and O. Croissant. 1977. "Papillomaviruses: A Possible Role in Human Cancer." In *Origins of Human Cancer*, ed. H. H. Hiatt, J. D. Watson, and J. A. Winsten. *Cold Spring Harbor Conferences on Cell Proliferation.* Vol. 4. Cold Spring Harbor, N.Y.: Cold Spring Harbor Laboratory.

93. Rapp, F., and F. J. Jenkins. 1981. "Genital Cancer and Viruses." *Gynecol. Oncol.* 12:25–41.

94. Howley, P. M., and T. R. Broker. 1985. *Papillomaviruses: Molecular and Clinical Aspects.* UCLA Symposia on Molecular and Cellular Biology, New Series. Vol. 32. New York: Liss.

95. Durst, M., L. Gissman, H. Ickenberg, and H. zur Hausen. 1983. "A Papillomavirus DNA from a Cervical Carcinoma and Its Prevalence in Cancer Biopsy Samples from Different Geographic Regions." *Proc. Nat. Acad. Sci.* 80:3812–3816.

96. Yang, Y-C., H. Okayama, and P. M. Howley. 1985. "Bovine Papillomavirus Contains Multiple Transforming Genes." *Proc. Nat. Acad. Sci.* 82:1030–1034.

97. Schwarz, E., U. K. Freese, L. Gissman, W. Mayer, B. Roggenbuck, A. Stremlau, and H. zur Hausen. 1985. "Structure and Transcription of Human Papillomavirus Sequences in Cervical Carcinoma Cells." *Nature* 314:111–115.

XI

EVOLUTION
OF THE GENE

CHAPTER 28

The Origins of Life

Throughout the preceding chapters, we have seen that the more deeply we look at the molecular level, the more similar all known life forms appear to be. The evidence is therefore overwhelming that life on this planet must have descended from a single ancestral life form or **progenote**. But what this progenote was like and how exactly it first arose are terribly difficult questions.

The theory has occasionally been advanced that the first life forms did not originate on Earth, but arrived here from somewhere else in the universe. Even if this were true, we would still want to know how life originated elsewhere. In this chapter, we will assume, as do the vast majority of practicing biologists, that life originated on Earth.

The past 30 years have witnessed tremendous progress in our ability to conjecture about the origins of life. During this time, geologists have used advanced radioisotope dating techniques to show that the Earth first condensed from an interstellar cloud about 4.6 billion years ago. Chemists have shown that many of the essential building blocks of life (sugars, bases, and amino acids) could have assembled spontaneously from chemicals thought to have been present on the primitive Earth before the first living organisms arose (a process called **prebiotic synthesis**). **Paleobiologists**, who study ancient rocks and fossil sediments, have shown that the first fossil cells (**microfossils**) evolved from the prebiotically synthesized building blocks of life about 3 to 4 billion years ago, when the primitive Earth was only 1 to 1.5 billion years old. And molecular biologists have established that the many forms of life on Earth must share a common ancestor, since all cells are so similar at the molecular level.

In this chapter, we will see how recent advances in our knowledge of gene structure, protein structure, and RNA processing enable us to speculate with renewed confidence about the origin of life on Earth. In particular, we will emphasize the role that RNA enzymes might have played in the evolution of the first replicating biological systems, as well as in the early evolution of the translation apparatus and the genetic code.

Molecular Fossils Can Tell Us How Life Began[1,2,3]

A fossil is what remains after the slow process of mineralization has turned living matter into stone. Only the most durable parts of the organism (shells, bones, feathers, wood, or leaves) tend to survive this process, and thus most familiar fossils are large compared to cells. But under favorable circumstances, the process of mineraliza-

(a)

(b)

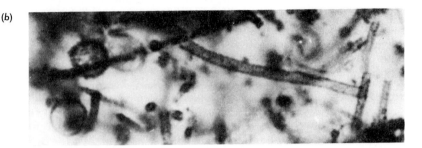

(c)

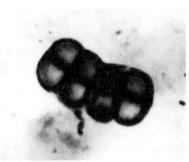

(d)

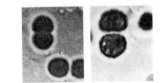

Figure 28-1
Fossil cells can be identified by using a light microscope to examine thin sections of ancient rocks. Procaryotic cells in rocks known to be (a) 3.5 billion years old from the Swartkoppie Formation in South Africa, (b) 2 billion years old from the Gunflint Iron Formation in Canada, and (c) 0.80 billion years old from the Spitsbergen Archipelago in Norway. The cells are much better preserved in the younger rocks. The cells in (b) represent a mixed colony consisting of both spheroidal and filamentous forms. Doublets in (a) and quadruplets in (c) are thought to be dividing cells. For comparison, (d) shows a dividing culture of the modern blue-green procaryotic alga *Aphanocapsa spheroides* at the same magnification as the specimen in (a). [See A. H. Knoll and E. S. Barghoorn, *Science* 198 (1977):396. Photographs courtesy of A. H. Knoll, Harvard University.]

tion can potentially preserve such small details as the wall of a single cell only a few microns in diameter. In the late 1950s, paleobiologists discovered that structures resembling modern algae and bacteria (microfossils) could be discerned by microscopic examination of extremely thin sections of certain ancient rocks (**stromatolites**) from Australia and South Africa (Figure 28-1). Stromatolites build up slowly in shallow seas when floating mats of algae or bacteria multiply and give rise to one layer of progeny after another.

There is strong evidence that the microfossils in stromatolites are really fossil cells, and not just bubbles or mineral deposits that happen to resemble a living organism (pseudofossils). First, deposits of blue-green algae resembling ancient stromatolites are still being formed in shallow bays today. Second, microfossils are comparable in size and shape to a variety of modern algal and bacterial cells. Third, different kinds of microfossils are often found together, as expected if different living species flourished side by side in a favorable environment (see Figure 28-1). And fourth, microfossils sometimes appear as doublets, suggesting that the cells perished in the act of division. Unfortunately, the process of mineralization preserves only the gross morphology of the cell wall, and most of the organic matter is destroyed as the rock undergoes cycles of heating and cooling. Thus, microfossils tell us little about the cell at the molecular level and provide no clues about the life forms that preceded the first microorganisms with tough cell walls.

We can reconstruct the earliest stages of evolution in either of two ways. One way is to ask what chemical or enzymatic reactions can occur when the conditions that may have prevailed on the primitive Earth are simulated in a modern laboratory. While these experiments cannot rigorously prove how life began, they can establish the plausibility of various scenarios.

The other way is to scrutinize modern cells for evidence of molecular fossils. **Molecular fossils** are structures (such as the placement of introns within a gene) or functions (such as the utilization of a particular metabolic pathway) that are shared by such diverse organisms that they must have been present in the first cells, as well as in earlier precellular forms of life. The abundance of such molecular fossils strongly suggests that living cells preserve within them a remarkably complete record of ancient evolution. Careful reading of this record may allow us to develop a coherent picture of primordial events which, until recently, were only fit for speculation. In this chapter, we will consider some current ideas about the origins of life, starting with the primitive "soup" and continuing to the present, with a special emphasis on what we can learn by reading the molecular fossil record.

The Formation of the Earth

Cosmologists agree that the solar system coalesced rather quickly from a whirling cloud of gas and interstellar dust about 4.6 billion years ago. As this nebulous cloud underwent gravitational collapse over a period of a few hundred million years, most of the mass of the cloud was concentrated in the sun. But for reasons that are still not well understood, some of the cloud did not fall into the sun and instead condensed into tiny **planetesimals** resembling meteorites. These planetesimals in turn aggregated to form the embryonic planets. The core of the primitive Earth melted from the heat of condensation, but the outer crust cooled quickly except where convection or volcanic eruptions brought heat and mass from the molten core to the surface. The primitive atmosphere was formed as various gases escaped from the interior of the earth (**outgassing**). However, it is a matter of contention whether outgassing produced an atmosphere that was highly reducing (H_2, NH_3, CH_4, and H_2O), mildly reducing (H_2O, H_2, N_2, CO, and CO_2), or nonreducing (H_2O, N_2, and CO_2). The crust and atmosphere of the primitive Earth cooled slowly, and the first shallow seas were formed as the less volatile gases, including water, rained back down to Earth. Lightning from these rainstorms, ultraviolet light from the intense solar radiation penetrating through the thin atmosphere, and alternating cycles of cold nights and hot days: These provided the chemical cauldron in which prebiotic synthesis created the essential building blocks of life.

Prebiotic Synthesis of the Building Blocks of Life

We normally think of evolution as the process by which new species of organisms arise from old, but we can extend the concept of evolution to include the process by which the first living cell evolved from

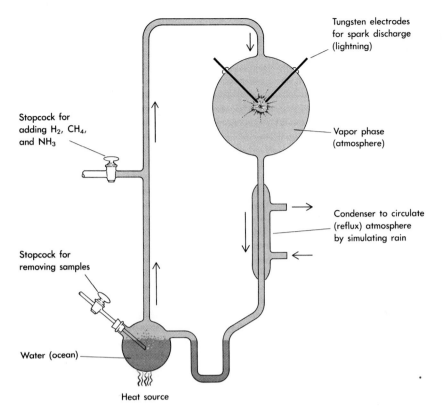

Stopcock for adding H₂, CH₄, and NH₃

Tungsten electrodes for spark discharge (lightning)

Vapor phase (atmosphere)

Condenser to circulate (reflux) atmosphere by simulating rain

Stopcock for removing samples

Water (ocean)

Heat source

Figure 28-2
A reflux apparatus used to simulate prebiotic synthesis of organic molecules under conditions thought to prevail on the primitive Earth. [After S. L. Miller and L. E. Orgel, *The Origins of Life on Earth* (Englewood Cliffs, N.J.: Prentice-Hall, 1974), Figure 7-1.]

the chemicals that were present on the primitive Earth. In this sense, the evolution of life began long before the first living organisms existed.

It now seems obvious that the first living systems could not have evolved unless the most essential building blocks of life (purines, pyrimidines, sugars, amino acids, etc.) had already been synthesized by prebiotic pathways, but this concept was revolutionary when proposed by Alexander Oparin in 1938. Prebiotic synthesis was first subjected to an experimental test in 1953. In an attempt to simulate possible conditions on the primitive Earth, high voltage sparks ("lightning") were passed through the gaseous phase ("atmosphere") of a refluxing mixture of CH_4, NH_3, H_2O, and H_2 (Figure 28-2). These four molecules had all been identified spectroscopically in clouds of interstellar gas and were probably among the major constituents of the Earth's early atmosphere. Samples of the aqueous phase ("ocean") were then withdrawn from the refluxing mixture over a period of days and were found to contain at least ten different amino acids as well as hydrogen cyanide (HCN) and various aldehydes. One possible pathway for amino acid synthesis under these conditions is shown in Figure 28-3.

In other laboratory experiments, the nucleic acid base adenine was synthesized by refluxing a concentrated solution of hydrogen cyanide in ammonia (see Figure 28-3). Polymerization of HCN also gives rise to orotic acid, which can be photochemically decarboxylated by sunlight to yield uracil. The synthesis of various sugars from formaldehyde (H_2CO) had been described as early as 1861, but was thoroughly reinvestigated in the 1960s as interest in prebiotic synthesis grew. Interestingly, the polymerization of formaldehyde results in ribose but not deoxyribose. As we shall see, this observation is but one of many suggesting that RNA preceded DNA in the evolution of life.

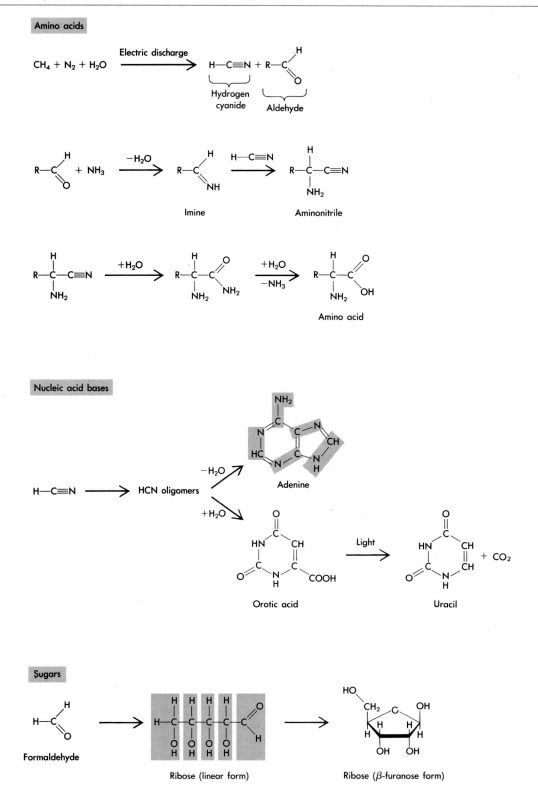

Figure 28-3

A few of the many proposed pathways for prebiotic synthesis of amino acids, nucleic acid bases, and sugars by natural condensation of simple molecules present in the oceans and atmosphere of the primitive Earth. Each of these reactions yields other products besides the biologically relevant molecules shown here. For illustrative purposes, the atoms in the backbone of the adenine ring have been grouped arbitrarily to emphasize that adenine can be thought of as a polymer of hydrogen cyanide monomers. Similarly, ribose is shown as a polymer of formaldehyde.

The kind of prebiotic reactions shown in Figure 28-3 could have taken place in either a **reducing** or a **nonreducing atmosphere**, but not in the modern, strongly **oxidizing atmosphere** containing 21 percent O_2. However, calculations suggest that the abundance of atmospheric oxygen today is due solely to the accumulated synthesis of O_2 by living systems. Thus, the Earth's early atmosphere would have been reducing until the emergence of the first photosynthetic procaryotes.

The pathways of prebiotic synthesis remain speculative because we have so little definite knowledge about the exact conditions prevailing on the primitive Earth. In fact, many of the proposed prebiotic reaction pathways are mutually incompatible. For example, at the concentrations required to generate sugars in alkaline solution, formaldehyde would react rapidly with the amino groups of both amino acids and nucleic acid bases. Nonetheless, local variations in temperature, chemical composition, pH, and other factors might have permitted incompatible reactions to occur in different places or at different times on the primitive Earth.

The proposed prebiotic pathways pose other problems as well. The yield of biologically interesting molecules is often poor, side reactions abound, and synthesis often requires unusually high concentrations of the reactants. However, it is possible that repeated cycles of freezing and thawing, or cycles of evaporation and rehydration, might have concentrated either the reactants or the products.

The conclusion we can draw from more than 30 years of laboratory experiments is that a multitude of alternative pathways may have contributed to prebiotic synthesis. Our confidence in the notion of prebiotic synthesis is also greatly strengthened by recent chemical analysis of meteorites that have fallen to Earth from elsewhere in the solar system. Amino acids and other building blocks of life are found deep within these rocks, where they cannot be regarded as earthly contaminants.

The smallest building blocks of life (nucleic acid bases, sugars, and amino acids) were probably assembled into still larger molecules by sequential **condensation** (removal of water molecules). Water molecules could have been removed physically by heating or chemically by reaction with a water-hungry condensing agent like a polyphosphate. Such condensations would generate the nucleic acid precursors by forming a glycosidic bond between a nucleic acid base and a ribose ring. Condensations would also create **polymers** by forming phosphodiester bonds between the nucleic acid precursors and amide bonds between amino acids. Thus, as the prebiotic cauldron cooled, the resulting soup of organic chemicals probably contained oligonucleotides, peptides, and oligosaccharides.

THE RNA WORLD

Living Systems Must Replicate and Evolve

The first cells were clearly alive, but what about the precellular collections of molecules that preceded them? Can we draw a line between the inanimate world created by prebiotic synthesis and the very earliest stages in the evolution of life? To answer this question, we must first ask ourselves, What is the simplest collection of molecules that would qualify as a "living system"? Opinions will vary, of course, but most biologists would agree that every living system must at the very least be capable of both replication and evolution. Replication enables the parental copy of the system to give rise to two identical (or nearly identical) daughter copies of itself. Evolution implies that the process of replication is always slightly imperfect, so that variant daughter copies may have the potential to replicate faster than the parent or to adapt better to changing environmental conditions.

The First "Living Molecule" Was Almost Certainly a Nucleic Acid

We have seen in Chapter 3 that the self-complementary structure of a nucleic acid double helix is perfectly suited for the task of replication. One strand serves as the template surface for polymerization of a new daughter strand. Polypeptide chains, on the other hand, are not self-complementary and cannot serve as templates for their own replication. This profound distinction between nucleic acids and proteins constitutes a powerful argument that the first replicating structures (genomes) were made of nucleic acids. Moreover, the universality of nucleic acid genomes in contemporary cells and viruses makes it difficult to believe that the ancestral living systems did *not* have either an RNA or a DNA genome.

Until very recently, reconstructing the origin of the earliest life forms appeared to be a hopeless case of "the chicken and the egg." Only nucleic acids seemed capable of storing and replicating genetic information, and only proteins were known to function as catalysts. Thus, it was exceedingly difficult to imagine how the first nucleic acid genome (whether DNA or RNA) could be replicated without a protein catalyst, and equally difficult to imagine how an efficient protein catalyst could arise without a nucleic acid genome to encode it. In 1981, a startling discovery presented a solution to this paradox. The ribosomal RNA intron of the ciliate *Tetrahymena* was shown to excise itself from the rRNA precursor without the benefit of a protein catalyst (see Figures 20-13 and 20-14). The ability of the rRNA intron to function as an enzyme implied that the very first living molecule might have been an **RNA replicase** that catalyzed its own replication without help from a protein (Figure 28-4). This hypothetical RNA would therefore have functioned as both the genetic material and the replication enzyme. The discovery in 1983 of a second catalytic RNA species (the RNA component of RNase P; see Figure 14-19) lent strong support to this notion. Since 1983, many new instances of catalytic RNA have been discovered. As we shall see, the common features of these catalytic RNAs suggest that they are molecular fossils whose history dates back to the earliest life forms. Studies of catalytic RNA therefore play a major role in shaping our current views about the origins of life.

How did the first RNA replicase arise? Presumably, random prebiotic condensation of mononucleotides formed longer and longer polymers. Most of these random RNA polymers would have been functionless, but an occasional RNA molecule may have been able to work as a very primitive, error-prone RNA replicase. This first RNA replicase would have been able to use either itself or another nucleic acid as a template for polymerization of nucleic acid precursors found in the prebiotic soup. Of course, no replicase can copy its own active

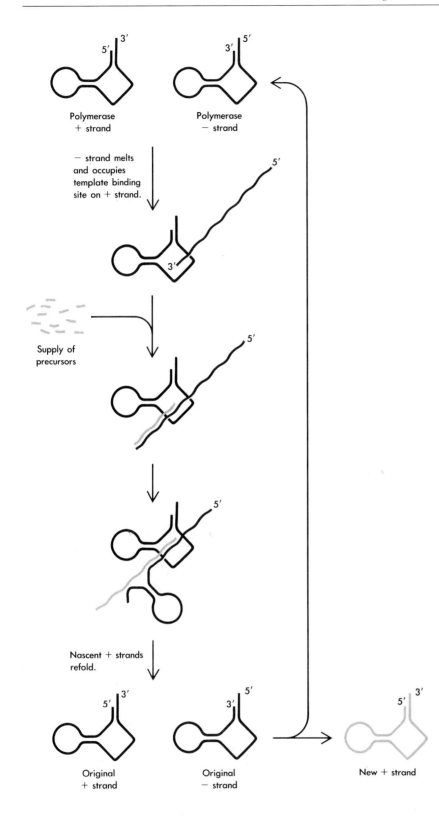

Polymerase + strand

Polymerase − strand

− strand melts and occupies template binding site on + strand.

Supply of precursors

Nascent + strands refold.

Original + strand

Original − strand

New + strand

Figure 28-4
Simplified scheme for an RNA replicase. The + and − strands of the RNA enzyme have similar secondary structures because the sequences are complementary. Only synthesis of a new + strand is shown, but − strands would be replicated in the same way. For clarity, several assumptions have been made in drawing the figure. The RNA polymerase is shown as a *processive* enzyme that remains bound to the template strand until a complete copy has been synthesized. However, the polymerase could also be *dissociative*, separating from the template after addition of each new nucleotide. The precursors are shown as mononucleotides, although we discuss the possibility that random oligonucleotides could have served the same purpose (see Figures 28-11 and 28-12). The nascent RNA strand is thought to refold as synthesis proceeds; otherwise, formation of a stable RNA duplex might block further replication.

site, so we must imagine that *two* RNA replicases arose simultaneously. This is not implausible because, as we shall see, very short RNA sequences can have distinct enzymatic activities (see Figure 28-7).

The hypothetical RNA replicase would not only replicate but also evolve, because some of the variant copies generated by error-prone replication would be able to replicate faster or more accurately than the parental molecule. We therefore regard this RNA replicase as the first "living system." In the section entitled The RNA World, we will see that recent progress in the field of RNA processing suggests that just such an RNA replicase was the not-so-distant ancestor of certain molecules found in living cells today.

Why RNA and Not DNA Was the First Living Molecule

As already discussed, new examples of RNA molecules with enzymatic activity are constantly being discovered (see Figure 28-19), but no enzymatic activity has ever been attributed to DNA. This is perhaps the most powerful argument that the first replicating nucleic acid was RNA rather than DNA. However, there are other substantial reasons for believing in the primacy of RNA:

- The difference between RNA and DNA is that the deoxyribose rings of DNA lack 2'-hydroxyl groups. Both the 2' and 3'-hydroxyl groups are known to play a structural role in tRNA, where they are involved in a number of unusual hydrogen bonds that fold the "cloverleaf" tRNA secondary structure into the final L-shaped tertiary structure (see Figures 14-5 through 14-8 and Figure 28-13). RNA enzymes, like protein enzymes, create a catalytic surface by folding into complex secondary and tertiary structures. Lacking 2'-hydroxyl groups, DNA may be less adept than RNA at folding into the complex three-dimensional shapes required for catalytic activity.

- The 2' hydroxyl of ribose also plays a direct catalytic role in some RNA enzymes. Evidence for this comes from studies of the self-splicing mitochondrial Group II introns (see Figures 28-19 and Table 28-1), as well as from theoretical consideration of the available functional groups in RNA (see Figure 28-6). Thus, DNA enzymes could never be as versatile as RNA enzymes.

- Although all cells today have DNA genomes, DNA precursors are always synthesized by reduction of RNA nucleoside diphosphate precursors using the highly conserved enzyme **ribonucleoside diphosphate reductase** (see Figure 28-20). This suggests that the biochemical pathway for ribonucleotide synthesis evolved first, when cells (or precellular living systems) had RNA genomes. Later, when DNA genomes replaced the more labile RNA genomes, the preexisting biochemical pathways were adapted for the synthesis of DNA precursors.

- Finally, ribose is much more readily synthesized than deoxyribose under simulated prebiotic conditions.

RNA Enzymes Work Like Protein Enzymes[4]

Regardless of whether it is made of protein, RNA, or even some other exotic material, every enzyme must have an active site. This site is a three-dimensional surface that binds the substrate and positions the functional groups of the enzyme around the bound substrate so that catalysis can proceed. When the crystal structure of tRNA was first

Table 28-1 Intron Boundary Sequences

Intron Type	5' Splice Junction	3' Splice Junction
Group I, mitochondrial + nuclear	U̲↓	G̲↓
Group II, mitochondrial	↓G̲U̲GCG	Y$_n$A̲U̲↓
Chloroplast mRNA (*Euglena*)	↓G̲U̲G^C_UG	Y$_n$A̲^U_C↓
Nuclear mRNA precursor (general)	^C_AAG↓G̲U̲^A_GAGU	Y$_n$A̲G̲↓
Nuclear mRNA precursor (yeast)	↓G̲U̲AUGU	Y$_n$A̲G̲↓

Underscored nucleotides are essentially invariant; other nucleotides represent a consensus. Arrows designate the actual 5' and 3' splice sites. [After T. R. Cech, *Cell* 44 (1986):207.]

determined in 1974, it became clear that RNA molecules could fold up into three-dimensional structures resembling those formed by proteins (see Figures 14-5 through 14-8). However, no one was sufficiently bold (or farsighted?) to suggest at the time that RNA might function as a catalyst. The reason for such universal hesitation was that only 4 relatively similar nucleotides (instead of 20 diverse amino acids) could be used to construct the active site of an RNA enzyme. Although we now know that RNA can function catalytically, we might still expect RNA enzymes (also called **ribozymes**) to be less versatile than protein enzymes, because RNA does not have as large a variety of functional groups.

Like protein enzymes, RNA enzymes almost certainly use **proton exchange reactions** to catalyze hydrolysis of RNA substrates (Figure 28-5). In fact, very similar mechanisms may be responsible for the hydrolysis of RNA by the protein RNase A and the catalytic RNA component of RNase P. Despite some superficial differences between the mechanisms (RNase A leaves a 3' phosphate and RNase P a 5' phosphate), in each case one functional group (a general base) removes a proton (H^+) while a nearby functional group (a general acid) adds a proton. A wealth of structural and enzymatic data suggest that the proton exchange reactions in RNase A are catalyzed by histidines 12 and 119. (Histidine is a particularly useful amino acid for catalysis because both nitrogens of the imidazole ring can bind protons.) Like proteins, RNA molecules also possess a variety of potential functional groups (Figure 28-6), but we are still far from understanding which groups in RNA are catalytically important.

Very Short RNAs Can Be Catalytic[5–9]

How realistic is it to suppose that random prebiotic condensation of ribonucleotides could actually fashion a crude RNA replicase? The answer to this question depends on the size of the smallest RNA that can function as a catalyst. If all catalytically active RNA molecules had to be as large and complex as the RNA component of RNase P (377 nucleotides in *E. coli*; see Figure 14-19) or the self-splicing intron of the *Tetrahymena* rRNA (413 nucleotides; see Figures 20-14 and 20-8), it would be very difficult to argue that even a very crude RNA replicase could arise by random condensation of ribonucleotides. However, recent progress in the study of plant viruses has demonstrated that as few as 52 nucleotides can act as an accurate RNA catalyst, and a crude catalyst might be simpler.

Hydrolysis of RNA by pancreatic RNase

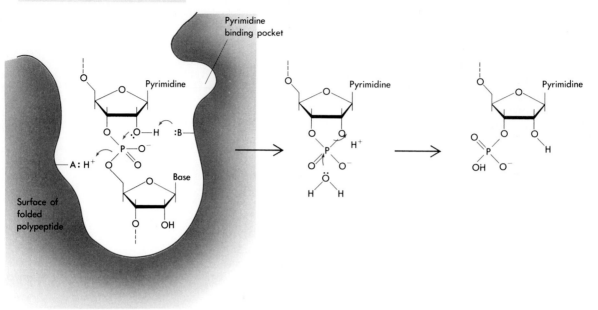

Figure 28-5
Both protein enzymes and RNA enzymes (ribozymes) catalyze hydrolysis of RNA substrates by donating and capturing protons in proton exchange reactions. Pancreatic RNase cuts after the pyrimidines U and C, producing a 3' phosphate and a 5' hydroxyl. The proton donor and acceptor groups (labeled A and B) are thought to be histidines 119 and 12 at nearly opposite ends of the 124-residue enzyme. Hydrolysis proceeds through a cyclic 2',3'-monophosphate intermediate that is enzymatically hydrolyzed in a second step. [After F. M. Richards and H. Wyckoff, in *The Enzymes*, Third Ed. (P. D. Boyer, editor), Vol. 4:647.] Hydrolysis of RNA by RNase P produces a 3' hydroxyl and a 5' monophosphate. The identity of the proton donor and acceptor groups in the catalytic RNA molecule is not yet known. [After N. R. Pace and T. L. Marsh, *Origins of Life* 16 (1985):97.]

Hydrolysis of RNA by RNase P

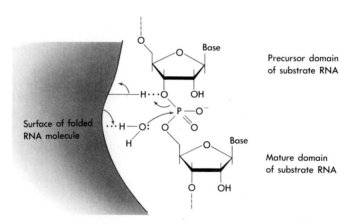

Many plant viral pathogens have single-stranded RNA genomes (see Figure 24-37). The genome of Lucerne Transient Streak Virusoid (LTSV), for example, is a covalently closed single-stranded RNA molecule of 324 nucleotides. The most stable secondary structure for the infecting + strand of the virusoid appears to be a double-stranded rod with many "bulged" bases and unpaired loops (Figure 28-7a). Surprisingly, the viral + strand is capable of accurate self-cleavage if the molecule is first heated to melt the most stable rod-shaped structure and then quickly cooled to "freeze" the molecule in a less stable (but enzymatically active) alternative secondary structure. Equally surprising was the observation that the viral − strand can cleave itself at a site only six nucleotides away from the + strand self-cleavage site. The role of self-cleavage in the viral life cycle remained a mystery until the demonstration that viral replication proceeds by a rolling circle mechanism (Figure 28-7b). Then it became clear that self-cleavage is required to excise monomeric viral RNAs from the multimeric RNA tails produced by the rolling circle mechanism. The reason why the self-cleavage sites in the + and − strand multimeric tails assume

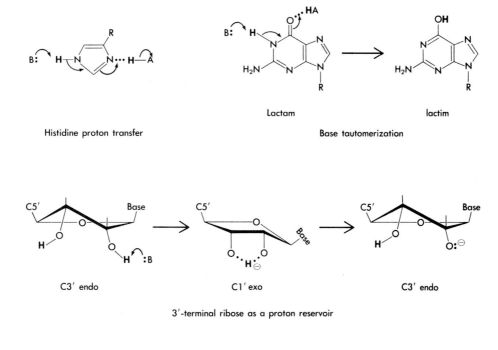

Histidine proton transfer

Base tautomerization

3'-terminal ribose as a proton reservoir

Base tautomer activation of an internal ribose 2' OH

the enzymatically active secondary structure (Figure 28-7c) rather than the more stable structure (Figure 28-7a) is not yet understood. Perhaps the binding of host cell proteins stabilizes the catalytic conformation.

Self-cleavage is a common and perhaps universal feature of the many different small, single-stranded RNAs that infect plants. Comparison of the cleavage sites in different viral RNAs initially suggested that self-cleavage does not require a large or complex RNA structure. This was verified experimentally by testing the ability of very short fragments of LTSV RNA to cleave themselves in the test tube. The minimal structural requirements for self-cleavage are three short stems meeting in a loop; within this structure, 17 highly conserved nucleotides are thought to play a role in catalysis, while the remaining nucleotides presumably establish the proper secondary and tertiary structures for self-cleavage (see Figure 28-7c). Unfortunately, we know so little about the interactions responsible for RNA tertiary structure that there is no way to predict the three-dimensional structure of the self-cleaving RNA domain from the two-dimensional base-pairing scheme (see Figure 28-7c).

Figure 28-6
Possible mechanisms by which RNA might catalyze proton exchange. The enzymology of RNA catalysis is not yet sufficiently advanced to distinguish between these (and other) possibilities. For comparison, the figure also shows proton transfer by the imidazole side chain of histidine, a reaction that is thought to occur at the active site of pancreatic RNase (see Figure 28-5). [After N. R. Pace and T. L. Marsh, *Origins of Life* 16 (1985):97.]

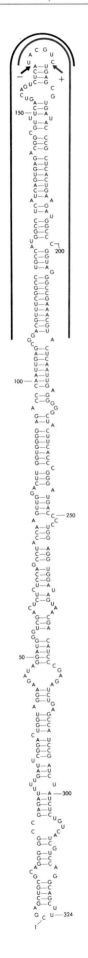

(a)

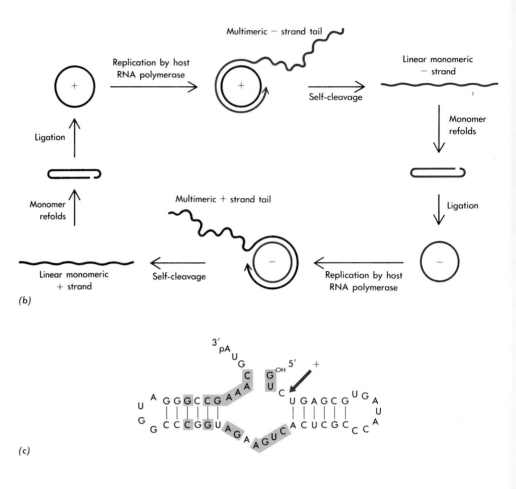

(b)

(c)

Figure 28-7

Only 52 bases are sufficient for self-cleavage of a plant virus RNA genome. (a) Secondary structure for the circular + strand of Lucerne Transient Streak Virusoid (LTSV) RNA. Sites of self-cleavage for the + and − RNA strands are denoted by color and black arrows, respectively, and the nucleotide sequence domains required for these reactions are indicated by the corresponding bars. (b) Rolling circle model for LTSV RNA replication. Both the + and − RNA strands must be capable of self-cleavage in order to excise monomeric units from the multimeric RNA tails of the rolling circle intermediates. Note how the secondary structure shown in (a) aligns the 5′ and 3′ ends of the linear monomer for ligation. (c) When the nucleotide domain indicated by the color bar in (a) assumes the alternative secondary structure shown here, it is enzymatically active. These 52 nucleotides alone are capable of cleaving the phosphodiester bond indicated by the color arrow. The boxed nucleotides are conserved in a large variety of other single-stranded RNA plant viruses and appear to be required for catalysis. [Courtesy of A. C. Forster and R. H. Symons, The University of Adelaide, Adelaide, Australia.]

The Self-Cleaving Transcripts of the Newt[10]

Virtually the same self-cleaving RNA sequence first discovered in plant viruses (see Figure 28-7) has recently been found in the newt *Notophthalmus viridescens*. The genome of this vertebrate has many blocks containing head-to-tail (tandem) repetitions of a 330-base-pair sequence. Some of these blocks of DNA are transcribed into very long nuclear RNA molecules that are subsequently exported to the cytoplasm. Unexpectedly, the cytoplasmic transcripts are all small RNAs whose length is an integral multiple of 330 nucleotides. Inspection of the DNA sequence revealed that the 330-base-pair DNA repeating unit contains the same conserved sequence elements as the plant viral self-cleavage site (see Figure 28-7c), implying that the long RNA transcripts were cleaving themselves at 330-base intervals.

How could a vertebrate have acquired tandem DNA copies of a 330-nucleotide sequence that is capable of self-cleavage as RNA? The most plausible explanation is that a circular, single-stranded RNA virus (like LTSV) infected an ancestor of modern newts. Virus replication then led to the production of a multimeric RNA tail (see Figure 28-7b) that was reverse-transcribed into DNA and inserted into the newt genome. The generation of pseudogenes (see Figure 20-30), processed genes (see Figure 20-31), and *Alu* sequences (see Figure 20-37) in mammalian genomes provides an attractive precedent for such a **reverse flow of genetic information** from cellular RNA into genomic DNA.

Insertion of the self-cleaving RNA sequence must have occurred long ago in the genome of an ancestral newt, since all species of the salamander family from Europe, North America, and Asia have homologous genomic sequences. A perplexing question then is why, after so many millions of years of evolution, a sequence that is now perpetuated as DNA has retained the capability for self-cleavage as RNA. In any event, the discovery of the 330-base-pair newt DNA sequence suggests that many more molecular fossils of catalytic RNA are lurking in unlikely places. Such fossils will continue to expand the library of important clues regarding the nature of the earliest living molecules on Earth.

Sequential Transesterification Reactions Promote RNA Self-Splicing[11, 12, 13]

We have seen that proton exchange reactions are responsible for cleavage of RNA by both protein and RNA enzymes (see Figure 28-5). Proton exchange reactions also account for the series of **phosphoester bond transfers**, or **transesterifications**, that lead to autocatalytic excision and subsequent reactions of the *Tetrahymena* rRNA intron (see Figures 20-13, 20-14, 28-8, and 28-19). Superficially, transesterification appears to be a far more sophisticated reaction than either the site-specific hydrolysis of tRNA precursors catalyzed by RNase P or the autocatalytic cleavage of plant viral RNAs. But mechanistically, both transesterification and hydrolysis involve nucleophilic attack of a hydroxyl group on a specific phosphoester bond. When the attacking group is a water molecule, the result is hydrolysis; when the attacking group is the sugar hydroxyl group of a ribonucleotide, the result is a transesterification (Figures 28-8 and 28-9). Since hydrolysis and transesterification are so closely related, small changes in the structure of an RNA enzyme could in principle convert a simple RNase (such as the plant viral self-cleavage domain; see Figure 28-7c) into an enzyme capable of catalyzing transesterifications, and vice versa.

The series of transesterifications that catalyze excision of the *Tetrahymena* rRNA intron all involve attack by the 3′ hydroxyl of a ribonucleotide on a phosphoester bond. The 3′ hydroxyl group of the rRNA

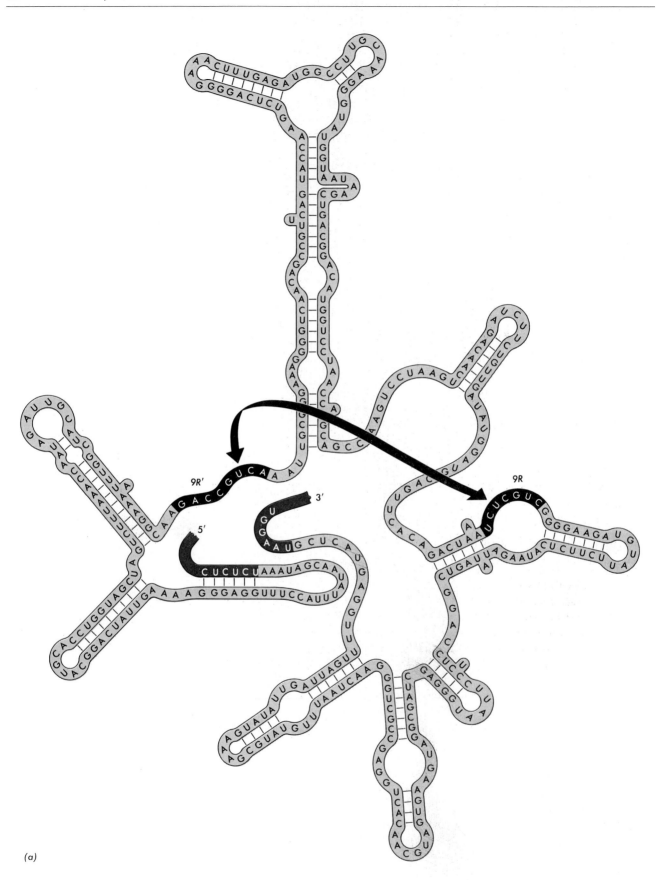

(a)

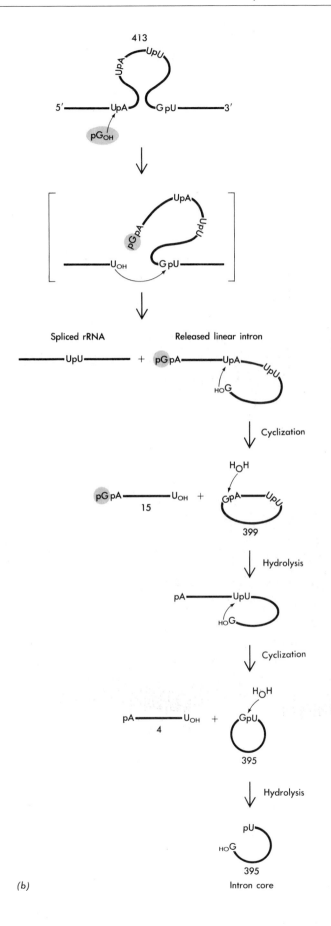

(b)

Spliced rRNA   Released linear intron

Intron core

Figure 28-8
(a) Structure of the *Tetrahymena* riboso-mal RNA intron. For simplicity, the intron is shown here in two dimen-sions, but long-range interactions such as base pairing between the sequences marked *9R* and *9R'* (black arrow) are known to fold the intron into a com-plex three-dimensional structure. [After T. R. Cech, *Sci. Amer.* 255 (1986):64.] (b) Schematic representation of the re-actions of the self-splicing intron. The initiating guanosine nucleotide is indi-cated by a shaded circle. Square brack-ets denote postulated intermediates that have not been isolated. Numbers repre-sent the length in nucleotides of the intron and its fragments. In both (a) and (b), exon sequences are shown in color.

Figure 28-9
Proton exchange reactions also catalyze RNA transesterifications. The elaborate secondary and tertiary structure of the *Tetrahymena* intron (see Figures 20-14 and 28-8) are represented here and in the next figure by a simple curved line. Once the free guanine nucleotide binds to the intron, its 3'-hydroxyl group is activated to initiate a nucleophilic attack on the phosphoester bond at the 5' splice site. Nucleophilic attack by guanosine initiates the series of transesterifications (phosphoester transfer reactions) that excise the intron as a linear molecule and ligate the two exons (see Figures 20-13 and 28-8). Experimentally verified hydrogen bonds between the guanosine nucleotide and its binding site on the RNA are indicated by dotted lines; the hydrogen bond donors and acceptors of the guanosine binding site could include ribose rings and phosphates, as well as the bases themselves. Note that catalysis by the protein enzyme pancreatic RNase proceeds by a similar mechanism, with the 2'-hydroxyl group of a ribose ring being activated to attack the adjacent phosphate (see Figure 28-5). [After B. L. Bass and T. R. Cech, *Nature* 308 (1984):820.]

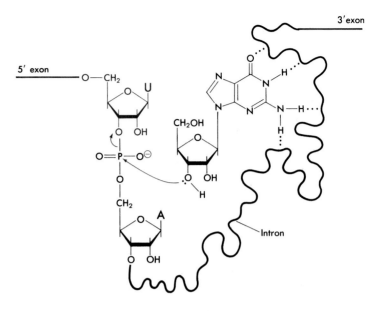

precursor itself cannot initiate the transesterifications, or the intron would become attached to the 3' end of the rRNA. Instead, the RNA enzyme binds a free guanosine nucleotide (GMP, GDP, or GTP), and the 3'-hydroxyl group of this mononucleotide initiates the sequence of transesterifications. As a result, all the products of the initial reaction are linear RNA molecules. (However, the 3' hydroxyl of the excised intron can attack an internal phosphoester bond to generate circular RNAs; see Figures 20-13, 28-8, and 28-19).

The transesterification reactions responsible for excising the ribosomal intron do not consume any high-energy bonds in the form of nucleotide diphosphates, triphosphates, or other high-energy dinucleotides such as NAD or FAD (see Figure 28-14). Instead, the same number of phosphoester bonds are made as are broken, and the energy of the existing phosphoester bonds is conserved. Thus, the intron autocatalytically rearranges existing phosphodiester bonds, rather than using high-energy intermediates to make new ones. In this sense, the intron can be thought of as an *RNA "rearrangease"* or *RNA isomerase*. As we shall see, the first RNA replicase may have been a similar RNA enzyme that was capable of rearranging prebiotically synthesized random oligonucleotides into a complementary copy of an RNA template strand.

Exploring the Fossil Evidence in the *Tetrahymena* rRNA Intron[14]

Studies of the *Tetrahymena* rRNA intron began in 1981 with the relatively modest goal of understanding how the RNA carries out catalytic reactions, but the work soon had important implications for the origin of life. In the next few sections, we will describe experiments showing that the excised *Tetrahymena* rRNA intron can function as a poly C polymerase (see Figure 28-11). We will then extrapolate from the evidence provided by this molecular fossil to the reactions and mechanisms of ancient RNA enzymes. Specifically, we will present plausible models for how a poly C polymerase might be further modified to function as an RNA replicase (see Figure 28-12) and perhaps even as a primitive tRNA synthetase (see Figure 28-17).

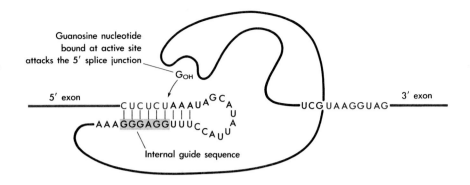

We hope in this extended discussion of the autocatalytic *Tetrahymena* rRNA intron to demonstrate convincingly that RNA molecules could have functioned as the first replicating (and therefore living) systems. We will also describe a scenario in which variants of this RNA replicase gave rise first to the **RNA world**, a rudimentary living world where RNA functioned as the sole catalytic molecule, and then later, with the advent of protein synthesis, to a more complicated world where RNA and protein shared the catalytic functions (the **ribonucleoprotein**, or **RNP world**). This RNP world was the immediate precursor of our own **DNA world**.

The Core of the Self-Splicing Intron Can Function as an RNA Polymerase[15–19]

An enzyme is defined as a molecule that can catalyze a chemical reaction without itself being permanently altered in any way. The intact *Tetrahymena* rRNA intron is therefore not a true enzyme, because it is chemically changed by the very reaction it catalyzes. However, as shown in Figure 28-8, the sequential transesterification reactions that excise the intron as a linear molecule do not destroy the active site of the RNA: The guanosine residue at the 3' end of the excised linear intron can still attack an internal site near the 5' end of the intron in an autocyclization reaction. The cyclization joint is subsequently hydrolyzed by the RNA enzyme to liberate a linear form of the intron, which can in turn undergo yet another cycle of autocyclization and hydrolysis, thereby fragmenting the original intron. The role of autocyclization and fragmentation may be to ensure that excision of the intron is irreversible under normal circumstances.

Even the second round of autocyclization and hydrolysis does not destroy the active site of the RNA. The final product is a linear molecule in which the active site of the intron remains intact. Once it is deprived of an internal substrate, this intron core can use other RNA molecules as substrates.

A remarkable series of experiments revealed that the intron core can function as a true enzyme that is capable of polymerizing poly C from oligo C. In the normal self-splicing reaction, the G-rich **internal guide sequence** of the intron selects the 5' splice site by forming complementary base pairs with the 5' exon sequence (Figure 28-10). This base-pairing scheme positions the 5' splice site at the active site. The guide sequence plays a similar role when the intron core functions as a poly C polymerase (Figure 28-11): The G-rich internal guide sequence selects oligo C precursors from a mixture of oligonucleotides

Figure 28-10

The internal guide sequence of the *Tetrahymena* ribosomal RNA intron selects the 5' splice site. A six-nucleotide internal guide sequence near the 5' end of the intron base-pairs with the 5' exon sequence just upstream from the 5' splice site. This aligns the active site of the intron with the phosphoester bond at the 5' splice site. The activated guanosine nucleotide with a free 3'-hydroxyl group (see Figure 28-9) is shown bound to the active site. Presumably, the 3' splice site is selected by a similar but more complicated mechanism that positions the 3' hydroxyl of the 5' exon next to the 3' splice site. Exon sequences are shown in color, and the internal guide sequence is boxed. The intron shown here is completely schematic; the actual structure is complex and brings the 5' and 3' exons close together (see Figures 20-14 and 28-8). [After G. Garriga, A. M. Lambowitz, T. Inoue, and T. R. Cech, *Nature* 322 (1986):86.]

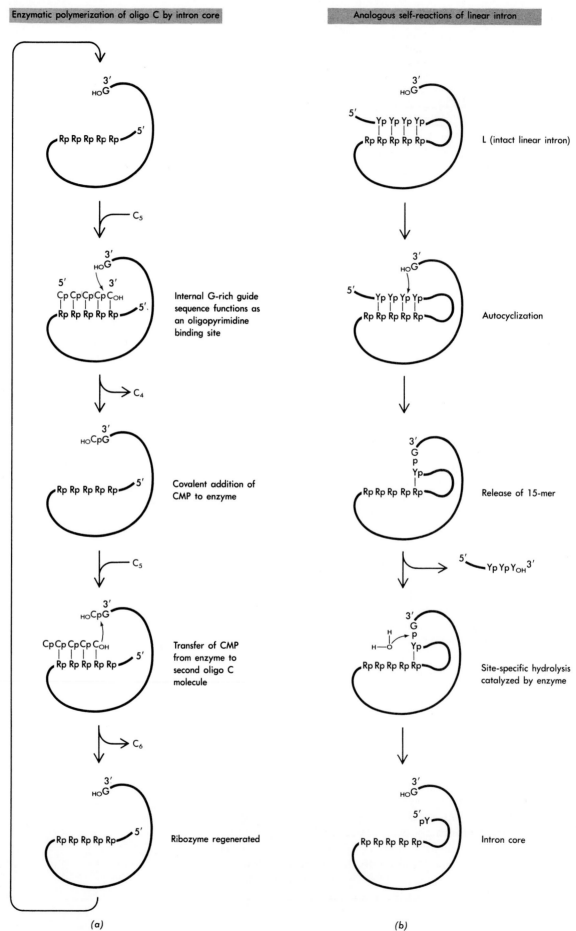

Enzymatic polymerization of oligo C by intron core

Analogous self-reactions of linear intron

Internal G-rich guide sequence functions as an oligopyrimidine binding site

Covalent addition of CMP to enzyme

Transfer of CMP from enzyme to second oligo C molecule

Ribozyme regenerated

L (intact linear intron)

Autocyclization

Release of 15-mer

Site-specific hydrolysis catalyzed by enzyme

Intron core

(a)

(b)

Figure 28-11
The *Tetrahymena* rRNA intron core can function as a poly C polymerase. Enzymatic polymerization of poly C from oligo C is shown in (a). The analogous self-reactions of the intact 413-nucleotide intron are shown in (b) to illustrate the strict correspondence between reactions that the intron can catalyze using itself or oligo C as substrate. The first round of autocyclization is shown as releasing a fragment 15 nucleotides long (15-mer); subsequent hydrolysis followed by a second round of autocyclization would release an additional 4-mer. Note that the G at the active site of the intron core is the 3' end of the original intron, not the guanosine cofactor that initiated excision. R represents a purine. [After A. J. Zaug and T. R. Cech, *Science* 231 (1986):470.]

in solution and properly positions them at the active site. The intron core then functions as a poly C polymerase by catalyzing the same transesterification reactions as in self-splicing; the only difference is that oligo C molecules take the place of the 5' splice site. The result is molecular rearrangement of the oligo C substrate: C residues are moved from one oligo C to another, so that some molecules become longer at the expense of others.

At first glance, it may strike some readers as false advertising to claim that the intron core functions as a poly C "polymerase." After all, the poly C "polymerase" merely rearranges existing oligonucleotides without creating any new phosphoester bonds, while modern RNA polymerases synthesize new RNA chains de novo by polymerizing nucleotide triphosphates one at a time. But modern polymerases require activated mononucleotide triphosphate precursors. These precursors could not have been plentiful in the primordial soup, because they are unstable and easily hydrolyzed to nucleotide diphosphates and monophosphates. Oligonucleotides, in contrast, are relatively stable and must have been able to accumulate under prebiotic conditions (how else could the first RNA replicase have arisen?). Thus, if the primordial soup was rich in random oligonucleotides but poor in activated mononucleotides, the ability to use oligonucleotides as substrates for RNA polymerization would have been a major advantage for the first RNA replicase.

A Detailed Model for the First RNA Replicase[20, 21]

Although the ability of the excised linear form of the *Tetrahymena* rRNA intron to function as a poly C polymerase could be fortuitous, a far more exciting possibility is that the intron core retains the ability to polymerize oligonucleotides because it is in fact a molecular fossil of the primordial RNA replicase. Viewed in this light, the essential task of the intron core originally was, and still remains, the polymerization of RNA; self-splicing is simply a fortuitous side reaction of the RNA polymerization activity.

Recent experiments have shown that the intron core can be changed from a poly C polymerase into a poly U polymerase simply by making the internal guide sequence A rich rather than G rich. This result suggests the tantalizing possibility that the internal guide sequence corresponds to the template of the original RNA replicase. If so, it may be possible to transform the intron core back into a template-dependent RNA polymerase by sophisticated genetic engineering:

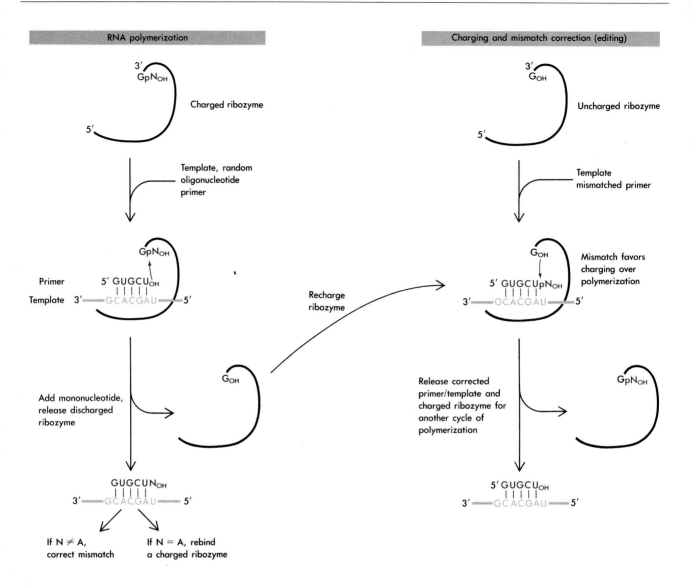

Figure 28-12

A scheme for RNA self-replication. Deletion of the internal guide sequence from the intron core might generate a template-dependent RNA polymerase able to copy an external template. Random oligonucleotides would serve not only as primers for RNA replication but also as the source of activated mononucleotides. Note that the polymerase must dissociate from the template after addition of each mononucleotide in order to be recharged. If the same or similar mechanism could add oligonucleotides to the growing RNA primer, the RNA enzyme might also function as an RNA ligase to join the randomly initiated RNA fragments into a full-length copy of the template RNA. The particular template sequence used here to illustrate the mechanism is arbitrary. [After T. R. Cech, *Proc. Nat. Acad. Sci.* 83 (1986):4360.]

The *internal* guide sequence of the intron core would have to be deleted and replaced by a separate single-stranded RNA template capable of functioning as an *external* guide sequence. The resulting template-dependent RNA polymerase might then be able to copy other RNA molecules (Figure 28-12).

Since the intron core already has the catalytic abilities of an RNA replicase, the major challenge in this genetic engineering project would be to facilitate an association between the catalytic domain of the intron core and the external guide sequence (template). The three-dimensional crystal structure of various tRNAs shows that folding of the negatively charged RNA backbone can create binding pockets for divalent cations such as Mg^{2+}, Ca^{2+}, and Zn^{2+} (Figure 28-13). This implies that the intron core could be redesigned as a **metalloenzyme** with several metal binding pockets on its surface. These bound metals would then hold the template RNA in place by forming ionic bridges to the negatively charged phosphates of the template backbone.

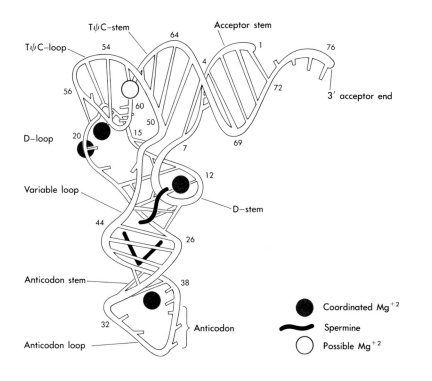

Figure 28-13
The three-dimensional structure of yeast tRNA^{Phe} showing how the phosphate backbone of the folded RNA chain creates specific binding sites for the cations Mg^{2+} and spermine. (Spermine, which contains four positively charged nitrogen atoms, is a polyamine with the structure NH$_3^+$–(CH$_2$)$_3$–NH$_2^+$–(CH$_2$)$_4$–NH$_2^+$–(CH$_2$)$_3$–NH$_3^+$.) The position of the cations was determined by refining the tRNA^{Phe} crystal structure shown in Figure 14-6 to a resolution of 2.5 Å. As in Figure 14-6, the sugar-phosphate backbone is drawn as a continuous ribbon, base pairs are shown as complete crossbars, and unpaired bases are indicated by short bars extending from the backbone. [After G. J. Quigley, M. M. Teeter, and A. Rich, *Proc. Nat. Acad. Sci.* 75 (1978):64.]

All Reactions of the *Tetrahymena* rRNA Intron Are Fundamentally Similar[12, 18]

Viewed superficially, the excision, circularization, hydrolysis, and poly C polymerization reactions of the *Tetrahymena* rRNA intron seem very different (see Figures 28-8 through 28-11). But on closer examination, these transesterification reactions can all be seen to proceed by the same molecular mechanism: Invariably, the 3' hydroxyl of the intron core is activated to participate in a proton exchange reaction (see Figures 28-5 and 28-9). Thus, this same 3'-hydroxyl group always serves as the active residue of the RNA enzyme. The only apparent exception is the initial cut at the 5' splice site during intron excision, but here, a free guanosine nucleotide simply substitutes for the 3'-terminal guanosine hydroxyl of the intron itself.

Since the intron activates its own 3'-hydroxyl group, the secondary and tertiary structures of the intron core must be designed to surround the active residue with an appropriate array of proton donor and acceptor groups. The activated 3'-hydroxyl group can then attack any phosphoester bond that is held in the proper position by the internal guide sequence. No significant energy is gained or lost in the resulting *isoenergetic* transesterification reaction, so the reaction can proceed forward and backward with equal ease. A direct consequence of the *reversibility* of the reaction is that the activated hydroxyl group not only can attack a preexisting phosphoester bond ("charging"), but also can *be* attacked when it forms part of a phosphoester bond ("discharging"):

Uncharged 5' intron-OH 3' + substrate 5' pNpNpNpN-OH 3' $\longrightarrow$
 charged 5' intron-pN 3' + substrate 5' pNpNpN-OH 3'

In most reactions of the intron, binding of the substrate RNA to the internal guide sequence positions a particular phosphoester bond at the active site for transesterification. Only when no RNA substrate is bound does hydrolysis of the charged enzyme occur. In this case, the active site inefficiently catalyzes the attack of water (H–OH) instead of a bound oligonucleotide (5' pNpNpNpN–OH 3') on the activated phosphoester bond (see Figures 28-8 and 28-11).

The ability of the *Tetrahymena* intron to catalyze so many different reactions may indicate that the active sites of RNA enzymes are generally more flexible than those of protein enzymes (also see section entitled Why Does RNase P Retain Its RNA?). This may provide a critical clue regarding the origin of protein synthesis, because it suggests that the activated phosphoester bond of a charged RNA replicase could react with chemical groups other than the 3' hydroxyl of RNA (see section entitled A Model for How a Charged Ribozyme Might Have Specifically Activated an Amino Acid).

Metabolic Pathways Evolved To Synthesize Scarce Molecules[22]

We have seen that many molecules of biological importance were synthesized prebiotically under the harsh conditions prevailing on the youthful Earth. However, despite the best efforts of astronomers, geologists, chemists, and paleobiologists, our knowledge of these primordial conditions remains uncertain and fraught with speculation. Indeed, it would not be an exaggeration to say that every expert in the field of molecular evolution has a different notion of what exactly was in the prebiotic soup. This uncertainty compounds the difficulty of evaluating various scenarios for the origin of life.

Any replicating system requires a source of precursor molecules. In the particular scenario we have presented for the origin of life, the prebiotic soup would have to be rich in random oligonucleotides to serve as precursors for the expanding population of RNA replicases. But even if the first replicating system were composed not of RNA but of some other substance (protein, carbohydrate, or even inorganic crystals), the replicating molecules would eventually begin to exhaust the local precursor supply. This would be true whether the molecules were replicating and evolving on a lake bottom, in a tidal pool, or at the edge of a geothermal vent on the ocean floor.

Initially, nutrient limitation would simply decrease the replication rate, and the "starving" population of replicating molecules would encounter scarce precursor molecules less and less often. But unless nutrients could be supplied to the expanding population faster than the replicating molecules were naturally destroyed (e.g., by spontaneous hydrolysis), the population would eventually decline. Some replicating molecules might escape from the starving population, perhaps by diffusion or by a physical catastrophe such as a mud slide, flood, or volcanic eruption. But even if such emigrant molecules colonized a new niche where molecular precursors were still abundant, the problem of nutrient limitation would inevitably recur.

The only real solution to the problem of nutrient starvation would be the evolution of a subpopulation of RNA molecules capable of accelerating (catalyzing) synthesis of those precursors in shortest supply. For example, it is possible that both D-ribose and the free nucleic acid bases were in relatively abundant supply, but that formation of the glycosidic bonds to generate nucleosides required catalysis to keep pace with the expanding population of replicating RNA molecules. Thus, the most ancient enzymes of **intermediary metabolism** may have evolved to accelerate the *slowest* steps in the prebiotic synthetic pathways. (Intermediary metabolism consists of the synthesis of biological molecules, called **anabolism**, and the breakdown of these molecules, called **catabolism**.)

Later, as living systems became more complex and made still greater demands on the nutrient supply, additional steps in the hybrid prebiotic/biotic synthetic pathways would become rate limiting. Carried to its logical extreme, this point of view predicts that modern biosynthetic pathways are the direct descendants of prebiotic pathways. In fact, many experts in molecular evolution have attempted to deduce the nature of prebiotic synthetic pathways from the enzymatic pathways responsible for intermediary metabolism today.

Nucleotide Cofactors May Be Metabolic Fossils of the RNA World[23]

The only proteins in the primitive RNA world were random polypeptides assembled by prebiotic condensation of amino acids (see Figure 28-3). Admittedly, some of these random polymers might have possessed rudimentary catalytic abilities. For example, a positively charged polymer like polylysine could interact with negatively charged oligonucleotides to facilitate their polymerization by an RNA enzyme. But in the absence of a working translation system (see Figures 28-17 and 28-18), there would be no way to inherit the information encoding such a potentially useful polypeptide. As a result, the polypeptide could not be synthesized repeatedly and could not be improved upon by selecting for more efficient catalysis. Eventually, the expanding population of replicating RNA molecules would outstrip the availability of those few random polypeptides that were catalytically useful, and the population of RNA molecules would starve and decline.

There was only one way that the enzymes of intermediary metabolism could have replicated and evolved, and thereby kept pace with the demands of a growing population of RNA replicases: The first enzymes of intermediary metabolism had to be made of RNA and had to be replicated by the very RNA polymerase population whose replication they would in turn accelerate. In the resulting *mixed* population of RNA molecules, the RNA enzymes of intermediary metabolism would be selected not only for catalytic efficiency, but also for efficient replication.

The RNA enzymes we have discussed so far (RNase P and the *Tetrahymena* ribosomal intron) use RNA as the substrate. What reasons do we have for supposing that an RNA enzyme could catalyze steps in intermediary metabolism using a substrate other than RNA? First, RNAs contain an abundance of potential catalytic groups and are capable of folding into complex tertiary structures (see Figures 28-6, 28-8, and 28-13). Thus, there is no intrinsic reason why an RNA enzyme could not be nearly as versatile as a protein enzyme. Second, the original enzymes of intermediary metabolism might have evolved most readily from the RNA replicase itself. For example, if the nucleotide binding site of the RNA replicase was able to recognize the cyclic form of the ribose ring, a variant RNA replicase might have catalyzed conversion of the linear form of ribose to the cyclic form (see Figure 28-3) simply by holding the linear molecule in proper conformation (see Figure 2-8). Later, as living systems evolved and made more complex demands on intermediary metabolism, the number of distinct RNA enzymes would inevitably increase.

Is there any evidence that the first enzymes of intermediary metabolism were actually made of RNA? Meticulous characterization of metabolic pathways by the pioneers of experimental biochemistry during the late nineteenth and early twentieth century led to the realization that the catalytic abilities of many proteins are augmented by **cofactors** bound either covalently or noncovalently to the active site. While protein enzymes may have been the first to take advantage of the catalytic versatility of cofactors, the intriguing possibility exists that RNA enzymes had already adopted this strategy in the RNA world. A remarkable number of such enzyme cofactors are closely related to the nucleotides (Figure 28-14). Although other interpretations are possible, one attractive hypothesis is that **nucleotide cofactors** are metabolic fossils of the enzymes of intermediary metabolism in the RNA world. This would explain why the enzymatically *active* end of the nucleotide cofactor generally lies opposite an enzymatically *inactive* adenylate or adenosyl group: Perhaps the adenylate or adenosyl group served as a convenient "handle" for an RNA enzyme to hold onto a nucleotide cofactor, much as the structure of the *Tetrahymena* ribosomal intron creates a specific binding site for a guanine nucleoside (see Figure 28-9). Later in evolution, the development of a working translation system would have allowed more efficient protein catalysts to assume the functions of most RNA enzymes. An evolving protein enzyme would then have retained the cofactor as a key part of its active site, but would have replaced the rest of the ancestral RNA (or RNP) enzyme with a polypeptide chain (Figure 28-15).

RNA Genes May Have Been Assembled from Smaller Pieces

Consider a living system composed of an RNA replicase and auxiliary RNA species that contribute to the efficient replication of the replicase and are in turn replicated by it. Initially, these auxiliary RNAs (which arose by random polymerization of nucleotides) would be extremely simple, each having only a single rudimentary function. But later, as the living system increased in complexity, the pressure for more efficient catalysis would select for complex auxiliary RNAs, capable of performing two or more functions in a coordinated fashion.

Figure 28-14
The nucleic acid bases serve as structural components of some coenzymes. NAD^+/NADH and FAD/$FADH_2$ mediate oxidation-reduction reactions, CoA carries acyl groups, and coenzyme B_{12} transfers hydride ions. Although the AMP residues in NAD^+, FAD, and CoA are not directly involved in catalysis, the 5'-deoxyadenosyl residue of coenzyme B_{12} is covalently bound to the central cobalt atom and participates directly in hydride ion transfer. Note also the unusual base 5,6-dimethylbenzimidazole in coenzyme B_{12}, as well as the structural resemblance between the nucleic acid bases and the catalytically active nicotinamide (NAD^+) and isoalloxazine (FAD) rings. The nucleic acid components of the coenzymes are shown in color.

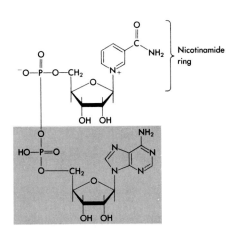

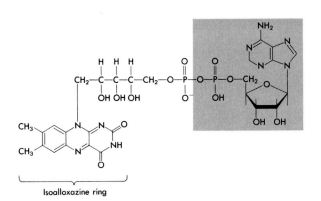

Nicotinamide adenine dinucleotide (oxidized form, NAD⁺)

Flavin adenine dinucleotide (oxidized form, FAD)

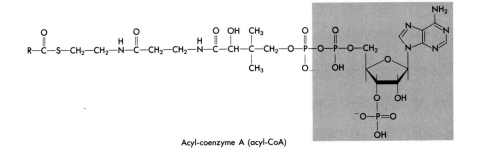

Acyl-coenzyme A (acyl-CoA)

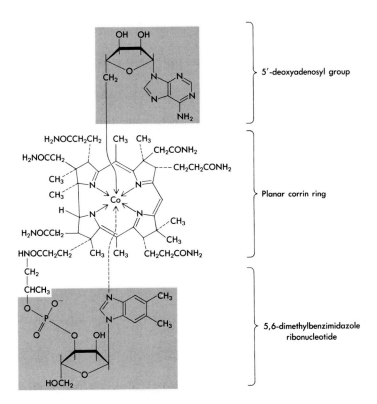

Coenzyme B₁₂ (5′-deoxyadenosylcobalamin)

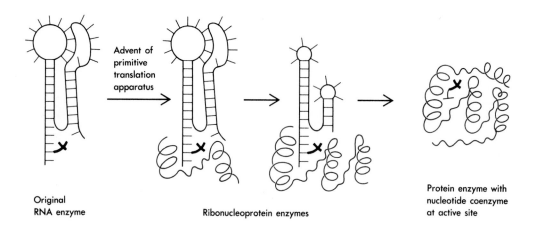

Advent of primitive translation apparatus

Original
RNA enzyme

Ribonucleoprotein enzymes

Protein enzyme with
nucleotide coenzyme
at active site

Figure 28-15
A model for the evolution of coenzymes from RNA enzymes. At first, the cofactor was base-paired to the RNA enzyme through its adenosyl group. Later, after a working translation system had evolved, catalysis by the original RNA enzyme was enhanced by interaction with small polypeptides. As the translation system became more reliable, larger polypeptides gradually took over all the functions of the RNA except for the critical role of the cofactor in catalysis. Ultimately, only the coenzyme remained at the active site of a protein enzyme. [Adapted from H. B. White, III, in *The Pyridine Nucleotide Coenzymes* (1982):1–17. Academic Press, New York.]

The most obvious way to create a multifunctional RNA would be to fuse two preexisting auxiliary RNAs (say, a pyrimidine binding RNA and a ribose binding RNA) into a single RNA chain with a new activity (perhaps the ability to catalyze formation of the glycosyl bond between ribose and the base). The only alternative to such a simple **gene fusion** would be to reinvent both preexisting activities at the same time as part of a single RNA chain, a truly remote possibility.

How could two RNA chains be fused into a single bifunctional molecule in the primitive RNA world? One possibility is that this reaction occurred by the same prebiotic pathways that synthesized the first oligonucleotides. Another possibility is that such fusions might have been catalyzed by the RNA replicase itself. In the model for the first RNA replicase (see Figure 28-12), as well as in actual experiments showing that the intron core functions as a poly C polymerase (see Figure 28-11), polymerization proceeds by removing mononucleotides from the 3' end of one molecule and adding them to the 3' end of another. The replicase in the model therefore functions as an *RNA rearrangease* or *recombinase,* and thus it might have joined and rearranged *poly*nucleotides as well as mononucleotides. This possibility could actually be tested by experiments with the *Tetrahymena* intron core.

The idea that complex RNAs could have been assembled from separate preexisting structural or functional domains of RNA is completely analogous to the idea that split genes encoding proteins can accelerate evolution in the DNA world: Many examples are known in which exons encoding distinct structural or functional protein domains have been joined into a composite gene by recombination at the DNA level between the adjacent introns (see Figures 20-20 through 20-23, Figure 28-22, and Table 28-2).

Compartmentation in Biological Systems[25]

No living system could continue to replicate and evolve unless all the enzymes and precursors were maintained at a high concentration and protected from destructive forces. Otherwise, the replicating molecules would diffuse away from each other or be chemically degraded faster than they could accumulate. For example, a mixed population of interdependent RNA molecules in a lakeside puddle might be hopelessly diluted by a torrential rainstorm or chemically destroyed by repeated evaporation to total dryness in the heat of a blazing sun.

Arguments such as these suggest that life must have evolved in a protected environment, but the nature of that environment is a matter for pure speculation. All we can say is that at some very early point in the evolution of life (perhaps even before the first RNA replicase), **compartmentation** was necessary to separate the living system ("inside") from its environment ("outside"). This compartment had to be sufficiently resilient to withstand abuse, sufficiently flexible to allow division without loss of contents, and sufficiently permeable to allow nutrients to diffuse in while wastes diffused out.

Today, all living systems are compartmentalized by membranes built of lipid bilayers (see Figures 2-26 through 2-31). Lipid bilayers separate the inside of every cell from the external environment and also define the internal compartments of eucaryotic cells (nucleus, endoplasmic reticulum, mitochondrion, chloroplast, etc.). Unlike most other biologically important macromolecules (nucleic acids, proteins, and carbohydrates), lipid bilayers assemble *spontaneously* from mixtures of lipids in aqueous solution when the lipid concentration exceeds a certain critical value. Below this concentration, the individual lipid molecules are soluble in the aqueous phase. At higher concentrations, the lipid molecules aggregate with one another to form huge sheets. More importantly, fragments of the sheets spontaneously break off and reseal themselves as closed balloon-like **vesicles** (Figure 28-16; see also Figure 2-31).

Lipid vesicles can fuse with each other to form larger vesicles, or they can divide irregularly by pinching off daughter vesicles. These spontaneous transformations of a lipid bilayer bear a striking resemblance to the budding of a yeast cell (see Figure 18-2), the movement of vesicles through the Golgi apparatus (see Figures 18-9 and 18-16), the budding of enveloped viruses from eucaryotic cells (see Figure 24-2), and the release of secretory granules after fusion of the granule membrane with the plasma membrane (see Figure 18-16). The implication is that living systems have devised mechanisms for stabilizing, accelerating, and directing the natural transformations of lipid sheets.

The spontaneous formation of closed lipid vesicles suggests that lipidlike molecules may have played a crucial role in providing a protected environment for primitive living systems. The budding of lipid vesicles also suggests a crude mechanism for "cell" division: Primitive living systems, each surrounded by a membrane bilayer, would simply divide unevenly whenever the vesicle became too large to withstand the hydrodynamic forces acting upon it. Similarly, vesicle fusion could have served as a primitive form of mating. If we suppose that useful genes arose in two different (and possibly competing) living systems, each enclosed in a separate lipid vesicle, vesicle fusion could bring the two genes together into a single living system. This would accelerate evolution by ensuring that any useful RNA sequence was made available to the entire "breeding" population of vesicles.

One problem in thinking of the first "cell" as RNA encapsulated in a lipid vesicle is that both RNA and *phospho*lipids are negatively charged (see Figure 28-16). Repulsion between these negative charges might then be expected to make encapsulation a very inefficient process. Recent experiments, however, show that the addition of positively charged (basic) proteins increases the efficiency of encapsulation by reducing the repulsion between negative charges. This suggests that prebiotically synthesized basic peptides may have facilitated the encapsulation of the first living systems in lipid

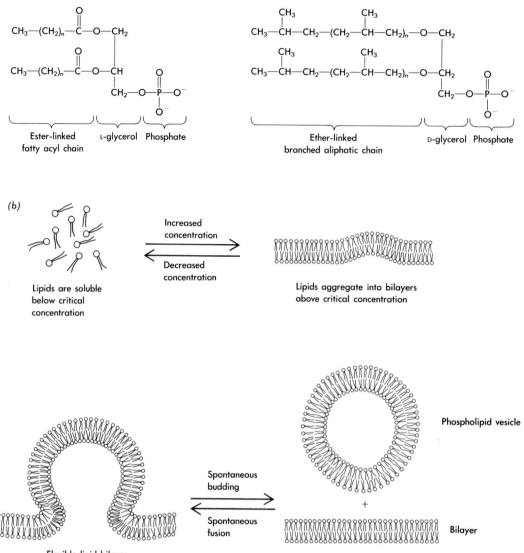

(a) The simplest phospholipids:

Ester-linked fatty acyl chain   L-glycerol   Phosphate

Ether-linked branched aliphatic chain   D-glycerol   Phosphate

(b)

Lipids are soluble below critical concentration

Increased concentration

Decreased concentration

Lipids aggregate into bilayers above critical concentration

Flexible lipid bilayer can pucker

Spontaneous budding

Spontaneous fusion

Phospholipid vesicle

+

Bilayer

Figure 28-16
Many lipids spontaneously form bilayers that are relatively impermeable to water and are capable of fusing and budding. (a) All phospholipids have a hydrophobic aliphatic chain and a hydrophilic glycerophosphate group. (b) Bilayers form spontaneously because the hydrophobic fatty acyl chains tend to associate with each other, while the hydrophilic phosphate groups remain on the outside of the bilayer because they prefer to interact with water. In archaebacteria, the phospholipids are ether-linked to D-glycerophosphate rather than ester-linked to L-glycerophosphate as in eubacteria and eucaryotes (see Table 28-5).

vesicles. Alternatively, the most primitive lipids may have lacked phosphate head groups.

Although lipid membranes are a particularly attractive means for effecting compartmentation, there are other possibilities for the most primitive (and therefore least demanding) living systems. For example, many polymers and even some minerals can form porous gels with the consistency of pudding or gelatin. Such **coacervates** might have retained the larger functional RNA molecules while allowing free diffusion of smaller molecules such as precursors and wastes.

Unfortunately, the earliest forms of compartmentation could not have been preserved as microfossils (see Figure 28-1). However, the synthesis of ester-linked lipids has been achieved under simulated prebiotic conditions, and today all living organisms use functionally similar ester-linked or ether-linked lipids (Table 28-5). Taken together, the evidence is strong that lipid bilayers are molecular fossils of the original form of compartmentation in biological systems. Perhaps all living systems, from the first RNA replicase to the most sophisticated neuron in the human brain, have been surrounded by lipid bilayers.

Table 28-2 Many Animal Proteins Have Sequence Segments in Common*

EGF-type segment
   Epidermal growth factor (EGF)
   Tumor growth factors
   Low density lipoprotein receptor
   Blood clotting factor IX
   Blood clotting factor X
   Protein C
   Tissue plasminogen activator
   Urokinase
   Complement component C9
   Notch gene product (fruit fly)
   lin-12 gene product (nematode)
C9-type segment
   Complement component C9
   Low density lipoprotein receptor
   (Fibronectin?)
   Notch gene product (fruit fly)
   lin-12 gene product (nematode)
Fibronectin "finger"
   Fibronectin
   Tissue plasminogen activator
Proprotease "kringle"
   Plasminogen
   Tissue plasminogen activator
   Urokinase
   Prothrombin

*The terms "finger" and "kringle" refer to specific disulfide-linked modular units of protein construction. Note that some proteins are a mosaic of at least three different modular units found in other proteins.
SOURCE: R. F. Doolittle, D. F. Feng, M. S. Johnson, and M. A. McClure, *Cold Spring Harbor Symp. Quant. Biol.* 51 (1986):447–455.

THE RNP (RIBONUCLEOPROTEIN) WORLD

Translation Evolved in an RNA World

Because proteins are more versatile catalysts than RNA, intense selective pressure forced early living systems to devise an efficient mechanism for the mass production of useful proteins. And since a living system would be at a serious disadvantage if it could not pass on the "accomplishments" of one generation to the next, selective pressure likewise ensured that the production of such proteins would be *heritable*. However, only nucleic acids can store and replicate the genetic information specifying proteins; the proteins themselves cannot (see the section entitled The First "Living" Molecule Was Almost Certainly a Nucleic Acid). Thus, very early in evolution, the appropriate molecular machinery arose to replicate RNA molecules encoding useful proteins and to translate the RNA code into the correct polypeptide.

The Modern Ribosome Is Very Complicated[26–31]

Ever since the realization in 1956 that DNA makes RNA makes protein (the central dogma; Chapter 3), two great questions have dominated all discussions of molecular evolution: What was the nature of the first replicating system, and how did the first primitive ribosome translate RNA sequences into protein? The recent discovery and characterization of enzymatically active RNAs quickly persuaded most molecular biologists that the first replicating systems were made of RNA. Can we use our picture of the RNA world and our knowledge of contemporary protein synthesis to make reasonable (and perhaps even testable) proposals about the origins of translation and the genetic code?

Today, translation is carried out by the ribosome, a supremely sophisticated molecular machine that contains about 4500 bases of RNA and 50 different proteins (see Figures 14-13 through 14-16). Such sophisticated machinery did not arise all at once, but developed over perhaps a billion years as evolution tinkered one step at a time with the basic design of the original ribosome. Thus, the crude translation machinery that gave birth to the RNP world, a world made of both RNA and protein, must have been radically simpler than the ribosome of today. How, then, can we know whether *any* parts of the modern ribosome are remnants (molecular fossils) of the original translation system, or whether the original molecular machine has been rendered unrecognizable by subsequent tinkering?

In our discussions of the primitive ribosome, we will make three simplifying assumptions.

First, we assume that the modern translation apparatus is a vastly improved version of the original apparatus, but does not differ from it fundamentally. We make this assumption because molecular evolution is generally *conservative*, modifying and adapting old pieces of molecular machinery whenever possible, instead of devising totally new ones.

Second, we assume that the original translation system was very crude and could not discriminate reliably between chemically similar amino acids such as lysine and arginine, or leucine, isoleucine, and valine (see Figure 2-16). Thus, translation of any particular template RNA (mRNA) would have produced a mixture of chemically similar but nonidentical polypeptide chains of varying length. The ability to start and stop translation precisely and to distinguish as many as 20 different amino acids would then arise step by step as the ribosome improved.

Third, and most importantly, we assume that the first tRNA synthetase and possibly the first ribosome were derived from the RNA replicase or from one of the RNA enzymes that depended on it for replication. This assumption follows logically from our first assumption that molecular evolution is a conservative process in which existing molecules are adapted to ever more sophisticated functions. The first RNA replicase must necessarily have been crude (see Figures 28-4 and 28-7), but the descendants of this RNA enzyme would have increased in complexity and diversity as populations of RNA molecules emerged that could replicate more rapidly. Thus, it seems likely that the original apparatus for polymerization of amino acids would have been a *variant* of the most sophisticated molecular machinery then available, rather than an altogether new invention.

Assuming that the mechanism of modern protein synthesis is a molecular fossil that still resembles the original translation apparatus, we should be able to divide our speculations about the origin of protein synthesis into two separate questions: How were the primitive tRNAs specifically charged with the correct (cognate) amino acid, and how were these charged tRNAs aligned prior to peptide bond formation?

The Adaptor Hypothesis Revisited

In 1955, Francis Crick had a crucial insight into how the translation apparatus aligns specific amino acids on the template RNA molecule (mRNA) before polymerization. The simplest model was that the mRNA folded up to create cavities on its outer surface, each cavity specific for a particular amino acid; the amino acids would then polymerize spontaneously, if they were already activated, or be polymerized by an additional enzymatic activity. Crick pointed out that it is relatively difficult to imagine how the hydrophilic bases in RNA could form hydrophobic pockets capable of interacting specifically with the aliphatic and aromatic side chains of amino acids such as leucine, isoleucine, valine, phenylalanine, and tyrosine (see Figure 2-16). Crick therefore argued that a class of *bifunctional* "adaptor" molecules must exist that would recognize *both* the mRNA sequence *and* the corresponding amino acid side chains (see the section in Chapter 3 entitled The Adaptor Hypothesis of Crick).

The discovery of the role that tRNAs play in protein synthesis dramatically validated Crick's adaptor hypothesis, but the nature of these RNA adaptor molecules was more complex than anticipated. Transfer RNAs are indeed bifunctional: The 3'-hydroxyl group of the acceptor stem is specifically coupled to an amino acid (see Figure 14-6), while the anticodon recognizes mRNA sequences. However, a modern tRNA is not a simple adaptor that can function by itself: The tRNA is unable to recognize the appropriate amino acid side chain directly and therefore requires a second adaptor molecule (a specific aminoacyl-tRNA synthetase) that can recognize both the tRNA and the appropriate amino acid (see Figure 14-9).

A Model for How a Charged Ribozyme Might Have Specifically Activated an Amino Acid[31]

Modern aminoacyl-tRNA synthetases are protein enzymes, but at the dawn of the RNP world, a tRNA synthetase could only have been made of RNA. Since the RNA replicase may well have been the most sophisticated enzyme at that time, let us examine the possibility that a variant RNA replicase might have served as the first tRNA synthetase (Figure 28-17).

We imagine the same series of reactions as in RNA polymerization by the RNA replicase (see Figure 28-12), but hypothesize that a particular amino acid has a fortuitous affinity for the active site of the charged enzyme. The 3' mononucleotide (NMP) of the charged enzyme could then be discharged by attack of the amino acid carboxyl group instead of the 3'-hydroxyl group of a polynucleotide (see Figure 28-19). This would generate an *activated* amino acid~NMP intermediate (AA~NMP), where the high energy of the broken phospho-

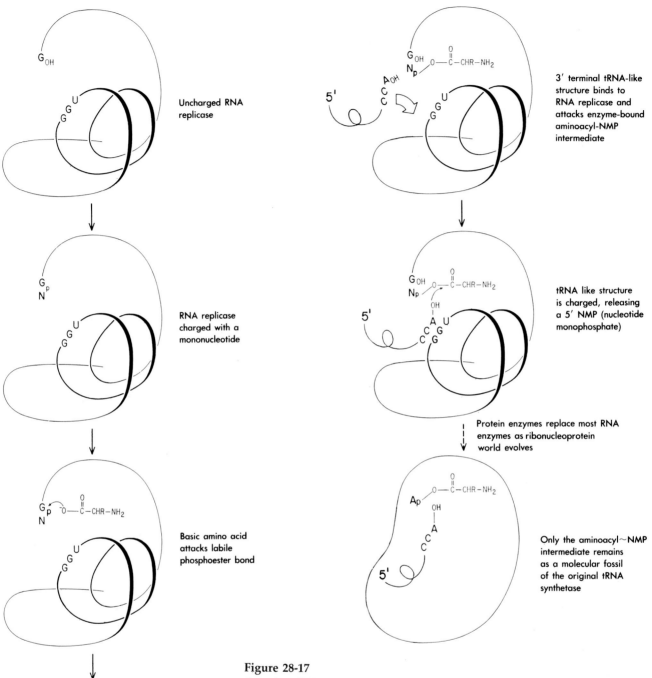

Figure 28-17

A variant RNA replicase may have given rise to the first tRNA synthetase. Since positively charged amino acids would be able to interact most easily with a variant RNA polymerase (see text), the hypothetical origin of lysine tRNA synthetase is shown. Later, as proteins gradually took over the function of most RNAs in the RNP world (see Figures 28-14 and 28-15), only the activated AA~AMP charging intermediate would remain as a molecular fossil of the original tRNA synthetase. [Courtesy of A. M. Weiner and N. Maizels.]

ester bond is conserved in the aminoacyl bond to the nucleotide phosphate. If the activated AA~NMP intermediate then remains bound to the enzyme, the resulting high-energy aminoacyl bond could be attacked by the 3'-hydroxyl group of yet another ribozyme with tRNA-like features. In order to promote stable binding of the tRNA-like ribozyme to the replicase, an internal guide sequence (3' UGG 5') would base pair with the 3' terminal CCA of the hypothetical tRNA (Figure 28-17). This series of reactions would charge the 3' end of the primitive tRNA with an amino acid (AA~RNA), using essentially the same sequence of reactions as are found in protein synthesis today (see Figure 14-9).

Prebiotically Synthesized Oligonucleotides as an Energy Source for Peptide Bond Formation[20, 21]

We assumed in this model for a primitive tRNA synthetase (see Figure 28-17), as we did in the model for the first RNA replicase (see Figure 28-12), that random prebiotically synthesized oligonucleotides served as the energy source for early biological polymerizations. But is this assumption energetically reasonable for protein synthesis? Does the phosphoester bond of the charged synthetase actually have sufficient energy to form the postulated high-energy AA~NMP charging intermediate? A simple calculation suggests that it does. Today, formation of the AA~AMP intermediate is driven by hydrolysis of a high-energy phosphoester bond in ATP (Equation 6-6):

$$\text{ATP} \longrightarrow \text{AMP} + \textcircled{P}\sim\textcircled{P} \qquad (\Delta G = -8 \text{ kcal/mol})$$

We have also seen that polymerization of nucleic acids from the nucleotide triphosphate precursors does not involve a large change in free energy (Equation 6-18):

$$\text{RNA } (n \text{ bases}) + \text{NTP} \longrightarrow$$
$$\text{RNA } (n + 1 \text{ bases}) + \textcircled{P}\sim\textcircled{P} \qquad (\Delta G = +0.5 \text{ kcal/mol})$$

Subtracting Equation 6-18 from Equation 6-6, we find that

$$\text{RNA } (n + 1 \text{ bases}) \longrightarrow$$
$$\text{RNA } (n \text{ bases}) + \text{NMP} \qquad (\Delta G = -8.5 \text{ kcal/mol})$$

This shows that the phosphoester bond of the charged synthetase contains at least as much energy as the high-energy bond in ATP that is hydrolyzed to form the AA~AMP intermediate today.

Protein synthesis is thus driven forward by *coupling* the synthesis of the low-energy peptide bond ($\Delta G = +0.5$ kcal/mol) to hydrolysis of a high-energy phosphoester bond ($\Delta G = $ about -8.5 kcal/mol). Although the ΔG for formation of the enzyme-bound AA~NMP intermediate may be positive, the large overall ΔG of about -8.0 kcal/mol favors protein synthesis (see Figure 6-3).

The First tRNA Synthetases May Have Activated Basic Amino Acids[31]

The very complexity of modern protein synthesis suggests that the first crude protein synthesizing apparatus would only have been able to synthesize short, irregular peptides. Since positively charged amino acids might preferentially bind to the negatively charged sugar-phosphate backbone of the tRNA synthetase, the first charging enzymes may have acted on basic amino acids like lysine, arginine, and histidine to produce basic AA ~ tRNAs. And since aminoacyl~RNA bonds are highly reactive, the aminoacyl~RNA bond of one such molecule could have been attacked by the amino terminal group of another aminoacyl~RNA. (As discussed later, even modern peptide bond formation may occur by such an "uncatalyzed" or "spontaneous" reaction.) The dipeptidyl~RNA resulting from spontaneous reaction of two aminoacyl~RNAs retains an activated aminoacyl bond, and so it could in turn be attacked by a third aminoacyl~RNA to generate a tripeptidyl~RNA, and so forth. The result would be random polymerization of basic amino acids into short positively charged peptides.

At the dawn of the RNP world, short basic peptides might have been useful in at least two ways. First, positively charged peptides could have neutralized the repulsion between negatively charged RNA chains, thereby stabilizing RNA enzymes or allowing them to bind other substrate RNA molecules more tightly. In this role, the basic peptides would resemble modern protamines, which help to pack DNA tightly within sperm heads by neutralizing the negative charges of the phosphate backbone and by forming crossbridges between adjacent DNA strands (Figure 21-7). Second, supposing the first lipids were *phospho*lipids, short basic peptides might also have stabilized lipid vesicles by interacting with the negatively charged phosphate head groups (Figure 28-16).

If indeed the ability to synthesize random basic peptides in an mRNA-independent and ribosome-independent reaction was sufficient to confer a selective advantage on the primitive living system, tRNA synthetases might have been the first components of the translation apparatus to arise, followed soon by the appearance of more sophisticated tRNAs specifically designed for protein synthesis. This would provide the driving force for the next great step in the evolution of protein synthesis: the invention of a ribosome with the ability to align the charged tRNAs on an mRNA, and to translocate the mRNA after each round of peptide bond formation.

The Elusive Peptidyl Transferase[26, 32]

What clues can the modern ribosome provide regarding the first template-dependent mechanism for polymerization of activated aminoacyl~tRNAs? In the modern bacterial ribosome, the 30S subunit binds the mRNA, while the 30S and 50S subunits contribute jointly to forming two tRNA binding sites on the surface of the 70S ribosome (Chapter 14). Evidently, the "fit" between the mRNA and the tRNAs on the ribosomal surface is so tight that only a tRNA with the correct anticodon can bind efficiently to the mRNA on the ribosome. This suggests that the task of the original ribosome might have been as simple as providing correctly oriented binding sites for the mRNA and the charged tRNA. Such binding sites would form a rigid "workbench" to accomplish the two essential functions of a primitive ribosome: to guarantee that the mRNA would base-pair with the tRNA anticodon rather than with some other part of the tRNA molecule, and to stabilize the transient base-pairing interaction between the charged tRNAs and the mRNA so that peptide bond formation could occur.

What is the mechanism of peptide bond formation? There have been numerous attempts to determine whether purified ribosomal proteins can catalyze polymerization of activated amino acids (AA~tRNAs), yet no single molecule has been identified as having peptidyl transferase activity. Of course, the explanation could be that the activity resides in several ribosomal proteins collectively, rather than in a single protein. But there are two other very real possibilities:

- The peptidyl transferase might be an RNA enzyme instead of a protein. This would imply that the large ribosomal RNA does not serve simply as a passive scaffold for ribosomal proteins, but actually catalyzes peptide bond formation. This hypothesis would help to rationalize the extraordinary conservation of ribosomal RNA structure and sequence (see Figure 14-15).

- The other possibility is that peptide bond formation is in fact "spontaneous." This would imply that the entire ribosome serves as a peptidyl transferase by aligning the activated AA~tRNA and the peptidyl~tRNA on the template mRNA, and not by chemically catalyzing the making and breaking of covalent bonds as is the case for most biological catalysts (see Figure 2-25).

The First Ribosomes Synthesized Short Peptides

The first primitive ribosomes (protoribosomes) probably arose to accelerate spontaneous peptide bond formation, perhaps by simply aligning two charged tRNAs and perhaps also providing peptidyl transferase activity. The protoribosome might have been an entirely new RNA enzyme, or it might have been derived from the tRNA synthetase itself by duplication of the tRNA binding site (Figure 28-18). (Duplication could be caused by a replication error, or by recombination between RNA molecules as discussed in the section entitled RNA Genes May Have Been Assembled from Smaller Pieces.) Originally, the pair of tRNA binding sites on the protoribosome might have interacted solely with the conserved CCA ends of the charged tRNAs; however, selective pressure for more stable and accurate alignment of the charged tRNAs would have led to additional interactions between the tRNAs and their protoribosome binding sites. The region on the tRNA that participated in these additional interactions would in effect become the anticodon, and the corresponding region of the protoribosome would function as a built-in mRNA (see Figure 28-18).

Coevolution of tRNA Synthetases and Peptide-Specific Ribosomes May Have Defined the Rudiments of the Genetic Code[31,33]

As long as there was only one kind of tRNA synthetase specific for a particular basic amino acid (say, lysine) the protoribosome would simply have made short lysine peptides. But eventually, the selective advantage of being able to make other useful peptides would have forced the first tRNA synthetases, the first primitive tRNAs, and the protoribosomes to diversify. For example, if a variant tRNA

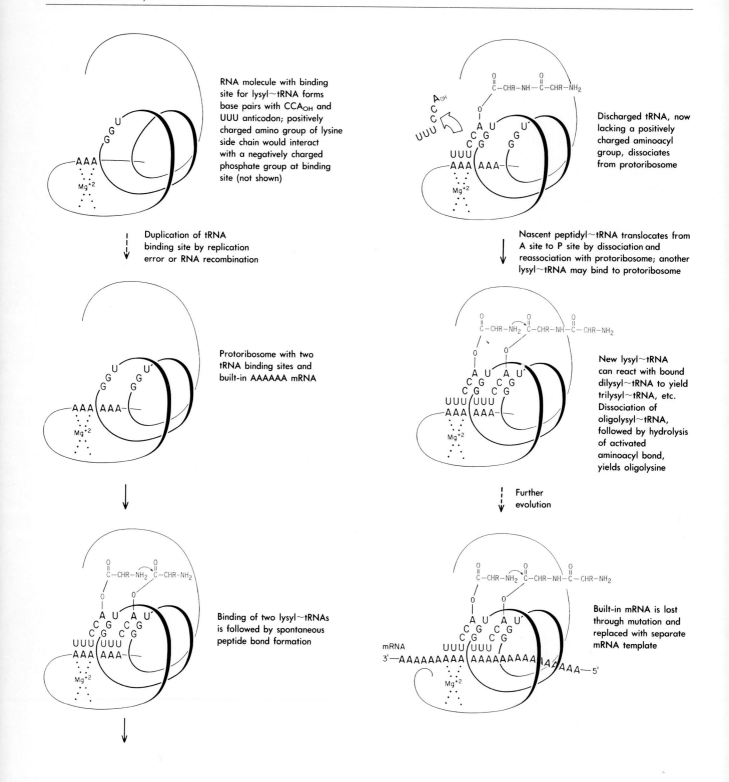

Figure 28-18

The first ribosomes (protoribosomes) may have been peptide-specific. For simplicity, the primitive tRNAs are shown as hexanucleotides with the anticodon immediately adjacent to the 3' terminal CCA$_{OH}$; the actual tRNAs would have been more complicated. Since the first tRNA synthetases are likely to have been specific for basic amino acids like lysine (Figure 28-17), a lysine-specific protoribosome is illustrated here. [Courtesy of A. M. Weiner and N. Maizels.]

synthetase arose that could activate arginine better than lysine, selective pressure would ensure that the primordial tRNA structure evolved into two slightly different tRNA species. The lysine- and arginine-specific protoribosomes would distinguish between these two tRNAs based on the ability of the anticodons to form base pairs with their built-in mRNAs (Figure 28-18); the two tRNA synthetases might recognize some more general feature of the primitive tRNA structures. The result would be two different peptide-specific protoribosomes, each with a different internal guide sequence serving as a "built-in" mRNA, and each with a corresponding tRNA synthetase. In this way, the rudiments of the genetic code would be defined as the population of tRNA synthetases and peptide-specific protoribosomes coevolved.

Although the rudimentary genetic code may have begun as a "frozen accident," reflecting the fortuitous affinity of RNA enzymes for particular amino acids and oligonucleotides (primitive tRNAs), the evolving code must soon have become subject to many different kinds of selection. Examples of such fine tuning today are the use of related codons for chemically related amino acids, and degeneracy in the third position of the code, as ways to minimize the potentially harmful effects of transcriptional and translational errors (see Chapter 15).

The Unexpected Diversity of Modern tRNA Synthetases[34, 35]

All modern tRNA synthetases are made exclusively of protein, and they all catalyze the same reaction (aminoacylation) using structurally very similar substrates (tRNAs). Before any of these tRNA synthetases had been characterized at the molecular level, the common expectation was that they would turn out to be structurally similar variants of a common ancestral protein—the first protein that was able to aminoacylate a tRNA-like species. Thus it was extremely puzzling to find that some tRNA synthetases are large, and some are small; some are monomers, and some are dimers; and no extensive amino acid homology among them is detectable.

We have seen that all tRNA synthetases may indeed share a common ancestor, but that this ancestor was probably an RNA enzyme (the RNA replicase) rather than a protein. As proteins gradually replaced most RNA enzymes during evolution of the RNP world (Figure 28-15), the stepwise conversion of each tRNA synthetase (made of RNA) into a tRNA synthetase (made of protein) would have had to take place *independently*. Assuming that there are many different ways of converting an RNA enzyme to a protein enzyme, this would account for the puzzling diversity of modern tRNA synthetases.

Loss of the Internal Guide Sequence May Have Made mRNA-Dependent Translation Possible

The dependence of peptide-specific protoribosomes on built-in mRNAs would have severely limited the variety of peptides that could be made. This limitation would then have provided the driving force for the evolution of a true ribosome capable of using a separate mRNA molecule as the template for protein synthesis. In principle, replacement of the built-in mRNA of a peptide-specific protoribosome with a separate mRNA template is completely analogous to transformation of the *Tetrahymena* intron into a template-dependent RNA replicase by replacement of the internal guide sequence with an external guide sequence (Figure 28-13). Assuming that the negatively charged backbone of the built-in mRNA was held in place by interaction with bound metals ions on the surface of the peptide-specific protoribosome, simple deletion of the built-in mRNA region may have sufficed to allow binding of a separate mRNA species.

The Ancestral Ribosome Remains a Puzzle

The simple model presented here for the origin of protein synthesis (see Figures 28-17 and 28-18) has two main strengths: It demonstrates that a primitive RNA enzyme could in principle accomplish activation and attachment of an amino acid to a primitive tRNA-like molecule. It furthermore shows how selection for the synthesis of small peptides on a very primitive ribosome (protoribosome) may have played a major role in defining the rudiments of the genetic code and in initiating the stepwise evolution of the modern protein synthetic apparatus. On the other hand, the model does not even attempt to address the nature of the first true ribosome that was capable of translocating the mRNA as well as initiating and terminating protein synthesis.

The best current hope for reconstructing the ancestral ribosome is to try to understand the three-dimensional structures of both the small and large subunits of the modern procaryotic ribosome (see the section entitled The Modern Ribosome Is Very Complicated). This ambitious goal, unthinkable only a few years ago, may now be within reach. The approximate positions of all 21 proteins in the small 30S subunit of the *E. coli* ribosome are already known (see Figure 14-39), and these data are currently being refined. In addition, techniques developed in studying the small subunit are now being applied to the more difficult task of locating the 34 proteins in the 50S subunit. Even more promising is the very recent crystallization of the 50S ribosomes from a salt-tolerant bacterium. The resulting crystals yield precise X-ray diffraction data at a resolution of 6 Å leading to the expectation that perhaps within the next two decades the 50S subunit of the ribosome (and conceivably the 30S subunit) will be understood at the atomic level. By learning how the modern ribosome performs the task of protein synthesis, we may eventually be able to identify molecular fossils of the ancestral translation machine.

Nuclear mRNA Splicing May Have Evolved from Self-Splicing Introns[36, 37, 38]

Unlike the self-splicing of the *Tetrahymena* rRNA intron, the splicing of nuclear mRNA introns requires the U1, U2, U4, U5, and U6 small nuclear ribonucleoprotein particles, or snRNPs (see Figure 20-18) and proceeds by a lariat mechanism (see Figures 20-19 and 28-19). The presence of an snRNA component in each of the five different snRNPs required for nuclear mRNA splicing suggests that introns of this type first appeared in the RNP world. But how could such an

but replicate through a double-stranded DNA intermediate; hepadnaviruses have circular single-stranded DNA genomes, but replicate through an RNA intermediate.

The extraordinary diversity of viral life styles suggests that most families of viruses evolved independently, so the origins of T7 may not be revealing about the origins of MS2. There is, however, one especially useful question that can be asked about *all* viruses, and that is whether the virus is old or new. A virus that evolved quite recently in a sophisticated host cell (perhaps resembling a modern procaryote or eucaryote) might be expected to have a life style resembling that of its host. In fact, many double-stranded DNA viruses may be of relatively recent vintage. Alternatively, a virus that evolved long ago in a simpler (perhaps even precellular) living system might be expected to retain certain hallmarks of the more primitive living world in which it arose. This suggests the fascinating possibility that some RNA viruses may be molecular fossils of the RNA or RNP world, and that retroviruses (whose RNA genomes replicate through a DNA intermediate) and hepadnaviruses (whose DNA genomes replicate through an RNA intermediate) may be fossils of a world in transition from RNA to DNA genomes. The implication is that viruses may be a rich source of information about early evolution, if only we can learn to decipher the molecular fossil record.

Segmented Viral RNA Genomes May Be Molecular Fossils of the RNP World

The genomes of many RNA viruses, including influenza and reovirus, are segmented (Table 24-1, Figure 24-16). Since there is good reason to think that genomes in the RNA and RNP world were also segmented, these viruses may be very ancient.

The DNA genomes of modern organisms contain thousands of genes lined up one after another. By recognizing transcriptional start and stop signals in the DNA, RNA polymerase subdivides these vast tracts of genetic information into more manageable units: mRNAs encoding only one or a few proteins, and noncoding RNA species like rRNA and the RNA component of RNase P. The translational apparatus then further subdivides the genetic information in mRNAs by recognizing the translational start and stop signals that identify protein-coding regions. In the relatively primitive RNP world, where neither the transcription nor the translation apparatus may have been sufficiently advanced to recognize *internal* sequences within an RNA molecule, it might have been advantageous to encode separate proteins (and functional RNAs) on separate RNA segments. End-to-end transcription followed by end-to-end translation would then yield the desired protein products.

The disadvantage of storing genetic information in segmented form is that progeny can only survive by inheriting at least one of each genomic segment. (This is as true for a modern eucaryote with multiple DNA chromosomes as for an RNA virus with a segmented genome.) Perhaps, by studying the (currently unknown) mechanism by which a modern RNA virus packages one copy of each genomic segment in the viral capsid, we can learn how the segments of RNA genomes might have been segregated to daughter cells in the RNP world.

THE DNA WORLD

DNA Fulfills the Need for a Stable Genetic Material[44]

Eventually, living systems became so complex that RNA was no longer a suitable genetic material. The reason is that RNA is far more vulnerable to "spontaneous" hydrolysis than is DNA: The 2'-hydroxyl group on RNA has the ability to catalyze cleavage of the adjacent phosphoester bond (see Figure 28-5), and this reaction is accelerated by the extreme conditions that may have been unavoidable on the primitive Earth (e.g., high pH, high temperatures, and the presence of certain divalent cations such as Zn^{2+}). In fact, nineteenth-century biochemists prepared DNA by heating cells in alkali under conditions where DNA survives but RNA is quickly reduced to oligonucleotides. Spontaneous hydrolysis therefore severely limited the complexity of RNA-based living systems.

The relative instability of RNA suggests that no organism as complex or at least as fast growing as a bacterium could evolve with an RNA genome, even if a major fraction of cellular metabolism were devoted to rapid repair of the spontaneous lesions. RNA is, however, a sufficiently stable material for structural components of the cell such as ribosomal RNA, tRNA, and snRNAs, which need not have a useful lifetime of more than a few cell generations.

When Did Genomes Become Double-Stranded?

Strictly speaking, we cannot know whether genomes first became double-stranded in the RNA, RNP, or DNA world. But there are two persuasive reasons for believing that double-stranded RNA genomes evolved before the birth of the DNA world. One reason is that RNA is subject to a high rate of spontaneous hydrolysis, but in a double-stranded RNA genome, such lesions could be healed by having the intact complementary strand serve as a splint for ligation of the two RNA fragments. An even better reason is that the existence of duplex genomes in the RNP world would have expedited the transition from RNA to DNA genomes.

Ribonucleoside Diphosphate Reductase Made the DNA World Possible

All organisms synthesize DNA precursors by reducing the 2'-hydroxyl group of ribonucleotide diphosphates (rNDPs) to make deoxyribonucleotide diphosphates (dNDPs), a reaction catalyzed by the enzyme ribonucleoside diphosphate reductase. Thus, the evolution of this enzyme was probably the key event that ushered in the DNA world (Figure 28-20).

But the actual conversion from RNA to DNA genomes would have been a complicated process, requiring at least four steps: the *synthesis of DNA precursors*, the *reverse transcription* of preexisting RNA genomes into DNA, the *replication* of the new DNA genomes by a DNA polymerase, and finally the *transcription* of the resulting genomic

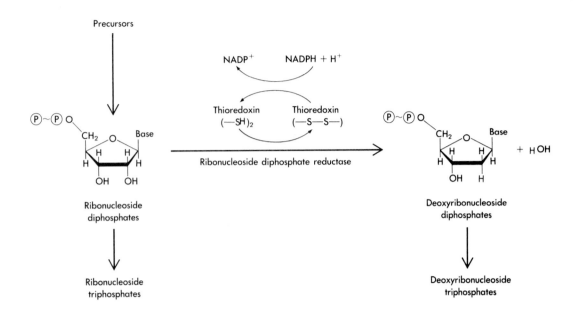

DNA segments back into functional RNA molecules. We cannot discount the possibility that four new enzyme activities—ribonucleoside diphosphate reductase, reverse transcriptase, DNA polymerase, and a DNA-dependent RNA polymerase—evolved simultaneously. But there is a simpler (and more attractive) scenario for conversion from RNA to DNA genomes. Once the RNP world reached a certain level of sophistication, the production of dNDPs by ribonucleoside diphosphate reductase might have enabled preexisting replication enzymes to copy RNA genomes into DNA and to transcribe the resulting DNA genomes back into functional RNA molecules.

How could preexisting enzymes copy RNA into DNA? If RNA genomes were already double-stranded in the RNP world, enzymes capable of replicating and transcribing double-stranded nucleic acids would already have existed. These enzymes would at first have used RNA precursors exclusively, but some of them may have been able to use DNA precursors, too, as these new molecules became available.

Figure 28-20
The enzyme ribonucleoside diphosphate reductase performs the key step in the conversion of RNA precursors into DNA precursors. The base can be either A, C, G, or U. The structures of the coenzyme NAD$^+$ and NADH are given in Figures 2-3 and 28-14; in NADP$^+$ and NADPH, the 2′ hydroxyl of the adenosyl residue has an additional phosphate. By reversible reduction of a cystine disulfide bridge to two cysteine sulfhydryl groups, the protein thioredoxin carries two reducing equivalents (hydrogens, in color) from NADPH to ribonucleoside diphosphate reductase (see Figure 2-19).

DNA Precursors Would Have Been Dangerous in an RNP World

The introduction of ribonucleoside diphosphate reductase into an RNP world was dangerous, because the incorporation of DNA precursors (lacking the 2′-hydroxyl group) into growing RNA chains would have seriously altered the structure and catalytic activity of the resulting RNA molecules. This implies that the polymerases responsible for producing functional RNAs had to be able to discriminate between RNA and DNA precursors. How could this be done?

In the RNA and RNP worlds, a clear distinction may already have existed between (double-stranded) genomic RNAs and (single-stranded) functional RNAs. In this case, the RNA replicase that copied genomic RNA may have been different from the RNA transcriptase that copied genomic RNA into functional RNA. If only the replicase, and not the polymerase, could accept DNA precursors, then the danger of having DNA precursors in an RNP world would have been diminished.

Why Does DNA Contain Thymine Instead of Uracil?[45]

Supposing that the first functional ribonucleotide reductase simply reduced each of the four RNA precursors to make DNA precursors, we would expect the original DNA genomes to have contained uracil. Why, then, does DNA today have thymine in place of uracil? Spontaneous deamidation of cytosine to uracil is known to occur at a low but significant rate (see Figures 12-9 and 12-10). The potentially harmful effects of these C to U mutations are prevented by the enzyme uracil-DNA glycosylase, which excises uracil (but not thymine) from DNA (see Figures 12-9 and 12-10). This suggests that uracil was replaced by thymine (5-methyluracil) after the initial conversion from RNA to DNA genomes, so that the increasingly efficient repair enzymes could recognize the hydrolysis products of cytosine in DNA.

We have seen that ribonucleoside diphosphate reductase converts the four rNDPs to dNDPs (Figure 28-20). Three of these dNDPs can be phosphorylated directly to yield DNA precursors (dATP, dCTP, and dGTP) but the fourth product (dUDP) must be broken down to dUMP before it can be converted to dTMP by the enzyme thymidylate synthase. (dTMP is subsequently phosphorylated to yield dTTP.) Thymidylate synthase is a dimer of identical subunits of molecular weight 35,000 daltons in organisms as varied as bacteria, bacteriophage, yeast, viruses, and vertebrates. The extraordinary conservation of this enzyme suggests that the transformation from uracil-containing to thymine-containing genomes occurred only once, very early in the DNA world.

Curiously, no RNA repair systems are known. Apparently the occasional deamidation of C to U in structural RNAs does little damage, perhaps because even the most stable structural RNA species are gradually diluted out by newly synthesized RNA as the cell multiplies.

Why Is the DNA Double Helix Right-Handed?

Two molecules are said to be **stereoisomers** if they are identical except for the geometric configuration of the four bonds surrounding a tetrahedral carbon atom. For a simple molecule like an amino acid, which has only one such asymmetric carbon atom, one stereoisomer will simply be the mirror image of the other (see Figure 5-11). Molecules with an asymmetric arrangement of bonds around a central carbon atom are able to rotate the plane of polarized light, a property known as optical activity. Thus, the two stereoisomers of a simple asymmetric molecule like glyceraldehyde are often referred to as either D or L, depending on whether the stereoisomer rotates polarized light in a right-handed (*d*extrorotatory) or left-handed (*l*evorotatory) direction (Figure 28-21). For convenience, complex molecules with more than one asymmetric carbon atom are sometimes designated as D or L to show how they are related to a simple D or L molecule (see Figure 28-21), but then the D or L no longer indicates whether the molecule is dextrorotatory or levorotatory when tested experimentally.

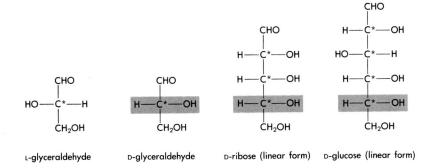

L-glyceraldehyde   D-glyceraldehyde   D-ribose (linear form)   D-glucose (linear form)

Figure 28-21

D-ribose, D-glucose, and D-glyceraldehyde are geometrically related. Asterisks indicate asymmetric (tetrahedral) carbon atoms with four different substituents. Horizontal lines denote bonds that emerge from the plane of the page; vertical lines denote bonds that recede into the page. The asymmetric carbon atom shared by D-ribose, D-glucose, and D-glyceraldehyde is shown in color.

Today, all living organisms have D-ribose rather than L-ribose in their nucleic acids, and as a result, the typical nucleic acid duplex is right-handed rather than left-handed. (Neither the phosphates nor the bases are asymmetric.) Similarly, all contemporary living organisms have L- rather than D-amino acids in their proteins. Yet, every scheme for prebiotic synthesis of sugars or amino acids produces nearly equal mixtures of the D and L forms. Why, then, are biological systems not built from L-ribose and D-amino acids (a mirror image of our world), or from all D forms or all L forms? The reason surely lies in the detailed enzymology of the first RNA replicase and the first working translation system.

Whether we are considering a sophisticated modern RNA polymerase or the first crude RNA enzyme, all RNA polymerases depend on base pairing between the template and the pool of potential nucleotide precursors to select the correct nucleotide for addition to the growing daughter strand. Since active sites necessarily have a handedness, the first RNA enzyme would have preferred either the D precursors or the L precursors. As a result, the handedness of the active site of this crude RNA replicase would have been heritable, although the enzyme itself might have been composed of both D- and L-nucleotides. In fact, not only would all the immediate descendants of this first enzyme continue to discriminate between D-ribose and L-ribose, but natural selection for the most rapid and accurate RNA replicase would tend to increase the initial level of discrimination between stereoisomers. Still later, when the first DNA genomes were copied from preexisting RNA genomes, DNA inherited the right-handedness of its RNA predecessors. Thus, the handedness of our nucleic acids appears to be a "frozen" accident of biological history.

Why Are Proteins Made of L-Amino Acids Rather Than D-Amino Acids?

If the handedness of DNA reflects the handedness of the active site of the first RNA replicase, a similar argument could be used to explain why proteins are made of L-amino acids and not D-amino acids. The first successful polypeptide polymerase (and therefore all of its descendants) might simply have preferred L-amino acids. But there is another, far more interesting possibility. Perhaps once the fateful decision had been made to use nucleic acids containing D-ribose, then the first proteins *had* to be constructed from L-amino acids.

Since translation arose after RNA replication, the first primitive tRNA-like molecules and synthetases would have been built from

D-ribose. If such RNAs were able to interact more easily with L-amino acids than with D-amino acids, we could rationalize the use of L-amino acids in living organisms today. This line of reasoning predicts that a mirror image of our own biological world (with L-nucleotides and D-amino acids) could exist, perhaps elsewhere in the universe, but that an all D or L biological world could not. Such intriguing but highly speculative ideas cannot be evaluated until we know much more about the origin of the translational apparatus.

DNA Metabolism Evolved in a World Rich in Protein Catalysts[46]

Early in the evolution of the RNP world, every sophisticated catalytic reaction had to be carried out by RNA; rudimentary proteins might have assisted the RNA catalysts, but were probably not catalysts themselves. As the RNP world became able to translate longer mRNAs accurately, proteins became more important (see Figure 28-15). Some RNA enzymes, such as those responsible for mRNA splicing and protein synthesis, eventually evolved into RNP machines that survive today (snRNPs and the ribosome); other RNA enzymes (the tRNA synthetases?) ultimately lost their RNA component. The complexity of RNP machines such as the ribosome and snRNPs suggests that the mature RNP world was very sophisticated.

Since proteins are more versatile catalysts than RNA (see Figure 28-6), accurate synthesis of long polypeptides enabled living systems to become far more complex. In addition, as new enzymes arose to meet new needs in the evolving RNP world, the superior catalytic abilities of protein guaranteed that most of these enzymes would be made of protein instead of RNA or RNP. Today, if an enzymatic function is carried out by an RNP complex, we can safely assume that the enzyme is a molecular fossil of the RNP world. But if a modern enzyme (e.g., the DNA replication machinery) is composed solely of protein, we cannot be absolutely certain that it began as a pure protein rather than as an RNP enzyme whose RNA was gradually replaced by protein (see Figure 28-15).

RNP Machines Survive in the DNA World[47]

Today, DNA replication is carried out by an efficient protein machine that evolved in a world rich in protein catalysts. In fact, DNA replication proceeds nearly as fast as deoxyribonucleotide triphosphate precursors can diffuse into the active site of the enzyme, about 750 nucleotides per second. The more ancient process of polypeptide chain elongation is carried out by the ribosome, a ribonucleoprotein fossil of the RNP world, and proceeds much more slowly, at a rate of about 15 amino acids per second.

The fact that nucleotides are polymerized 50 times faster than amino acids could be interpreted to mean that translation is inefficient compared to DNA replication and that RNP complexes are poor catalysts compared to proteins. But another interpretation is that translation is an intrinsically slower process than DNA replication because it must *decode* a nucleic acid rather than merely *copying* it. The immediate precursors of DNA are simply the four deoxynucleotide triphosphates; these small molecules diffuse rapidly and can be present in very high concentration without becoming a metabolic burden on the cell. The immediate precursors of protein synthesis are 20 different charged tRNAs; these larger molecules diffuse slowly, must be charged with the correct amino acid by the appropriate tRNA synthetase, and cannot be present in significantly higher concentrations than they already are (tRNA constitutes about 15 percent of total cellular RNA).

Recent calculations suggest that the peptide bond occurs very efficiently once the peptidyl-tRNA and the charged tRNA are correctly positioned on the ribosome. Thus, we should not mistake ribosomes and snRNPs for clumsy, inefficient fossils of the RNP world. Like the DNA replication machinery, these RNP complexes continued to evolve in a world rich in proteins and have undoubtedly been improved wherever possible by the addition of protein. The persistence of RNP machines in a DNA world where protein is the major catalyst therefore suggests that the RNA component of RNP complexes cannot easily be replaced by a polypeptide chain.

Why Does RNase P Retain Its RNA?[4]

Since the main task of ribosomes and snRNPs is to perform sequence-specific reactions using RNA templates and substrates, it is conceivable that RNP machines (which can recognize other RNA molecules by base pairing) might even be *more* efficient than equivalent machines made entirely of protein. Evidence for this comes from RNase P, the RNP enzyme responsible for recognizing and processing 30 to 40 different pre-tRNAs (see Figure 14-19). Although the catalytic function of the RNA component of RNase P is always the same, the sequence of the RNA is surprisingly different in different bacterial species. This implies that the RNA structure is subject to few constraints. But then why hasn't the function of the RNA been taken over by the protein component of this RNP enzyme? An obvious possibility is that RNA can do something that proteins cannot. Perhaps an RNA structure is more flexible than a protein structure and can more easily accommodate many slightly different pre-tRNA substrates.

The Great Antiquity of mRNA Introns[48, 49, 50]

The abundance of mRNA introns in vertebrate genes poses a straightforward question: Were preexisting protein genes broken up by the insertion of mRNA introns, or were mRNA introns gradually lost from an ancestral gene? Since mRNA introns are less abundant in lower eucaryotes than in vertebrates and are altogether absent in familiar bacteria like *E. coli*, we might be tempted to guess that these introns have been inserted into preexisting genes in higher eucaryotes. But given the strong evidence that bacteria and lower eucaryotes must streamline their genomes in order to reproduce as rapidly as possible (Chapters 18 and 20), it is impossible to tell whether these simpler organisms have discarded ancestral introns or have been under selective pressure to resist the insertion of new introns.

The only sure way to determine when and why mRNA introns arose is to trace the evolutionary history of an ancient gene encoding a ubiquitous highly conserved protein. This has been done for the enzyme triosephosphate isomerase, which carries out a basic step in both glycolysis (see Figure 4-16) and gluconeogenesis (the synthesis of glucose from smaller metabolites). Since the amino acid sequence of triosephosphate isomerase is very similar in both procaryotes and eucaryotes, this enzyme must have evolved into its final form in the progenote.

The gene encoding triosephosphate isomerase has no introns in *E. coli* or yeast, five introns in the fungus *Aspergillus,* six introns in chickens and humans, and eight introns in maize (Figure 28-22). Since the positions of five introns are strikingly conserved between maize (corn) and vertebrates, these introns must have been present before the divergence of plants and animals over a billion years ago. The position of the sixth conserved intron appears to have shifted by three amino acids, a process known as **intron sliding** (see Figure 20-23). However, the conservation of six intron positions between maize and vertebrates does not tell us whether the introns were inserted into the triosephosphate isomerase gene of the last common ancestor of plants and animals or whether they were present still earlier in evolution. Nor can this question be resolved by examining the triosephosphate isomerase gene from *Aspergillus;* the fungal gene shares only one intron with maize and chicken, and the remaining four introns in *Aspergillus* are so different that intron sliding cannot be distinguished from independent insertion.

Protein-Coding Genes Were Assembled by Exon Shuffling[50, 51]

The best evidence that introns may have been present in the very first triosephosphate isomerase gene comes from a comparison of the highly conserved three-dimensional structure of the enzyme with the intron positions in the various genes. Many of the introns are located between adjacent blocks of DNA encoding a basic unit of protein structure known as an α/β domain (see Figure 28-22). This strongly suggests that the first triosephosphate isomerase gene had introns because it was assembled from at least ten relatively small primordial exons, each encoding a tiny domain of protein structure. A corollary is that modern triosephosphate isomerase genes have been derived from the ancestral gene by differential intron loss. The alternative interpretation of the data—that introns were inserted into a preexisting triosephosphate isomerase gene as mutagenic agents—tends to be excluded because introns sometimes interrupt the codon for a highly conserved (and therefore functionally essential) amino acid like Trp_{169}.

Exons Correspond to Functional Domains of Proteins[50, 52]

Not long after the discovery of split protein genes, it became clear that modern exons sometimes encode relatively large domains of protein structure (Chapter 20). This observation gave rise to the hypothesis that the rate of evolution may be enhanced in the DNA world by **exon shuffling**: Recombination at the DNA level between different intron sequences could generate novel combinations of exons, and such

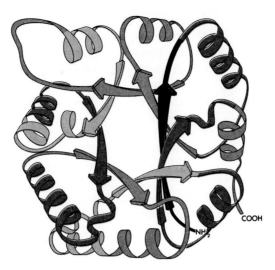

(a)

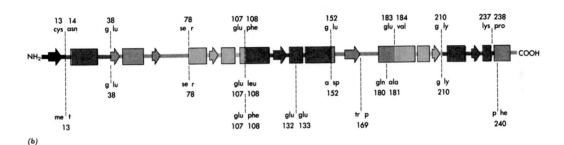

(b)

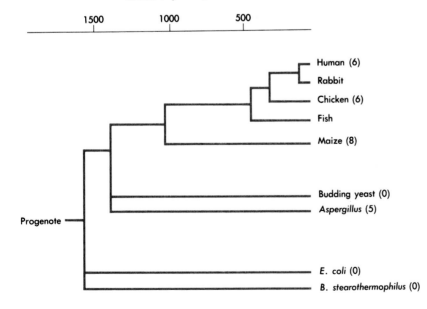

Millions of years ago

(c)

Figure 28-22
The ancestral triosephosphate isomerase gene may have had at least ten introns. (a) The three-dimensional structure of triosephosphate isomerase with coils representing α-helices (see Figure 2-22) and arrows representing β strands (see Figure 2-23). The eight β strands form a symmetrical β barrel (see Figure 5-21) surrounded by a sheath of α helices; the protein thus consists of eight similar α/β structural domains. Protein domains encoded by separate exons in maize are indicated by different colors. (b) The protein sequence with α-helical domains (arrows) and β-strand domains (rectangles) are colored as in (a). The positions of introns are shown in the protein sequence of maize (top), chicken (middle), and the fungus *Aspergillus* (bottom). (c) An evolutionary tree constructed as in Figure 28-24 from the amino acid sequence of triosephosphate isomerase; divergence times were calculated by assuming that amino acid substitutions accumulate in triosephosphate isomerase at a constant rate. Where known, the number of introns in the gene is indicated in parentheses. [After W. Gilbert, M. Marchionni, and G. McKnight, *Cell* 46 (1986):151.]

novel exon combinations will almost always produce translatable mRNAs because mRNA splice junction sequences are interchangeable (see Figures 20-20 through 20-23).

A single exon encoding a domain of epidermal growth factor provides an especially dramatic example of exon shuffling. This exon appears in four unrelated proteins including epidermal growth factor itself, the low-density lipoprotein receptor, and blood clotting factors IX and X (Table 28-2; see p. 1127). Dispersal of this exon must have taken place in the DNA world, since vertebrate proteins evolved comparatively recently. In contrast, triosephosphate isomerase is such an ancient protein that we cannot tell from the available data whether its gene was assembled from separate exons by RNA recombination in the RNP world or by DNA recombination in the DNA world.

Introns Can Be Precisely Deleted in the DNA World[53]

The gradual loss of ancestral introns over evolutionary time could occur by several different mechanisms. Intron loss caused by random deletion at the DNA level would generally be *imprecise*. As a result, a few amino codons would be gained, lost, or changed at the boundaries of the deletion. In fact, this might be one cause of intron sliding. *Precise* intron loss would most likely reflect the reverse flow of genetic information—that is, any process in which a spliced or partially spliced mRNA (or cDNA copy of the mRNA) served as the source of genetic information for generating the new gene copy. For example, an intron would be deleted if the mRNA (or a cDNA copy of the mRNA) served as the template for conversion of the parental gene (see Figure 20-29). Alternatively, introns would be lost if a cDNA copy of a spliced or partially spliced mRNA were integrated into the genome as a *functional* processed gene (see Figure 20-31). One of the two insulin genes of the rat appears to have lost an intron by just this mechanism.

The Red Queen Argument: Running Fast To Stay in Place[54, 55, 56]

In trying to understand evolution, we must avoid the simplistic notion that evolution always improves genes. Many genes need no improvement. Basic processes like DNA replication, transcription, protein synthesis, and metabolism are remarkably similar in all living organisms, and the proteins and RNAs that carry out these highly conserved **housekeeping** functions (page 704) are themselves often highly conserved. For example, triosephosphate isomerase has been essentially unchanged for more than 1.6 billion years; the enzyme works just the same in corn as it does in bacteria (see Figure 28-22). But over the course of evolution, mutations in one part of the triosephosphate isomerase molecule have necessitated compensatory changes elsewhere in the molecule to preserve protein function. RNA and protein sequences will therefore change with time (Chapter 20), even when the changes preserve gene function rather than modifying or improving it. The only exceptions will be gene products whose structure is so critical that the sequence cannot tolerate changes. The amino acid sequences of histones H3 and H4, for example, are identical in cows and garden peas!

Table 28-3 Some Human Proteins Are Much Older Than Others

Ancient Proteins
First editions. Essentially unchanged since the divergence of procaryotes and eucaryotes. Mostly mainstream metabolic enzymes. Example: triosephosphate isomerase (human and bacterial sequences are 46% identical).
Second editions. Human and procaryotic proteins have homologous sequences but apparently different functions. Example: glutathione reductase in human red blood cells and mercury reductase in the bacterium *Pseudomonas* (27% identical).

Middle-age Proteins
Proteins found in most eucaryotes but procaryotic counterparts are as yet unknown. Example: actin.

Modern Proteins
Recent vintage. Proteins found in animals or plants but not both. Not found in procaryotes. Example: collagen.
Very recent inventions. Proteins found in vertebrates but not elsewhere. Example: plasma albumin.
Recent mosaics. Modern proteins that are clearly the result of exon shuffling. Example: low density lipoprotein receptor.

SOURCE: R. F. Doolittle, D. F. Feng, M. S. Johnson, and M. A. McClure, *Cold Spring Harbor Symp. Quant. Biol.* 51 (1986):447–455.

Although many proteins have remained essentially unchanged over the entire course of evolution (Table 28-3, "first editions"), other proteins have slowly accumulated mutations that made it possible to put them to new uses ("second editions"). In addition, some novel proteins were invented relatively late in the course of evolution ("middle-age proteins" and certain "modern proteins") while still others were assembled from segments of preexisting proteins as we discussed previously ("recent mosaics").

The History of the DNA World Is Written in Gene Sequences[57–60]

The slow accumulation of mutations over time makes it possible to measure the evolutionary distance between organisms by counting the number of accumulated changes (mutations) in a gene encoding a highly conserved protein or RNA. The evolution of the organisms from a common ancestor is then represented by a branched pathway called a **genealogical tree** (also known as a **phylogenetic** or **evolutionary tree**). The branching pattern of the tree is calculated using the **principle of parsimony** to determine the minimum number of genetic changes required to derive the sequence of the gene in each organism from a common ancestor (Figure 28-23).

The evolutionary distance separating organisms on a phylogenetic tree is usually expressed in units of accumulated amino acid or nucleotide changes rather than in units of millions of years. The reason for this is that gene sequences change at different rates in different organisms, and some genes will change faster than others, depending on how sensitive the gene product is to mutation (Table 28-4). (We saw previously that the RNA component of RNase P evolves very fast because it is quite *insensitive* to mutation.) Strictly speaking, then, the principle of parsimony can only establish the branching pattern of the phylogenetic tree. To convert such evolutionary distances into chron-

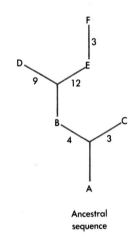

(a) The correct dendrogram minimizes total changes.

Ancestral sequence

(b) The total number of changes is greater for an incorrect dendogram.

Ancestral sequence

Figure 28-23
Evolutionary (phylogenetic) trees can be established using the principle of parsimony. First, the sequence of a particular highly conserved gene or protein is obtained for all the organisms under consideration (A, B, C, D, E, and F). Then the organisms are arranged in a branching diagram (dendrogram) that minimizes the number of changes required to derive all the sequences from a single ancestral sequence. Numbers along the branches indicate additional changes occurring during that time interval. For example, organisms D and E have two and five changes, respectively, relative to organism C, which in turn differs from the ancestral A sequence by three changes. Thus, D and E differ from A by five and eight changes, respectively.

ological time, it is usually necessary to examine the fossil record to see when organisms of a certain type first appeared. However, it is sometimes reasonable to assume that mutation rates are comparable for closely related organisms. In such cases, a highly conserved protein like globin or cytochrome *c* can be used as a **molecular clock** to measure how long the species have been diverging from each other.

All Bacteria Look Very Much Alike[61]

No one today would seriously suggest that bats and birds are closely related simply because they both can fly, but until 1977, most biologists and molecular biologists were unaware that the generally accepted division of all life forms into only two kingdoms (procaryotes and eucaryotes) was based on equally superficial characteristics.

Bacteria lack nuclei and have rigid cell walls about 1 μm across (only the irregularly shaped mycoplasmas lack cell walls; see Figure 4-25). Their length varies slightly, depending on whether they are bacilli (rod shaped), cocci (spherical), or spirochetes (spiral). The uniform size, simple design, and lack of internal architecture in the bacterial cell present a striking contrast to the eucaryotic cell. Eucaryotic cells are larger (about 10 μm) and are characterized by a sophisticated nucleus, extensive membrane systems, internal organelles, and often the ability to participate in complex developmental pathways.

For many years, the apparent simplicity of bacteria persuaded biologists that modern procaryotes are more similar to the first living systems than are modern eucaryotes. Moreover, it seemed reasonable to think that bacteria evolved into eucaryotes by stepwise changes, the most critical of which would have been the invention of a nuclear envelope (separating transcription from translation), loss of the cell wall, an increase in overall size, and the acquisition of internal organelles such as mitochondria and chloroplasts.

The Bewildering Biochemical Diversity of Bacteria

Although bacteria are remarkably similar in appearance, they are remarkably diverse in all other respects, and serious problems arise

Table 28-4 Different Proteins Evolve at Different Rates

Protein	Point Mutations per 100 amino acids per 100 million years
Pseudogenes	400
Fibrinopeptides	90
Lactalbumins	27
Lysozymes	24
Ribonucleases	21
Hemoglobins	12
Acid Proteases	8
Triosephosphate isomerase	3
Phosphoglyceraldehyde dehydrogenase	2
Glutamate dehydrogenase	1

SOURCE: R. F. Doolittle, D. F. Feng, M. S. Johnson, and M. A. McClure, *Cold Spring Harbor Symp. Quant. Biol.* 51 (1986):447–455.

when an attempt is made to lump them all into a single procaryotic kingdom. Bacterial species inhabit an enormous range of environments. **Thermoacidophiles** (heat- and acid-loving bacteria) such as *Sulfolobus* live in hot sulfur springs, where the temperatures (90°C) are near the boiling point of water and the surroundings are extraordinarily acidic (pH 2). Another thermoacidophile known as *Thermoplasma* was discovered under smoldering piles of coal tailings; like mycoplasmas, this organism has no cell wall and is protected from its inhospitable habitat by only a lipid bilayer. In contrast, **halophiles** (salt-loving bacteria) such as *Halobacterium* not only tolerate but actually require nearly saturating concentrations of salt, such as those found in the Great Salt Lake and the Dead Sea.

Bacterial species also differ dramatically in metabolism. Most molecular biologists are only familiar with bacteria such as *E. coli* and *Salmonella*, which are medically important and experimentally tractable, and have been intensively studied at both the genetic and biochemical level. These bacterial species generate energy by fermentation in the absence of oxygen and by oxidative phosphorylation in the presence of oxygen (see Figure 2-12). Oxidative phosphorylation produces energy by establishing a proton gradient that is coupled to the generation of ATP (the **chemiosmotic theory**; see Figure 2-14). In this process, hydrogen atoms (H) from food molecules are used to reduce molecular oxygen (O_2) to water (H_2O).

Other bacteria known as **methanogens** generate energy by an analogous reaction in which hydrogen gas (H_2) reduces carbon dioxide (CO_2) to methane (CH_4). Because molecular oxygen poisons the enzymes that reduce carbon dioxide, methanogens live in **anaerobic** (oxygen-free) environments such as bogs and lake bottoms, where they are responsible for the production of flammable "marsh gas." In fact, methanogens flourishing in modern sewage treatment plants can produce commercially marketable amounts of industrial gas. There are even bacteria that produce energy by using hydrogen gas to reduce elemental sulfur to hydrogen sulfide (H_2S)!

Obligate anaerobes cannot tolerate exposure to air. These bacteria are hard to work with in the laboratory and have only recently begun to receive the attention they deserve. **Facultative anaerobes** can survive in both the presence and absence of oxygen and have been better studied. For example, bacteria of the nitrogen-fixing genus Rhizobia will grow on the laboratory benchtop and can be analyzed genetically. However, Rhizobia are induced to differentiate into nitrogen-fixing bacteroids only by exposure to plant metabolites within the protected anaerobic environment of root nodules (see Figure 22-49).

A Universal Phylogeny Based on 16S Ribosomal RNA[62]

The enzyme cytochrome *c* is a component of the respiratory chain of all eucaryotes and many procaryotes and thus has been particularly useful for building phylogenetic trees. Some of the most interesting and peculiar bacteria lack cytochrome *c*, however, and even in those bacterial genera with the enzyme, the proteins are often too different to establish reliable relationships between them. To remedy this serious gap in our understanding of the evolution of the simplest living organisms, Carl Woese and his colleagues decided in 1969 to examine the sequence of 16S ribosomal RNA from as many organisms as possible, with special emphasis on unusual bacteria that had previously eluded reliable phylogenetic placement.

Woese chose 16S rRNA for construction of a phylogenetic tree because it is truly universal and is so highly conserved in structure and function that phylogenetic trees are relatively easy to construct. In addition, 16S rRNA is an abundant RNA that can be quickly purified and analyzed even from small samples of cells. The early stages of 16S rRNA sequence analysis culminated in 1977 with a radical hypothesis. Woese proposed that procaryotes should be divided into two groups, called the archaebacteria and the eubacteria, which are as different from each other as either is from the eucaryotes. The clear implication of this proposal was that archaebacteria, eubacteria, and eucaryotes had all descended from an earlier common ancestor that did not survive. This hypothesis met with substantial resistance within the biological community because it contradicted two common but unfounded assumptions—that all bacteria are closely related and that bacteria more closely resemble the first living cells than do any eucaryotes.

Archaebacteria Assume Their Rightful Place[61-65]

Despite widespread scepticism about the value of dividing procaryotes into eubacteria and archaebacteria, proponents of the hypothesis continued to refine the universal phylogeny based on 16S rRNA (Figure 28-24) and to amass supporting biochemical evidence (see Table 28-5). Today, there is no longer any doubt that all living organisms belong to three coequal kingdoms, or **lines of descent**, and that none of these three kingdoms can be thought of as having given rise to the others (see Figure 28-24 and Table 28-5). Instead, all three have descended from an earlier living organism, or progenote, whose nature we can only infer by asking what archaebacteria, eubacteria, and the eucaryotic nucleus have in common. (The eucaryotic nucleus is directly descended from the progenote, but as we shall see, eucaryotic organelles such as the mitochondrion and chloroplast were derived by endosymbiosis of oxygen-fixing and photosynthetic eubacteria.)

The universal phylogeny based on 16S-like rRNA reveals other startling conclusions. Human beings (*Homo sapiens*) are in fact more closely related to corn (*Zea mays*) than a Gram-negative bacterium (*E. coli*) is to a Gram-positive bacterium (*Bacillus subtilis*) (see Figure 4-8 for the significance of Gram staining). Thus, the evolutionary distance separating two different bacteria can be greater than the distance between a sophisticated plant and the most sophisticated animal. The 16S-like phylogeny also provides definitive evidence for the endosymbiont hypothesis that mitochondria and chloroplasts are descended from eubacteria (see the section entitled The Endosymbiotic Origin of Mitochondria and Chloroplasts).

The Progenote (First Cell) Differed from All Modern Cells

The universal phylogeny based on 16S-like rRNA tells us that the three great kingdoms of living organisms are all descended from a progenote. But what was this progenote like? The abundance of

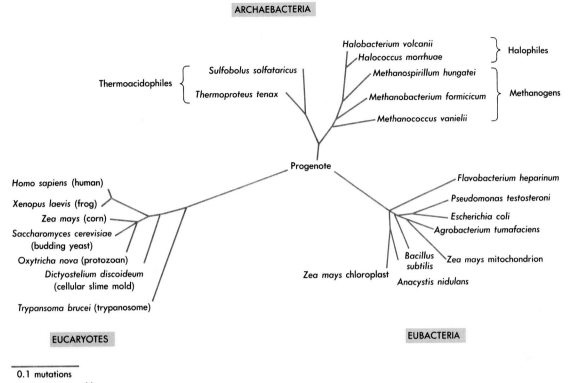

ARCHAEBACTERIA

Thermoacidophiles
- Sulfobolus solfataricus
- Thermoproteus tenax

Halobacterium volcanii
Halococcus morrhuae } Halophiles

Methanospirillum hungatei
Methanobacterium formicicum } Methanogens
Methanococcus vanielii

Progenote

Homo sapiens (human)
Xenopus laevis (frog)
Zea mays (corn)
Saccharomyces cerevisiae (budding yeast)
Oxytricha nova (protozoan)
Dictyostelium discoideum (cellular slime mold)
Trypansoma brucei (trypanosome)

Flavobacterium heparinum
Pseudomonas testosteroni
Escherichia coli
Agrobacterium tumafaciens
Zea mays mitochondrion
Bacillus subtilis
Zea mays chloroplast
Anacystis nidulans

EUCARYOTES

EUBACTERIA

0.1 mutations per sequence position

introns in archaebacteria and eucaryotes suggests that the progenote had introns but that these were lost during eubacterial evolution as the genome was streamlined for very rapid growth (see Table 28-5). Similarly, since eubacteria and eucaryotes have ester-linked unbranched lipids containing L-glycerophosphate, it is likely that the progenote did, too.

Table 28-5 A Few of the Known Differences Between Archaebacteria and Eubacteria

Archaebacteria	Eubacteria
Genomic rearrangements common	Genomes quite stable
Transposable elements often abundant	Few transposable elements
Some introns in rRNA and tRNA	No introns known
No peptidoglycan in cell wall	Peptidoglycan cell wall
Branched-chain fatty acids	Straight-chain fatty acids
Ether-linked lipids	Ester-linked lipids
Lipids contain D-glycero-phosphate	Lipids contain L-glycero-phosphate
rRNA, tRNA, and ribosomes share both eubacterial and eucaryotic features	
Larger multisubunit RNA polymerases resembling the eucaryotic enzymes	Simpler RNA polymerases
"Reverse" gyrases in thermophiles introduce + supercoils	Gyrases introduce only − supercoils
EF2 sensitive to diphtheria toxin as in eucaryotes	EF2 insensitive to diphtheria toxin

Figure 28-24
An evolutionary tree can be constructed by comparing the complete sequences of 21 different 16S and 16S-like ribosomal RNAs (rRNAs). The scale bar represents the number of accumulated nucleotide differences per sequence position in the rRNAs of the various organisms. Note that the scale bar cannot be recalibrated in billions of years without making the unjustified assumption that mutations accumulate in the DNA of all organisms at the same rate per unit time. [After N. R. Pace, G. J. Olsen, and C. R. Woese, *Cell* 45 (1986):325.]

But we cannot automatically assume that any trait shared by two of the three great kingdoms must reflect the nature of the progenote. Such a shared trait could also have arisen more than once as different organisms independently discovered its value. Independent evolution of the same characteristic in separate branches of a phylogenetic tree is called **convergent evolution**.

Bacteria Are More Highly Evolved than Higher Organisms

Efforts to deduce the nature of the progenote are also confounded by the fact that different organisms evolve at different rates. Although mutations arise in DNA throughout the life cycle of an organism, the effect of these mutations on fitness can only be tested in each *new* generation. As a result, rapidly multiplying organisms like bacteria and many lower eucaryotes have had a far greater opportunity to lose or modify the characteristics of the progenote than have more slowly growing higher organisms. This implies that many bacteria, although they are no more ancient than eucaryotes (see Figure 28-24), are actually more highly evolved.

The Endosymbiotic Origin of Mitochondria and Chloroplasts[66–70]

Eucaryotic cells contain a variety of internal organelles, each surrounded by a lipid bilayer. Many of these organelles (e.g., lysosomes, peroxisomes, and the endoplasmic reticulum) are relatively simple (see Figure 18-8). But two of them, mitochondria and chloroplasts, are about the same size as bacteria and, like bacteria, have circular DNA genomes (see Figure 15-17). Mitochondrial and chloroplast genomes encode the rRNA and tRNA components of the organellar translation apparatus, as well as mRNAs for organellar proteins that are synthesized within the organelle. The mitochondrial and chloroplast ribosomes are sensitive to antibiotics such as chloramphenicol, which kill many bacteria but do not affect the cytoplasmic ribosomes of eucaryotes.

The resemblance of mitochondria and chloroplasts to bacteria naturally led to the idea that these organelles began as free-living bacteria that had been engulfed by a primitive eucaryote (the **urcaryote**; see Figure 28-25). Once internalized, these symbiotic bacteria flourished within the host eucaryote as **endosymbionts**, while supplying the host with the ability to generate energy by oxidative phosphorylation and (in the case of plants) by photosynthesis. As the protomitochondrion and the protochloroplast slowly degenerated into specialized organelles, genes were transferred from organellar DNA to the nuclear genome of the host, leaving only a handful of essential genes behind in the organelle. As a result, most mitochondrial and chloroplast proteins are now encoded in the nuclear DNA, translated in the cytoplasm, and transported across the outer

Figure 28-25

A possible scheme for early evolution. [After J. E. Darnell and W. F. Doolittle, *Proc. Nat. Acad. Sci.* 83 (1986):1271.]

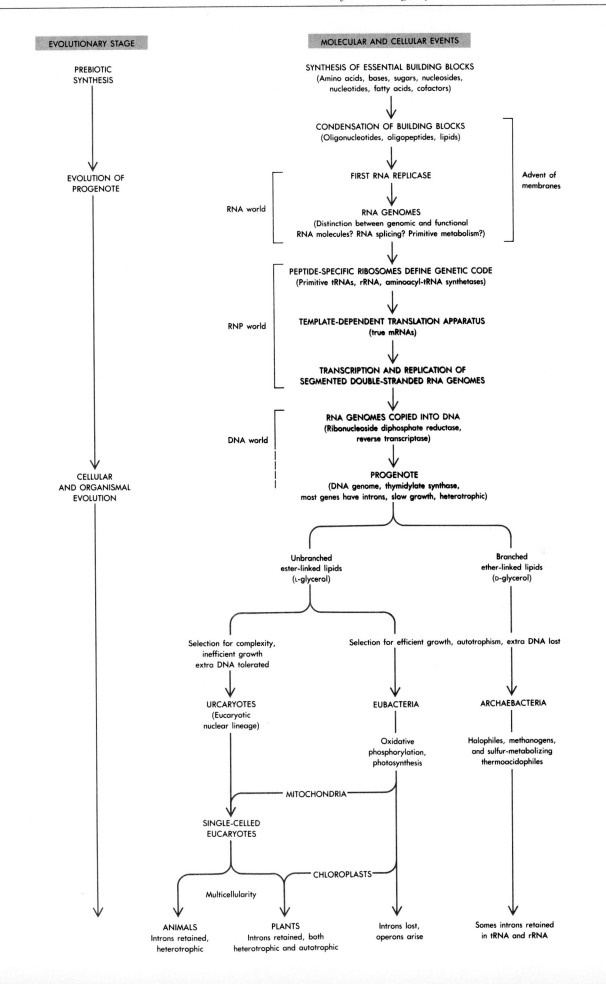

membrane of the organelle. Only those molecular species that cannot cross the outer membrane (rRNA, tRNA, mRNA, and some proteins) must still be encoded in the organellar genome. (Some RNA molecules *can* cross the organellar membrane, however, as shown by the recent discovery that the RNA component of a mammalian mitochondrial RNA processing enzyme resembling RNase P is encoded by a nuclear gene.) As fewer and fewer proteins were encoded within the evolving endosymbiont, the organellar translation system no longer had to be extremely accurate. Eventually, the organellar translation apparatus degenerated into an apparently minimal translation machine (see Figure 14-15) and the mitochondrial genetic code underwent some surprising changes (see Table 15-9).

Did mitochondria and chloroplasts evolve from eubacterial or archaebacterial progenitors? Comparison of bacterial and eucaryotic cytochrome *c* initially suggested that mitochondria might have descended from the purple photosynthetic eubacteria. Comparison of animal mitochondrial and eubacterial 16S rRNA sequences failed to prove this, however, because the animal mitochondrial rRNA sequences had diverged too extensively to permit a meaningful comparison. Fortunately, plant mitochondrial 16S rRNAs are less divergent, and in this case, the comparison led to a surprising result. Plant mitochondria descended from a group of purple eubacteria that includes rhizobacteria (see Figure 22-49), agrobacteria (see Figure 22-48), and rickettsias (see page 544). Even today, each of these procaryotes is able to live within or in very close association with eucaryotic cells. This makes the endosymbiont hypothesis all the more plausible and allows us to complete a tentative scheme for early evolution (Figure 28-25).

Several Bacteriophage T4 Genes Contain Self-Splicing Introns[71, 72]

No introns have ever been found in *E. coli*, the most intensively studied of all procaryotes. The discovery in 1984 of an intron in a bacteriophage of *E. coli* therefore came as quite a shock in 1984. Three different bacteriophage T4 genes are now known to have self-splicing introns resembling the *Tetrahymena* rRNA intron. Two of the genes encode enzymes that convert RNA precursors into DNA precursors (thymidylate synthase and the small subunit of ribonucleoside diphosphate reductase; see Figure 28-20). By expressing high levels of these two enzymes, T4 diverts the metabolic resources of the infected bacterium from making RNA to making DNA, thereby increasing the yield of DNA-containing progeny phage.

Why do self-splicing introns interrupt useful T4 genes? Although it is possible that the introns are harmless or hard to get rid of, another fascinating possibility is that the introns might actually contribute to efficient phage growth. Recall that self-splicing is initiated by attack of a free guanine nucleotide on the 5' splice site of the intron (see Figures 28-8 and 28-9). When high levels of guanine nucleotides are present early in infection, efficient self-splicing of the transcripts will produce mRNAs whose protein products catalyze conversion of RNA precursors into DNA precursors. As guanosine nucleotide levels begin to fall later in infection, the efficiency of self-splicing will decrease, and the rate of conversion of RNA precursors to DNA precursors will consequently slow down. T4 thus appears to use the dependence of self-splicing on the concentration of guanosine nucleotides as a regulatory device; the introns prevent the cellular ribonucleotide pool from being reduced to such a low level that the yield of progeny phage would suffer.

How Did Bacteriophage T4 Get Its Introns?[73, 74]

Similar self-splicing introns are found at three different sites in the bacteriophage T4 genome. Since complicated self-splicing introns are unlikely to have arisen independently at three different sites, the introns must be capable of transposing from one site to another. This suggests that T4 may have acquired its first intron by infecting a host cell whose genome already contained the intron. As previously described, mitochondria are descended from endosymbiotic bacteria (see Figure 28-24), and fungal mitochondria are known to contain self-splicing introns (see Figure 28-19). Thus, T4 bacteriophage may have acquired its introns by infecting the mitochondrion (or protomitochondrion) of a eucaryotic cell. Such **horizontal transmission** of a genetic element from one species of organism to another (as opposed to **vertical transmission** of an element from parent to daughter) may be responsible for the widespread occurrence of self-splicing introns in apparently unrelated organisms.

Fungal mitochondria have self-splicing introns (Table 28-1 and Figure 28-19) but the mitochondria of higher organisms do not. This is especially puzzling, because the eubacteria that gave rise to the mitochondria of higher plants (and possibly animals) lack introns today. One explanation would be that fungal mitochondria acquired self-splicing introns by horizontal transmission after endosymbiosis began. Alternatively, if eubacteria originally had introns, these introns may have been retained by the endosymbiont but lost by free-living eubacteria. Still a third possibility is that endosymbiosis occurred more than once among the eucaryotes. In this case, mitochondria would be **polyphyletic,** and fungal mitochondria might have descended from a different endosymbiont than the mitochondria of other eucaryotes.

Molecular Biology Is a Subdiscipline of Biology

The evolution of genes is necessarily complex because genes exist only within organisms, and organisms can interact with one another and with the environment in exceedingly subtle ways. Although molecular biology is concerned primarily with molecules, it must always be seen as a subdiscipline of biology. As the next two sections will illustrate, we simply cannot expect to understand the structure and function of genes unless we are prepared to understand the biology of the organisms in which those genes reside.

Wormy Mice in a Hybrid Zone[75]

Europe is inhabited by two distinct but interbreeding species of house mouse (Figure 28-26). *Mus musculus* resides in eastern Europe, and *Mus domesticus* in western Europe. Hybrid mice having traits derived from both species are found only in an extraordinarily narrow zone

Figure 28-26
A map of Europe showing the distribution of the two interbreeding species of house mouse, *Mus domesticus* and *Mus musculus*. The two species meet and hybridize in a narrow zone (heavy black line) extending from Denmark through central Europe to the Black Sea. Restriction maps were established for the mitochondrial DNA (mtDNA) of mice collected at each of the circled locations. Hollow circles indicate mice with *M. domesticus* mtDNA, color circles indicate mice with *M. musculus* mtDNA, and black circles indicate mice with *M. musculus* nuclear genes but mtDNA from *M. domesticus*. The data show that in parts of Scandinavia, mtDNA from the western European house mouse has invaded the eastern European house mouse and displaced the resident mtDNA. [After S. D. Ferris, R. D. Sage, C.-M. Huang, J. T. Nielsen, U. Ritte, and A. C. Wilson, *Proc. Nat. Acad. Sci.* 80 (1983):2290.]

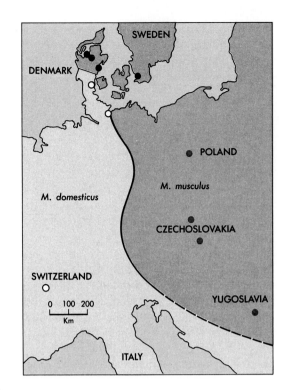

about 20 km (12 miles) wide running from the Adriatic Sea to the North Sea. Why is the hybrid zone so narrow? The zone does not follow any major geographic feature, such as a mountain range, that might affect the climate, so we cannot argue that mice having *M. musculus* traits are unable to survive west of the zone, while mice with *M. domesticus* traits are unable to survive to the east. Nor can we argue that the zone is narrow because mice seldom migrate more than 20 km; each parent species has in fact spread over a range exceeding 1000 km. Instead, it would appear that the zone is narrow because the hybrid progeny of these two interbreeding species are somehow *unfit* and therefore reproduce more slowly than either parental species.

Inefficient reproduction of hybrid mice explains why a stable boundary exists between the two species and also why the hybrid zone is so narrow. *M. musculus* cannot move westward without mating with *M. domesticus*, thus producing hybrid mice at the boundary zone. The hybrid mice are somewhat unfit and cannot continue westward out of the zone because they are unable to compete with the resident *M. domesticus*. A similar argument explains why *M. domesticus* cannot move eastward. The resulting hybrid zone is self-perpetuating and serves as a "genetic sink" to block gene flow between the two populations. The width of the zone depends on how unfit the hybrid mice are.

Why do the hybrid progeny of *M. musculus* and *M. domesticus* reproduce poorly? To answer this question, mice were caught in barns and houses from 30 different towns within and around the hybrid zone and then examined for abnormalities. Curiously, mice from the hybrid zone were frequently infected with intestinal parasites such as pinworms (nematodes) and tapeworms. Since both parental species are relatively resistant to these parasites, the implication is that reassortment of parental chromosomes in the hybrid mice lowers resistance to infection. This suggests that resistance is a **polygenic trait** conferred by several different genes working together. Reassortment would therefore break up groups of cooperating genes or produce incompatible combinations of parental genes. In either case, the fitness of the hybrid mice would decrease, and they would reproduce more slowly.

This example emphasizes that genes are part of genomes, genomes produce organisms, and organisms have complex life cycles. Thus, natural selection on the gene occurs at the level of the organism, not at the level of the gene itself. As a result, a lowly pinworm parasite can block genetic exchange between two potentially interbreeding species.

Mitochondrial and Nuclear DNA Are Evolving Separately[76, 77]

Further molecular examination of the European house mouse population led to another interesting conclusion. As shown in Figure 28-26, the two interbreeding species of house mouse, *M. domesticus* and *M. musculus*, meet and hybridize in a narrow zone extending from Denmark through central Europe to the Adriatic Sea. Since mammalian mitochondrial DNA (mtDNA) is only 16.5 kilobases in size (see Figure 15-17) and is easy to purify, it was possible to compare the restriction maps of mtDNA from mice collected at many different locations. This comparison revealed that in parts of Scandinavia, mtDNA from the western European house mouse (*M. domesticus*) has "invaded" the eastern European house mouse (*M. musculus*) and displaced the resident mtDNA. Why this has happened is a mystery.

The *M. domesticus* mtDNA might conceivably confer a selective advantage on mice living in Scandinavia, perhaps because of minor differences in the local climate or because mtDNA from *M. domesticus* but not from *M. musculus* confers resistance to some disease or parasite found exclusively in Scandinavia. Alternatively, since mtDNA encodes only a small fraction of the mitochondrial proteins (Chapter 18), it is conceivable that mutation in a nuclear gene of Scandinavian mice is incompatible with the presence of *M. musculus* mitochondria.

The inheritance of mtDNA in European mice demonstrates that a group of genes may be selected simply because the genes are physically linked on the same replicon or because they work together to confer a selective advantage on the organism. Even though natural selection ultimately takes place at the level of the organism, the organization of genes on chromosomes evidently plays a crucial role in determining the variety of genetic variants that are available for natural selection to act on.

One Eve or Many?[78–82]

The earliest examples of anatomically modern human fossils have been found in Africa, suggesting that a band of the earliest humans left Africa and that their descendants migrated throughout Europe, Asia, and finally the Americas. Although this theory that modern humans (*Homo sapiens*) first emerged in Africa is widely believed, the fossil evidence is unlikely to be complete, and there is nothing to exclude the possibility that modern humans evolved on another continent (where the fossil evidence has yet to be discovered) and then migrated to Africa.

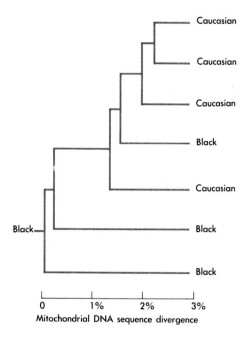

0   1%   2%   3%
Mitochondrial DNA sequence divergence

Figure 28-27

A human genealogical tree based on the nucleotide sequences of human mitochondrial DNA. The most likely tree was deduced using the method of maximum parsimony as described in Figure 28-24. The root of the tree lies in Africa, and there are two primary lines of descent; one leads to Africans only, the other to some Africans and to all other populations. [After R. L. Cann, M. Stoneking, and A. C. Wilson, *Nature* 325 (1987):31.]

Can we address the question of where *Homo sapiens* originated by examining the *molecular* fossil record that is written in the DNA sequences of present-day humans? We have seen that highly conserved molecules like cytochrome *c*, triosephosphate isomerase, and 16S ribosomal RNA evolve slowly and can be used to establish evolutionary trees that span hundreds of millions of years (see Figures 28-22c to 28-24). But radioisotope dating of fossil remains suggests that our own species diverged from primitive hominoids only 100,000 to 140,000 years ago. Molecules like cytochrome *c* or 16S rRNA change too slowly to be of much help in constructing an evolutionary tree describing the emergence of the human races over such a brief time. Instead, we need a "fast clock" molecule that evolves quickly, and for this purpose, mitochondrial DNA seems to be ideal.

The mutation rate of mitochondrial DNA is about ten times higher than that of nuclear DNA, primarily because mitochondria have no DNA repair systems. As a consequence of such rapid evolution, mitochondrial DNA exhibits much greater variation from one person to another than does nuclear DNA. This has made it possible to construct a human genealogical tree by comparing mitochondrial DNA sequences from different individuals (see Figure 28-27), just as the phylogenetic tree leading back to the progenote was constructed by applying the principle of parsimony to many different 16S rRNA sequences (see Figures 28-23 and 28-24). Whether the mitochondrial DNAs from 7 or 147 individuals of different racial extraction were compared, the conclusion was the same. Most modern Africans are closer to the ancestral mitochondrial DNA sequence than are any modern Caucasians, implying that humans first emerged in Africa and then emigrated to other continents.

The human genealogical tree also suggests that all present-day humans may have shared a common ancestral mother (Figure 28-27). Mitochondrial DNA is known to be **maternally inherited**: Although both the oocyte and the sperm contain mitochondria, progeny inherit mitochondrial DNA from the mother only. The fact that 147 different mitochondrial DNA sequences can all be arranged in a single tree implies that there was indeed a single original mitochondrial DNA sequence. Supposing that the mitochondrial DNA of our most ancient ancestors exhibited the same degree of individual variation as mitochondrial DNA does today, this would mean that we all share a common mother, our mitochondrial Eve. The molecular fossil record therefore has the potential to tell us about events as ancient as the origin of the first replicating system or as recent as the emergence of human beings from the higher primates.

Summary

Molecular biology is an experimental rather than a theoretical science. Whether we want to know why globin genes are expressed only in red blood cells or why the fruit fly has a segmented body, we expect every theory to be tested experimentally and ultimately proved or disproved. We can afford to have such high expectations of molecular biology because the living systems that we study can be manipulated experimentally to test each theory.

Questions of molecular evolution are far trickier, because

it is usually harder, and sometimes even impossible, to do the right experiments. If we want to know, for example, whether humans are more closely related to chimpanzees or gorillas, we would really like to examine the missing links between the species. But these transitional organisms have not survived, and we must instead compare DNA sequences in the surviving species.

Questions of molecular evolution become even more daunting when we try to speculate about the very earliest life forms. We have seen that even the best fossils preserve only the durable parts of an organism; the macromolecules that most interest us are lost. As a result, we must attempt to reconstruct ancient organisms by examining living organisms (molecular fossils) and by designing model experiments that mimic the conditions thought to prevail on the primitive Earth. Unfortunately, it is impossible to obtain direct proof for any particular theory of the origin of life. The sobering truth is that even if every expert in the field of molecular evolution were to agree on how life originated, the theory would still be a best guess rather than a fact.

In this chapter, we have presented one possible scenario for the origin of the first "living molecule" that could replicate and evolve. Our purpose has been to show how the recent discovery of enzymatically active RNA species has made it possible to speculate usefully about the molecular origins of life. Aside from the intellectual challenge of trying to imagine how life began, such speculation inevitably inspires specific experiments that sharpen and modify the speculations themselves. For example, the experiment showing that the *Tetrahymena* ribosomal RNA intron core can function as a poly C polymerase was the first in a series of experiments designed to test whether an RNA molecule could actually copy itself in the test tube today. This experiment also led directly to the model for how an RNA enzyme could function as an RNA replicase and even perhaps as a tRNA synthetase. Although the scenario we presented is most unlikely to be true in detail, it surely contains important elements of the truth. Showing how life could have begun is a first step toward learning how life actually began.

Bibliography

General References

Day, W. 1984. *Genesis on Planet Earth: The Search for Life's Beginning.* 2nd ed. New Haven, Conn.: Yale University Press. A lively but idiosyncratic treatment of the origins of life, drawing on data from such diverse fields as geology, prebiotic chemistry, paleobiology, and molecular biology.

Ferris, J. P., and D. A. Usher. 1983. "Origins of Life." In *Biochemistry,* ed. G. Zubay (coordinating author). Reading, Mass.: Addison-Wesley. A superb, eminently readable, and often witty introduction to the chemistry of prebiotic synthesis and the origins of metabolism.

Miller, S. L., and L. E. Orgel. 1974. *The Origins of Life on Earth.* Englewood Cliffs, N.J.: Prentice-Hall. A concise and immensely readable classic monograph by two leaders in the field, emphasizing the role of prebiotic synthesis in the origin of life.

Raff, R. A., and T. C. Kaufman. 1983. *Embryos, Genes, and Evolution.* New York: Macmillan. A delightfully written essay showing how changes in DNA that affect embryogenesis can ultimately lead to the evolution of new species.

Schopf, J. W., ed. 1983. *Earth's Earliest Biosphere: Its Origin and Evolution.* Princeton, N.J.: Princeton University Press. Although aimed almost exclusively at specialists, each of the advanced technical essays in this authoritative collection begins with a preface suitable for beginning students of the field.

Cited References

1. Knoll, A. H. 1985. "The Distribution and Evolution of Microbial Life in the Late Proterozoic Era." *Ann. Rev. Microbiol.* 39:391–417.
2. Knoll, A. H., and E. S. Barghoorn. 1985. "Archean Microfossils Showing Cell Division from the Swaziland System of South Africa." *Science* 198:396–398.
3. Dill, R. F., E. A. Shinn, A. T. Jones, K. Kelly, and R. P. Steinen. 1986. "Giant Subtidal Stromatolites Forming in Normal Salinity Waters." *Nature* 324:55–58.
4. Pace, N. R., and T. L. Marsh. 1985. "RNA Catalysis and the Origin of Life." *Origins of Life* 16:97–116.
5. Keese, P., and R. H. Symons. 1985. "Domains in Viroids: Evidence of Intermolecular RNA Rearrangements and Their Contribution to Viroid Evolution." *Proc. Nat. Acad. Sci.* 82:4582–4586.
6. Hutchins, C. J., P. D. Rathjen, A. C. Forster, and R. H. Symons. 1986. "Self-Cleavage of Plus and Minus RNA Transcripts of Avocado Sunblotch Viroid." *Nucl. Acid Res.* 14:3627–3640.
7. Keese, P., and R. H. Symons. 1986. "The Structure of Viroids and Virusoids." In *Viroids and Viroid-Like Pathogens,* ed. J. S. Semanicik. Boca Raton: CRC Press, pp. 1–48.
8. Buzayan, J. M., W. L. Gerlach, and G. Bruening. 1986. "Non-enzymatic Cleavage and Ligation of RNAs Complementary to a Plant Virus Satellite RNA." *Nature* 323:349–353.
9. Dinter-Gottlieb, G. 1986. "Viroids and Virusoids Are Related to Group I Introns." *Proc. Nat. Acad. Sci.* 83:6250–6254.
10. Epstein, L. M., and J. G. Gall. 1987. "Self-Cleaving Transcripts of Satellite DNA from the Newt." *Cell* 48:535–543.
11. Cech, T. R. 1986. "RNA as an Enzyme." *Sci. Amer.* 255:64–75.
12. Cech, T. R., and B. L. Bass. 1986. "Biological Catalysis by RNA." *Ann. Rev. Biochem.* 55:599–629.
13. Bass, B. L., and T. R. Cech. 1984. "Specific Interaction Between the Self-Splicing RNA of *Tetrahymena* and Its Guanosine Substrate: Implications for Biological Catalysis by RNA." *Nature* 308:820–826.
14. Cech, T. R. 1985. "Self-Splicing RNA: Implications for Evolution." *Internat. Rev. of Cytol.* 93:3–22.
15. Zaug, A. J., J. R. Kent, and T. R. Cech. 1984. "A Labile Phosphodiester Bond at the Ligation Junction in a Circular Intervening Sequence RNA." *Science* 224:574–578.
16. Garriga, G., A. M. Lamobwitz, T. Inoue, and T. R. Cech. 1986. "Mechanism of Recognition of the 5' Splice Site in Self-Splicing Group I Introns." *Nature* 322:86–89.
17. Been, M. D., and T. R. Cech. 1986. "One Binding Site Determines Sequence Specificity of *Tetrahymena* Pre-rRNA Self-Splicing, *Trans*-Splicing, and RNA Enzyme Activity." *Cell* 47:207–216.
18. Szostak, J. W. 1986. "Enzymatic Activity of the Conserved Core of a Group I Self-Splicing Intron." *Nature* 322:83–86.
19. Zaug, A. J., and T. R. Cech. 1986. "The Intervening Sequence RNA of *Tetrahymena* Is an Enzyme." *Science* 231:470–475.
20. Cech, T. R. 1986. "A Model for the RNA-Catalyzed Replication of RNA." *Proc. Nat. Acad. Sci.* 83:4360–4363.
21. Haertle, T., and L. E. Orgel. 1986. "Template-directed Synthesis on the Oligonucleotide d(C_7–G–C_7)." *J. Mol. Biol.* 188:77–80.
22. Horowitz, N. H. 1945. "On the Evolution of Biochemical Syntheses." *Proc. Nat. Acad. Sci.* 31:153–157.
23. White, H. B., III. In *The Pyridine Nucleotide Coenzymes* (1982):1–17. New York: Academic Press.
24. Gilbert, W. 1986. "The RNA World." *Nature* 319:618.
25. Jay, D. G., and W. Gilbert. 1987. "Basic Protein Enhances the Incorporation of DNA into Lipid Vesicles: A Model for the Formation of Primordial Cells." *Proc. Nat. Acad. Sci.* 84. In press.
26. Moore, P. B. 1985. "Polypeptide Polymerase: The Structure and Function of the Ribosome in 1985." In *Proc. Robert A. Welch Found. Conf. Chem. Res.* 29:185–214.

27. Crick, F. H. C., S. Brenner, A. Klug, and G. Pieczenik. 1976. "A Speculation on the Origin of Protein Synthesis." *Origins of Life* 7:389–397.

28. Woese, C. R. 1970. "The Problem of Evolving a Genetic Code." *BioScience* 20:471–485.

29. Eigen, M., and R. Winkler-Oswatitsch. 1981. "Transfer-RNA, an Early Gene?" *Naturwissenschaften* 68:282–292.

30. Crothers, D. M. 1982. "Nucleic Acid Aggregation Geometry and the Possible Evolutionary Origin of Ribosomes and the Genetic Code." *J. Mol. Biol.* 162:379–391.

31. Weiner, A. M., and N. Maizels. 1987. "3' terminal tRNA-like structures tag genomic RNA molecules for replication: Implications for the origin of protein synthesis." *Proc. Nat. Acad. Sci.* 84:7383–7387.

32. Burma, D. P., D. S. Tewari, and A. K. Srivastava. 1985. "Ribosomal Activity of the 16S·23S RNA Complex." *Arch. Biochem. Biophys.* 239:427–435.

33. Wong, J. T.-F. 1981. "Coevolution of Genetic Code and Amino Acid Biosynthesis." *Trends in Biochem. Sci.* 6:33–36.

34. Webster, T., H. Tsai, M. Kula, G. A. Mackie, and P. Schimmel. 1984. "Specific Sequence Homology and Three-Dimensional Structure of an Aminoacyl Transfer RNA Synthetase." *Science* 226:1315–1317.

35. Hoben, P., N. Royal, A. Cheung, F. Yamao, K. Biemann, and D. V. Söll. 1982. "*Escherichia coli* Glutaminyl-tRNA Synthetase. II. Characterization of the *glnS* Gene Product." *J. Biol. Chem.* 257:11644–11650.

36. Sharp, P. 1985. "On the Origin of RNA Splicing and Introns." *Cell* 42:397–400.

37. Cech, T. R. 1986. "The Generality of Self-Splicing RNA: Relationship to Nuclear mRNA Splicing." *Cell* 44:207–210.

38. Schmelzer, C., and R. J. Schweyen. 1986. "Self-Splicing of Group II Introns in Vitro: Mapping of the Branch Point and Mutational Inhibition of Lariat Formation." *Cell* 46:557–565.

39. Tazi, J., C. Alibert, J. Temsamani, I. Reveillaud, G. Cathala, C. Brunel, and P. Jeanteur. 1986. "A Protein That Specifically Recognizes the 3' Splice Site of Mammalian Pre-mRNA Introns Is Associated with a Small Nuclear Ribonucleoprotein." *Cell* 47:755–766.

40. Gerke, V., and J. A. Steitz. 1986. "A Protein Associated with Small Nuclear Ribonucleoprotein Particles Recognizes the 3' Splice Site of Premessenger RNA." *Cell* 47:973–984.

41. Zhuang, Y., and A. M. Weiner. 1986. "A Compensatory Base Change in U1 snRNA Suppresses a 5' Splice Site Mutation." *Cell* 46:827–835.

42. Sharp, P. A. 1987. "Splicing of Messenger RNA Precursors." *Science* 235:766–771.

43. Maniatis, T. and R. Reed. 1987. "The Role of Small Nuclear Ribonucleoprotein Particles in Pre-mRNA Splicing." *Nature* 325:673–678.

44. Reanney, D. C. 1986. "Genetic Error and Genome Design." *Trends in Genetics* 2:41–46.

45. Hardy, L. W., J. S. Finer-Moore, W. R. Montfort, M. O. Jones, D. V. Santi, and R. M. Stroud. 1987. "Atomic Structure of Thymidylate Synthase: Target for Rational Drug Design." *Science* 235:448–455.

46. Alberts, B. M. 1986. "The Function of the Hereditary Materials: Biological Catalyses Reflect the Cell's Evolutionary History." *Amer. Zoologist* 26:781–796.

47. Jacob, F. 1977. "Evolution and Tinkering." *Science* 196:1161–1166.

48. Marchionni, M., and W. Gilbert. 1986. "The Triosephosphate Isomerase Gene from Maize: Introns Antedate the Plant-Animal Divergence." *Cell* 46:133–141.

49. McKnight, G. L., P. J. O'Hara, and M. L. Parker. 1986. "Nucleotide Sequence of the Triosephosphate Isomerase Gene from *Aspergillus nidulans*: Implications for a Differential Loss of Introns." *Cell* 46:143–147.

50. Gilbert, W., M. Marchionni, and G. McKnight. 1986. "On the Antiquity of Introns." *Cell* 46:151–154.

51. Südhof, T. C., D. W. Russell, J. L. Goldstein, M. S. Brown, R. Sanchez-Pescador, and G. I. Bell. 1985. "Cassette of Eight Exons Shared by Genes for LDL Receptor and EGF Precursor." *Science* 228:893–895.

52. Brändén, C.-I. 1986. "Anatomy of α/β Proteins." In *Current Communications in Molecular Biology: Computer Graphics and Molecular Modeling,* eds. R. Fletterick and M. Zoller. Cold Spring Harbor, N.Y.: Cold Spring Harbor Laboratory, pp. 45–51.

53. Soares, M. B., E. Schon, A. Henderson, S. K. Karathanasis, R. Cate, S. Zeitlin, J. Chirgwin, and A. Efstratiadis. 1985. "RNA-Mediated Gene Duplication: The Rat Preproinsulin I Gene Is a Functional Retroposon." *Mol. Cell. Biol.* 5:2090–2103.

54. Van Valen, L. 1974. "Molecular Evolution as Predicted by Natural Selection." *J. Mol. Evol.* 3:89–101.

55. Zuckerkandel, E. 1976. "Evolutionary Processes and Evolutionary Noise at the Molecular Level." *J. Mol. Evol.* 7:269–311.

56. Smith, J. M. 1983. "The Genetics of Stasis and Punctuation." *Ann. Rev. Genetics* 17:11–25.

57. Britten, R. J. 1986. "Rates of DNA Sequence Evolution Differ Between Taxonomic Groups." *Science* 231:1393–1398.

58. Wilson, A. C. 1985. "The Molecular Basis of Evolution." *Sci. Amer.* 253:164–173.

59. Helm-Bychowski, K., and A. C. Wilson. 1986. "Rates of Nuclear DNA Evolution in Pheasant-Like Birds: Evidence from Restriction Maps." *Proc. Nat. Acad. Sci.* 83:688–692.

60. Vawter, L., and W. M. Brown. 1986. "Nuclear and Mitochondrial DNA Comparisons Reveal Extreme Rate Variation in the Molecular Clock." *Science* 234:194–196.

61. Woese, C. R. 1981. "Archaebacteria." *Sci. Amer.* 244:98–125.

62. Pace, N. R., G. J. Olsen, and C. R. Woese. 1986. "Ribosomal RNA Phylogeny and the Primary Lines of Evolutionary Descent." *Cell* 45:325–326.

63. Doolittle, W. F. 1985. "Archaebacteria Coming of Age." *Trends in Genetics* 1:268–269.

64. Hofman, J. D., L. C. Schalkwyk, and W. F. Doolittle. 1986. "ISH51: A Large, Degenerate Family of Insertion Sequence-Like Elements in the Genome of the Archaebacterium, *Halobacterium volcanii.*" *Nucleic Acid Res.* 14:6983–7000.

65. Kjems, J., and R. A. Garrett. 1985. "An Intron in the 23S Ribosomal RNA Gene of the Archaebacterium *Desulfurococcus mobilis.*" *Nature* 318:675–677.

66. Yang, D., Y. Oyaizu, H. Oyaizu, G. J. Olsen, and C. R. Woese. 1985. "Mitochondrial Origins." *Proc. Nat. Acad. Sci.* 82:4443–4447.

67. Chomyn, A., M. W. J. Cleeter, C. I. Ragan, M. Riley, R. F. Doolittle, and G. Attardi. 1986. "URF6, Last Unidentified Reading Frame of Human mtDNA, Codes for an NADH Dehydrogenase Subunit." *Science* 234:614–618.

68. De Block, M., J. Schell, and M. Van Montagu. 1985. "Chloroplast Transformation by *Agrobacterium tumefaciens.*" *EMBO J.* 4:1367–1372.

69. Shinozaki, K., M. Ohme, M. Tanaka, T. Wakasugi, N. Hayashida, T. Matsubayasi, N. Zaita, J. Chunwongse, J. Obokata, K. Yamaguchi-Shinozaki, C. Ohto, K. Torazawa, B. Y. Meng, N. Sugita, H. Deno, T. Kamogashira, K. Yamada, J. Kusuda, F. Takaiwa, A. Kato, N. Tohdoh, H. Shimada, and M. Sugiura. 1986. "The Complete Nucleotide Sequence of the Tobacco Chloroplast Genome: Its Gene Organization and Expression." *EMBO J.* 5:2043–2049.

70. Chang, D. D. and D. A. Clayton. 1987. "A Mammalian Mitochondrial RNA Processing Activity Contains Nucleus-Encoded RNA." *Science* 235:1178–1184.

71. Gott, J. M., D. A. Shub, and M. Belfort. 1986. "Multiple Self-Splicing Introns in Bacteriophage T4: Evidence from Autocatalytic GTP Labeling of RNA in Vitro." *Cell* 47:81–87.

72. Sjöberg, B.-M., S. Hahne, C. Z. Mathews, C. K. Mathews, K. N. Rand, and M. J. Gait. 1986. "The Bacteriophage T4 Gene for the Small Subunit of Ribonucleotide Reductase Contains an Intron." *EMBO J.* 5:2031–2036.

73. Michel, F., and B. Dujon. 1986. "Genetic Exchanges Between Bacteriophage T4 and Filamentous Fungi?" *Cell* 46:323.

74. Colleaux, L. d'Auriol, M. Betermier, G. Cottarel, A. Jacquier, R. Galibert, and B. Dujon. 1986. "Universal Code Equivalent of a Yeast Mitochondrial Intron Reading Frame Is Expressed into *E. coli* as a Specific Double Strand Endonuclease." *Cell* 44:521–533.

75. Sage, R. D., D. Heyneman, K.-C. Lim, and A. C. Wilson. 1986. "Wormy Mice in a Hybrid Zone." *Nature* 324:60–63.

76. Ferris, S. D., R. D. Sage, C.-M. Huang, J. T. Nielsen, U. Ritte, and A. C. Wilson. 1983. "Flow of Mitochondrial DNA Across a Species Boundary." *Proc. Nat. Acad. Sci.* 80:2290–2294.

77. Wilson, A. C., R. L. Cann, S. M. Carr, M. George, U. B. Gyllensten, K. M. Helm-Bychowski, R. G. Higuchi, S. R. Palumbi, E. M. Prager, R. D. Sage, and M. Stoneking. 1985. "Mitochondrial DNA and Two Perspectives on Evolutionary Genetics." *Biol. J. Linnean Soc.* 26:375–400.

78. Cann, R. L., M. Stoneking, and A. C. Wilson. 1986. "Mitochondrial DNA and Human Evolution." *Nature* 325:31–36.

79. Latorre, A., A. Moya, and F. J. Ayala. 1986. "Evolution of mitochondrial DNA in *Drosophila subobscura*." *Proc. Nat. Acad. Sci.* 83:8649–8653.

80. Wainscot, J. 1987. "Out of the Garden of Eden." *Nature* 325:13.

81. Wainscoat, J. S., A. V. S. Hill, A. L. Boyce, J. Flint, M. Hernandez, S. L. Thein, J. M. Old, J. R. Lynch, A. G. Falusi, D. J. Weatherall, and J. B. Clegg. 1986. "Evolutionary Relationships of Human Populations from an Analysis of Nuclear DNA Polymorphisms." *Nature* 319:491–493.

82. Jones, J. S., and S. Rouhani. 1986. "How Small Was the Bottleneck?" *Nature* 319:449–450.

Index

This index covers both volume I and volume II.
Page numbers in bold face refer to a definition and major discussion of the entry. F after a page number indicates a figure; T after a page number indicates a table.